EMERGENCY RESCUE

GLENCOE FIRE SCIENCE SERIES

Bryan: **Fire Suppression and Detection Systems**
Bush/McLaughlin: **Introduction to Fire Science,** Second Edition
Carter: **Arson Investigation**
Clet: **Fire-Related Codes, Laws, and Ordinances**
Erven: **Emergency Rescue**
Erven: **Fire Fighting Apparatus and Procedures,** Third Edition
Erven: **First Aid and Emergency Rescue**
Erven: **Handbook of Emergency Care and Rescue,** Revised Edition
Erven: **Techniques of Fire Hydraulics**
Gratz: **Fire Department Management: Scope and Method**
Isman/Carlson: **Hazardous Materials**
Meidl: **Explosive and Toxic Hazardous Materials**
Meidl: **Flammable Hazardous Materials,** Second Edition
Meidl: **Hazardous Materials Handbook**
Robertson: **Introduction to Fire Prevention,** Second Edition

ABOUT THE AUTHOR

Lawrence W. Erven, now retired, was battalion chief of the Los Angeles City Fire Department, and has over thirty years of fire fighting and rescue experience. He was a pioneer in fire science education. In 1955 he was instrumental in developing the first complete fire science curriculum in the country. From 1955 until 1969 he taught fire fighting, emergency rescue, and fire apparatus courses at Los Angeles Harbor College. He is the author of *First Aid and Emergency Rescue, Handbook of Emergency Care and Rescue, Techniques of Fire Hydraulics, Fire Fighting Apparatus and Procedures,* and numerous study guides, instructor manuals, and articles on fire science and emergency rescue.

EMERGENCY RESCUE

Lawrence W. Erven
Battalion Chief (Retired)
Los Angeles City Fire Department

GLENCOE PUBLISHING CO., INC.
Encino, California
Collier Macmillan Publishers
London

This book is dedicated to the members of emergency organizations, who daily risk life and limb to rescue the more unfortunate, and ease their suffering.

Copyright © 1980 by Lawrence W. Erven

Printed in the United States of America

All rights reserved. No part of this book may be reproduced or transmitted in any form or by any means, electronic or mechanical, including photocopying, recording, or by any information storage and retrieval system, without permission in writing from the Publisher.

Glencoe Publishing Co., Inc.
17337 Ventura Boulevard
Encino, California 91316
Collier Macmillan Canada, Ltd.

Library of Congress Catalog Card Number: 78-71736

1 2 3 4 5 6 7 8 9 10   83 82 81 80 79

ISBN 0-02-472640-0

Contents

Preface IX

Part 1.–RESCUE EQUIPMENT AND TECHNIQUES 1

Chapter 1.–Rescue Personnel 2

Emergency Medical Technicians 3
Personnel Qualifications 5
Training for Law Enforcement Officers 7
Professional Conduct 7
Civil Liability 11
Questions for Review 17

Chapter 2.–Rescue Equipment 19

Simple Machines 19
Lifting Devices 20
Cutting Tools 27
Rope 31
Manual Rescue Tools 37
Forcible Entry 37
Questions for Review 40

Chapter 3.–Transporting the Victims 41

Personnel Safety 41
Victim Care 42
Patient-Handling Complications 43
Rescue of Casualties 47
Methods of Transportation 56
Questions for Review 63

Part 2.–Rescue from Land and Air Vehicles 65

Chapter 4.–Handling Highway Accidents 66

Responding to the Scene 66
Size-up 67
Emergency Medical Care 70
Traffic Control 74
Automotive Fires 77
Questions for Review 81

Chapter 5.–Extrication from Crashed Motor Vehicles 82

Stages of Extrication 82
Training and Preparation 83
Environmental Factors 85
Size-up 86
Gaining Access to Victims 87
Administering Lifesaving Care 88
Disentanglement 89
Preparation for Removal 100
Transportation of the Injured 101
Bus Accidents 102
Questions for Review 105

Chapter 6.–Rescue from Aircraft Accidents 106

Approaching the Crash Scene 107
Fire Fighting 107
Aircraft Rescue 110
Forcible Entry 113
Military Aircraft 114
Questions for Review 117

Part 3.–Metropolitan Rescue 119

Chapter 7.–Rescue from Collapsed Buildings 120

Size-up 120
Rescue Operations 121
Signs of Building Collapse 123
Locating Trapped Casualties 126
The Four Stages of Rescue 126

Contents

Damage to Buildings   **127**
Hazards from Damaged Utilities   **130**
Rescue Techniques   **132**
Raising and Supporting Structural
   Elements   **139**
Debris Handling   **143**
Questions for Review   **143**

Chapter 8.–Rescue from Elevators   **145**

Psychological Care of Passengers   **145**
Qualified Elevator Mechanics   **147**
Elevator Construction   **147**
Electrical Power   **150**
Basic Elevator Rescue   **151**
Advanced Elevator Rescue   **153**
Emergency Exits   **157**
Emergency Service Provisions   **158**
Questions for Review   **162**

Chapter 9.–Rescue from Electrical Contact   **164**

Electrical Principles   **164**
Hazards of Live Wires   **167**
Medical Effects of Electric Shock   **170**
Electrical Burns   **172**
Rescue from Electrical Contact   **174**
Fire Fighting Hazards   **175**
Questions for Review   **177**

Chapter 10.–Rescue of Trapped Victims   **179**

Control at the Scene   **179**
Machinery Accidents   **179**
Impaled Victims   **181**
Trapped Child   **181**

Garbage Disposals   **182**
Questions for Review   **182**

Chapter 11.–Rescue from High Places   **184**

Accessibility of Victims   **184**
Pole Top Rescue   **184**
Hoisting the Victim   **186**
Rappelling   **187**
Lifesaving Nets   **188**
Removal of Casualties   **191**
Ladders   **191**
Questions for Review   **194**

Chapter 12.–Rescue from Burning Structures   **196**

Fire Behavior   **198**
Survival in a Structure Fire   **205**
Fire Rescue by Police Officers   **206**
Fireground Size-up and Tactics   **207**
Search and Rescue Procedures   **209**
High-Rise Structures   **218**
Questions for Review   **219**

Chapter 13.–Rescue Using Fire Apparatus   **221**

Aerial Ladder   **221**
Elevating Platforms   **225**
Questions for Review   **227**

Part 4.–Environmental Rescue   **229**

Chapter 14.–Handling Natural Disasters   **230**

Weather   **230**
Winter Storms   **231**
Thunderstorms   **235**
River Floods   **238**

Hurricanes **241**
Tornadoes and Windstorms **245**
Tsunamis **249**
Volcanoes **251**
Earthquakes **252**
Questions for Review **257**

Chapter 15.–**Rescue from Landslides and Cave-Ins** **258**

Cave-Ins and Landslides **258**
Mine Accidents **262**
Rescue from Tanks, Pits, and Holes **264**
Rescue from a Pressurized Tunnel **266**
Questions for Review **269**

Chapter 16.–**Rescue from Water** **270**

Drowning **271**
Submerged Automobiles **273**
Diving Accident Rescues **274**
Scuba Diver Rescue **277**
Scuba Diving Hazards **278**
Ice Rescue **288**
Underwater Search and Recovery **293**
Drownproofing **303**
Cold Water Survival **304**
Deep Water Rescue **307**
Questions for Review **309**

Chapter 17.–**Wilderness Search and Rescue** **310**

SAR Organizations **311**
Use of Aircraft in Searches **313**
Immediate Action **314**
Evaluation of the Rescue Situation **316**

SAR Suspension or Termination **320**
Exposure and Exhaustion **321**
Search for a Missing Person **324**
Mountain Rescue **327**
Cave Rescue **332**
Mine Rescue **335**
Desert Survival and Rescue **336**
Swamp, Marsh, and Slough Rescue **339**
Quicksand and Quagmires **339**
Questions for Review **340**

Chapter 18.–**Rescue by Helicopter** **341**

Fixed-Wing Aircraft **341**
Safety Rules **342**
Helicopter Landing Sites **342**
Air Ambulance Operations **345**
Search and Rescue Operations **347**
Fire Fighting Operations **349**
Questions for Review **351**

Part 5.–**Rescue from Hazardous Materials** **353**

Chapter 19.–**Respiratory Protection** **354**

Respiratory Hazards **355**
Human Respiration **357**
Conservation of Air **358**
Respiratory Protection in a High-Pressure Atmosphere **360**
Training Programs **361**
Safety Precautions **365**
Breathing Apparatus Testing and Approval **368**
Types of Breathing Apparatus **368**
Questions for Review **382**

Contents

Chapter 20.-**Managing Hazardous Gases, Liquids, and Chemicals** 383

Hazardous Materials 383
Gases 390
Heat 391
Smoke 392
Smoke Inhalation 393
Commonly Encountered Hazards 398
Pesticides 402
Fumigations 405
Riot Control Agents 406
Questions for Review 407

Chapter 21.-**Handling Nuclear Radiation Incidents** 408

Radiation 409
Search and Rescue 411
Transportation Incidents 413
Atomic-Weapon Hazards, Precautions, and Procedures 416
Nuclear Weapons Explosions 418
Atomic Structure 420
Radiation 421
The Fission Process 423
Problems of Radiation 424
External and Internal Radiation 425
Protective and Decontamination Procedures 432
Questions for Review 433

Chapter 22.-**Handling Bomb Threats and Explosives** 436

Purpose of Bomb Threats 437
Developing a Plan for Handling Bomb Threats 438

Evacuation of the Premises 439
Terrorist Activities 440
Bomb Search 440
Suspicious Object Located 444
Letter and Package Bombs 445
Bomb Explosion 445
Bomb Disposal 445
Explosives 446
How Explosives React to Heat or Fire 449
Size-up of Incidents Relating to Explosives 451
Fire Fighting Rules 452
Questions for Review 452

Glossary 454

Appendix 456

Index 458

Preface

When rescue personnel arrive at the scene of a traffic accident, fire, or other emergency, a rapid size-up and intelligent plan of action are required immediately. *Emergency Rescue* provides the student with the knowledge of rescue equipment and techniques needed to make these crucial decisions.

Emergency Rescue provides information on the correct use of rescue equipment and techniques, both for rescue practices students and for personnel currently involved in rescue work. This includes members of municipal and volunteer rescue squads, as well as ambulance attendants, fire fighters, lifeguards, and police officers.

Emergency Rescue was planned to reflect the practical as well as the theoretical aspects of rescue work. Much time was spent observing rescue incidents and discussing them with rescue squad personnel, fire fighters, emergency medical technicians, mountain rescue personnel, and military personnel. Technical information and assistance were generously provided by the manufacturers of rescue equipment. The result is that this text provides a breadth of coverage of rescue operations previously unavailable to instructors.

Organization. *Emergency Rescue* is divided into five parts, allowing instructors to present units of similar rescue procedures in the sequence they think best. Part One, "Rescue Equipment and Techniques," enables the student to understand rescue procedures by explaining the categories and activities of rescue personnel; the types of rescue equipment and the techniques for using them; and correct procedures for transporting the victims.

In Part Two the student learns the techniques for rescue from land and air vehicles. This includes chapters on handling highway accidents, extrication from crashed motor vehicles, and handling aircraft accidents.

Part Three is devoted to metropolitan rescue. It explains procedures for rescue from collapsed buildings, rescue from elevators, and rescue from electrical contact. Also included are chapters on rescue of trapped victims, rescue from high places, rescue from burning structures, and rescue using fire apparatus.

Preface

Part Four, on environmental rescue, begins by explaining how to handle natural disasters. It then moves on to describe rescue from landslides and cave-ins; rescue from water; wilderness search and rescue; and rescue by helicopter.

Part Five covers rescue from hazardous materials. The first chapter in this section describes in detail the various devices for respiratory protection and their use. This part also explains how to handle rescue incidents involving hazardous gases, liquids, and chemicals; nuclear radiation; and bomb threats and explosives.

Features. *Emergency Rescue* contains many drawings to communicate rescue techniques to students visually, as well as photographs of actual rescue incidents. The instructor can use the more than two hundred illustrations in several ways: to show what kinds of rescue situations may be encountered; to illustrate recommended rescue techniques and equipment for a particular situation; or to describe a rescue problem based on the illustration and then ask students for possible solutions.

Another feature of the text touches upon the responsibility of rescue organizations to educate the public about correct procedures to follow during a flood, hurricane, or other emergency. Throughout the text the student is informed of the general procedures to be followed by the public for their safety during an emergency, as well as the steps to be taken by rescue personnel for their own protection and survival.

Learning Aids. Every chapter contains multiple-choice Questions for Review. Answers to these questions are supplied in an appendix.

An illustrated glossary gives the student an easy way to review the functions of rescue tools described at the beginning of the book, and allows instructors to assign chapters out of sequence if they desire.

Acknowledgments. I would like to thank the reviewers of this text. Roger D. White, M.D., of the Mayo Clinic provided a detailed medical review of the manuscript, and Charles Keenan, director of the Emergency Medical Technology department at Phoenix College, contributed valuable suggestions on content.

I would also like to thank the rescue practices instructors who contributed their advice: Captain Roy Kline of Pasadena City College; William Lane of Allan Hancock College; Clayton Johnson of the Minnesota Department of Education; Joe Thomas of J. Sargeant Reynolds Community College; James Lindsay of Indiana Vocational Technical College; James Guthrie of Miami-Dade Community College; Gene Cobb of Miramar College; Bob Dawson of the College of San Mateo; Captain W.E. Ackerly of the Hialeah Fire Department; Mark Wallace of Aims Community College; Captain Jan Dunbar of American River College; and Tony Licata of Hillsborough Community College in Florida.

All rescue personnel know the satisfaction they feel when they save a life or prevent additional pain or injury. The greatest support rescuers responding to a call can have is to know they have the training and equipment to do everything possible for the victims.

PART 1 RESCUE EQUIPMENT AND GENERAL RESCUE TECHNIQUES

1 Rescue Personnel
2 Rescue Equipment
3 Transporting the Victims

CHAPTER 1

Rescue Personnel

Accidents are the leading cause of death among persons between the ages of one and thirty-seven years, and are the fourth most frequent cause of death at all ages. The tragedy of this high accidental death rate is that trauma kills thousands of persons who otherwise could expect to live long and productive lives; even people afflicted with malignancy, heart disease, stroke, and other chronic diseases usually die late in life. Thus, many more millions of productive working years are lost annually because of deaths from accidents than from chronic diseases among older persons. A report of the National Academy of Sciences charged that "thousands of lives are lost through lack of systematic application of established principles of emergency care." Until recently many of the ambulances throughout the country were converted station wagons and modified hearses that were poorly equipped and operated by a single person with little or no training.

Paramedic personnel—including fire fighters, police officers, ambulance attendants, lifeguards, and members of rescue squads and the highway patrol—can reduce appreciably the suffering and death that result from accidental injury or sudden illness. Many rescue groups direct their attention to the critical period between the accident or the onset of the illness and the time when definitive care is available from a physician with adequate facilities.

Military experience has shown the importance of well-trained aides who can properly perform resuscitation, administer emergency medical care, and promptly evacuate combat casualties to hospital facilities. Many of the problems associated with combat casualties are the same as those found in civilian accidents. Extensive training is required to give such emergency personnel sufficient judgment and technical competence to perform effectively.

All emergency service personnel have an obligation that extends beyond offering rescue from endangering environments, extrication from entrapping types of accidents, and administration of pre-hospital emergency medical care. Safety and survival are extremely important, both for rescuers and potential victims. Rescue personnel will rapidly change their status from an asset to a liability if they are not constantly alert to potentially hazardous situations. Understanding the importance of safety and survival should not be confined to emergency personnel, but must be continuously communicated to the general public. Emergency service organizations should spearhead programs to educate the public about whom to call and what to do in emergencies until help arrives, and about how to protect themselves in natural disasters.

Personnel of the emergency medical services (EMS) systems should conduct classes in prehospital emergency medical care for all persons in the community. The subjects should include, but not be confined to, what emergency services are available, how to summon aid in case of an accident or sudden illness, blood pressure screening programs, training citizens in the techniques of cardiopulmonary resuscitation (CPR), saving persons who are choking on food (cafe coronary), and other techniques for maintaining life in a victim until trained paramedical personnel can respond. Some cities (for example, Seattle) have greatly decreased the number of fatalities resulting from sudden heart attacks by teaching a large percentage of the residents how to administer cardiopulmonary resuscitation until the mobile coronary care units arrive.

EMERGENCY MEDICAL TECHNICIANS

Ambulance services that are established only to rush the sick and injured to a hospital are not in the best interests of the community. Sophisticated lifesaving methods, techniques, and equipment in the emergency department of a hospital are useless to individuals who are dead on arrival. Emergency medical care procedures, therefore, must be started at the scene of the emergency and continued during transport, to prevent the patient's condition from worsening irreversibly (Figure 1-1).

A number of governmental and medical organizations in the United States have recognized the need for training ambulance and rescue personnel in emergency medical care. Under the provisions of the Highway Safety Act of 1966, the National Highway Traffic Safety Administration has developed guidelines and recommendations for emergency medical services, which state that all ambulances should be equipped with certain lifesaving equipment and operated by at least two persons trained in specified areas of emergency care. The responsibility of ambulance services to provide more than transportation alone is clearly emphasized.

Training Standards

National and community leaders have developed, adopted, and registered National Apprenticeship and Training Standards for Emergency Medical Technicians (EMTs). These standards establish criteria of performance, knowledge, training, and experience to assure the development of truly professional EMTs at both the basic and advanced life-support levels. This program is a positive step forward. It will help ensure that communities have competent professional rescue and paramedic personnel capable of responding immediately to aid the victims of accidents and sudden illness. It is important that prehospital emergency care and transportation be properly recog-

RESCUE EQUIPMENT AND GENERAL RESCUE TECHNIQUES

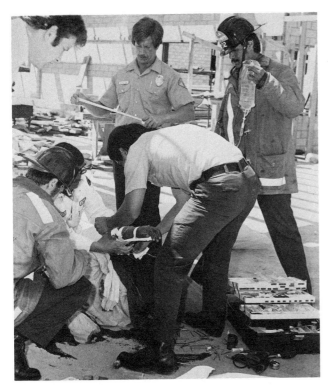

Fig. 1-1. Paramedics and fire fighters are administering prompt prehospital emergency medical care at the scene of a construction accident. Treatment consisted of covering the patient to alleviate the effects of traumatic shock, administering fluids intravenously, maintaining a clear airway for adequate ventilation, and continuing the emergency care during transport until hospital physicians could accept the victim. The ABCs of prehospital emergency medical care (**Airway, Breathing, Circulation**) designate the relative importance of these basic rescue procedures and the order in which they should receive attention. (Courtesy of Los Angeles County Fire Department.)

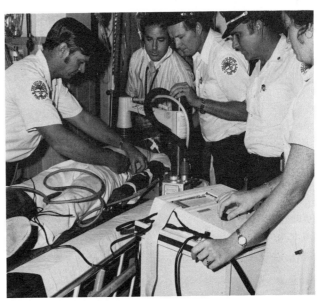

Fig. 1-2. Thorough training under the direction of competent medical personnel will assure the highest caliber of prehospital emergency care for the sick and injured. Under the close supervision of a physician and a nurse, emergency medical technicians here assist the hospital staff. The patient is receiving CPR mechanically, he has been intubated, an intravenous infusion lifeline has been established, and cardiac activity is being monitored. This type of training is highly beneficial to everyone concerned: the paramedics become proficient and the hospital staff becomes acquainted with the skills, techniques, and knowledge that is possessed by the EMTs. (Courtesy of Miami Fire Department.)

nized as a vital public service. Because of its lifesaving potential, prehospital emergency care should receive the same support and status as fire and police protection. A well-coordinated, medically directed emergency medical service program will necessarily involve integratsonnel and medical leaders. Such a project is a positive step toward the assurance of quality prehospital care of the sick and injured (Figure 1-2).

According to the National Apprenticeship and Training Standards for Emergency Medical Technicians, the recommended professional titles are defined as follows:

Emergency Medical Technician (EMT) includes both prehospital care and transportation personnel at several levels.

EMT-A shall mean Emergency Medical Technician—Ambulance.

EMT-P shall mean Emergency Medical Technician—Paramedic.

EMT-A Apprentice shall mean an applicant who has been employed to acquire the skills, knowledge, and ability to become an EMT-A.

EMT-P Apprentice shall mean the EMT-A or equivalent who has been employed to acquire the additional skills, knowledge, and ability required to become an EMT-P.

EMS shall mean Emergency Medical Services.

Terminology

Throughout the world the personnel who are trained for and provide rescue services and emergency medical care have various titles that may or may not reflect their level of training and experience. It is not practical to attempt to compare the competence or knowledge of personnel in various jurisdictions by their titles. In this book, to prevent redundancy and to eliminate any thought that certain techniques and procedures are restricted to personnel of a certain title, the terms emergency service personnel, paramedic, emergency medical technician (EMT), rescuer, and ambulance attendant shall be used interchangeably.

PERSONNEL QUALIFICATIONS

Rescue operations can best be performed by qualified personnel acting in a prudent and cautious manner. Personnel selected should definitely be interested in, and have a desire to do, rescue work. To continue to be an asset to the organization, every person in the crew must possess a constant enthusiasm for keeping abreast of the latest rescue techniques and procedures.

Rescue personnel should be able to act on their own initiative, determine what is to be done, and then do it. Rescue operations often preclude close supervision. All personnel must cooperate, so the crew can work together as a team.

Mechanical aptitude is mandatory. Rescue squads are called to many types of accidents. Emergency vehicles carry a large variety of forcible-entry and extrication tools and equipment, but very often the ideal tool cannot be obtained, so equipment and methods must be improvised.

Squad members must have enough education to converse intelligently with medical personnel. While it is not necessary to be able to use a profusion of medical terms, personnel should be interested in learning and using the terms in common use. Many of these terms will be encountered in this text.

Discipline

Discipline means a willing and dutiful submission to control. In the public safety and emergency care services, obedience and subordination to authority and rules of conduct are essential, because a delay in action or a reluctance to perform an assigned task could have disastrous results.

Rescue operations are similar to military maneuvers in that when discipline breaks down, the battle is lost. To save life and property in the midst of a chaotic accident or a hazardous emergency, a strong disciplinarian must control the activities on the scene. The chain of command in an organization has two purposes: to facilitate

RESCUE EQUIPMENT AND GENERAL RESCUE TECHNIQUES

successful operations, and to protect the individual employees. Rescue operations are dangerous and usually require a great amount of exertion. These difficulties are multiplied when rescue activities are attempted in an uncooperative and haphazard manner. The officer in charge, either personally or through subordinate officers, must be constantly in touch with all members of the rescue squad to ensure efficient operations and safety of personnel under constantly changing conditions. Individuals must discipline themselves to keep in touch with their individual officers and other rescue personnel at all times. Heat, smoke, fire, toxic fumes, and other hazards often make it difficult for crew members to carry out their assignments or hold their ground. It requires strong personal discipline to stand rather than run, yet all members must realize that in many situations the lives of fellow crew members may be in the hands of an individual. Many rescuers have lost their lives because of a break in the chain of command.

Training

Rescue operations often require that paramedics risk their lives to save others, so all rescue squad members must be prepared to protect themselves to keep from becoming casualties. The attitude and conduct of squad members must at all times reflect a sincere dedication to serving fellow human beings. Moral and ethical standards must remain high. Rescue personnel must seek always to increase their knowledge and skills, to perform to the best of their abilities (with full recognition of their own limitations), and to accept and benefit from constructive criticism and advice.

EMTs must be thoroughly familiar with the usefulness and limitations of their emergency vehicle, communications equipment, and every item of rescue supplies and equipment available.

All personnel must have a thorough knowledge of the use, protection, and limitations of the protective clothing and breathing apparatus with which they are equipped. They must be trained to recognize that safety clothing and equipment is provided for their protection and must be worn. The situation determines the amount of protection to be used. Crew members must have a thorough knowledge of the life hazards to which they may be exposed during rescue operations, and the precautions that should be observed.

Preplanning and practice can be a tremendous aid in all forms of rescue. Situations can be anticipated, outlined, and solved in practice sessions so that all personnel can gain experience and confidence in their own abilities.

Uniforms

Some type of uniform should be worn by all rescue personnel. Greater emphasis is now being placed on appearance and personal grooming, since the professional appearance of ambulance attendants and rescue crews will inspire confidence and respect in both patients and bystanders. Clean clothing is essential; under ideal conditions, ambulance attendants would wear the white clothing characteristically associated with hospital personnel. However, white uniforms are usually impracticable, since attendants are often not employed full time on an ambulance; they may be fire fighters, police officers, or volunteers. A white one-piece coverall garment will transform an individual into a more professional-looking person in a few seconds. Many crews wear caps and jackets that designate their organizations; these can be slipped on quickly. A badge or shoulder patch should be attached to the clothing; all uniforms should provide some means of identifying the members of the organization.

Cleanliness

A carefully groomed appearance promotes respect. Cleanliness and neatness are important—not only in preventing further infection of the victim, but also in stopping the spread of communicable diseases. Hands and fingernails should be kept clean; a neat haircut and a shave for male rescuers are important. A dirty, unkempt appearance must be avoided. Of course, an attendant must sometimes answer a call in working clothes, but this should be the exception rather than the rule.

One of the worst habits that ambulance personnel can develop is smoking while working on a patient. This practice is not only offensive to the victim, but also represents a possible source of contamination, if ashes should drop into an open wound.

TRAINING FOR LAW ENFORCEMENT OFFICERS

Law enforcement personnel of all ranks should be continually trained and advised so they can administer competent emergency medical care. Members of the police or highway patrol are usually on the scene of traffic accidents before the ambulance arrives—often before it has been summoned—which makes initial lifesaving measures their responsibility. The victim may suffocate or bleed to death if prompt measures are not taken. Well-meaning but ignorant attempts at removing an injured person from a wrecked vehicle may cause further injury, and untrained rescuers may not recognize the signs of a seriously ill patient or critically injured accident victim. A person with relatively minor injuries can suffer irreversible brain damage or even die, because no one thinks to turn the victim on one side or on the stomach, so mucus and blood can drain from a clogged airway.

Many prescription and over-the-counter drugs can affect judgment and behavior either favorably or unfavorably. That many drugs, particularly those prescribed by physicians, undoubtedly contribute substantially to highway safety by maintaining the health and well-being of their users is commonly overlooked. Erratic driving and collisions are commonly attributed to the driver's use of alcohol or drugs, but even if there is an odor of alcohol or some sign of drug addiction, the victim often deserves more consideration than to be jailed as a drunk. Police officers have sometimes failed to recognize medical signs and symptoms, and therefore have not summoned a physician or ambulance personnel so a proper diagnosis could be made. The National Highway Traffic Safety Administration has developed a training course specifically for the law enforcement officer, entitled "Crash Injury Management."

PROFESSIONAL CONDUCT

Whether paramedics are paid for their services or volunteer their time and effort, they should conduct themselves in a professional manner while offering emergency medical care or other services to a patient. The nature of the services provided by members of rescue squads and ambulances, and the conditions under which their work is performed, require that each member exercise particular care in personal conduct.

A professional manner is demonstrated by a person who is

RESCUE EQUIPMENT AND GENERAL RESCUE TECHNIQUES

clean and neat; who wears some type of uniform that sets off the paramedic from the crowd; who remains calm when those all around are excited; who is able to render emergency care in an efficient, confident manner; who will operate the emergency vehicle in a safe way; and who is sympathetic but firm in relations with the victim and the victim's family. A professional emergency medical technician will combine with medical skills and practices the attitudes and compassion of a good Samaritan in every endeavor involving patients, relatives, associates, and the general public.

Attitude

A serious, professional attitude must be maintained at all times; unnecessary loud talk and "horseplay" are out of place. Arrogance and superiority will be resented. A true professional must be courteous, pleasant, and considerate. People react differently to emergency situations; the ability to remain calm under all circumstances is not acquired easily. It takes time, training, and experience for rescue squad and ambulance personnel to develop a professional attitude. Kindness, tact, and good judgment must prevail in all dealings with the victim, the family, and all other people concerned with the emergency.

Personnel selected for a paramedical group should possess special qualifications, because the success of a rescue operation depends on the capability of every person assigned to it. It is not easy to look at the victims of an accident and, knowing that their chance for life is negligible, still remain calm. It is equally difficult to handle a small child afflicted with a serious illness without becoming emotionally involved. Many emergency incidents will include hideous sights, screaming victims, and utter confusion. No person who adds to the agitation by loud shouting or frantic gestures can be considered a true professional. A calm, efficient attitude on the part of all personnel will help the injured persons most. It takes a special kind of person to be an effective member of a rescue squad or ambulance crew—not everyone can qualify. For those who can, the often unspoken, yet perceptible, gratitude of patients and their families, and the satisfaction of doing the job well, provide rewards beyond description.

Emergency service personnel must have sympathy, compassion, and an understanding of the pain, worry, and anguish that more unfortunate citizens are subjected to. Criteria for these aspects of emergency medical care and victim rescue are rooted in common sense, understanding, and the awareness of others as human beings with distinct feelings and personal needs. The competent emergency medical practitioner must have the ability to recognize difficulties and to deal with them without subjecting a patient to undue mental or physical anguish (Figure 1-3).

It is imprudent and irresponsible to subject the public to the idiosyncrasies of well-meaning but psychologically unqualified personnel who fail to recognize the need for, or do not possess, the necessary attitude, temperament, disposition, compassion, and controlled constitution required for emergency rescue work. The supervisor must be aware of the psychological health of the crew, and be able to recognize these danger signs. A change in a rescuer's attitude should be carefully assessed. The supervisor must be willing to allocate the time, thought, and effort needed to evaluate, counsel, correct, and improve—or if necessary, terminate—members of the crew who do not maintain a professional attitude. The competent delivery of emergency care requires more than practical application of training.

Controlling The Emotions

Paramedics will often be confronted with situations that tax their ability to remain calm and to perform effectively. Great fortitude is necessary if one is to witness horrifying events and yet exhibit self-control and the ability to respond efficiently to the suffering of others. Such an attribute can be developed only through training, experience in dealing with all degrees of physical and mental distress, and especially through an unswerving dedication to serve humanity. Even the most experienced physician or combat-hardened medical corps member may sometimes find it difficult to submerge personal reactions and proceed without hesitation in their work—to release victims from life-endangering situations, administer life support measures to the mutilated, or recover the remains of those mangled in violent highway accidents, aircraft disasters, or explosions. The EMTs must be prepared to face these situations with equanimity and to fulfill their responsibility as members of the medical team.

Rescue personnel must remain calm, be sympathetic without becoming emotionally involved, and assure every patient that the worst is over and the help so desperately needed is already at hand. Everyone has emotions. Love, sympathy, fright, anger, depression, excitement, and many other feelings are integral parts of a person's personality. Controlling these emotions is not an easy task; yet effective ambulance attendants must exercise this control when dealing with patients and their families. They cannot be devoid of emotions, yet must not allow their emotions to gain control over their judgment and competence.

The manner in which the EMT cares for the patient may, in some respects, be as important as the emergency medical care measures themselves. The EMTs must be in complete control of their emotions, even though the scene around them may be a confusing chaos. If they allow the sight of blood, the cries of the injured, or the urgings of the onlookers to affect them, their emotions will be noticed by patients and bystanders alike, and they will be of little value in the emergency situation. It is not that paramedics should have no sympathy for the suffering victims, but rather that they should be able to control their feelings until after the emergency is over.

A good way for a paramedic to overcome emotional involvement when confronted with a gruesome, unnerving sight is to stop, take a good deep breath, and then become totally immersed in the treatment of the patient so that everything else is shut out of the mind.

Conversation

In many instances, the anxieties of a patient will be increased by discussing the injuries or illness; in other circumstances, the patient's anxiety may be increased by silence. The answer to the problem lies somewhere between these two extremes. There can be no standing rule that will apply to every patient, because personalities differ. One person will want to know the truth about the injuries, another will not. One person may be extremely upset and emotional, another may not. There is no simple statement or answer that will relieve all the patient's anxieties; the attendant will have to "play it by ear." The EMTs must not rely on a statement such as, "Everything is all right now," or "Everything is fine." Obviously, everything is not fine or the victim would not be on the way to the hospital (Figure 1-4).

RESCUE EQUIPMENT AND GENERAL RESCUE TECHNIQUES

Fig. 1-3. Rescue personnel must treat all victims with compassion and understanding. Mental trauma, anxiety, grief, despair, and other emotional problems will often incapacitate a greater number of people than physical trauma and sudden illness. Paramedics are shown transporting an adult and a child in a Stokes stretcher across a flooded river. (Courtesy of Los Angeles County Sheriff's Department.)

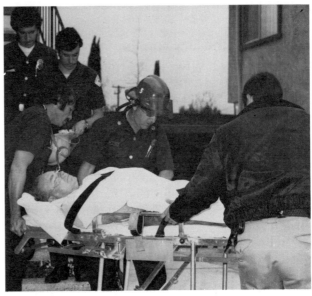

Fig. 1-4. Regardless of the state of consciousness, never discuss the patient's illness or injuries.

Never, under any circumstances, should paramedical personnel offer opinions about the extent of the victim's injury. This is not their responsibility; moreover, they are not qualified to offer medical opinions.

Emergency personnel should never, under any circumstances, attempt to estimate the time required for the recovery of the patient; this is the responsibility of the physician. If a well-meaning attendant convinces the patient that the recovery period will be short and a doctor determines otherwise, a great amount of unnecessary anxiety and disappointment often results.

Injuries or illness should not be discussed with the patient at all. If the patient insists, and all attempts to change the subject fail, the attendant can only give reassurances that the hospital staff will be able to help and that the patient will recover. It will be beneficial to the patient if, through reassurance and a calm, efficient demeanor on the part of the attendant, anxieties are reduced.

A patient should be spared trivialities or unnecessary conversation. Rescue personnel should use discretion in talking about the case in front of the patient or the family, or in furnishing information about the case to anyone else.

Tone of Voice

Many emergency scenes, especially those in which several persons have been injured, tend to become shouting matches and displays of utter confusion. Psychologists have found that the louder a person speaks, the louder others will respond until all are shouting and screaming at one another. Shouting and confusion transmit excitement, fear, and a sense of urgency to the patient; these actions magnify the victim's fears of the extent of the injuries. The EMT who can stay cool and calm under the stress of the emergency should demonstrate this attitude by speaking in a gentle, placid, normal voice, because this will relieve some of the apprehension of patients and their families. Such composure will also help the attendant to remain calm, and this is vitally important.

Cooperation With Medical Personnel

Paramedical personnel must establish and maintain effective liaison with the medical personnel with whom they come in contact, so all persons who treat the patient work as a team.

Rescue personnel should never assume the role of physician. If possible, they should immediately summon medical assistance when needed. They should not enter into technical discussions that might lead to arguments with a doctor or nurse. Although a nurse is usually operating under orders from a physician and is the physician's representative, good judgment must be used by EMTs in following such orders. If in doubt, and there is time, emergency personnel should speak directly with the physician or superior officer.

CIVIL LIABILITY

All emergency service personnel are legally obligated to perform certain services, and also have a moral responsibility to achieve these with caution and safety. Inappropriate actions are subject to scrutiny and criticism. Rushing blindly into a rescue or emergency care problem without taking all safety precautions is not only dangerous, but could create legal problems. Proper training and intelligent action are the best protection against legal involvement. If paramedics do all they can to rescue the victim from a hazardous location, intelligently support life in an injured or ill person, provide efficient and appropriate transportation to a medical facility, and, above all, use their common sense, the risk of legal liability is indeed small. Lay persons who have been asked whether they would give aid in an accident have stated their compassion is limited by the fear of a lawsuit; there is a widespread dread of being sued as the result of stopping and giving assistance to an injured stranger. The fear of legal entanglements should not discourage anyone's providing the assistance that is morally owed to every sick or injured person.

The following points of law apply to most states, and are general in application. It is generally held that no person is legally bound to offer or administer emergency care to any other person unless:
1. There is a relationship of *in loco parentis*; that is the person in question (teacher, guardian, game supervisor, camp supervisor, etc.) is acting as a parent to a minor or incompetent who is injured.
2. There is some specialized relationship between the victim and the person giving first aid, such as host and guest, employer and employee, etc.

In either of these situations, the law does not require that the uninjured person act in all situations; but if the uninjured person does not offer emergency care to the injured, civil liability may be incurred.

One is required to render emergency care or be liable for damages resulting from not rendering it under the following circumstances:
1. When the accident happened through any action of the uninjured person, and through the negligence or fault of the uninjured person.
2. In an automobile accident, even if the fault is that of the injured person, one must offer assistance.

Any assistance rendered need only be what can be expected of any "reasonable person" under the circumstances. A person need not do everything right, as long as what is done is not obviously wrong to a reasonable person. Following the first aid and emergency care manuals is more than most individuals would do in an emergency, so that following the usual EMS procedures is clearly more than enough to protect one from civil liability.

Travelers are under no obligation to assist an injured or sick stranger whom they meet on the way. But if they attempt to render assistance, they must do so with reasonable skill and care so as to avoid causing further injury or aggravation. "Reasonable care" means the amount of care that would be exercised by the "average prudent person" under the same circumstances. Other persons, even physicians, have no legal duty to aid a stranger; but there is a high moral and ethical obligation to offer aid when possible.

Volunteers are not obliged to pay for emergency service rendered by physicians and hospitals in the absence of an agreement to do so. Any individual may call a hospital, doctor, ambulance, or other emergency service and not be liable for the cost. A person may volunteer emergency care, but no one can order or compel anyone to help.

It is a police officer's duty to render or provide care for the injured; failure to do so may make the officer liable to prosecution. A police officer may question the patient, but not if the patient's condition will be worsened by the interrogation.

A person is not liable if standard first aid methods are used properly and the person does not utilize self-made or self-planned treatments. A first aider who allows a simple (closed) fracture to become compound (open) through carelessness or willful misconduct may be liable. A first aider, once assuming responsibility for emergency care, must remain in charge of the patient until the patient is turned over to a qualified person, such as a doctor or an ambulance crewman. First aiders must summon medical aid or make sure that it is called.

Every emergency medical technician should remember that his or her training is in emergency medical care, not the law. Therefore, all patient care decisions must be based on the standards of good medical care, not on legal implications. Any analysis based on avoiding legal problems, instead of considering only the medical circumstances, is dangerous for both the patients and the rescue personnel.

Under no circumstances should a patient be permitted to sign any papers unless the patient is fully conscious and lucid. No person is liable for a contract entered into before being fully conscious or without having complete use of all faculties.

Abandonment

Once anyone has begun to examine or assist an injured or ill person, he or she must remain with the patient until care is assumed by a nurse, physician, or other medically trained individual of at least equal competence. Leaving the patient before this could constitute abandonment.

Once any assistance is commenced, one must do everything possible to help the patient, or at least be sure that the injured person is not left in worse plight than before assistance was rendered. The basis for this rule is that the very presence of a rescuer at the scene might have prevented other persons from stopping to render aid, for many people will continue past an accident if they see that someone is already with the victim. If the first rescuer then leaves, the victim's chances of obtaining prompt attention may be seriously reduced.

Legal Terms

While it is not entirely out of the realm of possibility that emergency care personnel might be named in a lawsuit, most authorities consider it unlikely. Here are some legal terms that may be of interest.

Respondet Superior (let the master answer) is the legal term which, roughly translated, means that the employer is legally responsible for the harm caused by an employee as long as the individual is acting within the scope of employment. This implies that while the rescuer is on duty, the employing agency is legally liable for any errors and omissions performed in the course of emergency medical care and transportation of victims.

Institutionalized Liability refers to the fact that since the majority of EMTs receive all their instructions from a base hospital, the hospital would appear to be liable for errors in the prescribed method of treatment communicated to the paramedic.

Borrowed Servant. Generally, when an employer assigns an employee to duty for a third party and surrenders direction and control of the employee's work, the employee becomes the servant of the third party. Consequently, the primary employer is no longer liable for the actions of the employee. An example of this concept would be an EMT who is employed by a city, but who is trained and supervised by a hospital. This would also apply to a physician who might stop at the scene of an emergency and take over responsibility for victim care by assigning tasks to the EMTs. In this case, it would be advisable for the paramedics to consult with their base station to ascertain whether the physician on the scene should be allowed to take charge of the situation.

Under the doctrine of sovereign or governmental immunity, some state and municipal governments, and, to a lesser degree, the federal government, are immune from legal action unless permission is granted beforehand.

Good Samaritan Laws

Most states have passed "good Samaritan laws." The intent of these laws is to protect those who voluntarily undertake the care of injured persons at the scene of an accident from being sued as the result of their efforts. These laws are intended to relieve a good Samaritan of any fault or responsibility at law for errors or omissions in the care that is rendered. Most good Samaritan laws contain two requirements that will limit the legal protection of the person rendering care: (1) the emergency care rendered must not be negligent, and (2) the attention must be administered without payment to the good Samaritan. However, even if a state does not have such a law, no one should be deterred from stopping and rendering aid in an emergency.

Originally, the good Samaritan laws were written to protect medical personnel who might attempt to provide assistance at the scene of an accident. These laws recognize that the results of the emergency medical care may not always be perfect, or even satisfactory: needed instruments or other supplies may not be available, the physician or nurse may not be proficient in the required skills, or the victim may be too badly injured to be saved. This doctrine may afford some shelter for EMTs, but better protection is provided by rendering competent and intelligent care.

The good Samaritan laws grant an immunity, but not absolute protection by any means. The provisions of these laws differ widely from state to state. Most of the statutes provide that immunity is not granted when gross negligence or willful and wanton misconduct results in an injury to the victim. An individual paramedic could still be held liable for intentional or willful harm inflicted upon a patient.

Since payment is one of the key factors determining whether a person is protected by the good Samaritan laws, many authorities hold that personnel being paid to administer emergency medical care and to drive and operate ambulances should not consider themselves protected from civil suit by the good Samaritan laws for emergency care rendered while on the job. The exception, of course, is the off-duty paramedic who helps at the scene of an accident.

For further protection, the laws of nearly every state exempt emergency care from the medical practice laws. In other words, a person can give aid in an emergency without having to worry about being arrested for practicing medicine without a license.

Consent

Under normal circumstances, a person must give consent before medical treatment may be supplied. This permission must be the informed consent of a rational, lucid, conscious individual if it is to have any validity. Obtaining approval for the necessary emergency care from the victim of an accident or sudden illness is very often difficult, if not impossible. As a general rule, consent is implied if the patient is in immediate need of emergency care and is unconscious or too irrational to make an intelligent decision, or when the victim is a minor and a parent is not available to give consent. When a victim is unconscious, the law presumes implied consent to receive the necessary emergency care and agreement to be transported to a place where the patient can receive medical attention.

When a victim is too irrational to make an intelligent decision—whether drunk, under the influence of drugs, emotionally upset, mentally retarded, or suffering from any of the other factors that interfere with a person's ability to think clearly and intelligently—implied consent gives the attendant the right to care for the patient. A paramedic's job is to provide the most effective medical care for the patient, *not* to make the safest legal decision for himself or herself. If the victim is later declared to have been mentally competent (and therefore capable of giving consent at the time of the emergency), the EMT's error was made so that the best emergency care could be offered to that person. If a rescuer doubts a victim's mental competence, the police may be able to place the victim in protective custody; the victim then can be treated and transported under police orders. If the victim is under the supervision of a physician via radio, the rescuer should describe the situation and follow the physician's orders. Sometimes, when a victim's ability to consent to treatment is in doubt, the wife, husband, or other relative may agree that the patient requires further treatment.

If the patient refuses to be cared for or transported, is rational and lucid, and is clearly not in immediate need of medical aid, then those wishes must be respected. It is best to recommend that the patient be examined by the family physician or the local hospital, to rule out any possibility of hidden injuries. If the patient still refuses emergency care or transportation, it is a good practice to make a notation on the incident report that the patient refused aid, and to have the patient or witnesses sign it. The best rule to follow is to offer whatever emergency care is medically necessary at the moment, and then follow the patient's desires.

The law recognizes that a minor may not have sufficient wisdom, maturity, and judgment to give valid consent for certain procedures. In these situations, the right to consent is given to parents, or to individuals who are so close to the minor as to be treated as the equivalent of parents. If a bona fide emergency exists, the consent of the minor is implied. Since the parents possess the primary right to consent, an attempt should be made to reach them, but this endeavor need be only minimal. The parents' consent to emergency care may be assumed.

Right to Refuse Treatment

Many emergency patients are confused or are suffering from delusions. Under these circumstances, the refusal of treatment cannot be assumed to be a knowing rejection. On the other hand, competent adults who, for religious reasons, refuse specific kinds of treatment are generally within their legal prerogatives.

Emergency care personnel must determine as best they can whether the patient's mental condition is impaired. The primary purpose of emergency medical care is to render assistance to persons in need. Failure to provide necessary treatment to an individual creates a greater exposure to legal liability for paramedics than the administration of excessive emergency care to victims who neither require nor desire attention.

Confidentiality

Emergency care personnel have both an ethical and a legal duty not to disclose details of patient care to anyone except those immediately concerned in a patient's care. Disclosure of details without patient approval could bring a lawsuit.

Liability Releases

The question has often been raised whether rescue units should obtain a liability release before rendering aid. Court decisions do not generally favor the use of releases to avoid liability for future negligence, especially in situations where the person giving the release is in an emergency situation and has little choice in the matter. Rescue personnel should concentrate on assisting persons and property in danger to the best of their ability, realizing that the best defense against any allegation of negligence is to be able to show that a high standard of care and expertise was exercised when rendering assistance.

Negligence

Emergency medical care is not perfect. Despite advanced techniques of selecting and training paramedical personnel, there are conditions and circumstances at the scene of an emergency that create the opportunity for injury and serve to compound the irreducible element of human error.

It is very important that paramedics carefully explain the emergency procedures that they are employing to the victim and the relatives (if possible) so they will be constantly aware that the attendants are well trained and competent. If there is a physician who is in attendance or who is supervising emergency medical activities over the radio, this fact should also be conveyed. Often the victim may believe that injuries would have been less serious, or survivors may believe that a patient could have been saved, if a physician had been available.

Survivors often look for a scapegoat to blame for the death of casualties to lessen any feeling of guilt they might have. The parents of a child who drowned in a swimming pool or was crushed when running out into the street may blame themselves for not caring for the youngster better and preventing the death. But if the feeling exists that the child was alive when the paramedics arrived, and then died while en route to the hospital, the parents may attempt to shift part or all of the guilt to the rescuers, rationalizing that the death was really not their fault, but the result of poor emergency care. Automobile drivers and other persons who cause serious injuries or death to others may also use the same reasoning to avoid facing the consequences of their actions.

A paramedic may be sued for damages on a charge of negligence, even if competent, proper, and faultless care was provided to every patient. The best defense against a charge of negligence is to follow precisely the recommendations and standards of emergency care, and to document exactly certain information: the patient's appearance upon the paramedic's arrival, any prior treatment, and the emergency care rendered by the paramedic. The reports and records should thoroughly document every aspect of the incident, to refresh the paramedic's memory at a later date.

Standards of care may be imposed by statutes, ordinances, case law, administrative orders, or local customs. In many jurisdictions, violations of one of these standards is said to constitute "presumptive negligence." In these situations, negligence is assumed to exist if a violation of the standard can be shown. Persons rendering emergency medical care should familiarize themselves with the particular legal standards that exist in their state, and adhere to them religiously.

Negligence suits are referred to in legal parlance as torts. Tort law is concerned with civil wrongs committed by an individual against another person. The plaintiff (the person who sues) must prove that the defendant (the person who is sued) had a duty to the plaintiff; the plaintiff must prove that the duty was breached (violated); that injuries (damages) were suffered as a result of that breach of duty; and that there was a proximate cause (meaning that the violation of duty actually caused the injuries the plaintiff complains of). Damage in the legal sense is loss, injury, or deterioration caused by another's negligence. Of the four basic elements of negligence, the most elemental is that of duty. The duty of EMTs is to conduct themselves according to the standard of care expected of reasonably trained, prudent emergency care personnel.

By using only approved techniques and equipment, paramedics are protected from a charge of gross negligence. Qualified personnel must act in a prudent and cautious manner. A further obligation rests on the paramedics to periodically and properly examine and test their equipment and supplies. Proper records should document these tests. If the defendant is judged not to have acted prudently, and resultant damage can be proved, the accused is held liable.

The fate of most lawsuits for alleged negligence will depend on the issue of damages. In order for a person (the plaintiff) to collect in a suit for alleged negligence, he or she must allege and prove that he or she was harmed or injured in some discernible fashion. In reality, courts and juries do not give judgments for hurt feelings. It is very seldom that they will give an award for soft tissue injuries. The plaintiff has the burden of proof in demonstrating that there really are injuries that will support a judgment.

Driving Liability

States and communities have a variety of laws and ordinances that grant emergency vehicles the right of way. Many jurisdictions also have laws that protect drivers of emergency vehicles from civil suits in certain situations. All personnel must be thoroughly familiar with their legal protection and liabilities under emergency conditions.

Generally, an emergency vehicle operator is entitled to the right of way only if the other drivers and pedestrians relinquish their claims, if there is a bona fide emergency, and if the vehicle is continuously displaying its warning lights and sounding a siren.

Whether or not there is an emergency depends on three factors: (1) the nature of the call received, (2) the circumstances perceived in the mind of the operator while responding, and (3) the manner in which the operator drove in response to the call. Recent court decisions have rendered the opinion that whether the vehicle is being driven in response to an emergency call depends not on whether or not there is a bona fide emergency, but on the nature of the call received and the situation as presented in the operator's mind. In other words, on the basis of the facts then known to the driver, did the driver believe that an emergency truly existed?

The existence of a verified emergency may protect the emergency driver from liability for accidents incurred during the cautious and competent operation of the vehicle, but there are no laws to shield the driver against proven claims of gross negligence or incompetence.

QUESTIONS FOR REVIEW

1. What agency or organization has done the most to encourage the training of ambulance and rescue personnel in prehospital emergency medical care?
    a. Department of Health, Education, and Welfare
    b. National Highway Traffic Safety Administration
    c. Department of Urban Affairs
    d. American Medical Association

2. Which type of paramedic behavior and attitude will engender the most confidence among patients and bystanders?
    a. Calm and efficient demeanor
    b. Constant compassionate advice
    c. Loud shouting and frantic gestures
    d. Taciturn and uncommunicative silence

3. Which of the following patients has the most soundly based legal right to refuse prehospital emergency medical care?
    a. A child with no parents in attendance
    b. An unconscious pedestrian who was struck by a car
    c. A drunk person who is incoherent
    d. A lucid person who is refusing on religious grounds

RESCUE EQUIPMENT AND GENERAL RESCUE TECHNIQUES

4. The tragedy of accidental death is that trauma kills thousands of persons who otherwise could expect to live long and productive lives. How does accidental death rank among the most frequent causes of fatality among persons of all ages?
   a. First
   b. Second
   c. Third
   d. Fourth

5. Under what condition is a person obligated to aid a sick or injured stranger?
   a. When the person knows that beneficial help can be provided
   b. When no personal relationship exists
   c. There is a moral obligation, but no legal duty
   d. If the victim requests aid, no one can refuse to help

CHAPTER 2

Rescue Equipment

A wide array of tools and equipment is essential for effective emergency rescue operations. The list of equipment that might be needed at the scene of an emergency would be extremely long, and new, more efficient tools are constantly being designed. It is essential that all personnel have a high degree of mechanical aptitude and be acquainted with all the possible applications for the rescue tools and equipment with which their organization is equipped. It may be necessary to improvise or adapt equipment to uses for which it was not designed. Some of the more useful equipment and tools will be discussed in this chapter.

SIMPLE MACHINES

Ordinarily, a machine is thought of as a complex device, such as an automobile or a cement mixer. These are machines, but so are a hammer, a pry bar, and a saw. A machine is any device with which work can be accomplished. With great skill and ingenuity, large monuments such as the Pyramids and the Great Wall of China were constructed by using simple machines and human energy. Machines may be used to transform or to transfer energy, to increase speed, and to augment human energy. Rescue personnel are most interested in the ability of some machines to multiply force.

There are only six simple machines. They are the lever, the pulley, the wheel and axle, the inclined plane, the screw, and the gear. However, physicists recognize only two basic principles in machines: the lever and the inclined plane. The wheel and axle, the block and tackle, and gears may be considered levers. The wedge and the screw use the principle of the inclined plane. An understanding of the principles of simple machines is a necessary foundation for the study of compound machines, which are combinations of two or more simple machines.

RESCUE EQUIPMENT AND GENERAL RESCUE TECHNIQUES

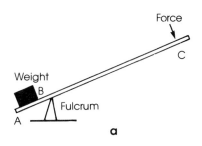

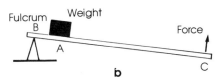

Fig. 2-1. The necessary force that must be exerted on a lever to lift a weight, and the mechanical advantage derived, depend on the ratio between the arms of the lever. To lift a very heavy weight by the exertion of a moderate force, the fulcrum should be placed as close as possible to the weight being lifted. Both of the above levers exert corresponding force.

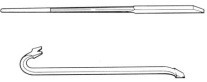

Fig. 2-2. Pry bars, claw tools, and wrecking bars all enable rescuers to exert leverage to multiply their strength. Levers are manufactured or improvised from on-the-scene materials in a great variety of sizes, shapes, and lengths when rescue operations involve lifting or moving heavy objects to remove trapped victims or to provide access into an area. Using a long wooden timber as a lever would probably release a person trapped by a fallen tree or wall faster than taking time to properly position a large jack, unless the load was excessively heavy. The bars and wrecking tools carried on rescue vehicles are useful for prying open the jammed or locked doors of automobiles or structures.

LIFTING DEVICES

Rescue operations often require lifting heavy objects to remove trapped victims or to provide access into an area. Most persons, unaided, can lift only about 80 percent of their own weight. Therefore, emergency service personnel must know how to effectively employ mechanical devices.

Both as a simple machine and as a component part of some complicated machines, the lever has many uses. Some of the more common examples of levers are the shovel, crowbar, pliers, hammer, scissors, and wrench. Essentially, a lever consists of a rigid bar capable of rotating about some point of support, known as the fulcrum or the axis.

The simplest lifting device used in rescue work is the lever or pry bar (Figures 2-1 and 2-2). Rescue personnel will use levers of various lengths and of several types (Figure 2-3). The force that a person must exert on a bar to lift a load and the mechanical advantage that is gained by using a lever depend on the ratio of the distance from the fulcrum to the worker to the distance from the fulcrum to the load. For example, if a person applies force on a bar 10 feet (3 meters) from the load and the fulcrum is 1 foot (30 centimeters) from the load, the worker would gain a 9-to-1 mechanical advantage. The advantage increases or decreases as the fulcrum is moved toward or away from the load. To lift a very heavy weight by the exertion of a moderate force, the fulcrum should be placed as close as possible to the weight (Figure 2-1).

Figure 2-1 illustrates two different types of levers: both will deliver a comparable mechanical advantage. The circumstances of an accident will determine where to place the fulcrum. In Figure 2-1, if force is applied on a bar 10 feet (3 meters) from the weight, and the fulcrum is 1 foot (30 centimeters) from the load, the worker will gain a 9-to-1 mechanical advantage. The advantage increases or decreases as the fulcrum is moved toward or away from the load.

Thus, as illustrated, distance AC is 10 feet (3 meters). Fulcrum B is 1 foot (30 centimeters) from the center of weight W. If the weight is 720 pounds (324 kilograms), the person must exert a force of 80 pounds (36 kilograms) on the bar to lift the weight. Mathematically, the problem is solved by the following proportion:

$$\frac{\text{Load}}{\text{Force}} = \frac{BC}{AB}$$

$$\frac{720}{F} = \frac{9}{1}$$

$$9F = 720$$

$$F = 80$$

The fulcrum must be made of some material that will not slip or crumble. If a metal beam is being lifted with a metal bar, a piece of wood placed between the bar and the beam will prevent the beam from slipping. Loads should be blocked up as they are lifted in order to take the weight off the bar, or to obtain an additional mechanical advantage in lifting the weight higher. Blocking will also prevent the load from dropping completely if slippage should occur.

Jacks

Jacks are used in lifting heavy loads and are of three general types.
Ratchet. The ratchet jack is a simple jack that raises a load by means

of a lever working against a ratchet that supports the load between each lifting stroke of the lever. Ratchet jacks are simple in construction and are manufactured in various sizes and lifting capacities. They are used frequently in rescue work.

Screw. The screw jack is operated by means of a lever that rotates a screw, which, in turn, raises the load. The screw jack is probably the safest type of jack, since all danger of slip-back is eliminated.

Hydraulic. The hydraulic jack consists of two cylinders containing tightly fitting pistons and a reservoir of oil. The smaller piston is attached to a lever and acts as a pump. This pump forces oil under pressure to the larger cylinder, forcing the piston outward and increasing the distance between the base of the jack and the top of the piston. The mechanical advantage is increased in the same proportion as the ratio of the areas of the two pistons. For example, if the area of the pump piston is 1 square inch (6 square centimeters) and the area of the ram piston is 100 square inches (600 square centimeters), the force exerted by the ram piston would be 100 times that placed on the pump piston. The mechanical advantage given the operator of a hydraulic jack is further multiplied by the leverage of the jack handle (Figure 2-4).

This type of jack is very useful since it can be made to lift heavy weights and is not unwieldly. However, a load should not be allowed to rest on a hydraulic jack for any appreciable length of time because the cylinder may leak and allow the load to drop.

The Use of Jacks. For safe and efficient use of jacks, the following rules should be observed:

1. When under a load, a jack should stand squarely on a heavy timber or other substantial footing so that it will not slip or sink into the ground. The footing must be dry and free from grease so that the jack will not slip.

Fig. 2-3. Concern is mirrored on the faces of rescuers as they labor with pry bars to remove the trapped victims of an accident. Pry bars, which are long metal levers, enable human energy to be multiplied. Mechanical advantage is gained by using a bar; the advantage depends on the ratio of the distance between the worker and the fulcrum to the distance between the load and the fulcrum. Levers can often be improvised from lumber or metal at the scene when a heavy weight must be lifted or moved. (Courtesy of the Chicago Fire Department.)

Fig. 2-4. Large hydraulic jacks are used to raise a truck off a crushed car so the driver can be removed.
(Courtesy of the Los Angeles City Fire Department.)

RESCUE EQUIPMENT AND GENERAL RESCUE TECHNIQUES

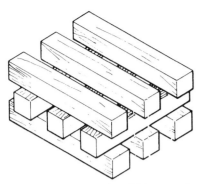

Fig. 2-5. Cribbing should be used to stabilize vehicles and other objects rescue workers are raising, to prevent the load from dropping completely if slippage should occur. A crosstie crib is made up of timbers stacked up in alternate directions to support jacks, heavy objects, or unstable weights, or to provide any other kind of bracing or shoring. It is important to use timbers of equal thickness and to level the ground where the cribbing will be placed.

2. As the weight is lifted, supports of solid material should be placed under it to prevent damage if the jack should fall. The weight of raised sections of walls and floors should not be allowed to rest entirely on the jack while rescue workers crawl underneath. Cribbing, a type of support, should be set under heavy weights or section of floors to prevent settling after the lifting operation is completed.
3. Jacks should be inspected and tested regularly. They should be kept clean, with working parts well oiled and greased.
4. When using several jacks under one load, all jacks must lift or lower together, so the load will not tip or too much weight be placed on one jack.
5. When lifting a metal object, a wood plank should be placed between the jack and the object to prevent slipping.
6. Be sure the "up" and "down" latch dogs are not cracked on ratchet jacks.
7. Be sure the jack handle fits the handle socket and always remove it when not in use.
8. Since the base of the jack is comparatively small, a plank or board should be placed underneath to obtain a greater bearing surface on soft ground.

Cribbing and Wedges

Cribbing and wedges, preferably made of hardwood, should be carried by every rescue squad to stabilize vehicles and support lifting and spreading devices. The most versatile cribbing is made of 2-inch by 4-inch and 4-inch by 4-inch lumber and is between 18 inches (45 centimeters) and 24 inches (60 centimeters) in length. To facilitate handling, it is a good idea to drill a hole through the width of the block at one end and splice a short length of light rope into a loop through the hole to form a handle. These rope loops will not interfere with stacking, and they will allow one person to carry many blocks. Cribbing can also be accomplished with timbers of various sizes ranging from 1 to 8 feet (0.3 to 2.4 meters) in length. The size of timber to use depends on the weight to be supported and the available working space. The ground should be level at the point where the cribbing is to be set; it is better to level off high spots than to fill in low ones. The crosstie crib is frequently used, because it is a safe method of supporting heavy weights (Figure 2-5).

Wedges are most durable if made of hardwood (Figure 2-6). They should have a long taper and be 2 to 4 inches (5 to 10 centimeters) thick at the large end. Wedges are used when a thin space requires filling, to snug up and steady an object being stabilized, or to help secure a wrecked vehicle.

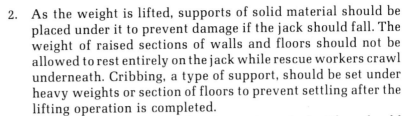

Fig. 2-6. Wedges can be used to exert a powerful lifting or spreading force. The more gradual the angle of the wedge, the greater the force that is exerted when the wedge is driven into a thin space.

Block and Tackle

The block and tackle is used to lift or move heavy weights, beams, sections of floors, and heavy timbers. Like levers and jacks, this device provides a mechanical advantage, enabling rescue personnel to move otherwise immovable obstructions. A block is composed of a frame made either of wood or metal, and a pulley or sheave, usually made of steel or hardwood. Blocks may be single, double, treble, and so forth. They are designated by the number of sheaves they contain. The lifting power is multiplied in ratio to the number of sheaves used. A block and tackle increases lifting power but reduces lifting speed (see Figures 2-7, 2-8, and 2-9).

For heavy work with a three- or four-sheave block, the fall (rope) should be rove (passed through the blocks) with the hauling part leading from the middle sheaves of the blocks instead of the outer sheaves. This reduces the chances of slewing the blocks in their straps. The hauling part of a tackle should be kept as nearly parallel to the other parts as possible, to prevent loss of power.

Safe Load. Generally, the safe load for a well-made block is in excess of the strength of the rope it will reeve. However, this is not always true of the hook. The hook is usually the weakest part of the block; its strength invariably measures the strength of the block. In lifting, the pull on the support of the fixed block is greater than the pull of the weight being lifted. This point is often overlooked. If it becomes necessary to determine the tension on a support to which the fixed block is made fast, the tension on the hauling rope must be added to the direct downward pull of the weight.

Lifting Power. Rope tackles are valuable not only to multiply power, but also to help apply power smoothly. Rope tackles permit an object to be lowered easily and exactly, because what is gained in power is lost in speed. A block is identified by the number of sheaves it contains, and the number of pulleys determines the power gain. Generally, the weight that can be lifted is equal to the applied force times the number of ropes leaving the movable block (Figure 2-10).

The rule for determining the lifting power of tackle is that the power at the movable block is to the power on the hauling part (rope) as the number of parts at the movable block is to 1:

$$\frac{\text{Power movable block}}{\text{Power hauling part}} = \frac{\text{Number parts at movable block}}{1}$$

However, this is not an exact calculation, because a certain amount of power is lost at each sheave through friction. Experience shows that this loss is represented by a 10 percent increase in the load for each sheave of the tackle. To find the power required to lift a given weight, add 10 percent friction at each sheave and divide by the number of parts (ropes) at the movable block. The above rule is for average working conditions and will vary according to changing friction at various speeds.

Example: How much power is required to lift 100 pounds with two double blocks?

10 percent × 4 (number of sheaves) = 40 percent
40 percent × 100 pounds (45 kilograms) weight = 40 pounds
                                             (18 kilograms)
100 pounds (45 kilograms) + 40 pounds (18 kilograms) = 140 pounds
                                                     (63 kilograms)
140 ÷ 4 (ropes leaving movable block) = 35 pounds
                                      (16 kilograms)

Chain Hoist

Chain hoists are often found in garages, factories, wreckers, and tow cars. They are usually used for raising and lowering weights because they are not too effective when used horizontally. The gear ratio of the hoist mechanism determines the mechanical advantage of the pull on the lifting chain. Chain hoists may be difficult to operate at night.

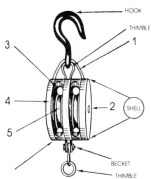

Fig. 2-7. The components of a block. Shell or Frame — Part which holds sheave or wheel. 1. Strap — Iron or rope passing around the shell forming attachments for making fast the "hook" at one end and the "eye" at the other. 2. Pin — Axle upon which the sheave turns in shell. 3. Swallows — The space between the sheave and frame of the block through which the rope passes. 4. Cheeks — The side pieces of the frame of a block.
5. Sheave — Wheel or grooved pulley over which rope runs.
6. Breech — The end of the block opposite the swallow.

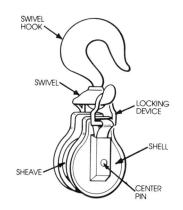

Fig. 2-8. A snatch block is used to change the hauling direction of a rope or cable. The side may be opened so the entire length of the rope will not have to be threaded through the block.

RESCUE EQUIPMENT AND GENERAL RESCUE TECHNIQUES

Fig. 2-9. The simplest pulley system consists of a rope passing around a single wheel mounted on an axle fixed to some stationary support. There is no increase in power. The gain is only in height or lift or change in direction of pull.

Winches

Winches on tow trucks, wreckers, and rescue squad vehicles are extremely useful for lifting victims from ditches, canyons, and other excavations. A snatch block may be needed to allow the pull to be directed in the correct direction.

Come-Along

Come-alongs, sometimes referred to as hand winches, will permit one person to create a force of up to several tons, depending on the model. These ratchet devices enable a cable or chain to create a force that can be directed around corners to stabilize vehicles, bend steering columns, pull car seats off the tracks, and perform many other operations at the scene of rescue incidents (Figure 2-11).

Hurst Power Tool

The Hurst power tool is a hydraulic spreading and pulling device that is completely portable, requires only one operator, and is ideally suited for extricating persons trapped in any type of wreckage requiring the use of great force (Figure 2-12).

This tool consists of a 5-horsepower engine which drives a hydraulic pump. Powerful jaws are connected to the pump by 16-foot-long (5-meter-long) hoses, which attach to the tool with quick-disconnect fittings. The arms of the working device are activated by thumb controls located on each handgrip. When the arms are fully extended, they span a distance of 32 inches (80 centimeters). Two types of jaws can be fitted to the arms and held in place by retaining pins. One type is for spreading and lifting; it is commonly used at automobile accidents, elevator incidents, building collapses, and other rescue scenes in which heavy materials require moving. The other jaws are used for cutting—for example, at aircraft accidents. The Hurst power tool operates with a maximum output of 10,000 pounds (4,500 kilo-

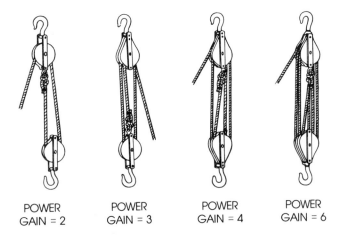

Fig. 2-10. Block and tackle combinations and power gains. When friction is ignored, the weight that can be lifted is equal to the applied force times the number of ropes leaving the movable block. In the above illustrations, the shown power gain is obtained when the higher block is fixed and the lower block is movable. If the block and tackle is reversed, and the upper hook is fastened to the movable load, the power gain will be increased by one.

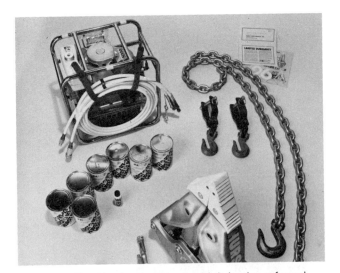

Fig. 2-12. The Hurst power tool, which is also referred to as the jaws of life, uses a gasoline engine to develop sufficient hydraulic pressure to exert five tons of force at the jaw tips. Using various snap-on jaws, the tool can spread, pull, cut, shear, or pierce. Courtesy of Hurst Performance, Inc. The "Jaws of Life" is a registered trademark of Hurst Safety Products.

grams) of force at the tips of the jaws; extreme caution, therefore, must be observed while operating this tool. The operator should wear full protective equipment, including gloves and face mask or safety goggles.

The engine should not be started until all hose connections are made up and checked. Allow the engine to warm up, then open the choke. The controls should be operated through several cycles until the arms operate smoothly through their full range of movement from open to closed. In extreme emergencies, the tool can be place in operation immediately. To open or spread the arms of the tool, the thumb controls are moved in a direction away from the cylinder body. To close the arms, move the controls in toward the cylinder body. If the controls do not operate the arms in this manner, the hoses have been incorrectly connected either at the tool or the power unit. The engine is stopped by depressing the grounding switch. Before shutting down, the jaws should be left about 1/2 inch (15 centimeters) apart to allow for changing the jaws without starting up the engine.

Fig. 2-11. A come-along hand winch will enable one person to exert tremendous force. This device can operate either vertically or horizontally at rescue incidents to release trapped victims.

Hydraulic Rescue Equipment

A remote controlled hydraulic jack has been developed that will greatly simplify and speed up many rescue operations. This jack consists of a hydraulic unit with a pump (source of power) separated from the ram (power end) by a section of long, flexible hose. A wide range of rams and attachments provides countless combinations for pushing, pulling, lifting, pressing, bending, spreading, and clamping. The ram end can be used in any position (Figure 2-13).

The jacks vary in size from a 4-ton (3.6 metric tons) light duty set to a powerful 50-ton (45 metric tons) unit. The light duty unit is

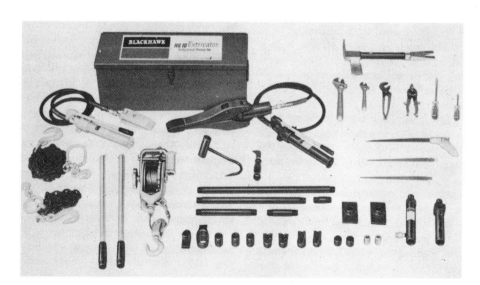

Fig. 2-13. A complete hydraulic rescue set which contains two Porto-Power® units for extricating victims from wrecked cars, lifting heavy objects, forcible entry into buildings, and other applications in which the powerful hydraulic rams can be used to push, spread, or brace heavy objects. In addition, the set contains a heavy-duty come-along, metal saws, pry-axe, chains, and tools which are necessary to cope with automotive and industrial accidents. (Courtesy of the Blackhawk division of Applied Power, Inc.)

RESCUE EQUIPMENT AND GENERAL RESCUE TECHNIQUES

ideal for removal of trapped victims from automobile wrecks, for forcible entry into buildings, and for similar situations where casualties may be caught or pinned. The unit can be used near gasoline, gas, or any other flammable vapors that would preclude the use of an acetylene cutting torch. Remote control of the hydraulic ram, plus the extra long hose, enable the rescuers to work at a safe distance when necessary.

The larger sets are valuable at train, bus, and truck accidents, mine cave-ins, building collapses, and other incidents where casualties may be pinned under concrete slabs, sections of collapsed buildings, street cars, or heavy machinery. There are many situations where such an extremely powerful, compact, and versatile power tool can be used.

Air Bags

Air lift bags, which can quickly be inflated with compressed air from a bottle or other source, are extremely useful in all types of lifting and spreading operations. These bags are manufactured in a variety of sizes, with a corresponding lift capacity that may range up to several tons. The top and bottom pieces are constructed of laminated neoprene and nylon fabric about 3/8 inch (9.4 millimeters) thick to resist cuts, tears, and punctures. The sides are of nylon and neoprene; they are thinner to permit flexibility (Figure 2-14).

Fig. 2-14. Air lift bags that can be quickly inflated are extremely useful in all types of lifting and spreading operations. As they have a large bearing surface, they will readily conform to soft and uneven surfaces.

Air lift bags require only 1-1/2 inch (37.5 millimeters) clearance for placement, an advantage they have over other types of jacking equipment. A maximum of 6 pounds per square inch (psi) of air pressure is needed for inflation. As there is a large bearing surface both top and bottom, these bags are ideally suited for lifting soft-sided vehicles such as buses, cargo vans, house trailers, and aircraft. They also readily conform to soft and uneven surfaces, such as muddy ground, snow, and other terrain where jacks with small bearing surfaces may sink. The bags can be used in a horizontal position to expand shoring in a collapsing ditch or to spread the bars on a door or window.

The necessity for blocking or otherwise stabilizing an object to be lifted is an important part of the accident scene size-up. One must attempt to visualize and try to anticipate the direction and force of movement that take place when lifting an object causes a change in its center of gravity. Rescuers must also consider the ultimate or immediate disposition of the object being lifted. Air bags have no lateral stability while being inflated. Thus, for example, a vehicle lying on a sloping surface might easily be lifted into a rolling position, which would introduce the possibility of further damage to itself and to other property or persons below. Such movement could be prevented or limited by blocking or using a guy line to a fixed object. Following the removal of a trapped victim, the vehicle is then lowered again to its original position. Vehicle wheels should be chocked to prevent movement during a lift.

Air lift bags may tend to pop out from under the load as the lift commences, because the angle between the bearing surfaces increases as the raising progresses in a rolling lift. It is definitely advisable to manually guide the bags, keeping the top and bottom pieces centered on each other until full contact is made on both bearing surfaces. If the bags still tend to pop out, it may be necessary to secure them with a rope loop or blocking. In some cases, a two-stage lift may be necessary.

Theoretically, it is possible to stack two or more bags on top of each other to increase the maximum height of the lift. However, this is definitely discouraged, because the lack of lateral stability will cause them to pop out of position long before any work is accomplished. This bag displacement would allow the object being raised to drop without warning. If it is necessary to exceed the maximum lift of one bag, the load should be lifted in two or more stages, using cribbing to hold the object while the bag is being repositioned. If the distance between the ground and the object is great, it is a good idea to place cribbing under the bag before the lift is started.

Manual Lifting

Average persons should not attempt to lift more than 80 percent of their own weight. If heavy manual lifting is necessary, the following procedure should be used: squat; balance properly; grasp the object; keep the back straight—not necessarily vertical; keep the feet together; lift evenly; let the legs take the load, keeping the load clear of the feet (see Figure 2-15). Reverse the procedure for lowering.

CUTTING TOOLS

Rescue workers may have to cut through heavy wood, iron beams, or steel girders, or force openings in masonry walls and concrete floors. Tools and equipment may include axes, bolt cutters, pipe cutters, hand and power saws, and portable oxygen-acetylene cutting equipment.

Axes

Axes, both flat-headed and pick-headed, are probably the tools most often used in rescue (Figure 2-16). They can be used for cutting, prying, and striking. They should be kept sharp, ready for instant use. It is best if the handles are maintained in a smooth condition and treated with boiled linseed oil or other wood conditioner, because painted handles have a tendency to blister the workers' hands during prolonged usage.

Fig. 2-15. Manual lifting. If heavy manual lifting is necessary, the person should squat; balance properly; grasp the object; keep the back straight — not necessarily vertical; lift evenly; let the legs take the load; keep the load clear of the feet. By allowing the strong leg muscles, instead of the more fragile spine, to exert the forces necessary to lift heavy weights, serious back injuries and ruptured intestines are avoided.

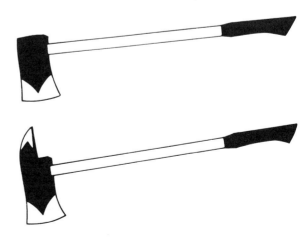

Fig. 2-16. Axes, both flat-headed and pick-headed, are extensively used in rescue operations for forcible entry and other cutting procedures.

RESCUE EQUIPMENT AND GENERAL RESCUE TECHNIQUES

Cutting Torch

An oxygen-acetylene cutting torch operator should have training and practice in the use of this tool, since it is dangerous if not handled properly. When handling the torch, the operator must make sure that in turning the flame away from the work, it is not turned on a fellow worker. The rubber hose, like most of the cutting apparatus, must not be exposed to heat, sparks, oil, or grease. These materials can easily be set on fire by oxygen under pressure, injuring rescue personnel, damaging equipment, and disrupting operations.

Portable torch working time is given in Table 2-1. The proper flame for cutting is attained by adjusting the valves on the torch. The two inner flames visible in the cone should be adjusted to become one. The metal to be cut is then preheated until it becomes cherry-red in color. Preheating is accomplished by holding the torch about 1/16 inch (1-1/2 millimeters) away from the metal. When the metal is properly heated, the torch should be moved about 1/2 inch (12-1/2 millimeters) farther from the metal. The high pressure oxygen valve is then slowly depressed, and cutting begins.

Table 2-1. PORTABLE TORCH WORKING TIME

METAL THICKNESS (in.)	METAL THICKNESS (cm)	OXYGEN PRESSURE (psi)	ACETYLENE PRESSURE (psi)	TOTAL TANK CAPACITY (minutes)
0-1	0-2.5	35	7	15
1-2	2.5-5	60	7	9

Caution: To avoid excessive discharge of acetone, in which the acetylene is dissolved, an acetylene cylinder should not have the valve opened or closed when the valve is lower than the body of the cylinder.

Objects to be cut must be braced to prevent them from falling on the victim, the operator, or other rescue workers. Further, rescue workers must be sure that in cutting a beam or support, they are not releasing additional debris or allowing other portions of the structure to collapse on themselves or trapped persons.

When cutting in a confined place, rescuers must be sure that adequate ventilation is provided. If cutting near trapped persons, they should be protected from the flame, sparks, fumes, and glare by using asbestos blankets or some other material such as wool blankets or tarpaulins soaked in water.

The operator must be especially careful that explosive fumes, gases, and liquids are not present and that the torch does not ignite combustible materials. When flammable liquids are present, the cutting torch should only be used to free trapped persons after all other methods are deemed impractical. In this case, there must always be an extinguisher or charged hose line available.

Oil, grease, graphite, or other hydrocarbon derivatives must never be permitted around or on any part of cutting apparatus. Operators should never stand directly in front of oxygen gages or regulators when opening the valves. Whenever possible, a spark lighter should be used to ignite the torch; if it is necessary to use matches or some other ignition source, they should not be held in the bare hand. The proper goggles and gloves should always be worn when using a cutting torch.

Caution must be used when it is necessary to cut into containers; they may have held flammable or explosive substances. Cutting any metal against a concrete surface should be avoided, since extreme heat causes concrete to spall (chip).

The torch tip must not touch molten metal, nor should it be used to knock out slag; the operator will have to use a hammer or chisel, because slag cannot be burned out since the metal is already oxidized.

No foreign matter can be allowed on the threads or seat; otherwise, a perfect seal may be prevented when changing tips. Only the tip cleaner provided should be employed to clean tips; tip holes are metered and must not be enlarged or damaged.

Proper speed and uniformity of motion are essential to the successful cutting operation. Too little speed results in an accumulation of slag on the underside of the cut; too much speed reduces the cutting efficiency.

The cutting lever should only be used when cutting metals containing carbon, such as iron and steel. The preheat flame may be used, in emergencies only, to melt nonferrous metals such as aluminum, copper, brass, and bronze. High carbon steels, such as are used in bumper bars and spring leaves, are more difficult to cut. They must be heated along the line of cut until the color begins to change, and then preheated to cherry-red at the starting point before proceeding with the cutting operation.

To cut cast iron, the torch should be moved from side to side along the line of cut, resulting in a cut about three times as wide as when cutting steel. More slag is produced when cutting cast iron; therefore, a wider cut is necessary to prevent the slag from accumulating and closing up the cut.

The oxygen valve should be shut off before shutting off the acetylene valve. If this procedure is reversed, a "pop" usually will occur. This "pop" throws carbon soot back into the torch and may eventually cause a flashback.

Caution: If there is spilled fuel in the area, it should be thoroughly washed down and away before any comprehensive rescue attempts are made. Fire fighters should be standing by with charged hose lines whenever rescue personnel are using flames or spark-producing equipment. Both rescuers and victims have been killed or maimed when gasoline has been ignited after a collision.

Saws

There are several different types of gasoline-powered saws that are extremely valuable for forcible entry, rescue, and ventilation work.

Rescue Saw (Figure 2-17). Rescue saws are powered with two-cycle gasoline engines; they are very portable, and various types of carbide-tipped blades and abrasive discs can be obtained to cut practically every type of material. These saws can do a tremendous amount of work in a very short period of time when used properly. But they can be unsafe and dangerous unless precautions are taken; since they have cutting blades of 12 or 14 inches in diameter, these tools should be used with caution.

When using any type of saw, observe the "buddy system"; one person concentrates on operating the saw while the other, the buddy, stands by to warn of imminent hazards, such as high-tension wires, unstable footing, or any other condition that might be a safety hazard. Both users should wear full protective clothing, including safety goggles.

A common mistake in using a rotary rescue saw is the failure to maintain normal blade speed. Allow the blade to feed into the work only as fast as the material is cut. Forcing the blade into the work in a vain effort to expedite the cutting process only lugs the engine, slows

RESCUE EQUIPMENT AND GENERAL RESCUE TECHNIQUES

down the blade speed, and reduces the cutting effectiveness; in addition, it increases the danger of breaking the abrasive discs.

To avoid binding the blade, cut in a straight line. A wavy kerf indicates a failure to hold the saw steady and the blade intermittently binds and slows down; this decreases the cutting speed and increases the danger of breaking the blade or abrasive disc. Binding can be avoided by holding the saw steady, starting a cut slowly, and then letting the saw find its own path.

When cutting ferrous metal, the abrasive disc throws out a heavy stream of sparks. Therefore, an extinguisher or a charged hose line should be ready whenever there are combustibles or flammable liquids in the area. If the saw is being used to extricate a victim from automobile wreckage, cover the person with a blanket or sheet for protection from the shower of sparks. If there is a choice, cut in the direction that will keep the flow of sparks away from both the victim and the flammables.

When opening up a roof, remove the tar paper and stones from the area in which the cut is to be made in order to prevent the tar from gumming up the blade; rocks can attain a tremendous velocity when hit by the disc.

Never stand in line with the blade unless protected by a guard. Pieces of the blades develop high speeds when they disintegrate. Test run a new blade at high speed before cutting to be sure it is safe. Always start a new cut if blades are changed, as the old cut may be too narrow for the new blade; this will cause it to bind and it may shatter or kick.

Fig. 2-17. Gasoline-powered rescue saw being used on a wrecked automobile. These rescue saws can be fitted with various types of blades to cut a wide variety of materials, including metals, masonry, wood, roofing materials, and other substances. The type of materials that can be cut is limited only by the selection of blades available. Remember that any gasoline-powered tool or appliance is a ready source of ignition whenever flammable gases, vapors, or liquids are present. (Courtesy of the Denver Fire Department. Photo by Terry Brennan.)

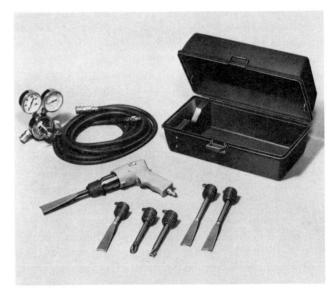

Fig. 2-18. Pneumatic rescue kit consists of an air hammer, several versatile cutting tools, pressure regulator, and hose. (Courtesy of the Ziamatic Corporation.)

Chain Saws. Chain saws are extremely valuable in providing sources of ventilation in burning buildings; openings should be made through the roofs and walls. These saws are capable of making much deeper cuts because their blades are longer than those of other saws. To reduce the risk of electrocution when cutting into a structure, always wear rubber electrician's gloves under the regular gloves; electrical wires could accidentally be cut into when making blind cuts.

One manufacturer furnishes a carbide-tipped chain that will cut brick, masonry, soft metals, and tar and gravel roofs with little trouble. These chains do not necessitate the frequent resharpening of the teeth that conventional chains require.

Pneumatic Chisel. A pneumatic chisel is one of the safest, lightest, and most economical tools for extricating automobile crash victims; in addition, it may be used for forcible entry into buildings, since it can cut openings in sheet metal doors. Body repair workers and muffler shops have been using air chisels for years; they can be safely operated in the presence of gasoline vapors (though this is not recommended), because they are free from sparks and flames. In addition to cutting sheet metal, air chisels can cut heavy door posts and can shear rivets (see Figure 2-18).

Pneumatic chisels are air hammers with chisel tips; they cut through metal with a reciprocating motion like a jackhammer. To operate, punch a small hole in the metal, hold the chisel at a 30° angle, and hold the trigger down while applying moderate, steady pressure. These tools are convenient and easily portable since the compressed air may be drawn either from a breathing apparatus air bottle with a pressure regulator or from a fitting on any vehicle equipped with air brakes. The air chisel requires about 4 cubic feet of air per minute at 100 to 150 psi pressure.

As the material cut becomes tougher, the air pressure at the regulator should be increased to give the needed power to the hammer. For example, for cutting thinner metals such as in car roofs, the regulator need only be set to approximately 90 to 100 psi. When cutting through the frame or bumper, the setting should be in the 250 to 300 psi range. Because every material being cut will differ, due to such factors as rust, rigidity, and supporting frames, it is impossible to give exact recommended air settings.

All personnel on rescue squads must thoroughly familiarize themselves with this tool in practice sessions; this will allow the correct feel of the hammer to be gained and the crew will obtain some knowledge of the appropriate air pressures. It is important to remember that the air pressure should never be set higher than needed to do a particular job. Air consumption increases as air pressure is raised and the air supply is, therefore, exhausted sooner.

There are a few safety rules that should be observed: Operators should wear full personal protection, including gloves and face mask or safety goggles; the air should be turned off or disconnected while changing chisels because an accidental triggering could launch a projectile; the cutting tool should always be held firmly against the work before starting and while making the cuts to eliminate the chance of its flying out of control and injuring someone.

ROPE

Rope is one of the most important tools of the rescue service. Rescue squads use steel-wire and fiber ropes. Ropes can be used to erect

RESCUE EQUIPMENT AND GENERAL RESCUE TECHNIQUES

derricks, jibs, and booms; to lift materials, tools, and equipment; to rescue injured persons; to join equipment such as ladders; to lower stretchers; and for lifelines. Practice and drill are the only methods through which rescue personnel can become proficient in the use of rope.

Listed in Table 2-2 are tensile strength and property comparison charts covering rope available for lifelines, and for rescue and drill operations. Note particularly the mildew resistance and relative strength of synthetic rope.

Inspection of Rope

If the age and condition of a rope is not known, the rope should be thoroughly inspected and tested before it is approved for service. The efficiency and safety with which rope can be used depends to a great extent on its receiving proper care and maintenance.

A length of rope may appear badly worn on the outside and still be in much better condition than another length which looks good on the surface. Chafing occurs in rope when the inner fibers rub or chafe against each other, as happens when a rope bends going through a pulley. Many of the central fibers become broken and

Table 2-2. ROPE COMPARISON CHARTS

AVERAGE TENSILE STRENGTHS

	½ in.	⅝ in.	¾ in.
Braided Cotton	1,000	1,200	1,400
Manila	2,650	4,400	5,400
Nylon	6,200	10,000	14,000
Dacron	4,400	7,300	10,000
Polyethylene	4,100	5,200	7,400
Polypropylene	4,200	5,800	8,200

PROPERTY COMPARISON CHART

	BRAIDED COTTON	MANILA	NYLON	DACRON	POLYETHYLENE	POLYPROPYLENE
Relative Strength	1	2	6	5	3	4
Order of Relative Weights	6	4	3	5	2	1
Working Elongation at 20% load	1	2	6	3	4	5
Relative Resistance to Impact Loading	1	2	6	3	4	5
Mildew Resistance	Poor	Poor	Excellent	Excellent	Excellent	Excellent
Acid Resistance	Poor	Poor	Fair	Excellent	Excellent	Excellent
Alkali Resistance	Poor	Poor	Excellent	Excellent	Excellent	Excellent
Sunlight Resistance	Poor	Fair	Fair	Good	Fair	Fair
Organic Solvent Resistance	Poor	Good	Good	Good	Fair	Fair
Melting Point		380°F (193°C)	480°F (249°C)	500°F (260°C)	265°F (129°C)	300°F (149°C)
Floatability		Non	Non	Non	Indefinite	Indefinite

Numbered Ratings: 1 lowest to 6 highest.

crumble into dust and small pieces. Because of this, ropes should be checked each time after being used. This inspection may save a life. In view of the many conditions that can affect the strength of rope, and since only a part of the rope may be affected, an examination of the entire length should be made at least every six months. If, upon examination, any of the following conditions are found to exist, and if the condition of the rope is such that there is any doubt as to whether or not it is safe to use, the rope should be replaced.

On the Surface of the Rope. Abrasions (broken fibers), cuts, extreme softness (when badly worn rope is extremely soft and has lost its stretch), decay, or burns caused by high temperatures or chemicals indicate the necessity of replacement.

Central Fibers of the Rope. To examine the inside of a rope, separate the strands at three foot intervals and observe the inner parts for broken fibers, fine powder, or a change in color of the fibers, which would indicate the presence of grit, mildew, or mold. Such conditions indicate the necessity of replacement.

Care of Rope

For easy handling, new manila rope should be stretched its entire length before use; this will reduce its tendency to twist and kink. The rope ends should be whipped or at least temporarily knotted to prevent fraying.

Rope should be kept as dry as possible. If it gets wet it should not be dried by heat, but should be stretched on a ladder or suspended from supports so that air can get to the fibers. When dried by heat, manila rope loses its natural oil and becomes brittle.

Rope strands frequently get cut by sharp edges. If it becomes necessary to pass a rope over a sharp edge, the edge can be padded with a board or a sandbag. Rope should not be dragged along the ground; sand or grit will work into the rope, cutting it and causing abrasions of the inner fibers. Rope should not be carried in a truck compartment that is damp or that contains sharp edged tools. After use, rope should be carefully inspected for cut strands. Never step on a rope, since this will grind dirt and grit into the fibers.

In storage, ropes should be off the ground, free from extreme temperatures, and away from contact with materials containing acids or strong alkalis. Oil or grease must not be applied to rope, since these substances allow dust and grit to collect on the rope and prevent air from reaching the fibers.

Where practical, splicing is considerably stronger than knotting for a rope join. A knotted line has only 50 to 60 percent of the full strength of its fibers; a spliced rope has 80 to 95 percent.

When new ropes are loaded the first few times, their fibers compact and the ropes become slightly longer (2 to 7 percent) permanently. After the break-in period, however, good lines will maintain their new length unless subsequently overloaded.

The strands of a nylon rope can be prevented from unraveling by whipping the end with a thin cord or by melting the end—for example, with a soldering iron.

Wash dirty rope with clean water and dry thoroughly before storing. Dirt on the surface and imbedded in rope acts as an abrasive on strands and fibers.

Protect rope from kinks. Rope repeatedly twisted in one direction will develop kinks, unless twists in the opposite direction are repeatedly thrown in. Kinks pulled through a restricted space, such as a tackle block, will seriously damage the rope.

RESCUE EQUIPMENT AND GENERAL RESCUE TECHNIQUES

Fig. 2-19. The thumb knot.

Fig. 2-20. The figure-of-eight knot.

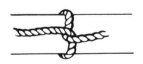

Fig. 2-21. The half hitch.

A line may be extremely difficult to break with a pull, yet can be easily broken with a snap; this is especially true when it has had considerable use. Sudden jerks on tackle may result in breaks that would never have occurred with a steady, natural pull.

When a rope is subjected to a strain, it will stretch; when the load is relieved, elasticity will cause the line to snap back to its original length, as a rubber band would. Whenever a rope is under a strain, all personnel must avoid being in the line of flight in case of breakage. Many persons have been killed and maimed by being struck by a rope that broke while under a heavy strain.

Knots, Hitches, and Bends

Humans have been tying knots far longer than historical records have been kept. Over 5,000 years ago, at opposite ends of the earth, Chinese boat people on the Yangtze river and Egyptian sailors on the Nile river expertly spliced and braided rope and made fast lines with the same knots being used today. With the same basic materials—rope twisted of cotton or hemp—the ancient Greeks, Romans, Incas, and American Indians all tied square knots, figure-of-eight knots, overhands, and clove hitches. Until recently, neither the knots nor the ropes they used had changed substantially since those ancient times.

Most of the common knots are more properly called bends. A bend is a way of fastening a rope to itself or to another rope. A hitch is a method of attaching a rope to a post, hose line, or other object. Another common term in the use of rope is the word *bight*. A bight is made when the rope is turned back on itself to form a partial circle.

Although there are many knots that can be used in rescue service, those listed in the following paragraphs should be sufficient for most rescue needs. Rescue personnel should be able to tie knots automatically and in pitch darkness.

Thumb or Overhand Knot (Figure 2-19). This knot may be used temporarily on the end of a rope to prevent fraying; it is sometimes tied on the end of a rope to prevent the rope from running through a block. To tie the knot, form a loop, making sure that the running part of the rope crosses the standing part, then pass the end around the standing part and through the loop.

Figure-of-Eight Knot (Figure 2-20). The figure-of-eight knot is an ideal basic knot for use at the end of a line to prevent a rope from slipping through a block. Make an underhand loop, then bring the free end over the standing part and pass it under and through the loop. This knot is larger, stronger, and easier to untie than the thumb or overhand knot. It is equally secure, but is less damaging to the rope fibers.

Half-Hitch (Figure 2-21). The simplest hitch is that known as the half-hitch. It is easy to tie and although it is not often used alone, it is frequently combined with other knots and hitches. This is the basis of many knots, and is one method that is used to attach a rope to a pole or a hook. To tie the knot, pass the short end of a rope around a pole or another rope, and then over the standing part so that when the rope is pulled, one part of the rope binds on the other. Two half-hitches are often placed on a pole or another rope since they will not slip while under a strain.

Clove Hitch (Figure 2-22). The clove hitch is useful for hoisting timbers and rescue tools aloft, as well as for fastening a rope onto another line that is stretched between two points. It may be used at the end or in the middle of a rope. To tie this knot, make two half hitches; or pass the running part of the rope around the object and bring the end out underneath the standing part; then pass the running end over the standing end and around the object a second time. Then bring the running end under itself to tighten, pulling on both the running and the standing ends. To tie in the middle of a rope (Figure 2-23), form two counterclockwise loops, one in the left hand and one in the right hand, the latter loop being passed in front of the left hand loop. Pass both ends over the object and draw tight.

Single Sheet Bend or Becket (Figure 2-24). This knot is useful for joining two ropes of different sizes together; it will not slip under conditions of varying tension. To tie this know, make a loop in the end of the larger rope, pass the end of the other line through it and form a half hitch around the loop of the first rope.

Double Sheet Bend or Double Becket (Figure 2-25). This knot is more secure than the single sheet bend and is used for joining ropes when there is a great difference in their sizes. It is formed in a manner similar to a single sheet bend except, after forming the first half hitch, the short end makes another turn around the two thicknesses of the thick rope and toward the bight.

Bowline (Figure 2-26). A bowline is a good securing knot; it also makes a nonslipping loop in the end of a rope. It is used to tie a lifeline to a rescue worker and is used frequently in raising and lowering heavy objects and tools. To tie a bowline, make a loop in the rope, then pass the free end through the underside of the loop, around the standing part of the line, and back through the loop.

Bowline-on-a-Bight (Figure 2-27). This knot is useful when raising or lowering a stretcher or a ladder horizontally since its double loop provides greater safety than a single loop. Also, it may be used to safely raise or lower a rescuer since the person may place one leg in each loop. To tie, form a bight in the line by grasping it at a distance from the end so that the rope is doubled. Form a small loop in the standing part as when tying a bowline; a loop 3 or 4 feet (about 1 meter) from the end of the bight will form loops sufficiently large for a person's feet to pass through. Then pass the free end through the loop (Figure 2-27a). Spread the doubled end and slide it over the large loops and up the standing part of the line until it is above the upper small loop (Figure 2-27b). Pull on the two large loops, while holding the small loop and the bight end in place, until the knot tightens up (Figure 2-27c).

Timber Hitch (Figure 2-28). This is a quickly-made hitch for lifting spars, poles, and planks; it will grip objects tightly if the objects are generally cylindrical in shape and have a rough texture. When hoisting timbers, a half hitch is commonly looped around the upper end of the timber so that it will stand on end. The timber hitch is formed by passing the running end of the rope around the object and then making a half hitch on the standing part of the rope, then twisting the long running end back around itself. The timber hitch should not be used on a metal pipe or on a plank that is small in relation to the rope.

Square Knot (Figure 2-29). To tie a square knot, hold both ends, one in each hand. Lay one rope on the other, and then pass one end around the other rope. Repeat this operation a second time, using the opposite rope on the top; draw the knot tight.

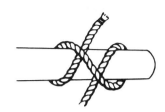

Fig. 2-22. The clove hitch.

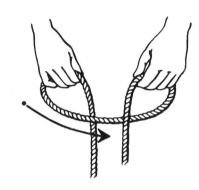

a

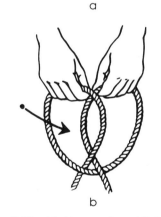

b

Fig. 2-23. Making a clove hitch in the middle of the rope.

RESCUE EQUIPMENT AND GENERAL RESCUE TECHNIQUES

Rescue Coil

When not in use, a rope should be coiled and ready for any emergency. The rescue coil (Figure 2-30) is a commonly used method of coiling heavy rope for ease in carrying and for dropping from high places. It will readily uncoil without kinking and will drop quickly and accurately. The coil is made on a special frame which is approximately 22 inches (55 centimeters) in diameter, or can be coiled on the end of a ladder. Many ropes used for rescue purposes, particularly those of larger diameters, have a loop or an eye at one end; this is generally formed by an eye splice.

Start with the eye end, allowing the eye to protrude approximately its own length. If there is no eye, leave about an arm's length of rope for a handle. For a 100-foot (30-meter) length of ¾-inch (2-centimeter) rope, coil seven turns of line around the frame in the first layer and six turns in the second layer, making sure that the rope is reasonably tight and that the second layer falls neatly into the grooves formed by the first layer of rope. Next, coil the rope at right angles to the two layers so prepared (Figure 2-30a). When the coil is finished, remove it from the frame. Then double the remaining portion of the

Fig. 2-24. The single sheet bend or becket.

Fig. 2-25. The double sheet bend or double becket.

Fig. 2-26. The bowline.

 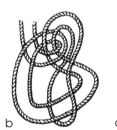 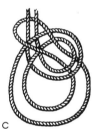

Fig. 2-27. The bowline-on-a-bight (French bowline).

Fig. 2-28. The timber hitch.

Fig. 2-29. The square knot.

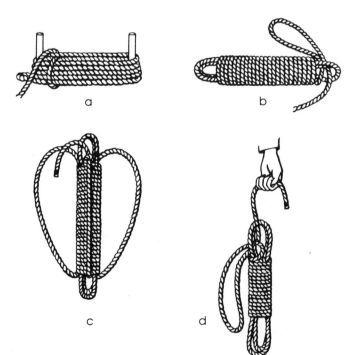

Fig. 2-30. Steps in making a rescue coil.

rope and pass it through the loop opening in the end of the coil (Figure 2-30b), and then through the loop opening in the opposite end of the coil. This makes two loops to go over the rescuer's shoulders so the coil of rope can be carried on the back (Figure 2-30c). Adjust the shoulder straps by loosening or pulling on the double rope. When the straps are adjusted, bring the remaining single rope up the side of the coil and tie a becket in the small loop formed by the shoulder straps as they protrude through the coil loop (Figure 2-30d).

To uncoil, untie the becket and the shoulder carry. Then holding the loop end, pull two or three lengths of rope from the coil so that it will uncoil freely as it is dropped.

When making a rescue coil with rope of other lengths or thicknesses, the number of turns required around the frame can best be determined by trial. After a proper length has been established, a piece of friction tape or a paint mark may be placed on the rope to indicate when enough has been coiled on the frame.

Wire Rope or Cable

Wire rope is used for lashings and slings and on the winches of rescue vehicles and wreckers for hoisting and dragging heavy objects and pulling down walls. It is about nine times stronger than manila rope of the same size.

Wire rope should be straight, without loops, before any pull is exerted. A pull on a loop will cause a kink, damaging the strands. After each use, wire rope should be cleaned and oiled to prevent rusting. It should be kept away from fire as even a slight burn will weaken the rope and render it dangerous for further use. When coiling for storage, care should be taken to avoid bending the rope too sharply.

MANUAL RESCUE TOOLS

There are many rescue tools that are manually operated. These appliances vary from simple axes and crow bars to multiple-use extrication and forcible-entry devices.

QUIC K-BAR-T® Rescue Kit is a tool that consists of a slide hammer that drives either a chisel or a cutting blade through practically every type of material (Figure 2-31). It will allow one person to cut open the roof or door of a vehicle to release a trapped motorist, pry open locks and jammed doors, breach brick and block walls, and accomplish many other forcible-entry tasks. Generally speaking, this tool does the work of seven other devices: power saw, bolt cutter, tin snips, fire axe, crowbar, wrecking bar, and sledgehammer.

QUIC BAR® and MINI QUIC BAR® are two multipurpose rescue tools for instant forced entry through almost any door or window. The QUIC BAR is a combination that can be used as a crowbar, pick, sledgehammer, or claw tool (Figure 2-32). The MINI QUIC BAR is a miniature version of the larger tool, made to be carried easily at all times by rescue personnel in a special holder.

FORCIBLE ENTRY

The purpose of locks and fasteners is to protect property against intruders, but these devices may become obstacles to rescue personnel responding to an emergency. A locked building requires forcible entry. However, it must always be remembered that doors and windows may not be locked; they should always be tried before force is used.

When human life is endangered or when time is a major factor, the fastest means of forcible entry must be employed, regardless of property damage. Entry is forced with as little damage as possible when human life is not a factor or immediate entry is not required. Forcing entry into a locked building does not involve much theory; it is mostly a matter of applying leverage and is best learned from demonstration and by observation. This does not mean that forcible entry is not an important subject; it can have considerable influence on the outcome of a fire or a rescue incident. Considerable skill is required to force entry in a manner that will save time and lessen damage.

Opening Doors

If the door is locked, examine it to determine which way it swings and which method of forcible entry will prove most effective. Residence doors generally open inward; outside doors in public buildings usually open outward. It is often advisable to find other means of entry if possible. In many instances, less damage may be done by breaking glass near the lock, if any exists; then a rescuer can reach through the hole and open the door from the inside.

Forcing Doors

There are several types of doors, and the methods of forcing them vary with the type of door involved. A large variety of tools have been devised to aid in forcible entry; most of these operate on the pry bar principle. Axes, both flat-head and pick-head, are very useful for such rescue operations. A lock may sometimes be removed or destroyed with less damage and expense than might occur if the door is forced.

If done properly, forcing locked doors calls for more skill than force. Consideration must be given to the type of door and the way it opens when deciding on the method and tools to be used to force it. Doors are usually forced in the direction they normally open—in, out, or up.

Hinge Pins. Some swinging doors have exposed hinges with removable pins. Removal of these hinge pins will often allow the door to be pried open from the hinge side.

Door Opening Out. Single hinged doors that swing out may be opened with a pry bar or an axe. The blade should be inserted between the jamb and the door, just above or below the lock. By prying with

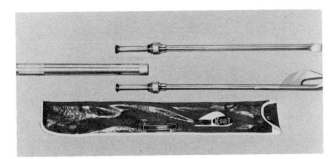

Fig. 2-31. The QUIC K-BAR-T® rescue kit consists of a slide hammer that will drive either a chisel or a cutting blade. This tool is used to cut open the sheet metal on crashed automobiles to release trapped victims.
(Courtesy of the Ziamatic Corporation.)

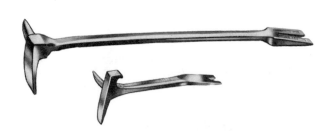

Fig. 2-32. QUIC BAR® and MINI QUIC BAR® forcible entry tools are multipurpose rescue tools useful for gaining access through locked or jammed doors or windows. (Courtesy of the Ziamatic Corporation.)

the handle to one side away from the door, the jamb can usually be sprung enough to allow the lock bolt to pass the keeper. The door should then be pulled open or pried clear with another tool when the lock has been freed from the keeper. Two tools may be used together: Alternately pry and hold with one, and get a purchase with the other.

Door Opening In. Swinging doors that open away from the rescue personnel present greater difficulties (Figure 2-33). Insert the blade of the tool between the stop and the jamb or the door and the stop, whichever is easier. Where there is a separate doorstop, it may be better to remove the stop completely. With a full bite behind the door, pry the door away from the jamb until the bolt passes the keeper. This type of door may be more easily forced by using two tools.

A method that works well when the door is on a street entrance is to place the end of a ladder against the lock side of the door with the end of one beam above the lock and the end of the other beam below. Several persons should grasp the ladder, holding it parallel to the ground, and push it steadily against the door. With this type of force, the door will usually spring open. The ladder should be pushed steadily and not rammed or banged against the door, because such use might damage the ladder.

Double Swinging Doors. This type of door can be forced with most pry tools by prying the two doors apart sufficiently at the lock to permit the lock bolt to pass the keeper. If a piece of molding is used to cover the crack between the doors when they are closed, it should be removed so that the blade of the tool can be inserted.

Tempered Glass Doors. The breaking characteristics of tempered plate glass are quite different from those of ordinary glass; this difference is due to the heat treatment that is given to the glass during tempering, which increases its strength and flexibility. This glass is several times stronger and more resistant to shock. When broken, the sheet of glass suddenly disintegrates into relatively small pieces.

Tempered glass doors are usually locked by dead bolts that fit into sockets in the sill; some are latched both top and bottom. Sometimes there is sufficient room between the door and the frame that claw tools or pry bars placed under the door can force it upward sufficiently to open. Some of the latches are sufficiently soft that they can be cut with a power saw or rotary cutting blade. It may be possible to drive the lock cylinder inward with the point of a fire axe or punch and then turn the lock mechanism with a screw driver to open the door. There are so many different types of doors and locks that no one method of forcing this type of door can be universally used.

Every other method of gaining entry into the building should be attempted before trying to force a tempered glass door. Whenever the glass in such a door must be broken, it can be shattered most easily by the pick point of a standard fire axe or other sharp pointed tool of hardened steel. Blunt-faced tools, such as sledgehammers, have relatively little breaking effect. The glass will fragment into small granules with relatively blunt points and dull edges.

Opening Windows

The same precautions to be observed when opening doors should also be observed when forcing windows, and the same tools and methods should be used. It is impossible to open many types of windows forcibly and it may be quicker and less damaging to break out a pane of glass and then reach inside to unlock the window.

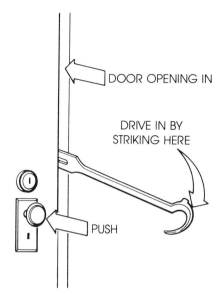

Fig. 2-33. For a door that opens inward, insert the blade of an axe or a pry bar between the door and the doorjamb. Pry the door away from the jamb until the bolt will pass the keeper. Then forcefully push on the door. For a door that opens outward, pry out on the door with the tool.

RESCUE EQUIPMENT AND GENERAL RESCUE TECHNIQUES

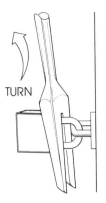

Fig. 2-34. When entry is barred because of a lock, the least damage will be caused if the hasp or chain is cut, broken, or twisted off. The simplest method is to place the claw of a crowbar or other tool over the hasp or lock and twist.

Double-hung windows may be opened by prying upward on the lower sash rail. If they are locked on the check rail, the screws of the lock will give and the sash will open. Prying should be done at the center of the sash to prevent breaking the glass.

If necessary, break the glass. Since replacement of glass is inexpensive, it may be broken without hesitation, rather than damage the window or door frames. Clean out the shards of glass to prevent injury.

When confronted with leaded glass, insulating glass, or tempered glass, judgment should be exercised before breaking the glass, since its value may be greater than the value of the window or door frame.

Forcing Hasps and Locks

It will probably cause less damage to break or twist off the hasp than to attempt to break a padlock. This may be accomplished with practically any type of lever or pry bar (Figure 2-34).

QUESTIONS FOR REVIEW

1. Which of the following is the most important requisite when a rescue organization is suddenly confronted with a highly complex victim extrication situation?
    a. That rescue personnel possess a high degree of mechanical aptitude
    b. That the rescue squad be equipped with a wide variety of tools and equipment
    c. That, at least two of the personnel be trained emergency medical technicians so that immediate care can be started while the rest of the crew is extricating the victims
    d. That a fire department pumper be requested to stand by in case the wreckage should become ignited

2. Which one of the following is not a machine?
    a. A crowbar
    b. A block and tackle
    c. A wedge
    d. A block of wood

3. Rescuers are using a 12-foot-long, 4 inch × 4 inch length of lumber to pry a 500-pound iron beam off the leg of a victim. If the fulcrum is placed 1 foot from the beam, how much pressure must be exerted on the end of the lever to raise the beam?
    a. 33.65 pounds
    b. 39.70 pounds
    c. 45.45 pounds
    d. 51.30 pounds

4. A single sheave snatch block is affixed to a rafter and a line is passed through the pulley. One end of the rope is fastened to a heavy weight and several rescue personnel pull on the loose end of the line to lift the load. Which of the following statements is least correct about this pulley system?
    a. There is no increase in power.
    b. Since there are two ropes leaving the block, power is multiplied by two.
    c. There is a gain in height of lift over manual methods.
    d. There is a change in direction of pull.

CHAPTER 3

Transporting the Victims

A basic principle of emergency medical care is that every sick or injured person must be given essential treatment and the victim's condition must be stabilized before any movement is attempted. Unless the situation endangers the lives of casualties and rescue personnel, transportation of patients must be orderly, organized, and unhurried so that injuries cannot be further aggravated. Authorities report that more harm results from improper transportation of trauma victims than from any other aspect of emergency assistance. Careless or rough handling may not only increase the seriousness of the illness or injury, but may also be fatal.

PERSONNEL SAFETY

An emergency medical technician (EMT) must perform several basic tasks when handling patients; these include lifting, lowering, pulling, pushing, holding, and carrying individuals and equipment. If any one of these tasks is conducted improperly, discomfort and additional injury may result for the patient and, possibly, even for the EMT. If the correct procedures are followed, the possibility of injury will be reduced and the efficient lifting and moving of the patient will be ensured.

The muscles responsible for bending the body forward are mainly in the walls of the abdomen and are not capable of exerting great amounts of force. Some of the complex muscles involved in straightening the body act on the bony protuberances of the vertebrae and form a number of small levers. These muscles and small levers produce less powerful movements than the limbs. Therefore, paramedics should avoid lifting or moving patients when they are themselves in postures that make it necessary for them to bend and then straighten their backs. Discomfort or injury may result if undue strain is placed on the back when lifting a patient or if frequent bending and straightening of the back takes place. Even a moderately bad posture over a period of time can result in muscular fatigue and an aching back.

RESCUE EQUIPMENT AND GENERAL RESCUE TECHNIQUES

The muscles related to lifting and moving are composed of bundles of long, thin cylindrical fibers. In order for these muscle fibers to contract, energy must be released through chemical reactions triggered by nerve impulses. In addition to this energy, unwanted and sometimes poisonous by-products are produced that must be removed. This is achieved by the action of oxygen stored in the red muscle fibers themselves or in the blood. If the supply of oxygen from these sources is inadequate, the by-products of energy production will build up and cause a sensation of pain or muscular fatigue. This may result even if the force exerted is quite small, provided the muscle is contracted for a long period of time. Pain builds up and will subside only when the muscle is relaxed and blood is allowed to flow through it again. Consequently, significant amounts of force should not be sustained over a long period; all muscular activity should be intermittent.

Several types of injuries and discomfort may result if patient-handling tasks are performed improperly. EMTs are liable to experience muscle strain, sprained ligaments, dislocations, cramps, fatigue, and even permanent impairment if lifting and moving are not performed in a manner that properly utilizes the body's natural system of levers. In order to protect an EMT and help move patients safely, certain principles should be followed.

1. Paramedics should keep in mind their physical capabilities and limitations and not try to handle too heavy a load. When in doubt, they should seek help and not attempt to lower a patient if they are not sure of being able to handle the load.
2. EMTs must keep their balance when carrying out a job, maintain a firm footing, and keep a constant grip on handrails, stretcher, and patient.
3. Lifting and lowering should be accomplished by bending the legs, not the back. Keep the back as straight as possible at all times, bend the knees, and lift. Set the pelvis by tensing the muscles of the abdomen and buttocks.
4. When holding or carrying, keep the back straight and rely on shoulder and leg muscles.
5. When performing a task that requires pulling, keep the back straight and pull with the arms and shoulders.
6. All jobs should be carried out slowly, smoothly, and in unison with other workers. Move the body gradually and avoid twisting and jerking when handling patients.
7. When handling a patient, attempt to keep the arms as close to the body as possible in order to maintain balance. Do not keep the muscles contracted for a long period of time; rest frequently and whenever possible.
8. If it can be arranged, slide or roll heavy objects instead of lifting them.

VICTIM CARE

The life of an injured person may well depend on the manner in which transportation is accomplished. Rescue operations must be performed quickly, but unnecessary haste is both futile and dangerous. Unless conditions at the scene prevent it, movement of patients should be performed by at least two attendants. When additional help is required, volunteers from among the bystanders may provide valuable assistance if they are carefully selected and controlled. Untrained assistants must be thoroughly instructed before any operations are started and the recruits must not be allowed to advance into any hazardous location.

If there is an immediate danger from drowning, or if life is endangered by fire, heat, smoke, steam, electricity, poisonous or explosive gases and vapors, danger of falling walls, or collapsing structures, the victim must be moved only far enough to be out of immediate jeopardy, be given proper emergency medical care, and then be transported. After the casualty has been removed from the immediate hazards, all further transportation must be accomplished in a manner that will not aggravate injuries.

Since rescues will be conducted under almost every conceivable adverse condition, the method employed for casualty removal will vary according to the location of the victim and the type of injury that has been sustained. In some rescue operations victims will have to be lowered from upper floors of buildings; in others they may require hoisting up the face of a cliff or through a hole in the floor. The possibilities are endless.

After rescue, many patients will have to be carried over piles of debris, uneven ground, or other terrain where it will be difficult to obtain secure footing. Speed in transporting injured casualties is important, but it should be consistent with safety. The method used will depend on the immediate situation, the victim's injury and condition, and the available personnel and equipment.

Occasionally, one lone rescuer must lower an unconscious or injured person from a bed or other raised position to the floor, so the victim can be dragged from the building. The patient's head and shoulders are cradled in the rescuer's arms for support and protection while sliding the victim to the floor. With a little practice, this maneuver can be accomplished smoothly and with minor effort.

Immobilization of fractures before the victim is moved will assist considerably in retarding shock, relieving pain, and preventing other complications. A patient with extensive rib injuries will be most comfortable lying on the injured side, since this position tends to immobilize and support the broken ribs. Semiconscious and unconscious patients are best transported face down or with the head turned to the side, so that secretions will drain from the nose and mouth instead of being inhaled.

PATIENT-HANDLING COMPLICATIONS

A patient's condition should be stabilized as much as possible before transportation is attempted. Successful treatment is directly related to the rapidity with which a functional spontaneous cardiac or respiratory rhythm can be restored, or fractures be immobilized. It should be clear to all rescue personnel that in cases of cardiac emergency, restoring adequate, spontaneous circulation at the scene of the accident is the most important factor in the victim's survival. Every effort must be made to treat and stabilize the patient at the scene, since it is difficult to perform cardiopulmonary resuscitation (CPR) effectively during transportation. Once the patient is stabilized, it is reasonable to transport the person to a medical facility. Stabilization includes the following measures:
1. Assuring effective ventilation, either spontaneous or assisted.
2. Maintaining a stable cardiac rhythm and effective circulation, utilizing drugs as indicated.
3. Maintaining a functioning ECG monitor and an intravenous lifeline, if the unit is so equipped.

RESCUE EQUIPMENT AND GENERAL RESCUE TECHNIQUES

4. Establishing and maintaining communications necessary for consultation, transportation, and admission to a hospital or other medical facility.

Several relatively common conditions that complicate the basic handling procedures are routinely encountered by paramedics. Among these problems are accident victims with spinal injuries, unconscious or vomiting patients, persons requiring life-support techniques and equipment, and some psychiatric patients or hard-to-control persons.

Suspected Spine Injury

No accident victim requires more care in handling than a person with a suspected spinal injury. One of the main functions of the spinal column is to protect the spinal cord contained within it. Once the spine is fractured or vertebrae are displaced, the spinal cord is in great danger of being pinched or severed at the points of injury; this could result in paralysis or even death. Very few victims of spinal injury immediately suffer the severe consequences of an injured spinal cord; such injuries are typically the result of improper or careless patient handling, particularly at the time of the initial emergency treatment.

A patient is likely to have a fracture of the spine or neck, and should be handled accordingly, if the person has suffered a severe fall, diving accident, whiplash in a head-on or rear-end traffic collision, or face or head wounds; has been hit in the back by a heavy weight; or has been involved in any type of violent accident or collision. In most cases, the victim should not be moved until after immobilization, especially if there is any chance that a neck or spinal injury may have been incurred. If possible, immobilize the casualty in the position in which found. It is best to avoid straightening or rearranging the victim's body or limbs. A spinal injury will occur most often in either the neck region (cervical fracture) or in the lower (lumbar) area of the spine. Injuries of this type require that the head, neck, and trunk be kept immobile and in the alignment in which they are found. A backboard is an excellent device for immobilizing these types of injuries.

Injured Limb

All types of severe wounds and fractures of the victim's arms and legs render direct handling of the extremity painful and often harmful. Avoid any direct contact with the injured body part and move the patient as little as possible. Maintain adequate support of the injured limbs during treatment, immobilization, and transport (Figure 3-1).

After splinting and bandaging the fracture or wound, the EMT should determine whether the injured limb can be further protected by loosely securing it to the patient's body or to an uninjured limb. If this is not advisable, one paramedic can support the injured extremity while the victim is being transported to a stretcher. When lifting an injured limb, two hands should be used; place one hand above and the other hand below the injured area.

Unconsciousness or Vomiting

The unconscious patient is in a state of complete unawareness and has no idea what is going on in the area. The victim is unable to make any meaningful or purposeful movements. The rescuer should be prepared for this and treat the patient accordingly. The primary objective when handling an unconscious casualty is to keep the air-

Fig. 3-1. The victim of a violent accident should be completely stabilized before transportation. This motorcyclist required the traction splinting of his fractured leg, IV therapy, and other advanced life-support measures. (Courtesy of the Los Angeles County Fire Department.)

way free of obstruction. If the victim is positioned in the ordinary supine position on the stretcher, the tongue will probably drop back into the throat and may create an obstruction to breathing. Also, vomiting may occur during or following unconsciousness; this is a great threat to an open airway. The EMT can reduce the risk of respiratory obstruction by making certain that the patient's head is positioned at a slight downward tilt while lying down, preferably on the side. This positioning of the victim will tend to drain out any fluid in the air passages through the mouth and nose. If, because of injuries, the patient must lie on the back, a suction unit should be used to remove any blood, mucus, or vomit that may have accumulated. The patient should lie facing the side of the stretcher that will be more accessible to the attendant in the ambulance, so that close observation can be maintained.

Hard-to-Control Patients

In most cases, the psychiatric patient will not be difficult to control and should be handled like any other person. Occasionally, psychiatric patients will have been sedated by a physician prior to the time that they must be handled by ambulance personnel. However, on rare occasions paramedics are confronted with a violent or otherwise physically unmanageable person. The patient may be psychotic, intoxicated, or have some other problem which causes uncontrollable actions. In these circumstances, paramedics may have to use physical force and special restraining devices; such measures should always be used with discretion and only when absolutely necessary for the patient's safety and for the protection of others. Handling hard-to-control patients requires good judgment more than almost any other type of situation EMTs will encounter. Procedures must be adapted to fit the needs of the situation.

If conversation and the usual handling methods fail to bring the patient under control, place the wheeled stretcher at its lowest position (where it is most stable) and place the casualty face down, so that wrist and ankle restraints can be applied. Fasten the wrists together with restraints behind the patient's back and secure the ankles to each other. The patient is covered in the usual manner, and then secured to

RESCUE EQUIPMENT AND GENERAL RESCUE TECHNIQUES

the stretcher with one strap across the back just below the shoulders, an abdominal strap across the small of the back, and a third safety strap across the legs. Remember that these measures should be taken only as a last resort.

Cardiopulmonary Resuscitation During Transportation

Once cardiopulmonary resuscitation is begun, it must be continued until the patient recovers or until a physician takes over. Unless ventilation and sternal compression are carried out without interruption, irreparable brain damage resulting from lack of oxygen will occur.

Effective ventilation and sternal compression should be well established before the victim is moved. If the patient must be moved for definitive therapy while CPR is continued, it is wise to solicit the help of available bystanders. The ventilating operator may work from the head end of the stretcher and the chest compressing operator from the side, while three or more bystanders carry the stretcher. The most experienced rescuer will act as team leader. If necessary, the patient's shoulders may be raised using a rolled blanket to maintain a backward tilt of the head. When moving the patient through narrow passageways and other difficult places, manual resuscitation will have to be improvised, but it must not be interrupted for more than a few seconds at a time.

Patients Requiring Life-Support Equipment

In many emergency situations the patient's critical condition requires the implementation of life-support measures that cannot be interrupted, even when the patient is being carried to the ambulance or transported to the hospital. The usual life-support procedures are oxygen ventilation, heart-lung resuscitation, and the intravenous (IV) administration of fluids. The use of life-support equipment makes many of the patient-handling jobs more difficult for the EMTs, but for the most part these difficulties are readily overcome with the aid of a third person.

Patient Receiving Oxygen. When administering oxygen to a patient, the ideal procedure is to have a third person hold the oxygen cylinder or resuscitator while the attendants are transferring the patient to a stretcher. This allows the patient to receive oxygen without interruption; otherwise, the oxygen must be discontinued for the moving and resumed once the person has been placed on the stretcher. During transfer, the EMTs must be careful to keep the oxygen tubing free of kinks and entanglement.

After the patient is comfortably positioned on the stretcher, the oxygen cylinder or resuscitator may be placed between the person's legs, on the stretcher beside the victim (resting on the stretcher frame or headrest under the mattress), or in a special cot-mounted holder.

All jobs will be conducted as for any other patient. The casualty should be supine, covered, secured, and transported in the usual manner. A person receiving oxygen will usually be most comfortable with the stretcher headrest elevated between 30 and 45°. The oxygen cylinder should remain uncovered so the pressure can be observed.

Heart-Lung Resuscitator. Oxygen-powered, mechanical, cardiopulmonary resuscitatory devices play an integral and vital role in the lifesaving techniques of external compression and respiratory ventilation. While the patient is being manually maintained with cardiopulmonary resuscitation, the base of the mechanical resuscitator should be positioned under the person's body; the resuscitation unit is then applied to the chest and activated as soon as possible. It is best to secure both the patient and the machine to the stretcher or backboard to avoid any shifting, which could displace the compressor from the correct point on the sternum. While carrying or wheeling the stretcher, it is best to have the patient's head in a lowered, extended position to ensure an adequate supply of blood to the brain.

When transporting the patient down stairs, chest compression will be temporarily interrupted. At landings, or whenever the stretcher can be set down on a level surface, CPR support must be resumed for awhile before negotiating the next flight of stairs.

Attached IV Equipment. When the patient requires administration of liquids by an intravenous apparatus, the tasks of transferring the patient to a stretcher and moving the loaded stretcher are somewhat complicated. Once a patient has started receiving an IV, the bottle must always be held well above arm level to ensure a gravity flow of the liquid; a third person is almost a necessity. To transfer a patient onto a stretcher, the EMTs should position themselves on the patient's side away from the IV connection and lift the person, while a third attendant supports the patient's IV arm and holds up the bottle. Ideally, the patient is placed on a wheeled stretcher that has a cot-mounted bottle holder; otherwise, one rescuer must carry the bottle during the entire transport. The patient's IV arm is placed on an extra pillow and left exposed for observation.

Confined Areas. Quite often, paramedics will find that stairways, narrow corridors, or other confined areas prohibit them from getting the wheeled stretcher close enough to the patient for a direct transfer to the cot. In such cases, a portable stretcher, stair chair, or long backboard will prove useful in carrying the patient to the wheeled stretcher or ambulance. The patient's condition must be carefully considered when choosing a method; the long backboard would undoubtedly be the best method for transporting unconscious persons and those with spinal injuries or fractures of the lower extremities.

RESCUE OF CASUALTIES

While it is always desirable to employ the latest models of wheeled stretchers and other carrying devices for transporting victims, there are occasions when human energy and brute strength must be used.

If there is immediate danger from a spreading fire, toxic gases, basement flooding, or building collapse, casualties may have to be removed before emergency medical care can be administered or transportation provided. Victims should be moved only far enough to be out of immediate danger, then given emergency care and transported properly.

When victims are only slightly injured or must be moved immediately in spite of injuries, one of several methods may be used. To the extent possible, the type of carry or drag least likely to aggravate the patient's injuries should be selected.

Fire Fighter's Carry

One of the easiest ways for one rescuer to carry an unconscious person is by means of the fire fighter's carry (Figure 3-2). This method is effective if there is only one rescuer, if there is not too much smoke and heat to allow an upright position, if the victim is uninjured and of slight build, and if the distance to be traveled causes the various drags to be unpractical.

The rescuer turns the casualty face down and kneels on one knee facing the person's head. The rescuer then passes both hands

RESCUE EQUIPMENT AND GENERAL RESCUE TECHNIQUES

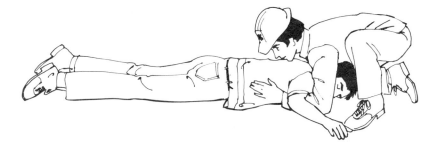

a. Turn the casualty face down on the ground and kneel on one knee facing the victim's head. Place both hands under the casualty's armpits and gradually work them down the side and across the back.

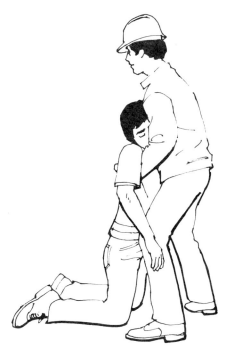

b. Raise the casualty to the knees.

c. Then take a firmer hold across the casualty's back and raise victim to standing position.

d. While holding the casualty around the waist with your right arm, grasp the casualty's right wrist with your left hand and draw the victim's arm over your head. (If the victim's injury is such that carrying from the opposite side would be preferable, simply substitute right for left in the instructions.)

Fig. 3-2. Fire fighters carry.

under the victim's armpits, slides both hands down the victim's sides, and clasps them together across the back.

Next the victim is raised to the knees so the rescuer can get a better hold across the back. When the victim is raised to a standing position, the EMT then thrusts the right leg between the casualty's legs, grasps the person's right wrist in the left hand, and swings the victim's arm around the back of the rescuer's neck and down the left shoulder.

Placing the right arm between the victim's legs, the rescuer then stoops quickly and pulls the victim across the shoulders. Then the casualty's right wrist is grasped with the right hand. The rescuer may now stand up and proceed to a safe environment. This procedure sounds long and complicated, but a little practice will simplify everything.

Transporting the Victims

e. Bend at the waist and knees and pull the casualty's right arm down over your left shoulder so that his body comes across your shoulders. At the same time, pass your right arm between the legs and grasp the right knee with your right hand.

f. The casualty is lifted as you straighten up.

g. Then grasp the casualty's right wrist with your right hand, leaving your left hand free. This is the position of carry. A person can carry another person some distance in this manner.

Fig. 3-2. (continued)

h. i.
The procedure for lowering the casualty to the ground is the reverse of the lifting procedure.

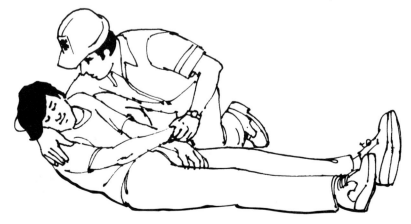

RESCUE EQUIPMENT AND GENERAL RESCUE TECHNIQUES

The procedure for lowering the casualty to the ground is the reverse of the above.

Seat Carry

The normal seat carry requires two rescuers (Figure 3-3). They raise the casualty to a sitting position, each steadying the victim by placing an arm around the back. Each rescuer then slips the other arm under the casualty's thighs, grasping the wrist of the other rescuer. One pair of arms forms a seat rest, and the other pair a back rest. Both rescuers then slowly rise in unison, lifting the victim from the floor. Another variation of the seat carry, the four-hand seat carry, also requires two rescuers. This is an excellent method of carrying conscious, and not too seriously injured persons a short distance.

A third variation, the saddle-back carry, is a simple and effective method of moving a conscious person a short distance when the injuries are not very serious.

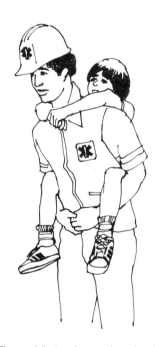

a. The four-hand seat carry is an excellent method for carrying conscious, not too seriously injured persons a short distance.

Fig. 3-3. Seat carry.

b. This is an effective method of carrying an unconscious injured person a short distance provided it is not necessary to keep the casualty flat.

c. The saddle-back carry is a simple and effective method of moving an injured person a short distance when the injuries are not serious.

Pack-Strap Carry

The pack-strap carry may be used by one rescuer to carry an unconscious victim if the heat and smoke do not preclude walking in an upright position (Figure 3-4). The rescuer makes a loop of any convenient material, such as triangular bandages, sheets, belts, or rope. Then the casualty is turned face up and the loop is passed around the victim's back and chest at the armpits. The rescuer then lies flat on top of the casualty, face up, and slips both arms through the loop, which should fit snugly. The rescuer then rolls over so both persons are snugly together, face down. The rescuer may then arise to an erect position with the casualty affixed to the back.

The victim may also be positioned back-to-back with the rescuer. In either case, both hands of the rescuer will be free to ascend or descend a ladder or to perform other tasks. The dangling legs of the victim may prove awkward while descending a ladder, but travel will be possible by observing every safety precaution. If sufficient bandage or other material is available, the casualty's hands may be tied together.

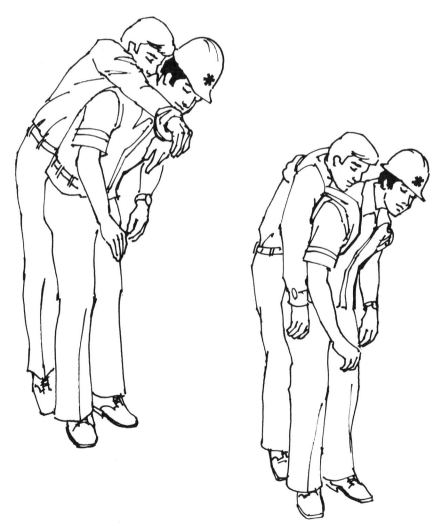

Fig. 3-4. The pack-strap carry may be used to carry an unconscious or nonambulatory person when the atmosphere will permit an upright position. The victim may be positioned either back-to-back or in the same direction as the rescuer. It is best if the victim is positioned high enough on the rescuer's back that the feet dangle clear of the floor. This carry should not be used when the patient has any serious injuries or breathing difficulties.

RESCUE EQUIPMENT AND GENERAL RESCUE TECHNIQUES

Three-Person Lift and Carry

The three-person lift and carry is often used to handle a severely injured casualty (Figure 3-5). Even with trained personnel, this technique does not offer adequate neck and back support. When there is any suspicion that spinal injuries may be present, a fourth paramedic should support the patient's head.

Three rescuers line up on one side of the casualty, preferably with the tallest person at the shoulders, another at the hips, and the third at the knees. The team kneels on their knees nearest to the victim's feet. The paramedic at the victim's shoulders works both hands under the victim's neck and shoulders; the person at the hips places both hands under the pelvis and the hips; and the EMT at the knees is responsible for lifting the knees and ankles.

At the command *lift*, usually given by the rescuer at the shoulders, the crew gently lifts and rests the victim on their knees. At a

Fig. 3-5. The three-person carry is often used to carry a seriously injured victim or to pick up a casualty and place him or her on a stretcher. Three rescuers line up on one side of the patient, preferably with the tallest at the shoulders, another at the hips, and another at the knees; from this position, they can move with the casualty through narrow spaces and down winding stairs. The rescuer nearest the victim's head is responsible for continuously checking the casualty's respiration and other vital functions.

second command, they slowly turn the victim toward them until the casualty rests in the bends of their elbows. At a third command, the crew rises together to a standing position.

When a stretcher or cot is not immediately available, these procedures allow the crew to move with the patient through narrow spaces and down winding stairs. Depositing the victim requires the reverse of this procedure. These steps are also recommended for picking up and placing a victim on a stretcher or cot. The casualty is raised only high enough to clear the stretcher; the stretcher is then slid under the victim, and the victim is lowered.

Chair Carry

The chair carry (Figure 3-6) is effective where sharp turns or a steep stairway would preclude the use of a stretcher or backboard. A commercial stair chair or an ordinary household straight-back chair may be used. It is a good idea to test the chair before use to make sure that it is solid and will support the patient.

After the victim has been placed in the chair, the rear paramedic tilts the chair back to enable the other crew member in front to get into position. If the victim is too large to be picked up and placed in the chair, then the person is placed face up. The legs are elevated and the chair back slid beneath the buttocks until the patient is in a seated position, although still lying on the floor. Straps may be used to secure the patient to the chair. The chair is then tilted upright and the victim is ready for transport.

Fire Fighter's Drag The fire fighter's drag, which enables one person to move a patient with relative ease (Figure 3-7), is often used to move an unconscious casualty in an emergency. This method is recommended for use in tunnels and other limited spaces, or in atmospheres

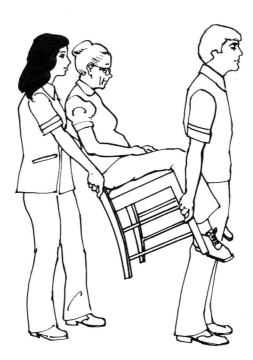

Fig. 3-6. Chair carry. It is very easy for two EMTs to carry a nonambulatory patient down winding stairways or through narrow passageways by placing the patient on an ordinary straight-back chair. The EMTs then grasp the chair in the most convenient positions. This method is good to know, as straight-back chairs are almost universally available.

RESCUE EQUIPMENT AND GENERAL RESCUE TECHNIQUES

where it is necessary to remain close to the floor in order to obtain an adequate supply of breathable air. The disadvantage of the fire fighter's drag is that the victim's head is unsupported and unprotected.

The rescuer kneels, turns the victim face up, and then ties the person's wrist together with a short length of rope, a soft belt, a necktie, or some other suitable material. This forms a loop of the victim's arms. In the event of patient injuries, it may be more desirable to loop a cravat bandage, belt, rope, or other convenient substitute over the casualty's head and under the arms. This method may be more comfortable for many rescuers because the victim can be raised higher off the floor by tightening up the loop.

After kneeling astride the victim, the loop of material (or the patient's arms) is passed over the rescuer's head and under the arms to support the victim. By raising the upper part of the body, the rescuer lifts the victim's shoulders clear of the floor. Crawling on the hands and knees will drag the casualty while allowing both persons to remain close to the floor where the best respirable atmosphere will be found.

While descending a stairway, the paramedic should reverse positions and descend down the stairs backwards to prevent the victim's head from striking the steps.

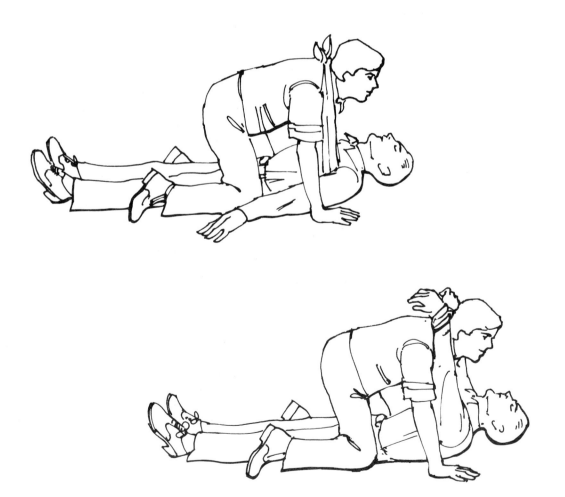

Fig. 3-7. The fire fighter's drag allows one person to rescue an unconscious victim. This drag is recommended for use in tunnels and other limited spaces, or in atmospheres where it is necessary to remain close to the ground in order to obtain an adequate supply of breathable oxygen.

Transporting the Victims

Incline Drag The incline drag may be used to ease an unconscious casualty down a stairway or incline (Figure 3-8). The victim should be positioned face up, head downward. The rescuer crouches at the victim's head and grasps the person under the armpits. The victim's head must be supported in the rescuer's arms, and the head kept as close to the stairs as possible. The wrists of an unconscious victim should be tied together to prevent injury.

Blanket Drag The blanket drag can be used to move a person who, because of injuries, should not be lifted or carried by one rescuer alone (Figure 3-9). When properly used, this drag will allow one rescuer to safely and effectively move an injured or unconscious victim. The blanket is placed lengthwise close to the patient with one-half gathered next to the body. The victim's arm opposite the blanket is extended overhead and the person is rolled to that side. While holding the patient on the side, the gathered folds of the blanket are pulled close to the body. The casualty is then rolled onto the blanket face up and wrapped snugly in the blanket with the arms at the sides. The blanket is then arranged to support the patient's head and neck. The victim is always pulled head first with the head and shoulders slightly raised, to keep the head from bumping against the floor. If it is necessary to descend stairs, care must be exercised to protect the victim's head. A mattress can also be used, but will require a minimum of two rescue personnel.

Clothing Drag

A casualty may be dragged by sturdy clothing (Figure 3-10). The clothing must be carefully snugged up around the victim's head and neck, but be careful that they do not become so tight around the neck that respiration is interferred with. The casualty's head is supported by the rescuer's forearm; the head must be kept as close to the floor as possible while traveling so the neck is not bent excessively.

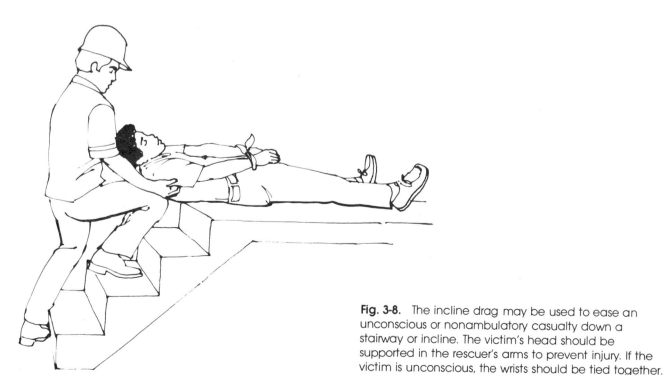

Fig. 3-8. The incline drag may be used to ease an unconscious or nonambulatory casualty down a stairway or incline. The victim's head should be supported in the rescuer's arms to prevent injury. If the victim is unconscious, the wrists should be tied together.

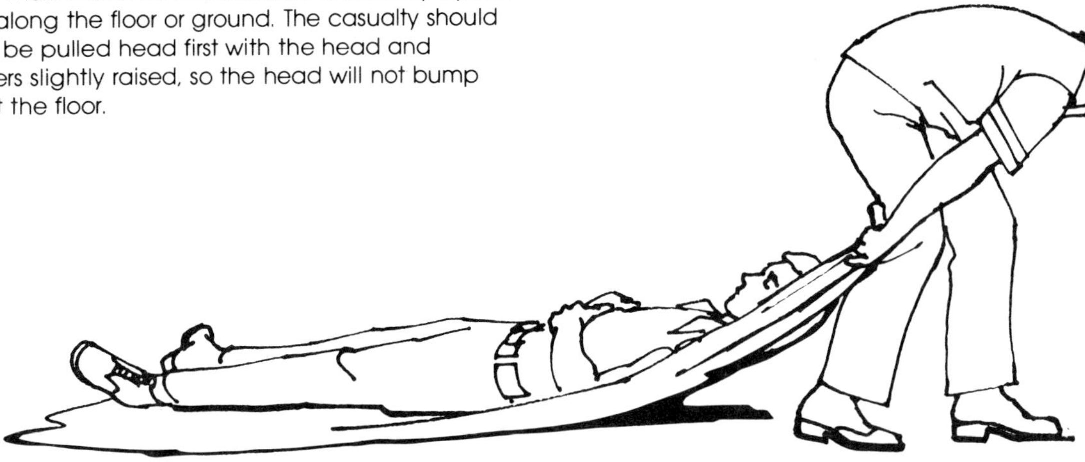

RESCUE EQUIPMENT AND GENERAL RESCUE TECHNIQUES

Fig. 3-9. The blanket drag can be used when one rescuer must move an unconscious or seriously injured victim along the floor or ground. The casualty should always be pulled head first with the head and shoulders slightly raised, so the head will not bump against the floor.

Fig. 3-10. Clothing drag. An unconscious victim dressed in sturdy clothing can be dragged from a structure by grasping the attire at the neck and pulling the victim towards the exit. Arrange the clothing carefully around the head and neck to offer support and prevent the head from dragging on the floor. This drag is especially valuable for removing victims from a hot, toxic, nonbreathable atmosphere, where survival is possible only by staying close to the pavement.

METHODS OF TRANSPORTATION

A multitude of problems may arise when sick and injured persons must be conveyed to medical attention. It is of utmost importance that the victim receive the best emergency care possible before transportation is attempted, that close attention be devoted to the patient during movement, and that all activities be conducted in the smoothest and most efficient manner possible to keep casualty morbidity and mortality at a minimum.

Backboards

Backboards, which are also called spine boards, are designed primarily for immobilizing accident victims with suspected spinal injuries. Backboards are available in two sizes. The long board is used for immobilizing the prone or supine patient; the short board serves as an intermediate device for immobilizing patients in a seated or cramped position, which would prohibit the direct movement of the victim onto a long board. The short board can be used to immobilize an

Transporting the Victims

automobile accident victim before removal from the wreck; it also provides a firm surface for administering external cardiac compression.

Backboards are typically constructed of plastic, aluminum, or plywood. Wooden surfaces are well sanded, varnished, and waxed to permit easy sliding of the patient onto the board. Nine-foot-long (2.7-meter-long) straps are used to secure the patient to the boards; other accessories include neckrolls, head and chin straps, a footrest, and other body straps. All models permit X-rays to be taken of the victim without removal from the board.

All accident victims should be immobilized before they are extricated or transported to a hospital, even if there is only a slight chance that they have suffered spinal injuries or fractures. A unique feature of the short backboard is its easily maneuverable size, which is just sufficient to support the victim's entire spine, but not the legs. After immobilization on the short backboard, the victim should be transferred, board and all, to a stretcher or long backboard before transportation. Once a spine board has been secured to a patient, it must not be removed until so directed by a physician (Figure 3-11).

The long backboard is an excellent carrying device for all victims with serious broken bones, especially those with fractures of the back and neck, but also for those with fractures of the pelvis and legs. This board is also useful as a stretcher for carrying patients down stairs and through narrow passageways.

Patients may be moved onto the long backboard by a method known as the patient roll, or log rolling. This operation requires a

Fig. 3-11. Patients may be moved onto the long backboard by a method known as the patient roll, or log rolling. Rescuers kneel at the victim's least-injured side, grasp the clothing on the far side, and at a signal roll the victim towards them. The backboard is placed flat on the ground (or tilted slightly upward on the ground behind the victim) and, at a signal, the casualty is rolled onto the board. Caution must be taken to hold the victim's head in line with a slight amount of traction. (Courtesy of Ferno-Washington, Inc.)

RESCUE EQUIPMENT AND GENERAL RESCUE TECHNIQUES

minimum of four people. The four rescuers should kneel at the victim's least-injured side; raise the arm that is the closest to the EMTs above the head; grasp the victim's clothing on the far side and, at a signal, roll the casualty toward them. The victim's head must be held in line with a slight amount of traction. Do not release the traction until the patient is fully immobilized. Place the backboard flat on the ground (or tilted slightly upward on the ground behind the victim) and, at a signal, roll the person onto the board, returning the arm to the side (Figure 3-12).

The patient may be slid lengthwise onto the spine board; this method is advantageous when the board cannot be placed alongside, because it requires very little direct handling of the victim. It can be performed with only three rescuers, though more help is advisable. Center the board at either the head or the feet of the victim, whichever is more accessible and convenient. Then, working in unison, slide the patient lengthwise onto the board. A rope loop under the victim's arms will afford an effective means of pulling on the patient, but the head should receive special consideration. The EMT caring for the head should keep it in line with the body and must apply a slight traction at all times until the head and neck are immobilized.

Stretchers and Litters

A regular ambulance cot or stretcher should be employed if one is available. If an improvised stretcher must be used, it is essential that the device be strong enough to bear the weight. Always test a stretcher before a patient is placed on it, if there is any doubt of its dependability. To do so, place an uninjured person, weighing as much or more than the patient, on the stretcher and lift it. The stretcher should be positioned as close as possible to the casualty so lifting and carrying is reduced to a minimum.

Fig. 3-12. The short backboard is most frequently used for stabilizing victims who are sitting or lying in an awkward or cramped position. This spine board is useful for immobilizing accident victims before extrication from the wreckage. The victim's head should be supported prior to and during immobilization, with slight traction in the upward position. With careful manipulation, the board may be slipped between the patient's back and the seat of the car with a minimum of disturbance to the spinal column. (Courtesy of Ferno-Washington, Inc.)

There must be enough personnel to carry the stretcher so that there is no risk of dropping the casualty. Whenever possible, the stretcher is brought to the victim instead of the victim's being carried to the stretcher (Figure 3-13).

As a general rule, an injured person is moved while lying on the back. However, in some instances, the type of trauma or location of the injury will necessitate the use of another position. A person having difficulty breathing because of a chest wound or other respiratory distress may be more comfortable if the head and shoulders are

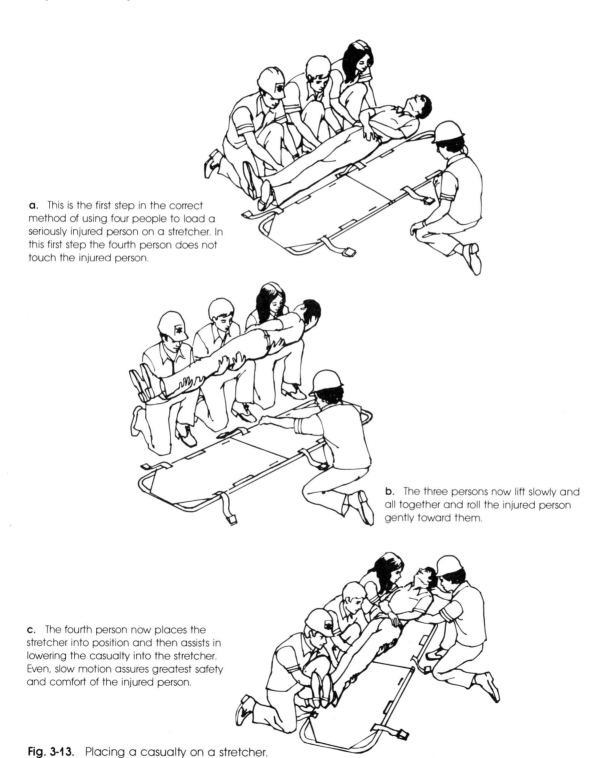

a. This is the first step in the correct method of using four people to load a seriously injured person on a stretcher. In this first step the fourth person does not touch the injured person.

b. The three persons now lift slowly and all together and roll the injured person gently toward them.

c. The fourth person now places the stretcher into position and then assists in lowering the casualty into the stretcher. Even, slow motion assures greatest safety and comfort of the injured person.

Fig. 3-13. Placing a casualty on a stretcher.

RESCUE EQUIPMENT AND GENERAL RESCUE TECHNIQUES

slightly raised. A victim with a broken bone should be moved very carefully so that further injury is avoided. A casualty with a severe injury on the back of the head should be kept lying on the side while being transported. Placing an unconscious or vomiting patient face down or on the side, with the head turned so blood and vomit may drain from the nose and mouth, will help prevent asphyxiation from a blocked airway. In all cases, the victim is placed in the position which will best prevent further injury (Figure 3-14).

The person must be securely fastened to the stretcher, and it must be carried so that the casualty will be moved feet first. The stretcher may be padded with blankets or garments.

If a motor vehicle must be used to transport a seriously injured victim, a passenger car should be avoided except in an extreme emergency. It is difficult to place an injured patient in a car without causing further injury and pain; and it is almost impossible to keep the person lying flat. The best means of transportation is, of course, an ambulance. When no ambulance is available, a truck or station wagon is a fairly good substitute.

If it is absolutely necessary to use a passenger automobile to transport a seriously injured person, no attempt should be made to put the victim in the car until the proper location within the vehicle and the best way to fit the patient into it has been thought out. An injured person is in no condition to be used as a measuring stick. The patient must not be placed anywhere in the car until rescuers are certain that undue bending, twisting, or turning will not be necessary.

A casualty must not be turned over to anyone without the rescuer's giving a complete account of the situation. The person taking over must know what caused the injury and what emergency treatment has been given.

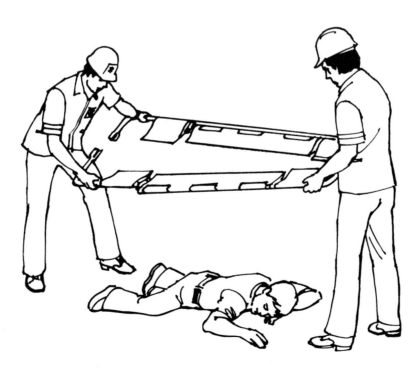

Fig. 3-14. A Scoop Stretcher enables EMTs to gently position the patient onto the stretcher through a scissorslike leverage action. This permits accident victims to be moved in the position found, thus minimizing the possibility of complicating the original injury. (Courtesy of Ferno-Washington, Inc.)

Transporting the Victims

Army Litter or Stretcher. This stretcher consists of two long poles with a bed, usually of canvas, between them, and crosspieces to keep the long poles apart and thus stretch the canvas. The poles are long enough to afford hand holds for the bearers at each end of the stretcher.

Stokes Navy Stretcher (Figure 3-15). The Stokes Navy Stretcher is a woven wire basket made to conform to the human body. The patient, after having been strapped to the basket, may be placed in a vertical position and lifted out of confined quarters. The stretcher should be padded with blankets and the victim made as comfortable as possible.

Improvised Stretchers. If regular stretchers are not available, it may be necessary to improvise some way of carrying the injured. Shutters, doors, boards, and even ladders may be used. All stretchers of this kind, of course, must be very well padded and great care must be taken to see that the casualty is fastened securely in place.

Pole and Blanket Stretcher (Figure 3-16). A blanket, shelter half, tarpaulin, or other material may be used for a litter bed. The poles may be made from such objects as strong branches, tent poles, skis, and so forth. A blanket should be opened, laying one pole lengthwise across the center, and folding the blanket over it. The second pole should be placed across the center of the new fold, then the free edges of the blanket folded over the second pole.

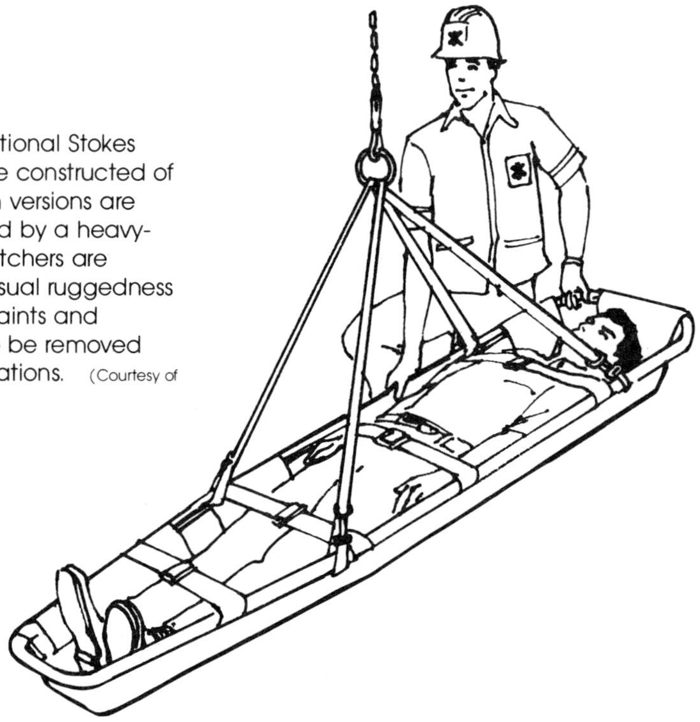

Fig. 3-15. A modern version of the traditional Stokes Navy basket stretcher. Early models were constructed of chicken wire and metal tubing; modern versions are constructed of durable plastic supported by a heavy-duty aluminum frame. These basket stretchers are valuable in rescue situations where unusual ruggedness and stability are necessary. Patient restraints and hoisting straps allow accident victims to be removed from a wide variety of endangering situations. (Courtesy of Ferno-Washington, Inc.)

a. Open a blanket, lay one pole lengthwise across the center and fold blanket over it.

b. Place the second pole across the center of the new fold.

c. Fold the free edges of the blanket over the second pole.

Fig. 3-16. Improvised stretchers.

RESCUE EQUIPMENT AND GENERAL RESCUE TECHNIQUES

Blanket [Figure 3-17]. Sometimes a blanket may be used as a stretcher. The casualty is placed in the middle of the blanket on his back. Three or four personnel kneel on each side and roll the edges of the blanket toward the victim. When the edges are rolled tightly and are large enough to grasp securely, the casualty should be lifted and carried. This type of stretcher must not be used for seriously injured persons, particularly those with fractures.

Rough Terrain Transportation

Situations where rescue may be necessary in rough terrain include removing casualties from hilly or mountainous regions, flooded areas, or other areas where conventional vehicular road travel is impossible. These conditions may be aggravated by snow, ice, and rain. On difficult terrain, even a minor handicap can render a person incapable of traveling without help. Rescuers must not underestimate the amount of physical effort required to transport an injured person. When available personnel and equipment are limited, it may be better to provide the patient with emergency care and to send for additional assistance.

The major considerations in rough terrain rescue are first, to locate the injured victim; second, to provide such immediate care as is necessary; and third, to have the knowledge, personnel, and equipment to evacuate the victim. Ski patrols and organized desert and mountain rescue organizations usually possess four-wheel-drive vehicles, snowmobiles, toboggans, and other especially designed or adapted equipment that is capable of operating in that particular section of the country.

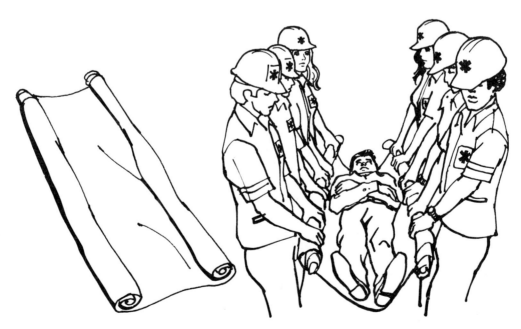

Fig. 3-17. If a regular stretcher is not available to transport an injured person, a blanket may be used to improvise a carrying device. This type of stretcher must not be used for a seriously injured person, particularly a person with fractures, because of its nonrigid construction. This technique requires the services of three or four rescuers on each side of the blanket.

When transporting a casualty in a jeep, truck, or station wagon, great ingenuity is sometimes required to suspend or pad the stretcher so the person obtains a reasonably smooth ride. One technique is to partly inflate a number of inner tubes or air mattresses to act as cushions. A thick sponge rubber pad is also effective. Swinging a stretcher on straps is not recommended; excessive swaying and bouncing occur because the straps have no cushioning effect.

An improvised sled may be constructed from an injured skier's skis and poles. A travois may be built by laying two small trees about 18 inches (45 centimeters) apart and then tying short cross braces between them. The patient is then laid upon this litter. Under ideal conditions, one or two persons can travel by lifting one end of the travois and dragging it along the ground. Stretchers may also be improvised from materials available at the scene.

QUESTIONS FOR REVIEW

1. Which of the following is the most common cause of additional trauma among accident victims during the treatment phase?
   a. Not recognizing a clogged airway
   b. Excessive pressure on the chest during cardiopulmonary resuscitation
   c. Improper transportation
   d. Not recognizing the presence of a fractured extremity

2. Paramedics should use which of the following groups of muscles when lifting a patient onto a stretcher?
   a. Back
   b. Legs
   c. Arms
   d. Shoulder

3. Immobilization on a backboard is least likely to be required for which of the following accident victims?
   a. A worker who was struck on the head by a falling beam
   b. A young person who became unconscious while diving into a pool from a springboard
   c. An automobile accident victim who received a cut forehead by striking the windshield
   d. A baseball player who was struck in the abdomen with a hard-hit ball

4. Which of the following is least recommended when it is necessary to control a combative psychiatric patient?
   a. Sedation
   b. Restraining devices
   c. Physical force
   d. Calm and sympathetic concern by the EMTs

5. When is the earliest time that a spineboard may be removed from an accident victim?
   a. When the paramedics determine that no spinal injury exists.
   b. When the patient refuses treatment
   c. When the victim can walk and talk clearly
   d. When so directed by a physician

PART 2 RESCUE FROM LAND AND AIR VEHICLES

4 Handling Highway Accidents
5 Extrication from Crashed Motor Vehicles
6 Rescue from Aircraft Accidents

CHAPTER 4

Handling Highway Accidents

Most traffic accidents are minor, but occasionally a vehicle collision will resemble a disaster scene. The efficiency with which rescue operations are carried out will largely depend on the plans and training of the rescue organizations and personnel that will be involved.

Generally, law enforcement personnel will be responsible for traffic control and investigation of the accident; ambulance attendants or rescue crews will administer emergency medical care and transport the victims to the hospital; and fire fighters will provide fire protection and flush away spilled flammable liquids. However, the first personnel to arrive on the scene, whether one person or a complete crew, will be responsible for all emergency service activities until additional help arrives (Figure 4-1).

RESPONDING TO THE SCENE

A safe but rapid arrival on the scene is necessary. The best method is to plan the approach over the quickest route, not necessarily the shortest. Drivers must know their area thoroughly; this includes shortcuts, detours, alternate routes, construction zones, and areas of congestion. A few seconds spent thinking out the best route may save minutes of hazardous high-speed driving.

Prompt arrival may allow the lifesaving emergency medical care that will save a critically injured victim. Timely appearance of rescue personnel will prevent poor public relations and critical comments from the persons in the accident and bystanders on the scene. To those involved in an accident, each minute they wait seems much longer than it really is. They wonder (and sometimes audibly express their thoughts) what is keeping the emergency personnel. Moreover, witnesses (who may be of great value) often become impatient and leave the scene once their curiosity is satisfied, unless rescue personnel are present to question them about the accident.

SIZE-UP

Whenever possible, the rescue crew must be given all the available information about the accident when they first receive the call, so a preliminary size-up can be made while they respond to the incident. The dispatcher or other officer who receives the call should obtain all pertinent facts from the person reporting the emergency and pass these on to the personnel who will be sent to the accident. The crew will be better prepared to evaluate the emergency when they arrive if they have been forewarned about the number of vehicles involved and their types; the number of victims and the seriousness of their injuries; whether hazardous or flammable cargoes are involved, so special precautions can be taken; and other facts that could take valuable time to learn after arrival on the scene.

Highway accidents are often the scene of confusion. A simple collision with no casualties will seldom cause any problems, unless the participants are seeking an arbitrator. When multiple vehicles and victims are involved in the accident, a number of complications may result. There may be hazards from fire and flammable liquids; casualties may be more serious; and survivors and bystanders may become a problem. The moans and cries of the injured and the devotion to duty of rescue personnel may cause them to rush to the aid of the victims without taking in the complete picture.

The person in command of the first unit on the scene is responsible for sizing up the situation. The first duty is to decide whether emergency units are required, and, if they are, to see that enough of them have been dispatched. It is important to request at the earliest possible moment all fire-fighting crews, ambulances, rescue squads, and other personnel and equipment needed to handle the situation. Taking a few seconds to make a deliberate and methodical assessment of the situation could save minutes or hours of effort later and spare the victims needless injury and death.

Priority of Concern

When there are injured victims, the first person on the scene must rapidly survey the victims to detect any life-endangering conditions. In certain instances, emergency medical care will be the most important action, and must be given without hesitation or delay. When injuries are minor, protection of the scene may warrant top priority. When a high-speed highway is blocked in an area of low visibility, prevention of additional crashes must be considered. Placing flares and directing traffic while a person bleeds to death or suffocates would undoubtedly be incorrect, but so would bandaging slight injuries before protecting the personnel on the scene with traffic control. Each situation must be considered on its own merits.

Searching for Victims

A thorough search of the entire accident scene must be conducted to be sure that all victims have been accounted for. Ask the other casualties, survivors, and bystanders how many were involved. Victims are often thrown out of the car on impact; a small child can easily be overlooked; a dazed or confused victim may wander off; or another vehicle may have transported patients to a hospital (Figure 4-2).

The bodies of infants and small children are often mistaken for bundles of rags or other debris, and no search is launched until one of the automobile occupants regains consciousness and asks questions. Passengers may be transported in the luggage compartments

RESCUE FROM LAND AND AIR VEHICLES

Fig. 4-1. Bystanders and police are giving emergency care to the victim of an automobile accident while awaiting the arrival of an ambulance. The ABCs of prehospital emergency medical care must be observed at all times. First, check to be sure the airway is clear so respiration is possible; second, be sure the patient is breathing; third, examine the patient's circulation to be sure there are no cardiac irregularities. After these steps, hemorrhage, shock, and other trauma should receive consideration. In this photo, the rescuers have laid the victim on his side so choking will not result from blood and vomit in the breathing passages while they are caring for the other injuries. (Courtesy of the California Highway Patrol.)

Fig. 4-2. A thorough search of the entire accident scene must be conducted to be sure that all victims have been accounted for. Victims are often thrown out of a car on impact; a small child could easily be overlooked in the wreckage; the bodies of infants and small children are often mistaken for bundles of rags or other debris. It is essential that every part of the wreckage and debris be thoroughly and methodically searched to be sure that there are no overlooked victims of the accident. (Courtesy of the California Highway Patrol.)

of cars or the freight sections of trucks and buses; these areas must be searched. Bodies have been discovered days after an accident in thick brush or deep canyons. In one case, a pedestrian was hurled by the force of the collision onto the roof of a nearby building. Some vehicles travel for hundreds of feet past the point of impact; occupants may have fallen from the car before it finally stopped.

An examination of the clothing and luggage in an automobile may offer some hint of the number and sex of the occupants. Toys will hint that a child may have been riding in the car. Emergency units should not leave the scene of a serious accident without making every effort to be sure that all of the occupants have been accounted for.

Positioning the Rescue Vehicle

The rescue vehicle should be parked safely and conveniently upon arrival at the scene. Consideration must be given to the safety of the crew, the vehicle, and other traffic. Do not park opposite the accident, so rescuers must cross the traffic stream every time it becomes necessary to return to the vehicle; parking on the same side of the highway will be safer and more convenient. Properly positioning the vehicle will let warning lights be used to alert other traffic, will make equipment more accessible, and will prevent potential looters from stealing tools and appliances.

Rescue personnel arriving first on the scene should position their vehicle to maintain efficient traffic control and flow. Some protection may be offered by parking in front of or behind the collision area, instead of alongside; in this way, only a minimal amount of roadway is blocked. Placing the emergency vehicle between the oncoming traffic and the wreckage will allow the warning lights to be used for alerting the other drivers, and the headlights will illuminate the scene at night. In such cases, protect the emergency vehicle from being driven into the accident scene by a rear-end collision by setting the parking brake and chocking the wheels in front. No matter where the vehicle is positioned, it should be parked far enough away from the rescuers and victims to avoid additional damage or injuries if it is struck by another vehicle or if flammable liquids around the wreckage are ignited.

Gasoline Washdowns

When gasoline or other fuels must be flushed from the highway with large quantities of water to avoid the possibility of its becoming ignited, consider where the fuel and water will run, and what additional action, if any, is necessary. Even though the water will flush away the flammable liquid, combustible vapors, being heavier than air, will flow along close to the ground and may create a hazard at a remote location. It is not unusual for lighted cigarettes or a warning flare hundreds of feet from the wreckage to ignite the vapors; the resultant fire may flash back through the vapor cloud to the accident scene.

Storm drains and ditches are especially hazardous because the flammable liquids and vapors can be channeled through them for a long distance and still remain dangerous. To reduce the threat of water pollution, an oil boom or other protection may be needed to contain the materials where the ditch or drain enters the ocean, lake, or river.

If flammable liquids are flushed into a sewer system, serious hazards may be created at the sewage disposal plant. Flammable vapors may back up into homes and businesses.

Truck Accidents

Truck and van accidents can involve rescue personnel in extremely hazardous and complex operations, because of the wide range of hazardous materials and flammable liquids transported on the highways. In spite of laws regulating the posting of warning placards, there often is no prior warning of the possibility of a dangerous condition. Emergency personnel must remain alert to these possibilities because explosions, fire, and detonations have occurred 30 minutes or more after a collision, with no forewarning or indication of disaster. The vehicle should have a cargo manifest in the truck cab; this should be checked at the first opportunity (Figure 4-3).

Because of the variety of hazards and the lack of knowledge regarding the cargo, rescuers are unable to preplan for a specific incident. When responding to a truck emergency, approach the wreckage with caution and attempt to identify the cargo. Any signs of vapor or smoke should alert all personnel that further precautions are necessary. If possible, it is best to approach the scene from upwind and upslope. Stop the rescue vehicle at a distance and approach on foot, taking advantage of any natural protection and buildings. If there is the slightest chance that toxic gases or vapors may be encountered, self-contained breathing apparatus must be worn, because some of the most lethal substances cannot be detected by any of the human senses. Keep the wind at the back and avoid any low spots in the terrain where gases or vapors may pocket or flow. Evacuate the area of all bystanders and rescue personnel who are not absolutely necessary to the operation.

RESCUE FROM LAND AND AIR VEHICLES

Reports of Accidents

Law-enforcement officers will attempt to obtain all necessary information for their reports, such as vehicle license numbers, the makes and the models of the cars, and the names of the drivers and occupants. These reports are generally designed to provide information about the causes of the accident in order to establish driver responsibility, negligence, or criminal intent. There will be an exchange of names and addresses between drivers, whenever possible, so that their insurance companies can be notified. These reports and other transmittal of information are certainly important, but ambulance and rescue personnel should be certain that this data collecting does not delay immediate emergency medical care and transportation of the seriously injured victims.

EMERGENCY MEDICAL CARE

It is generally recognized that emergency care at the scene of a highway accident often leaves much to be desired; in many cases, it directly contributes to the injuries and fatalities resulting from the crash. Many of the problems are created because relatively naive and inexperienced individuals, who are often first on the scene,

Fig. 4-3. Gasoline and other flammable liquids must be covered with a fire fighting foam or flushed from the highway with large quantities of water to avoid the possibility of accidental ignition. The gasoline spill from this overturned tank truck was covered with a thick blanket of foam by the first arriving fire fighters, to reduce the immediate fire hazard. While the wreckage was being turned upright, streams of water flushed the gasoline away from the vicinity as fast as it leaked out of the tank. Remember that flushing flammable liquids does not render them less hazardous, but merely transfers the dangerous fuels and vapors to a more remote location. It is not unusual for smokers or a warning flare hundreds of feet from the wreckage to ignite the vapors; the resultant fire may flash back through the vapor cloud to the accident scene.

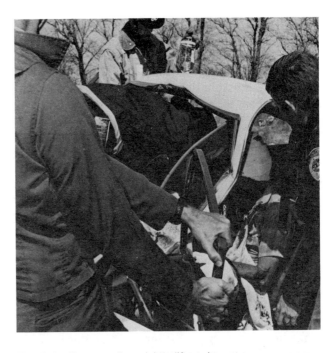

Fig. 4-4. Paramedics giving lifesaving emergency care to an accident victim who is still trapped in the wreckage. Prehospital emergency medical care must start as soon as an EMT can reach the victim. Every effort must be exerted to treat and stabilize the patient until other rescuers can bend, twist, cut, and otherwise remove the crushed metal from around the victim. Remember that the wreckage must always be removed from the victim, instead of manipulating and twisting an injured person to achieve extrication. In this photo, paramedics have used intravenous infusion and bandaging to preserve life in the victim until he can be removed from the wreckage. (Courtesy of the California Highway Patrol.)

undertake well-meaning but inappropriate actions—dragging the victims from the wreckage or propping the head up to make an injured person more comfortable. Possible neck and back injuries are generally disregarded.

Traffic control at the scene of a highway accident is necessary to prevent further accidents and injuries and to help assure a safe working environment for rescue personnel and equipment. A quick visual check of the victims and an experienced surveillance of traffic conditions will allow the first emergency personnel on the scene to determine which of the two functions is the more important—rescue or traffic control.

Emergency medical care for all victims of a highway accident must be immediately rendered by the first emergency service personnel to arrive on the scene, whether they are police, highway patrol officers, fire fighters, ambulance attendants, or lifeguards. Although emergency care procedures do not contribute directly to the process of extricating and disentangling victims from a wreckage (at least not in terms of their physical removal from the vehicle or debris), they are essential for saving human lives. Limited access to the victims may control the effectiveness of emergency care and treatment, but all attempts must be made to ensure that life is maintained. The time lapse between the actual crash and the administration of first aid measures must be reduced to a minimum. Once access is gained to the injured person and well-trained personnel with adequate equipment arrive on the scene, better care can be provided. Not only must the life-threatening medical problems be resolved, but those that could cause unnecessary permanent injury or needless suffering should be treated as early as possible.

Priority of Care

The simplest form of emergency medical care begins with a general survey of the situation to ascertain whether there are any serious injuries. Two factors are of major significance; the extent of the injuries and the position of the victim (Figure 4-4).

The sequence of emergency medical care activity follows a well-established pattern. Airway, breathing, circulation, bleeding, fractures, and shock should be considered in that order. Remedial action should take place before removal of the victim to an ambulance, unless the accident site is in an area of imminent danger. Experience has indicated that the immediate removal of victims from crashed vehicles is only warranted in a small percentage of cases, when fire, submergence, toxic fumes, or some other hazard exists.

The preliminary examination of the victim must be first directed toward two life-threatening injuries: interference with respiration and severe hemorrhage. Respiratory embarrassment, whether from obstruction, sucking chest wound, flail chest, or other causes, demands priority, because otherwise death may occur within very few minutes. Hemorrhage does not usually present an immediate threat to life, but it must be controlled to prevent shock. Once respiration and hemorrhaging are under control, the rescuer can examine the victim to determine the extent of injuries. This examination must identify not only the injuries involved, but also the problems that may arise as a result of the position of the victim.

Respiration

Maintenance of an opened airway is a key element to victim survival at any accident. Emergency physicians have stated that many accident victims with relatively minor injuries have died or suffered irreversible brain damage because no one at the scene was able to recognize immediately that the casualty had a blocked airway that required clearing. Providing oxygenation to the brain and heart via the lungs and circulatory system is the first priority in victim care. Loss of consciousness presents the potential for suffocation and requires the immediate attention of the first person to reach the accident scene.

Opening the airway by using the manual method of jaw lift or jaw thrust will facilitate the breathing of most uncomplicated cases, with manual positioning of the tongue necessary in some cases. The head should not be hyperextended or moved from side to side when there is any possibility of a cervical spine fracture; hold the head in a neutral position to prevent spinal cord trauma.

Neutral positioning of the head may not ensure adequate ventilation in the event of a cervical fracture or serious facial trauma. Using mechanical means—such as oral or nasal pharyngeal airways, an esophageal obturator airway, or a nasal endotracheal tube—will facilitate ventilation without head movement. Intubation using an oral endotracheal tube, except in very experienced hands, requires hyperextension of the head.

Lung expansion is assured by an intact thoracic wall. This means no sucking chest wounds, no flail chest, and no pneumothorax. It is of little help to deliver to the hospital a patient with an open airway but with flail chest, no ability to move air, hypoxia, and resultant cerebral and myocardial damage. Evaluation of lung expansion requires removing the victim's clothing and examining the entire chest by looking, listening, and feeling (if there are indications of body trauma).

Cardiac Evaluation

Evaluation of the cardiac status means to assure good cardiac output. Initiation of CPR in trauma situations may certainly be indicated. Myocardial damage resulting from a crushed chest may present the same problems as myocardial damage from occluded or partially occluded coronary arteries. Such patients should be closely monitored for cardiac disrhythmias, especially those of ventricular origin.

Bleeding Control

Injuries resulting from an auto accident usually involve some amount of bleeding, due to the immediate proximity of large amounts of glass and torn sheet metal. Safety designs have tended to reduce direct exposure to these hazards in the vicinity of the occupants, although lacerations will continue to occur as a result of dynamic contact with the interior surfaces of the vehicle. Severe or massive blood loss must be controlled immediately—but this does not refer to slight bleeding from minor lacerations. It refers to hemorrhaging that, if not controlled, may cause the victim to lose significant blood volume in a short period of time. Blood loss is generally controllable by applying direct pressure with the hand or a compress on the wound, and by continuing pressure until the bleeding stops.

While disentanglement and extrication operations proceed, efforts to render emergency medical care should continue as additional body areas become accessible. In the case of pinned victims, EMTs should be alert to the possibility that a severe bleeding potential may have been temporarily controlled by the pressure of some automobile structure, such as the steering wheel. When this pressure is released, additional bleeding may occur; any excessive hemorrhaging must be immediately controlled.

Stabilization of Fractures

Fracture stabilization before removal of the victim requires that the fracture site be immobilized. This can be done with in-line traction, using the hands of the rescuer or mechanical means such as air splints. In general, the application of rigid splinting material before extrication is not mandatory, except if the casualty has a cervical spine fracture or if there is any possibility of spinal damage.

Cervical, lumbar, and femoral fractures make up a significant portion of the injuries incurred in automobile accidents, and their immobilization requires a high level of preparation so the necessary equipment will be at the scene. The type of splint used must be reliable; the ease and rate of application under adverse psychological and environmental conditions must be considered. Logical planning and responsible action is only possible if training has sufficiently emphasized the time-proven adage "splint them where they lie." Although this reminder was originally meant to apply only to bone fractures, immobilization also provides needed protective support to soft-tissue injuries. The tension and torsion created by the process of extrication can add nothing but additional trauma and pain to victims with ruptured livers, spleens, or other vital organs; lacerated muscles; perforated intestines; and other serious injuries.

Unless a seat belt is snug, a collision impact can cause the knees to strike the dash, resulting in a fracture of the femur in midshaft. Fractures of the femur can result in lengthy disability due to nerve or vascular damage, nonunion, or infection. The immobilization of these fractures with a traction splint before the victim is moved will counteract spasms of the injured muscles, which may lead to arterial or neural involvement. If the victim is in a position that does not allow enough working room to apply a full-length traction splint, then apply a short splint between the legs and tie the legs together with bandages above and below the break. The victim can then be gently pulled from the restrictive area. As soon as the patient's legs are accessible, tie a bandage around both calves; an additional tie around both feet at the metatarsal arch will prevent the fractured limb from rotating. It is not necessary to remove any of the immobilization ties in order to apply a traction splint (Figure 4-5).

All victims of violent accidents, whether or not they are unconscious or complaining of neck or back injuries, should be fully immobilized and handled as if spinal fractures were a certainty. It is a good practice to immobilize all accident victims (except those with obviously minor injuries) on backboards before removing them from the wreckage (Figure 4-6).

Any pulling of the victim should be performed along the long axis of the body. The clothing may be grasped at the shoulder and hip to assist with lifting. This is often safer and more comfortable than placing the hands under the person. Using the patient's clothing for lifting will also protect the rescuers from cutting their hands on broken glass.

Shock

Accident victims are very susceptible to shock, which may accompany many different types of injuries. Casualties with fractures

RESCUE FROM LAND AND AIR VEHICLES

Fig. 4-5. EMTs sliding a backboard under a victim for transportation after a broken leg has been immobilized with a traction splint. *(Courtesy of the California Highway Patrol.)*

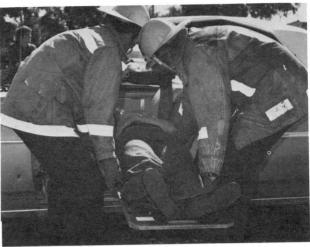

Fig. 4-6. Rescue personnel immobilize a collision victim with a possible spinal injury on a backboard before extrication from wreckage.

or massive bleeding may develop shock at an early stage of emergency care. This subject is discussed extensively in other sections of this book.

Neurological Status

Emergency medical technicians should evaluate the neurological status of every accident victim, especially if there is any indication of a head injury, by checking for pupillary size and equality along with evaluating the state of consciousness. To evaluate the state of consciousness, check for alertness, response to vocal stimulation only, response to painful stimulation only, and unconsciousness. A neurological status involving a deteriorating level of consciousness indicates a worsening condition. Changing pupillary size indicates the possibility of increasing intracranial mass or pressure. Indications of a deteriorating neurological status will dictate that the victim be rushed to a hospital as quickly as possible. Neurological assessment must be repeated every few minutes, in order to promptly recognize potentially lethal changes in neurological status.

Attention to Victims

Accident victims should be watched closely in case their conditions begin to deteriorate. The need for constant casualty surveillance is one of the reasons an ambulance should never be overburdened with excessive extrication or other equipment, which might be better transported by a rescue squad or by the fire department. There is a tendency for crews to leave the victim unattended while they are gathering up and loading the equipment back onto the ambulance. The same applies to police-operated ambulances, although police officers have many responsibilities at the traffic accident scene, such as the completion of accident report forms, obtaining statements from witnesses, traffic control, and measurement of skid marks.

TRAFFIC CONTROL

The main objective of traffic control is to forewarn other drivers. It is important to keep all vehicles moving in an orderly manner, to provide clear access for other emergency units, and to allow the casualties to be transported to a hospital with the least delay. Many police agencies give higher priority to traffic control than to the administration of emergency care to the critically injured. Their rationalization is that it makes little sense to devote all energy to saving one life while there is great potential for additional collisions, which would maim and kill other motorists. However, most authorities advocate the opposite viewpoint. They recommend that the initial action of the first person on the scene must be to rapidly check the victims to ascertain if there is any life-endangering trauma, and to care for it. Both actions are certainly necessary. When warning oncoming drivers is required to give them a chance to slow down before running into obstructions and to channel traffic into a safe path around the accident, it may be necessary to use volunteer bystanders to assist. If their aid is enlisted, they should be told what to do; otherwise, their undirected efforts may hamper rather than help the situation.

Traffic control at the scene of an emergency must receive high priority, because it is a protective activity that separates the involved casualties and rescue personnel from subsequent collision with other vehicles. Besides defending the accident scene from harm by limiting and guiding the flow of traffic in and around the area, efficient traffic control will reduce inconvenience to the other drivers and permit orderly access of other emergency vehicles. When drivers become irritated or annoyed because of conditions beyond their control, they often become irrational and angry. Their impatience could cause them to operate their cars recklessly and unsafely near the accident scene.

The methods of directing traffic depend mainly on the complexity and severity of the situation, but generally it is better to keep traffic flowing smoothly and safely around the area than to halt it completely. Obvious exceptions would exist when all lanes are blocked, when electrical wires constitute a hazard, or when flammable liquids and vapors represent a danger.

Flares

As soon as possible, appropriate warning devices such as flares, traffic cones, or flag signalers should be placed at a distance far enough from the accident to warn oncoming traffic from both directions. When constructing the flare pattern, it is necessary to guide the traffic away from, not into, the hazards. It is important to consider the traffic speed and any hindrances to visibility, such as curves, hills, fog, or rain. Under the best of conditions, the farthest warning device should be placed at a minimum of 100 feet (30 meters) from the wreckage when the traffic is proceeding at 30 miles (48 kilometers) per hour. The recommended distance is more than tripled when the speed is doubled. Any hindrances to clear visibility will require increased distances to achieve maximum safety (Figure 4-7).

Under safe conditions, the penetrating illumination of the flare is an effective aid to traffic control, even during adverse weather. However, several precautions must be observed. The fumes are toxic and the burning particles that are emitted may cause serious injury if they come into contact with the flesh. Avoid positioning the burning flares where the smoke may drift into the vicinity of the victims and rescuers. Never use flares in a confined area. Burning flares should not be held in the hand as a signaling device, because the arm motions

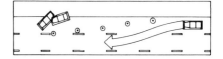

Fig. 4-7. After checking to be sure there is no spilled fuel, place flares about 15 to 20 paces apart. Angle them toward the side of the road as far back as possible. The object is to form a safety section by guiding approaching traffic around the accident.

RESCUE FROM LAND AND AIR VEHICLES

may cause personal injuries and clothing damage. Flares will offer a ready source of ignition if there are spilled flammable liquids, combustible vapors, or explosive chemicals at the scene. In case the presence of hazardous materials is known or suspected, it is best to dispense with the use of the flares entirely, and to rely on electrical warning lights and reflective safety markers for traffic control.

Every highway accident must be individually evaluated to determine the best method of regulating traffic. While ensuring the safety of all persons at a collision scene is of the utmost importance, it is also essential that the motoring public be delayed or hindered as little as possible. The flow of vehicles can be controlled in three ways:

1. *Blockade.* If all lanes are impeded to some degree by wreckage, debris, or victims, or if the passing traffic will endanger rescue personnel and victims, a complete closure of the roadway is warranted (Figure 4-8).
2. *Lane control.* When the entire roadway is not blocked, it is usually possible to direct traffic into the open lanes. This might mean channeling traffic from two or more lanes into a single line or alternately allowing the passage of automobiles from each direction.
3. *Alternate route.* When possible, the best procedure may be to detour the traffic onto alternate routes, instead of completely halting the vehicles. The availability of a convenient off ramp or intersection will largely govern the practicability of this method (Figure 4-9).

Hand Signals

Traffic control is often regulated by one person using hand signals. The most important aspect of directing traffic is to ensure that the drivers and pedestrians know exactly what is expected of them. Because verbal directions are seldom reliable at the scene of an accident, effective gestures must be delivered in a uniform, highly visible, meaningful manner.

When directing traffic, face the vehicles that need controlling and maintain a serious, businesslike demeanor. Look at the driver until eye contact establishes that attention has been gained. Point to the driver with an outstretched arm and finger or flashlight. Do not step in front of the oncoming vehicle. Check the position of the front wheels and the turn signals to detect any contradictory intentions of the driver. If possible, check the speed, distance, and route of other vehicles to avoid directing the responding driver into the path of another vehicle.

To halt traffic, point at the driver until eye contact is made, then open the hand and display the palm to the driver until the stop has been made.

To start traffic, point at the driver and when attention has been gained, open the hand and swing the arm with the palm up in a beckoning motion.

To direct a turn, gain the driver's attention and point in the desired direction.

Before any person can effectively direct traffic, hand signals and gestures must be adopted so that the desired intentions are clearly and authoritatively communicated to the drivers.

Handling Highway Accidents

Bystanders

Spectators and bystanders gather and crowd around emergency scenes with extreme fervor. Crowds of people are attracted to any unusual incident. Many stop with the intention of offering whatever aid they are capable of extending to the casualties. These good Samaritans are invaluable in helping the victims until the emergency personnel arrive. As a rule, most of these helpers are cooperative and will step back when the trained crews arrive.

The curious and morbid members of the public hurriedly park their cars in close proximity to the wreckage and then crowd around the injured. Usually these automobiles and spectators will complicate the rescue operation. As soon as personnel are available to do so, the bystanders should be requested to leave the scene.

AUTOMOTIVE FIRES

A vehicular fire can easily and rapidly change a minor traffic accident into a flaming disaster, so every precaution must be observed to prevent the ignition of any automotive vehicle involved in an accident. Because there are many dangers associated with car and truck fires, it is necessary to conduct all operations with extreme prudence. When no lives are in jeopardy, all activities must be cautious and deliberate.

Fig. 4-8. All traffic should be stopped when the entire roadway is occupied by victims, wreckage, debris, and emergency units. (Courtesy of the California Highway Patrol.)

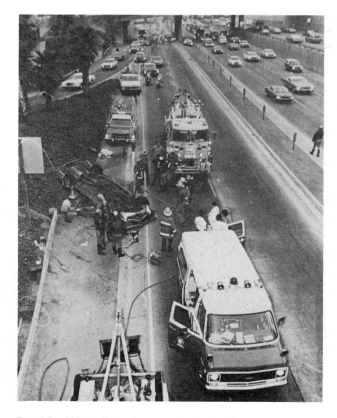

Fig. 4-9. When there is a convenient off ramp or intersection, divert all traffic onto alternate routes. Note the officer in the center divider preventing the opposite lanes of traffic from slowing down. (Courtesy of the California Highway Patrol.)

RESCUE FROM LAND AND AIR VEHICLES

Rescue personnel and fire fighters without extensive experience in combatting automotive fires have a tendency to underestimate the amount of heat that is generated, the rapidity with which these fires can spread and expand, and the number of devices and components that may suddenly rupture or explode. Therefore, novice or unskilled personnel tend to combat these blazes with small hose lines or fire extinguishers. It is true that the vast majority of automotive fires are easily extinguished with a minimum of expertise or effort; however, often, when least expected, the entire fire fighting capability of one or more engine companies will be required to control the situation (Figure 4-10).

Knowledgeable authorities usually recommend that no hose lines smaller than 1-1/2 inches (3.75 centimeters) be depended upon at the scene of motor vehicle incidents, whether or not ignition has occurred. It is also recommended that all fire fighting personnel remain dressed in their full complement of protective clothing, including helmet and gloves, while operating at or standing by a collision scene. In all cases, hose lines must be laid, charged with water, and constantly supervised whenever rescue personnel are engaged in activities around crashed or stalled vehicles where spilled flammable liquids are a possibility. Paramedics and other rescue personnel have been injured and killed when the wreckage suddenly became ignited during extrication procedures, and the trapped victims were burned to death.

Trapped Victims

When a vehicle is burning, a prompt attack on the fire must be launched so the trapped victims can be removed quickly and safely. Fire streams or the blast from carbon dioxide or dry powder extinguishers must be used to drive the flames away from the victim. A prompt, massive attack on the blaze is usually the best method of protecting the casualties. If substantial fire fighting equipment is not present and small fire streams or extinguishers must be depended upon, then using the extinguishers to form a shield between the victim and the flames offers the most effective protection. Many victims have perished after rescue personnel arrived on the scene because rescuers were unable to quench the fire before the casualties were extricated from the wreckage.

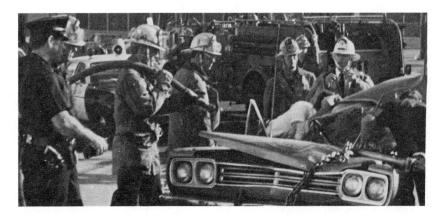

Fig. 4-10. Fire department personnel with loaded hose lines must protect EMTs against ignition of flammable liquids during extrication operations.
(Courtesy of the Los Angeles City Fire Department.)

Flammable Liquids

The most serious hazard to consider in a vehicle fire is the fuel tank, whether the car is operational, wrecked, or an abandoned derelict. The fuel may be gasoline, liquefied petroleum gas (LPG), or diesel fuel. In varying degrees, these could all be deadly and their hazards must be respected. Many fire fighters have been killed and injured by exploding fuel tanks while combatting relatively insignificant automobile fires. Because of the danger of explosion, immediate attention must be devoted to keeping the fuel tanks cool when they are or could be exposed to heat and flames.

Fuel tanks on most passenger automobiles are located in the rear, where they are susceptible to rear-end collisions. Many foreign cars carry the gasoline tank in the front, sometimes in the engine compartment. Although the positions of fuel tanks on trucks, buses, and other vehicles vary widely, they are readily located. Tanks are manufactured in various sizes and shapes of relatively thin metal. They are usually made by sealing two halves horizontally at the center. Caps and vents that previously permitted the flow of vapors and air into and out of the tanks have been replaced by an air pollution arrangement to control the emission of vapors into the atmosphere. Under this system, air is permitted to enter the tank to replace used fuel, but no vapor is allowed to escape. The heat during a fire will cause vaporization of the fuel, which results in a rise in pressure in the tank because lack of a vent prevents the escape of excessive pressure. If the tank ruptures during a fire, the almost instantaneous ignition of the vapors and fuel will resemble an explosion. When this occurs, the blast can cause a ball of fire and spray burning liquids for a considerable distance. The direction these flames will take is unpredictable, because it is impossible to know where the tank will burst.

On the other hand, suddenly cooling a tank that has been exposed to fire or heat with a stream of water may cause a quick drop in internal pressure. This, in turn, could allow outside air and flames to be drawn into the tank to form a flammable mixture. An explosion then becomes quite likely.

Liquefied petroleum gas cylinders are extremely hazardous, since they are normally pressurized at ambient temperatures. They may rupture when exposed to excessive heat or an impinging flame.

Diesel tanks are relatively safer than those filled with the more volatile fuels such as gasoline or LPG, but they still can be dangerous. Tanks that rupture because of heat can generate a mass of flames with explosive suddenness.

Rescue personnel must also remain alert to detect signs of leaking flammable liquids or a concentration of combustible vapors, because these could render a large area hazardous. Flammable fuels flowing from broken fuel lines, punctured tanks, and overturned vehicles may run down gutters, storm drains, or other low spots. When the vapors are exposed to a source of ignition, which may be some distance away, an extensive fire could occur. Whenever there is the possibility of a flammable-liquid spill, a fire department pumper should be requested to provide a gasoline washdown and to stand by in case of ignition.

Engine Compartment

Many automobile fires begin in the engine compartment, from a backfiring carburetor, a broken fuel line, or an electrical short.

RESCUE FROM LAND AND AIR VEHICLES

Raising the hood may cause an acceleration of the blaze or release a sudden burst of flames, because of the increased supply of oxygen. The hood should not be opened until a hose line or extinguisher is positioned to extinguish the flames. The fire may be safely controlled before raising the hood by directing a stream of water into the engine compartment through the front grill. Water passing over and through the radiator may knock the fire down sufficiently to allow the hood to be safely raised. When operating the hood release, keep the head turned to one side so the face will be protected from flames and hot gases.

Trunk Fires

Fires in locked luggage compartments may be difficult to extinguish when the keys are not available for easy opening. At times, the best procedure is to cut a hole in the back of the rear seat and spray water into the compartment until the main volume of fire is controlled. Before disconnecting the battery, check to see whether there is an electrical trunk release. It is usually difficult to force or pry doors or compartments open, but punching the lock cylinder on through the compartment door with a pointed tool and then inserting a screwdriver into the hole may allow the trunk to be unlatched. Be cautious when opening a luggage compartment, since flames and hot gases may erupt when the lid is raised.

Bursting Vehicle Components

There are many devices on modern automobiles and trucks that could rupture violently or catch on fire when subjected to excessive heat and flames. Any of the hydraulic or pneumatic components of the vehicle could cause serious problems. Even such innocuous items as bumper shock absorbers have injured fire fighters when the absorbers exploded from excessive exterior heat.

Any of the systems that contain confined liquids or air could rupture from heating. Such sections of the automobile include the cooling system, the air conditioning system, air or hydraulic brakes, and the various lines and hoses associated with these units. Any of these systems could result in serious injury if the contained pressure is suddenly and unexpectedly released, spraying bits of metal and rubber in the area.

Batteries

The electrical energy stored in batteries can cause injuries or start fires. One of the first actions to take on arriving at the scene of a burning or wrecked motor vehicle is to disconnect the battery so that a spark from it will not ignite flammable vapors. This action may involve no more than opening a switch, cutting a cable, or disconnecting a battery terminal. When uncoupling battery cables, disconnect the ground cable first to prevent a spark's resulting from shorting the battery terminal against the car frame or engine with a tool.

Many luxury automobiles use electrical power instead of cranks and levers to adjust seats, open and close windows, operate door lock, and unlatch hood and trunk locks. Check the vehicle over and use the electrical power to move the seat down and back, unlatch all doors and other locks, and open all windows before disconnecting the electrical power. Without a source of current, most of these operations are impossible.

Occasionally, battery cases will break apart with explosive force when subjected to excessive heat or flames. If this occurs, the battery acid solution is sprayed in all directions.

Burning Tires

Vehicle fires are often caused by fires that begin in a tire; or sometimes a car blaze may spread to the tires. Fully inflated tires that become ignited may burst, causing particles to fly in all directions with some force. Some heated wheels made of cast metal may shatter if they are suddenly cooled by a hose stream or a carbon dioxide extinguisher. Magnesium wheels may burn violently and could react with explosive intensity when struck by a stream of water. Serious burns or other injuries can result if fire fighters are struck by pieces of hot or burning tires and wheels.

When tires are threatened by flames, protect them by adequate cooling with water spray. If the tires are burning, water should be applied from a distance until the main volume of fire is knocked down.

QUESTIONS FOR REVIEW

1. When responding to the scene of a highway accident, the best plan is to approach over the
   a. shortest route.
   b. route that will be free of traffic congestion.
   c. quickest route.
   d. route that will not be used by other emergency vehicles.

2. Rescue personnel should position their vehicle at the scene of a night accident on a freeway so that it will
   a. illuminate the accident scene.
   b. protect the rescue personnel.
   c. provide the most accessible position for the rescue equipment.
   d. not interfere with the flow of traffic.

3. The correct sequence of initial emergency-care activity should always be
   a. airway first, then breathing and circulation.
   b. breathing first, then circulation and airway.
   c. circulation first, then breathing and airway.
   d. prompt, but not in any special order.

4. The victim of a violent automobile accident is fully conscious, but has a serious bruise on the forehead. There are no other obvious injuries. While extricating the patient from the wreckage, the rescuers should consider that the victim
   a. is uninjured.
   b. may possibly have a slight concussion.
   c. may become combative from the aftereffects of shock.
   d. could have spinal injury.

5. The most hazardous component of an automobile during a fire is (are) the
   a. fuel lines and pump.
   b. fuel tank.
   c. internal pressure in the engine cylinders.
   d. passenger compartment.

CHAPTER 5

Extrication from Crashed Motor Vehicles

Successfully extricating victims from crashed vehicles means moving the injured in such a manner that further trauma is not added to the injuries already sustained. Proper disentanglement will influence both the degree and duration of physical disability (Figure 5-1).

In most accidents, the drivers and passengers will already be removed from the vehicle, either by their own initiative or with the help of uninjured occupants or bystanders. However, surviving a high-speed collision may be directly related to remaining in the vehicle, since the structural reinforcements engineered into modern automobiles offer a great amount of protection. Experience has shown that victims who are thrown from their cars have a very high mortality rate. In many cases, though, the injured person may be pinned or trapped in the wreckage, not readily accessible in the small enclosed area of the interior of the automobile, or in an awkward, unnatural position, so that help is needed to escape from the confines of the vehicle.

STAGES OF EXTRICATION

The cardinal rule that no injured person should be moved until emergency care has been administered can be disregarded if the victim's or rescuers' lives are endangered. The victim must be transported out of danger as soon as possible if operations are jeopardized by fire, flammable vapors and liquids, the vehicle's teetering on the brink of a cliff, or any other immediate danger. Normally, extrication of victims from crashed vehicles is divided into five stages.

Gaining Access to Victims

Until a trained emergency medical technician (EMT) or paramedic can gain access to the victim, no life-support or emergency medical care can be supplied. Access must be rapid, using a minimum of mechanical tools. All that is required for initial access is a passageway large enough for the rescuer to examine the injured person and stabilize the victim's condition.

Administering Lifesaving Emergency Care

The sequence of emergency medical care activity follows a well-accepted pattern. Airway, breathing, circulation, bleeding, fractures, and shock should be considered in that order. Remedial action must be administered prior to removal of the victim unless the location of the accident is in imminent danger. The establishment of an airway, maintenance of a palpable pulse or blood pressure, and control of severe bleeding are further defined as immediate problems in which speedy handling directly influences survival.

Disentanglement

Disentanglement of the victim from the immediate surroundings is accomplished after access has been gained and the patient's condition has been stabilized.

Preparation for Removal

Once the structure and other impediments have been disentangled from the victim and other physical restraints in and around the vehicle have been removed, the patient should be carefully prepared for removal from the vehicle and subsequent transportation to the ambulance. This preparatory activity is directed toward facilitating removal and protecting the injured person from further trauma during transport.

Transportation

This phase consists of removing the casualty from the vehicle and transporting the injured person to the ambulance. This activity may consist of merely carrying the victim to the ambulance on a stretcher, or it could be as complicated as transporting the patient up the side of a cliff or across a flooded river (Figure 5-2).

TRAINING AND PREPARATION

Rescue squads, ambulance attendants, and all other paramedic groups must be prepared for any type of accident. Although every wreck is different, certain similarities can be expected. Training must be extensive in every area of vehicle rescue operations. It also must be continual if proficiency in rescue skills and techniques is to be maintained. Rescuers, machines, equipment, and methods must all be coordinated in order to be effective and efficient. The traditional procedure of "grab and pull" must be abandoned; effective rescue involves thorough and intelligent training, effective planning, efficient equipment, and modern techniques.

Sophisticated medical techniques and highly skilled hospital personnel are of little use so long as poorly trained and inadequately equipped rescuers continue to complicate injuries at the scene of collisions.

Outside Agencies

Training should be made available not only to paramedical groups, but to other agencies that might be called upon to help as well: electrical utilities to repair downed power lines, chemical companies to unload wrecked trucks carrying hazardous materials, vacuum trucks to clean up spilled flammable liquids, police and highway patrols to maintain traffic control, and other agencies that have the

RESCUE FROM LAND AND AIR VEHICLES

Fig. 5-1. Finding the best method of extricating trapped victims will often require the rescuers to use great ingenuity. In this accident, large hydraulic jacks lifted the truck sufficiently to allow the victims to be extricated. While the truck was raised, the crushed automobile was removed. The right front wheel of the truck offered a complication, since it was over the bridge railing. If there is a large crane in the vicinity, it will usually offer the fastest and most practical solution to lifting problems. (Courtesy of the Los Angeles City Fire Department.)

Fig. 5-2. Rescuer guiding a basket stretcher that is transporting an injured person up a steep cliff.
(Courtesy of the California Highway Patrol.)

equipment or knowledge that might be required at the scene of a highway incident.

Wreckers

Wreckers and tow trucks can offer valuable assistance. Their personnel should be well trained because they will often arrive on the scene even before the rescue squads or ambulances; in many cases, the initial extrication of victims will be handled by the drivers of these trucks. Preparation in practice sessions can greatly increase the victim's chances of survival at the scene and will reduce needless injuries caused by clumsy attempts to remove an injured person from the wreckage.

Incorporating local wreckers into rescue operations requires planning, training, and a working understanding among all participating parties. A thorough knowledge of the wrecker crew's potential to help is vital; consideration should be given to each wrecker's capabilities, to the size and capacity of the crew, to the accessibility of night crews, and to the number of personnel available. Some wreckers are equipped with twin booms and winches for side pulls or to stabilize the vehicle. The winch system is the wrecker truck's most important tool. The winch and booms may be used for lifting heavy weights, pulling seats, forcing impinging vehicle body parts, removing steering columns, and stabilizing materials or vehicles that are in a dangerous position.

Practice Sessions

Probably no single aspect of emergency rescue depends more on training and practice than learning the proper methods of extricating victims from automobile collisions. Emergency medical technicians and other paramedical groups should devote considerable time, labor, and thought to the problem of gaining access to and disentangling and extricating injured persons from damaged vehicles. Local wrecking yards are good places to obtain wrecked cars on which to practice. Invite local tow truck personnel to participate in these sessions; in this manner, paramedics can become acquainted with the potential uses and limitations of wrecker trucks, while the garage personnel can learn more about rescue techniques. Many accidents will demand all the personnel and equipment available; therefore, everyone should train together to form one large homogeneous rescue force.

Untrained Personnel

Unfortunately, the injuries and fatalities associated with vehicular accidents have often been greatly increased by the first aid given to the victims by well-meaning, but ignorant and misguided persons first to arrive on the scene, or by untrained and poorly qualified ambulance attendants. In the past few years this trend has reversed itself. Attaining the status of a qualified EMT, paramedic, or ambulance attendant requires an imposing amount of education, training, and experience. To some extent, this increased knowledge is being conveyed to the general public, making people more aware of the actions that should be taken when they are confronted with a sick or injured person.

Needless injuries to accident victims are frequently caused by ill-informed rescuers who attempt to free the victim by rolling the overturned vehicle right-side up. With the exception of instances in which imminent danger or other special circumstances exist—in which case an EMT must be present in the vehicle to stabilize victims while the vehicle is being uprighted—crews should be trained to remove persons from wreckage exactly as they find it. Training will provide personnel with the experience and confidence needed to gain safe entry into the vehicle regardless of its position.

Another common error of untrained rescue personnel is to rip and tear a car apart with tow chains or cables attached to trucks. In one case the crew chained an automobile against a power pole to steady it and attached other chains to a truck with the intention of spreading the vehicle back to its original dimensions. The jerking and tugging did enable the casualty to be disengaged, and somehow the victim survived the ordeal, but there are better methods by far.

ENVIRONMENTAL FACTORS

The environmental factors at the accident scene may be sufficiently hazardous to override all recommended emergency medical procedures. Employing the disfavored "grab and pull" method of handling the victims, instead of treating the injuries and immobilizing suspected fractures before transportation is attempted, could add trauma to the physical damage already incurred. However, when the alternative is death or critically increased hazards, discretion must be used.

At any accident scene the paramedics must decide immediately whether imminent disaster threatens, making immediate rescue imperative. An accident accompanied by fire demands immediate removal of the victims. A vehicle that is submerged or sinking requires fast action. A car teetering on the brink of a cliff may be stabilized; but again, this will take time to gather the necessary equipment, so "grab and pull" will be acceptable under some conditions. An automobile surrounded by a large flammable liquid spill that requires only a slight spark or flame to ignite the vapor cloud is a potential holocaust. A collision on a high-speed highway where the hazard of a second accident is imminent may require extraordinary techniques if protection and traffic control cannot be provided immediately.

However, most accidents occur within an environment that allows calm, deliberate, planned disentanglement and extrication of victims. Therefore, procedures must be adopted, personnel trained, and equipment obtained that will allow the emergency rescue operations at a traffic accident to be performed in the most professional and scientific manner.

SIZE-UP

The duties of an EMT or any other rescue personnel at the scene of a traffic accident are to size up the situation, gain access to the injured persons, administer immediate emergency medical care, disentangle and extricate the victims, and remove and transport them to a hospital.

No paramedic should begin operations until the situation has been realistically evaluated. In assessing a vehicular accident, an EMT must consider three things:
1. The number and types of vehicles involved and the extent of damage; this will provide an opinion as to whether the assigned personnel and equipment are prepared and capable of handling the task.
2. The number of persons injured and the extent of their injuries; this will inform the EMT about what emergency measures are required.
3. The hazards related to traffic, in addition to the other immediate and potential dangers that might be encountered.

Following this assessment, the rescuer must determine whether the initial force is capable of coping with the situation. Remember, the size-up procedure is a continuing process. Emergency medical technicians should be constantly alert to any changes in conditions so that alterations in operations can be made accordingly.

One- and two-car accidents present few problems unless there are unusual circumstances. If the wreck involves a tank truck, chemical carrier, school bus, or other vehicle that carries either a large number of passengers or a hazardous cargo, the rescuers will probably need additional assistance. Other resources may be requested, such as utility crews, heavy construction equipment, wreckers, police officers, extra ambulances, fire apparatus, and so on.

A survey of the situation should be made to determine whether special hazards are present. Hazard control will affect the safety of rescuers as well as victims. Safety precautions should be initiated if there is a threat to life or limb during the process of extricating the victim. The vehicles or cargo may require immediate stabilization before the attempt is made to rescue the injured persons. Perhaps an automobile crashed into a building and the entire structure is in jeopardy of collapsing. In some cases, landslides or snow masses may cause continuing hazardous land movement; stabilization then serves

Extrication from Crashed Motor Vehicles

as a protective measure for both the rescuer and the victim. Occasionally, a vehicle will be in a precarious position, such as hanging over a bridge, teetering on the brink of a bluff, or balanced in such a manner that further movement is imminent. Stabilization of such vehicles should be achieved through the use of jacks, chains, cables, cribbing, wedges, and other anchoring or blocking methods before initiating rescue operations. If the vehicle is unstable, stabilize it by blocking or cribbing. The less movement of the wrecked auto by prying or jolting, the better.

When there are downed high-voltage wires, walk around the vehicle to be certain that none of them is close to the car or actually touching it. If there is a chance that the auto is energized (regard all wires as potentially hot), request a utility crew and wait for them to declare the area safe. Order the occupants to remain in the car. Do not jeopardize the lives of rescuers by attempting to clear the downed wires with inadequate equipment. Special precautions should be taken if a car has hit a street light standard; exposed wires may be touching the frame of the car at an unseen point and making it hot. Instruct the occupants to remain calm and to stay in the car.

Gasoline from a leaking or ruptured fuel tank can pose a serious threat to both victims and rescuers. If a fire-department pumper has not been dispatched, request one. Do not climb into a car that has been soaked in gasoline or that is lying in a pool of fuel; one spark and it may ignite.

It is imperative that EMTs search the scene of an accident to locate all victims. Conduct a painstaking examination of the accident scene; question the victims and bystanders to find out how many persons were involved. Check the path of the car between the collision point and the wreckage, since victims are often thrown from the automobile. Examine the shrubs and weeds along the roadway. Injured persons may be transported to nearby hospitals or homes by other motorists or nearby residents. A small child or infant could easily be overlooked in the debris. Shock may cause a victim to become confused or disoriented to such a degree that the person wanders off unnoticed.

GAINING ACCESS TO VICTIMS

Gaining access to victims usually presents no problem. In the great majority of automobile accidents the injured person has been thrown out in the crash, has removed himself, or has been removed by uninjured passengers or other motorists.

In far too many of the more complicated accident situations the rescuers devote their time, skills, and energy to one continuous operation designed to reach the victim, administer emergency medical care, and remove the person from the wreckage at the earliest possible moment. Combining all activities at the scene into one complex conglomerate operation is time consuming, and valuable minutes that could be better used to provide the initial lifesaving care to the victims are wasted. If the *gaining access* and the *disentanglement* phases of the operation are well coordinated, but separated, the time between the crash and the implementation of emergency measures will be substantially reduced. In fact, a well-executed plan of operation can mean the difference between life and death at many traffic accidents.

The term *gaining access* describes the process in reaching victims who are trapped in a wreckage. A passage is necessary to

RESCUE FROM LAND AND AIR VEHICLES

enable an EMT to quickly evaluate the condition of the injured persons and administer emergency care to stabilize their condition until they can be extricated from the wreckage and removed to a hospital. Getting near a crash victim may be a major problem. The accessibility of a casualty depends on the location and condition of the vehicle after the crash. Gaining access may be accomplished as easily as opening a door on the car, or it may entail long and complicated operations requiring many personnel, much equipment, and an extraordinary length of time. In essence, a hole is made in the wreckage just large enough for the EMT with a kit to reach the victim's side. The position or orientation of the car, damage to the vehicle, and the location of the victim in relation to the automobile are important factors that must be carefully evaluated before any action is attempted.

A passage to a trapped casualty may be provided by establishing an entrance through a door, a window or windshield, or an opening in the vehicle body. It is generally recognized that forcible-entry attempts should be made in this order. Access to a victim might involve prying open the doors, removing or breaking out the windshield or windows, cutting a hole in the roof, jacking up the vehicle, stabilizing the car so it will remain safe and steady, or any other of the multitude of operations that will be described in more detail later in this chapter, in the section on disentanglement. Under some conditions, access may be gained by simply opening one of the doors in the normal manner.

Remove all boxes, seats, and other loose materials to ensure that the best possible access can be obtained. If the victim is still in the front seat, slide the front seat backward to allow more room to work; it may at times help to slide the front seat forward when there are victims in the rear. In some vehicles the seats can be lowered or raised to provide more space. When the seats are electrically powered, delay disconnecting the power until the seats have been repositioned.

The gaining-access phase of rescue operations includes finding and accounting for all occupants and pedestrians involved in the accident, in addition to providing a passage to injured persons who are still trapped in the wreckage.

In addition to medical care, the paramedics can also provide the casualty with protection from broken glass, flying objects, sparks, heat, and other hazards generated by subsequent disentanglement operations. A blanket or heavy coat can be used to shield the injured person so no further trauma is sustained.

If the person is conscious, the EMT must be considerate of the casualty's emotional state. Explaining the operations being conducted and warning the person before any loud noises or vibrations are generated by the rescue techniques will help reduce apprehension. Trapped victims often worry about fire; if the operations will entail use of tools that cause heat or smoke, such as a cutting torch or gasoline engine-powered saw, the person must be forewarned.

ADMINISTERING LIFESAVING CARE

The effectiveness of lifesaving emergency medical care given to trapped victims of a highway accident is dependent upon the accessibility of the victim, the training of the personnel, the equipment available, and the condition of the injured person. Limited access to the casualty may prevent the most effective care and treatment, but all attempts must be made to ensure that life is maintained. Not only must life-threatening medical problems be resolved, but trauma that may cause unnecessary permanent injury or needless suffering should be treated as soon as possible. The general rule is that no patient should be removed from the wreckage until life-threatening conditions are corrected and the victim's body is immobilized as one complete unit. Emergency medical care of sick and injured persons is covered in other sections of this book.

DISENTANGLEMENT

Disentanglement of the victim from the immediate surroundings is accomplished after access has been gained. Emergency medical care may be rendered before, during, or after disentanglement. However, it is generally recognized that the best care can be offered by the EMT who has gained access if all distracting extrication operations are suspended until the victim's condition has been stabilized. Establishing an airway, obtaining the blood pressure and other vital signs, starting IV therapy, controlling serious hemorrhages, splinting fractures, and performing other necessary emergency care procedures is sufficiently difficult at any accident scene. The jolting, vibrations, and noise that accompany extrication will further complicate an almost impossible situation.

This phase of vehicle rescue operations may be viewed in one of two ways, either as the disentanglement of the victim from the wreckage or as the disentanglement of the wreckage from the victim. It is far easier to bend the human body around an obstruction than it is to remove an obstruction from the human body, but such action may be hazardous to the casualty. Careful consideration must be given to the victim's known and suspected injuries. If it seems likely that the trauma may be complicated by manipulation, rescuers must forego this procedure in favor of disentangling the wreckage from the victim. Techniques of disentanglement may include removal of the steering wheel or column, seat removal, pedal displacement, forcing doors, opening door panels, and other techniques that involve severing, distorting, displacing, or disassembling the vehicle's components (Figure 5-3).

Tools and equipment may be required to free a victim from entrapment or entanglement with the vehicle. However, this is not always the case. The victim in the car might be trapped by a seat that can be manually removed. In some cases, careful movement of the body as a unit—preferably after preparing the victim for removal by using splints, backboards, and other supports—will offer freedom. In other instances, vehicular deformations are such that only the physical removal or forcible displacement of portions of the automobile will allow the casualty to be freed.

Disentanglement Procedures

Methods of gaining access to the victims and disentangling vehicle parts and debris from the victims require alternate approaches and a certain amount of ingenuity in the use of available methods and tools. The primary problem is that no two accidents are exactly alike, although some similarities do exist. Disentanglement, as well as access to the victims, depends to a great extent on the skill, training, and imagination of the EMTs. If the rescuers are well trained and experienced, they will be able to avoid adding to the injuries of the

RESCUE FROM LAND AND AIR VEHICLES

Fig. 5-3. The type of accident will determine the best method of disentangling the victims from the wreckage. When the vehicle is beyond the capacity of any available jacks, cranes, or other lifting appliances, then the only practical solution may be to drive the machine off under its own power. If at all possible, the victim should be extricated from the crushed vehicles before any other movement is attempted. Otherwise, moving the vehicle on top may result in additional pressure on the victims, increasing their injuries.
(Courtesy of the Los Angeles City Fire Department.)

trapped person. Creative thinking on the part of the rescuers will invariably determine the success or failure of the extrication.

Great care must be exercised during the disentanglement process because it is during this time that the tools and equipment are in closest proximity to the victim; their effects upon the person must be carefully considered. The effects of excessive heat, pressure, and force upon the victim must be minimal; evident and possible injuries must be taken into account. Most EMTs feel that it is best to use devices such as hydraulic rescue units or pneumatic chisels that do not create sparks, heat, or flame. This precaution is preferred because of the danger of spilled fuel in the vicinity. Both rescuers and victims have been killed or maimed when gasoline has been ignited after a collision. The construction of the vehicle and the deformation of the wreckage are of primary importance; these factors directly affect the manner in which the victims are trapped or pinned, and determine the correct means of disentanglement.

Be aware that cutting into a car body with an axe or bar can result in blows from the tool being transmitted directly to the injured person. A slow, steady application of force, or the low-impact vibration of an air gun, will minimize further injury and pain.

There may be a number of victims in a wrecked auto, some more seriously injured than others. Three procedures should be followed for extrication:
1. If only one door can be opened, it may be best to remove those victims who are closest to the door first and then gradually

Extrication from Crashed Motor Vehicles

work toward those farther away, regardless of the gravity of their injuries.

2. Remove those who can move on their own power first and as more space becomes available, handle those victims who require more room for treatment.
3. It may be imperative to treat the most seriously injured victims first (if they can be reached) because their lives may be in imminent danger.

The various operations an EMT must perform to gain access into the wreckage and to allow extrication of the injured victims require adopting the tools, equipment, and tactics necessary for three types of forcible entry.

Distortion. Distortion of the automobile is intended to move portions of the vehicle structure back toward their original construction, to move them to a more open position, or to create clearance or space for withdrawal of the victim. Distortion is a highly desirable method because it usually does not result in the production of sparks, heat, or flame. This technique basically involves forcing energy back into the structural system in a manner opposing that which occured during the crash. Great amounts of force may be required.

Operations may require bending, prying, spreading, enlarging, expanding, squeezing, and compressing. The most appropriate tools for generating such great force are hydraulic tools, pry bars, jacks, and winches.

Severing or Rending. Severing or rending is an obvious necessity when it is essential to provide a hole or to slice the structure. Generally speaking, the forces required to produce this severing, rending, or dividing action are relatively moderate, since only local distortion of the metal is required. Operations will include cutting, shearing, chipping, chiseling, sawing, grinding, puncturing, tearing, burning, breaking, and drilling. The key to success is to find the weak points in the vehicle and to attack them, leaving the distorted main structure with its energy frozen in. The main disadvantages are that some of the severing methods create sparks from impact, and sharp edges will be created. Cutting wheels, saw blades, and cutting torches are useful. Also included in this category is breaking, particularly breaking glass to gain access.

Tools and equipment will include bolt cutters, shears, pneumatic chisels, gasoline and electrically powered rotary blade and reciprocating saws, chain saws, hacksaws, axes, and oxyacetylene cutting torches.

Displacement. Displacement means moving out of the way parts of the vehicle, or items impinging upon it, by lifting, lowering, pushing, pulling, or rotating them. Also, the vehicle itself may be lifted off a person trapped underneath it. If a car is lifted, cribbing blocks should be inserted between the ground and the vehicle to support it and to prevent the car from dropping back down on the victim in case the jack or bar slips. Parts within the vehicle itself may be displaced relative to the main frame structure to provide clearance or an opening for access.

Tools and equipment will include human energy, vehicles, wreckers and tow trucks, winches and cables, blocks and tackle, rope, chain, jacks, pry bars, and hydraulic rescue tools.

Opening Locked or Jammed Doors

If the vehicle is more or less intact and is upright, the best entry is through the doors. Locked or jammed doors and closed windows impede rapid and simple access to the injured persons. Attempt to open the door in the normal manner. Check all doors; one may not be locked or jammed. A dazed or disoriented victim within the vehicle may be coaxed into unlocking a door.

If the occupant does not appear to be badly injured and the automobile is not excessively damaged, it is usually best to avoid needless property damage. A piece of stiff wire may be used between the glass and the channel to snare the locking button. Entry for the wire may be made through a vent window or by displacing the glass with a screwdriver. Perhaps the locking knob can be lifted if a small hole is made in the window glass. In some cases, a slight assistance from bars or axes will be all that is required to force open a jammed door.

After the locking mechanism of a door is exposed, the lock can be manually released. The most commonly suggested method is to cut a U-shaped piece of sheet metal from around the door handle to expose the safety lock mechanism. The exact spot to cut the hole varies slightly on front doors of different makes of cars. On the rear doors of many automobiles, the safety lock assembly is above the door handle. As different makes and models of automobiles have entirely different lock mechanisms, familiarity will come with observation and experience in practicing on junked cars. Do not skimp on the cutting; remove as much metal as is practical. Fold the sheet metal section back and pull up on the lock release bar with a gloved hand or a screwdriver. This action will usually open a door, unless it is jammed. If the lock cannot be released in any other way, shear the rivets holding the lock together and disassemble it (Figure 5-4).

Fig. 5-4. After one EMT has gained access to the vehicle and is administering lifesaving emergency care to the victim, other rescue personnel are using a slide hammer-powered cutter to force open the door.
(Courtesy of the Ziamatic Corporation.)

In some cars, breaking the door handle, that also contains the lock cylinder, completely off with a hammer or pry bar will expose the locking mechanism so it can be tripped.

The lock mechanism can be driven inward with the point of a fire axe, punch, or other pointed tool, or it can be pulled out by inserting a sheet metal screw into the key slot as firmly as possible and then yanking the cylinder out with an automobile body shop dent puller or by grasping the screw with a pair of vise grips and prying outward with a bar.

When the door is really jammed, it is unlikely that it can be forced with crowbars and hand tools until the safety lock is released. Cars built after 1966 are equipped with a safety latch that is made to withstand a force of over 4,000 pounds (1,800 kilograms). These devices are designed to prevent the doors from springing open in case of a collision. If the frame half of the latch can be reached, the rivets in the locking device can be cut once the bolt has been exposed.

If the door is crushed or the car is a late model, it is nearly impossible to force it open without hydraulic power tools. Perhaps the most powerful and versatile unit for forcing doors is the Hurst rescue tool. However, its use requires that an opening exists to begin with. Use a crowbar or pry bar between the door and the door frame to make room to insert the jaws of the device. In some cases, the jaws may be punched through the thin body metal. Another method of providing a gap between the door and the pillar is to position the Hurst tool vertically in the door window; operation of the device will force the upper door frame upward and push the lower door downward so that a sufficiently wide gap is created to allow insertion of the jaws. After the door is opened, it may be completely removed by repositioning the device near the top hinge to break it, and then near the bottom hinge to complete the job (Figure 5-5).

Hydraulic rescue sets composed of spreaders, wedges, extenders, and other appliances will also force doors. The best method is to use these devices in pairs so that as one is spread, the other can be repositioned. If only one is available, then a crowbar must be used to hold the gap open until the unit can be repositioned (Figure 5-6).

Removing the Windshield

When a door cannot be readily opened, the fastest entrance may be through the front or rear window. If the glass is already broken, clean out the shards of glass to prevent injury.

Older cars have the glass set in a U-shaped rubber channel. An intact windshield can be cleanly and rapidly removed by ripping off the metal trim with a hook, axe, screwdriver, or other flat tool; trim may be pulled away by grasping the strip with a pair of pliers or vise grips. After the trim is removed, cut the rubber channel on the top and both sides with a sharp hook-shaped knife such as a linoleum knife, and the windshield can be lifted out from the bottom.

If the glass has been cemented in, or if the rubber is vulcanized to the frame, the windshield can be forced out intact if an EMT can enter the car and push it out with both feet.

Breaking Out Glass

In modern automobiles there are two different types of safety glass, laminated and tempered. Laminated glass is manufactured by placing a thin sheet of plastic between two plates of glass and bonding the three layers together at high temperature and pressure. When

RESCUE FROM LAND AND AIR VEHICLES

Fig. 5-5. Using a Hurst rescue tool to remove a jammed door so the victim can be removed.

Fig. 5-7. Bending the cut-out roof section down to the ground to allow extrication of victims from a wrecked vehicle. (Courtesy of Ajax Tool Works, Inc.)

Fig. 5-6. Using hydraulic spreaders to force open the door of a wrecked automobile. These units may also be used to free trapped arms or legs, and in other situations where deformed metal or heavy objects have pinned down accident victims. (Courtesy of the Blackhawk Manufacturing Company.)

broken, the adhesive flexible plastic layer will retain the sharp shards of glass. Tempered glass is subjected to a hardening process to make it extremely tough. Tempered glass does not fracture into sharp shards when broken; instead, it shatters into small pieces that resemble popcorn. The cutting edges of ordinary glass are eliminated. Laminated glass is used in windshields and tempered glass is normally used in the side and rear windows.

Breaking glass can be hazardous. Glass can cause further injury to the patient, not only from the initial breaking of the window, but from glass that falls inside the car near the victim. This creates additional danger when removing the victim from the vehicle.

The best method of breaking and removing a laminated glass windshield is to drive a baling hook through the glass in one corner and then pull out on the glass one piece at a time. Another method is to drive the hook through the glass and then break the glass into large sections by forcefully moving the hook with a sawing motion. Laminated glass may also be removed by chopping through the windshield with an axe, pneumatic chisel, or other cutting tool. It is best to cut close to the frame so the pieces of glass will be supported and will not be driven about with great force.

Tempered glass is more difficult to penetrate because of its toughness. The best method is to sharply strike the window in a bottom corner with a sharp tool such as the point of a fire axe, Halligan tool, or punch. After a hole has been made, then a rescuer can enlarge the opening by pulling on the glass with a gloved hand.

Access Through the Roof

When the car is on its side, or the top has been flattened so entrance cannot be made through the doors or windows, access may be gained by cutting a hole in the roof. The car, of course, should be stabilized with cribbing blocks and wedges before any cutting takes place. Then cut a large U-shaped section out of the roof with a pneumatic chisel, rescue saw, or panel cutter. The usual practice is to make three cuts and then fold the flap either toward the ground or away from the side of the car where most of the activity will take place. Do not skimp on the cutting; in fact, it is always a good idea to expose the back seat as well as the front seat. The cut should be made as close to the roof edges as possible, because the steel in this area is more rigid and will cut more easily (Figure 5-7).

When using a pneumatic chisel, the best technique when the car is on its side is to start the cut at the lowest point and then slice the metal up, across, and back down in one continuous operation. If the automobile is upright, two rescuers can speed up the operation by teamwork. The first member starts the cut and proceeds as far as possible; the tool is then handed to the member on the opposite side who slices the top to the far side, changes direction toward the front or rear, and then starts the cut back toward the original side; the tool is handed back to the first member when the center of the roof is reached so the cut can be completed.

A flat-headed axe and sledgehammer can be used to cut the thin metal, especially if the car is on its side. Drive the axe into the roof near the top and then pound the axe vertically downward so the metal is sliced. The sledgehammer may be slid against the roof surface so a smooth contact is made. After the two vertical cuts are completed, the top horizontal cut may be made.

Panel cutters designed to resemble large can openers will quickly shear metal. These tools may vary in design, but they all operate similarly. Other cutting and shearing tools are improvised from sharpened automobile spring leaves; they are powered by pounding them along with a sledgehammer.

After the U-shaped section has been completely cut through on three sides, fold the section downward and flatten it to the ground. When cutting the access hole, make it as large as possible, to allow easy removal of victims. This section should not be entirely cut out lest it leave sheet metal with razor sharp edges on the ground. If the thin sheet metal vibrates excessively, it may be stiffened up by inserting a screwdriver or pinch bar into the cut close to the chisel.

Rods and braces under the roof will be exposed. Some may snap out easily; others must be cut out with bolt cutters or an air chisel. The head lining may now be torn out, creating a large, convenient opening.

Removing the Roof

Under some conditions it may be better to completely remove the roof and expose the entire interior of the automobile, instead of cutting a hole. The best method is to cut the front pillars first. While other team members are supporting the roof, cut the rear pillars. Whenever possible, a pinch bar should be used to remove chrome molding from the cutting area. With a car on which the roof is supported by four thin pillars, the roof may be entirely separated by cutting the pillars and then lifting off the top. In some cases, there may be thick reinforcements toward the rear of the car that are difficult to sever; the recommended procedure would be to cut the front pillars completely and the rear pillars partway, and then fold the roof upward and backward. It is sometimes dangerous to cut only two of the four roof supports and then fold the roof back. If the tools and equipment are available, all supports should be cut completely and the roof removed.

When a rescue saw is used, precautions must be taken to protect the victims from the sparks and to avoid igniting flammable vapors.

RESCUE FROM LAND AND AIR VEHICLES

A common hacksaw can cut the pillars surprisingly fast if blades are frequently replaced so they remain sharp. When using an air chisel, several increasingly deeper cuts should be made, instead of attempting to slice through the thick post in one operation; this prevents the tool from becoming stuck. A Hurst rescue tool or power shears will tear the car apart with little effort (Figures 5-8 and 5-9).

Access Through the Floor

When a vehicle is upside down with the roof and pillars compressed so that access cannot be made through a window opening, the best entrance may be made through a hole cut through the floor. Before starting any operations, the car must be stabilized to prevent further movement and all precautions must be taken against further injuring the occupants by accidentally contacting them with tools or igniting leaking fuels (Figure 5-10).

Structural members of the frame and running gear will dictate the location and limit the size of the opening. An air chisel, reciprocating saw, rescue saw, or other such tool will quickly cut through the thin metal. The best practice is to cut a small hole in the center of the desired panel, reach through and determine that no part of the victims or the car will interfere. After the limits of the cut are ascertained, the largest possible opening should be made.

Access Through the Trunk

Entrance into the car body through the trunk should only be attempted after the alternatives have been tried because this method can be slow and difficult. Before attempting forcible entry, check to determine if the keys in the ignition lock can be reached or if there is an electrical trunk release in the glove compartment. If the trunk lid has not been crushed down and jammed, it may be opened by driving the lock cylinder inward with the point of a fire axe or punch, or by pulling it outward by turning a sheet metal screw into the key slot and yanking the cylinder out with a dent puller or by grasping the screw with a

Fig. 5-8. Extrication of automobile collision victims may be simplified by lifting off the entire top after the door posts and other supports are cut with a pneumatic chisel or a rescue saw. (Courtesy of Superior Pneumatic and Manufacturing, Inc.)

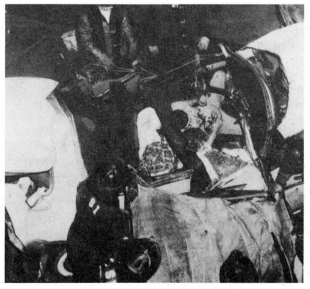

Fig. 5-9. Rescue squad removes automobile top so accident victims can be given emergency medical care. (Courtesy of the Chicago Fire Department.)

Extrication from Crashed Motor Vehicles

Fig. 5-10. Gaining access to injured persons trapped in an overturned car may be accomplished by jacking up the vehicle or by cutting a hole through the floor pan.
(Courtesy of the California Highway Patrol.)

pair of vise grips and prying with a bar. Another method is to cut the sheet metal around the lock mechanism, reach in, and pull or turn the lever or other mechanism to open the lock.

After the trunk has been opened, remove the spare tire and other articles and take out the back of the rear seat.

Removing or Forcing the Seat

Under some circumstances, the front seat may be pinning a victim against the steering wheel or dashboard. At times it may be desirable to adjust or completely remove the seat to gain working space. The front seats will travel forward and backward on tracks when a lock is released by the seat adjustment lever. Some automobile seats also have controls so that seat height and back angle can be adjusted. Many luxury cars are equipped with electrically controlled seat adjustments; these should be rearranged before the battery is disconnected.

Unless the vehicle body has been seriously deformed, the seat adjustment mechanism will probably work in the normal manner. To slide the seat backward, one or two team members should force the seat manually while the adjustment lever is lifted. Caution must be observed that the patient does not suffer any additional trauma or worry from this movement. Stabilization of the victim is required during this maneuver because the entire weight of the person could be supported by the seat. If the legs are fractured and impinging in the floorpan area, having the seat suddenly break free will painfully jolt the injured person. To force the seat back, two rescuers sit on the seat, place their feet against the dashboard, and use their powerful leg muscles to force the seat backwards.

A hydraulic rescue unit or a Hurst tool may be used to apply force between the door frame and the seat. Always remember to operate the seat adjustment lever. Another method is to remove the rear window and pass the winch cable through from a rescue vehicle or tow truck. Wrap the chain or cable around the solid part of the seat, not around the seat tracks that are secured to the car floor. The

RESCUE FROM LAND AND AIR VEHICLES

Fig. 5-11. Rescuer cutting fastenings with a pneumatic chisel so the seat can be removed.
(Courtesy of Superior Pneumatic and Manufacturing, Inc.)

chain should be centered to create an equal pull on both sides of the seat at the same time, so it will not bind on the tracks. A come-along can also be used; pass one end of the cable through the rear window and fasten the other end of the come-along to a sturdy structure on the underside of the vehicle. Shoring blocks placed under the cable or chain at appropriate locations will prevent the chain from sinking into the soft sheet metal and causing a snag. Inform the victim about the procedure before starting. With very little effort and one person holding back the seat release handle, the few remaining teeth on the sliding mechanism can be stripped and the seat pulled off the tracks.

To completely remove the seat, use the hydraulic rescue devices or the Hurst tool to accomplish the job simply and rapidly. Place the spreader jaws between the floorboard and the seat frame. Opening the spreader jaws will force the seat upward until the fastening bolts are torn loose. It may be necessary to insert a piece of wood or metal between the jaws and the thin metal to prevent slipping or tearing. A pneumatic chisel or rescue saw may be used to cut the bolts or track. When using an air chisel, always bear in mind that the item being cut must be rigid. If the body metal is rusted and weak, the bolts will tear out from their original anchoring. It is impossible to cut bolts under these conditions (Figure 5-11).

Removing or Displacing the Steering Wheel

The driver of a crashed automobile may be pinned between the seat and the steering wheel, have part of the body thrust between the spokes and rim of the wheel, be firmly trapped by a deformed steering column, or be suffering from any one of the various injuries and predicaments that can happen to a person who has been thrown forward with great force during a collision. A victim wedged between the seat and the steering wheel may find respiration difficult or impossible, even if the other injuries are not serious. Critical internal trauma that is not apparent on external examination of the casualty may be present. Before an EMT can perform even the fastest and most cursory examination of the victim, the person must be disentangled from the wreckage (Figure 5-12).

Fig. 5-12. Using a Hurst tool and chains to force the steering wheel and column away from a trapped victim. (Courtesy of the Los Angeles City Fire Department.)

When bending the steering column forward, pay careful attention to attaching the cable or chain to the steering column itself and to placing blocks on the hood at the junction of the windshield and hood. This allows an initial upward force on the steering column and prevents the cable or chain from merely crushing the body metal. If the attachment is not properly placed, the steering column may slip and in its rebound hit the patient, producing further injury. Because of the strength of the column, it is best not to remove the chains and tension of the steel cable until the patient is completely free of the steering column area.

Care must be taken that the extrication operations do not place any additional pressure on the injured person, causing further pain and trauma. Displacing or removing the seat or steering wheel could apply pressure in the wrong direction. Forcing the steering column upward and forward may cause the bottom of the wheel to move toward the rear while the top of the wheel is moving forward. All of these possible movements must be anticipated and compensated for before operations are started.

Cutting the steering wheel rim and spokes can be accomplished very rapidly with hand tools. A hacksaw will cut through both the steel core and the plastic covering. Large bolt cutters will sever the rim and spokes easily; if the cutters are not large enough to bite into the wheel, break off the plastic with pliers or some other tool, and then cut the metal core.

The wheel and steering column may be forced upward and forward with any of the hydraulic rescue devices or jacks. Use them between the floor and the steering column; place a block of wood on the floor so that the base will not slip or be forced through the thin metal.

Another method is to remove the windshield and pull the column forward with the cable from a rescue squad vehicle or tow truck equipped with a winch. If a winch is not available, use a come-along or Hurst tool to exert the force. Fasten one end of the cable or chain to a sturdy frame or axle member under the car and the other end of the cable to the steering column. Place wooden blocks at locations where the cable touches the body metal so that the force will be exerted in the correct direction (Figure 5-13).

Fig. 5-13. Using a powerful come-along to pull the steering column forward to extricate an injured automobile driver. (Courtesy of the Blackhawk Manufacturing Company.)

RESCUE FROM LAND AND AIR VEHICLES

Pedals and Other Devices

The driver or other occupant of a crashed automobile can be trapped or caught by a clutch, brake, or accelerator pedal; can be forced up under the dashboard and become entangled in the maze of wires and cables; or can otherwise have an extremity or body held forcefully in place so the victim cannot be simply slid out of the wreckage. There is seldom enough room to completely assess the situation or to work easily. Under no circumstances should the injured person's body parts be manipulated in an effort to free the casualty, because additional trauma can be caused.

Take time to obtain a clear and concise opinion of exactly what is trapping the victim and in what direction the entrapping device must be moved to relieve the pressure on the person. Because of the confined space within the vehicle, care must be constantly observed to avoid causing additional pain or trauma.

A pneumatic cutter may be used to cut open the front quarter-panel for access to the victim's feet when they are wedged and cannot be freed. Removal or spreading of the pedals, floorpan, and dash portions of the automobile that may impinge on the driver's feet, legs, and pelvis is an important maneuver. This can be the most frustrating part of automobile extrication. Sometimes the bulky power-driven hydraulic spreader or wedge cannot be maneuvered into the proper position and the manually powered tools are slow, requiring additional maneuvering and extension. Power tools often bend the metal adjacent to the patient's body, but do not completely free the body.

It may be useful to cut the victim's shoestrings and, possibly, the shoes that are impinging or caught. By removing the shoestrings, the entire foot can be slipped free, leaving the shoe behind.

Passing a light rope around the vehicle part and pulling in a direction away from the victim may move the device sufficiently to free the person. At times, the spreader jaws of a hydraulic rescue unit may be necessary.

PREPARATION FOR REMOVAL

After the vehicle components and other entrapping objects have been disengaged from the casualty so the person is completely untangled, the victim must be carefully prepared and supported before any initial movement from the automobile is attempted. This preparatory activity has a twofold purpose: (1) to protect the person from further injury during removal from the wreckage or during transportation to medical attention, and (2) to facilitate the removal of the victim from the vehicle.

Preparation for the removal will include the maintenance of an airway, the dressing of all serious wounds, the control of severe bleeding, and the immobilization of all broken bones or suspected fractures, as well as the correction of all life-threatening problems.

Preparing for the removal of a victim from a crashed vehicle involves adequate stabilization of all fractures. This does not necessarily dictate that rigid immobilization of an arm or leg is necessary. It is sometimes much easier to move an injured person with manual stabilization than with rigid immobilization, because the rigid or traction splints needed for total stabilization are awkward and bulky. Under some conditions, manually applying traction and support to a fractured extremity will allow extrication faster and with less trauma than taking time to apply splints in a confined area.

Immobilizing the injured limb by binding it to a body part will offer far greater support than attempting to stabilize the extremity manually. Simply binding the upper limbs to the victim's trunk and the lower limbs to each other will suffice until the injured person is removed to a place with sufficient space for adequate emergency medical care.

Short and long backboards are the most useful all-around devices for packaging victims as units to facilitate their removal from the wreckage. A rigid backboard is necessary whenever there is the slightest possiblity that the victim may have a neck or spine fracture. In addition, extrication is much easier because there is a firm solid base to which the person is attached; this aids lifting and handling. The backboard is first attached to the victim's body; the head should be secured only after this is done. If the head is attached to the board first, any movement of the body or board will act as a lever and the transmitted force could accidentally cause further injury (Figure 5-14).

The short spine board is used most frequently for stabilization of a sitting victim or one that is doubled up. A cervical collar should be placed around the victim's neck while the victim's head is supported and immobilized. The short board is then positioned behind the body and the head and torso are secured to the board by a headband, chin strap, and two long straps applied across the chest and around the thighs. An injured person can be lifted from the wreckage with little risk after being immobilized.

TRANSPORTATION OF THE INJURED

Transportation refers to the shifting of the victim from inside the wreckage to a point outside and the movement of the patient from the immediate vicinity of the accident to the ambulance. Just as gaining access and disentanglement are two separate and distinct operations, so are removal and transfer. These two operations may often be uncomplicated but sometimes they will be complex (Figure 5-15).

In the "grab and pull" type of rescue operation, rescuers simply manhandle the casualty from the wreckage. In a properly conducted rescue procedure—one in which the victim has received emergency medical care, has been successfully extricated from the wreckage, and has been treated with splints or a backboard—the removal operation is tailored to the requirements created by the accident.

Transfer may involve nothing more than carrying the "packaged" patient a few feet to the waiting ambulance, or it may consist of moving the victim hundreds of feet up a steep ravine. When rescue squads operate in difficult terrain, they should constantly train in techniques of overcoming the adverse conditions they may encounter. For difficult transfers, the most successful and least strenuous procedure seems to be using the winch and cable from a rescue squad vehicle, wrecker, or ambulance with a rigid carrying device such as a basket stretcher (or with a long backboard, if the victim is well secured). The essential requirements are that there be a place to attach a cable or lifeline to the head of the litter, so that it will withstand the stress of the patient's weight, and that there be enough handholds along the sides of the litter so that rescuers can support it and dodge obstructions. An additional lifeline may be lowered along each side of the main line to help the rescuers keep their balance while climbing. It is possible to use a fire department

RESCUE FROM LAND AND AIR VEHICLES

Fig. 5-14. The largest single cause of permanent disability resulting from back and neck injuries occurs **after** the accident. Trauma resulting from improper handling and transportation creates needless pain, crippling, paralysis, and death. Victims of violent accidents should always be immobilized before they are moved.

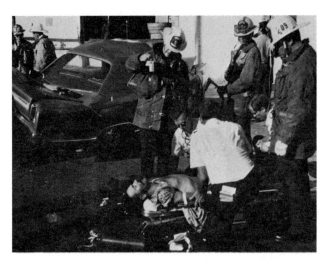

Fig. 5-15. The injured victim has been disentangled from the automobile wreckage and is being prepared for transportation to a hospital. Beginning intravenous infusion, administering oxygen therapy, splinting fractures, and checking to see that respiration and circulation are adequate will stabilize the victim's physical condition, so that treatment by the physician will have the best chance of success. (Courtesy of the Los Angeles City Fire Department.)

aerial truck as a crane, if the apparatus can be positioned close to the edge of a bluff or cliff.

In remote mountainous, desert, or seashore locations where there is adequate landing space, a helicopter may be the fastest and most comfortable method of transporting the victim to a hospital.

While transporting the victim, continue emergency medical care and maintain a constant watch on all life-threatening injuries related to respiration, circulation, hemorrhage, fractures, and shock.

BUS ACCIDENTS

Because of the large number of passengers in such vehicles, bus accidents or fires often present serious problems. The construction and location of the various components of a bus vary according to the model and manufacturer, but they follow a general pattern. A transcontinental transport will be described, but most of the points under discussion should be considered at all incidents involving large buses.

Opening Doors and Windows

To gain access to the interior of the bus in case the engine is not running, there are at least three manual releases for the door. One is located in front and to the left of the driver; another is positioned to the right just inside the door. Both of these controls can be operated from inside the coach. In case there is no one inside the bus able to work these releases, there is one located under the door just in front of the right front wheel that can be operated from the exterior.

There is no emergency door. In case of an accident's damaging the entrance door, rescue must be effected through the windows or skylights. Many coaches are equipped with pop-out windows; one person can remove them by simply placing the hands in the lower corners and pushing firmly from the inside. These windows are marked on the inside by a metal plate and on the skylights by decals. On older buses, the windows may be opened from the outside by placing a large screwdriver or pinch bar under the metal frame to force the window out. With the newer positive-latching system, the passenger windows must be opened from the inside; the windows can no longer be pried from the outside with a screwdriver or other rescue tool.

To open the front windshield, remove the rubber locking strip that runs around the glass. This will allow easy removal of the windshield and leave a smooth opening for entrance into the bus and for the removal of injured passengers.

A passenger trapped in the restroom can be rescued from the outside by opening the large window. It may be necessary to prop or hold the window open if it is hinged from the top.

Stopping the Engine

Most large buses are equipped with a diesel engine that does not need electrical current to keep running after it is started. If fire department or rescue personnel arrive at the accident scene and find that the engine is still running, there are several methods of stopping the engine. If the switch key will not turn off the engine, the emergency stop button located on the driver's switch panel should be used. If the emergency stop switch cannot be reached or fails to work, there are two battery shutoff switches located in the engine compartment that are identified by red paint; simply give them a half turn to the right. If the mentioned procedures fail to stop the engine, then break or cut the fuel supply line going to the fuel pump. If for any reason the engine still cannot be stopped, discharging a carbon dioxide (CO_2) extinguisher into the air intake will do the job. This method should be used only after all other procedures have failed. It is important that only CO_2 gas be used, since other extinguishing agents may damage the engine and injure rescue personnel.

Disconnecting the Batteries

Each coach is equipped with two batteries, located one on either side just to the rear of the back wheels. However, it should not be necessary to disconnect them, because the two battery shutoff switches in the engine compartment accomplish this. The individual circuits of the electrical system are protected by circuit breakers.

Fire

The most vulnerable places for fires to start in most coaches are in the brakes on the rear axle; in the transmission; and in the engine compartment. These blazes may be extinguished with water spray, dry chemicals, or CO_2. When combatting a fire around the transmission, attack the blaze over the wheels on either side with a fog stream.

Hazardous Areas

Since the coach has no frame, it is equipped with jack plates located under the body behind the rear wheels. If it is necessary to jack or block elsewhere under the bus, jack under a solid bulkhead. These bulkheads are located to the front and rear of the wheels. A jack may also be placed under the short engine supports. Always position a support under the body before jacking, since these bodies are made of aluminum and tear easily.

Instead of springs, many vehicles use an air suspension system. This system consists of a rubberized nylon bellows at each wheel, inflated through air chambers from the compressor and equalized by leveling valves. Because the bellows may be damaged in an accident or fire, block the coach body securely before allowing rescue personnel to go under the vehicle. Should a bellows collapse, the coach body can drop to within three inches of the ground.

The framework of the bus is made of tubular steel rods. These rods go all the way up and over the top for support. The fuel lines travel under the center of the floor from the tanks to the engine; be careful not to rupture a fuel line when cutting into the bus structure.

The air conditioning unit is located in front of the engine or to the rear of the baggage compartment. Discretion must be observed when cutting in this area, because of the danger of releasing the refrigerant.

It is very important to be careful when opening the rear engine compartment doors, because there are no safety guards around the fan blades. Their rapid rotation makes them difficult to see.

Removing Victims

Because of the narrow aisles provided in buses and other transportation vehicles, it is extremely difficult to remove a patient from the passenger section if the person has been immobilized. The best method is to place all the seat backs in an upright position, immobilize the patient on a long backboard or basket stretcher, and then slide the packaged victim along the seat tops.

School Bus Accidents

School buses are equipped with emergency exits to evacuate the occupants when the main door is inoperative; these exits may be either a door in the back or right rear side of the coach, or an extra large window that is easily opened. A rear-end collision may jam the emergency exit and, if the vehicle overturns onto the right side, the main door will also be blocked. Even without rear-end damage, small schoolchildren are unable to open the emergency exit door if the hinge side is up.

Before starting any forcible entry, block up the bus to stabilize it. This will prevent swaying and turning. Attempting extrication operations while the vehicle is unsteady and precarious could cause further injuries to the victims or rescuers. If the bus is still upright and all exits are jammed, the easiest procedure is to remove the windshield. When the coach is on its side, cutting a hole in the roof is the best way to remove injured passengers.

QUESTIONS FOR REVIEW

1. In which of the following accidents are victims most likely to receive lethal injuries from a traffic accident?
   a. The roof is crushed downward to the seat backs.
   b. The driver door is forced inward against the seat.
   c. The driver impacts the steering wheel with great force.
   d. The passenger is thrown from the car with moderate force.

2. When an automobile accident victim is trapped in a wrecked car, extrication must be performed by
   a. gently manipulating the person's body so extrication is accomplished without excessive further injury.
   b. disentangling the car from the victim rather than disentangling the victim from the car.
   c. carefully cutting away the body metal while gently manipulating the victim's body.
   d. removing the victim from the wreckage before initiating any emergency care.

3. The disfavored "grab and pull" method of extricating an injured victim from a wrecked automobile is encouraged only when
   a. the victim is not seriously injured.
   b. the victim is so seriously injured that the fastest method of extrication must be used.
   c. rescue personnel do not have the correct equipment available.
   d. imminent disaster threatens.

4. The first-arriving rescue personnel should size up the situation at a serious highway accident before operations are started. Which of the following should receive the least consideration?
   a. Vehicles involved and extent of damage
   b. Number of casualties and extent of their injuries
   c. Amount of rescue personnel and equipment presently on the scene
   d. Immediate and potential hazards that might be encountered.

5. When a badly wrecked automobile is still somewhat intact and upright, the best method of entry will probably be through the
   a. doors.
   b. windows.
   c. windshield.
   d. roof.

CHAPTER 6

Rescue from Aircraft Accidents

Sooner or later every airport experiences an aircraft accident on its premises or in the surrounding community. The crash of a passenger airplane resembles a head-on collision between a gasoline tank truck and a large bus. There will be mutilated wreckage, trapped people with life-threatening trauma, large quantities of flammable liquids, and either a major fire or the imminent threat of one. For a large airport, this could mean a significant disaster, involving several hundred casualties with varying degrees of injury. The fire fighting and rescue services of the entire area could quickly be overwhelmed.

In the propeller era, a relatively higher proportion of passengers would survive an accident—apparently because of the slower speeds—than in the early years after the introduction of jet aircraft. However, with the introduction of new cabin materials in compliance with stricter crashworthiness regulations, there are fewer flammable substances to jeopardize the occupants. With the introduction of wide-bodied jets, which allow considerably stronger construction and greater structural absorption of crash forces, the proportion of survivors in accidents on and around airports has increased.

Every accident will differ in time, location, severity, and a multitude of other factors. With this assumption as a basic premise, it must be clearly understood that no standard operating procedure can cover all situations or contingencies. However, preplanning and practicing will greatly improve the capabilities of the fire and rescue services to control any type of disaster that might occur. The successful handling of every emergency incident will depend on the judgment, initiative, prudence, and training of those concerned, particularly during the early stages of operations. The quality of the provisions for prompt extrication of victims from the wreckage, efficient emergency medical care, and effective fire fighting will largely determine whether the passengers, crew members, and occupants of adjacent ground structures will perish or survive (Figure 6-1). An airport disaster plan integrated with the facilities of the surrounding community must be adopted, to reduce the potential of a serious disaster to a minimum.

Rescue from Aircraft Accidents

APPROACHING THE CRASH SCENE

Always approach the scene of an airplane crash from upwind to avoid entering a downwind cloud of vapors from spilled flammable liquids. Carefully watch for aircraft occupants when approaching along the same path that the plane took after first hitting the ground, since survivors or the bodies of victims may have been thrown out of the wreckage. If the area is overgrown with brush or high grass, select another route or dismount from the rescue vehicle and approach on foot until the terrain has been carefully checked. Search the entire area for survivors who may have been thrown free of the aircraft. They may be wandering about in a dazed condition.

FIRE FIGHTING

Fires may occur at any time during the operation or servicing of aircraft, but are especially critical following a takeoff or landing crash. This type of fire spreads rapidly and, because of the unusual fuel dispersion and flame intensity, presents a severe hazard to the lives of persons inside the plane and those in adjacent structures. Combatting an aircraft fire with conventional apparatus is similar to structural fire fighting (Figure 6-2).

If there is no fire in the wreckage, rescuers must take every precaution to guard against the ignition of the flammable liquid spill. If there is fire, rescuers must determine to what extent the fire has threatened or spread to adjacent structures. The proportions and nature of the fire will greatly influence any action to save lives. It must be assumed that casualties are injured and need to be released and removed from the wreckage.

Size-up

The officer in charge must immediately conduct a preliminary size-up of the situation to determine the primary objectives in the rescue operation. The highest priority at any aircraft accident is the rescue of persons trapped in the wreckage or in any buildings that might be involved. This involves assessing the extent of the incident, the

Fig. 6-1. Rescuers search through the still-smouldering debris of an apartment complex that burned to the ground after being struck by a Navy jet in flames.
(Courtesy of UPI Photo.)

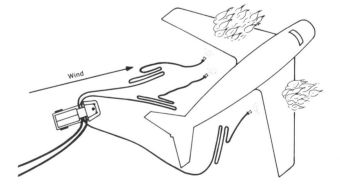

Fig. 6-2. Combatting an aircraft fire with conventional apparatus is similar to fighting a structural fire. If possible, lay lines while approaching the fire ground; otherwise, attack the fire with the booster tank and depend on other pumpers to provide the water supply. If practicable, attack from the windward side, approach along the fuselage, and drive the flames away from the passengers. Keep the fuselage wet and cool, and protect the rescue exits of the occupants.

probable amount of personnel and equipment that are responding, and the innumerable ways in which the accident might develop. The first action should be to radio the pertinent facts to the dispatcher and the other units responding to the crash scene. Request additional help if there is any doubt about the capability of the personnel and equipment on the scene to control the situation. When requesting assistance, consider the need to protect structures threatened by the accident, as well as the rescue operation itself. Proper fire stream procedures during aircraft rescue operations may, however, require forcing burning flammable liquids and vapors toward uninvolved structures.

When a plane has crashed into a structure, be aware that a fire accelerated by the fuel spill may be in progress or may be imminent as a result of delayed ignition. Extreme caution must be observed at all times.

Fuel

Fuel is the principal danger around any type of aircraft, because of the fire hazard it presents. There are two types of fuel that may be encountered: aviation gasoline and jet propulsion (JP) fuel. The fuel loads can vary from 50 gallons (190 liters) on small planes to thousands of gallons on large aircraft.

Most aircraft carry fuel in tanks incorporated into the wing structure. Even relatively minor collisions between the airplane and the ground or other structures will result in the release of fuel from the wing tanks. As aircraft become larger, the fuel systems become more complex and this, in turn, creates more opportunities for accidents and fires. When an aircraft crashes, the force of the accident may rupture a fuel line or a tank.

Oxygen Systems

All high-altitude aircraft are equipped with breathing oxygen systems. This oxygen is stored either as a compressed gas or as liquefied oxygen (LOX). Most modern airplanes use the LOX system. The oxygen may increase both the fire and explosion hazards of the aircraft, because oxygen intensifies burning. The danger of an explosion is extreme when liquefied oxygen comes in contact with flammable materials such as JP fuel and oil. When flammable liquids and LOX come together, one small spark can cause a violent explosion. Liquefied oxygen can also cause serious personal injury, since the liquid will freeze the tissue if it is spilled on a victim.

Tire Fires

Even though many aircraft tires are equipped with fusible plugs designed to release the air pressure at temperatures above 225°F (107°C), fire fighters run a serious danger of personal injury from bursting tires when combatting a fire involving the landing gear of an airplane. When the tire is not actually burning, the danger of a tire blowout can actually increase after the aircraft has stopped moving due to the transfer of heat from the hot brake shoes to the tire.

Caution. Approach the landing gear from the forward or aft direction if you suspect the brakes have overheated or if you are fighting a wheel fire; avoid the side tire area. If it is necessary to work in the hazardous area, safety cages should be installed around the wheels. Use dry-chemical extinguishers only if the tires are pressurized. If all tires are deflated, fine water spray or carbon dioxide (CO_2) may be used.

Fire Stream Techniques

The number of hose lines and the quantity of water will be determined by the availability of personnel, equipment, and water. All hose lines should be stretched and made ready, regardless of the fire situation upon arrival at the scene.

The first action should be to evaluate whether the fire can be extinguished immediately with the available personnel and equipment. Consider the trapped occupants' chances for survival under the existing conditions of fire and heat. Operations directed toward fire extinguishment are preferable, if the size-up justifies this approach. When there is any doubt that the trapped victims can withstand the heat, flames, and smoke exposure for the amount of time necessary to accomplish extinguishment, the fire streams must be used to protect the victims while they are being extricated.

Select an entry point into the aircraft (normally a door in the fuselage), and protect this opening from fire. Using spray streams of water or foam, direct cooling streams on the fuselage and work from the fuselage outward to force the flames away from the victims. Blast a path through the heated area, so rescues can be effected.

Nozzle operators should work together, keeping the fire streams low and parallel with the ground to avoid stirring up puddles of fuel that would increase the intensity of the fire. Rescue personnel and victims, as well as the aircraft fuselage, should be kept wet to reduce injuries.

When fuel is burning around the wreckage and victims are still trapped in the fuselage, approach the aircraft with flowing spray or foam streams along the fuselage from the most advantageous direction. When there is a perceptible breeze, approach from the windward side. The slope of the terrain should also be considered. If the aircraft is resting on the side of a hill, approach from the higher ground in order to take advantage of gravity in washing the fuel away from the critical areas.

The entire fuselage skin should be wetted down as soon as possible. The location of survivors and the concentration of the fire should determine where to apply the water first. Keep in mind that if the entire fuselage surface can be kept wet, the heat input from the exterior flames will be greatly diminished.

Normally, fire streams can be directed along the fuselage, usually from the tail or nose position. Efforts should be concentrated on driving the flames and heat away from the fuselage while making a rescue entry path. If the fire is very extensive, the advance must be rapid so that cooling water streams can be directed on the fuselage before occupants are suffocated or fatally burned.

When more than one hose line is available, direct all fire streams to bear on a similar line of approach from the same general direction to drive the flames and heat away from the fuselage. Opposing crews advancing their hose lines from opposite directions will probably force heat and flames toward each other. Be careful not to allow the fire streams to force flammable liquids and flames under or toward the wreckage. Advancing multiple lines simultaneously will develop a greater amount of extinguishing power. However, never delay operating the first hose line; advance the initial fire stream immediately and bring the others into play as they become available.

Although saving lives is the principal objective of crash fire fighting, the protection of adjacent and involved property must be considered. As soon as rescue operations are completed, all efforts should be concentrated on the protection of exposures and the extinguishment of structure fires started by the accident.

AIRCRAFT RESCUE

Speed is the watchword in saving lives; seconds can spell the difference between success and failure when rescuing trapped or injured victims of an accident. Inefficiency or lack of training cannot be tolerated; there is too much at stake. The rescue problems following an aircraft crash require skill, courage, teamwork, physical agility, and the utmost in mental alertness; this is highly specialized work.

The inherent fire hazards of aircraft have severe implications in the event of ground accidents. There are two distinct types of accidents to consider.

High-Impact Crashes

High-impact forces are generated by power-on ground collisions. Major aircraft structural failure results in almost certain instantaneous death to all occupants, distorted wreckage, debris, and probably fire. Rescue of any occupants is not likely and fire control is sought principally to protect exposed properties, to permit identification of the victims, and to preserve evidence to aid investigators in determining the cause of the accident (Figure 6-3).

Low-Impact Crashes

In the low-impact crashes, the fuselage remains relatively intact with the possibility that a well-trained and efficient crash rescue crew can save lives. Impact forces are low and it can be expected that survivor rates will be high and the majority of occupant injuries nonfatal, if fire does not block escape. Rescue comes first. Whenever possible, attack aircraft fires quickly and at least hold the fire in check until the occupants can be removed (Figure 6-4).

Crash Hazards

Aircraft occupants are exposed to several different types of hazards. Large amounts of heat and toxic gases result from a fire. Occupant survival depends on the ability of fire fighters to knock down the bulk of flames affecting the fuselage and, in some cases, to gain firefighting entry into the plane to cut off internal flame spread. The rescuers, aircraft fuselage, and exposed victims should be protected with spray streams or foam while in the heated area.

Victims may be trapped by structural disintegration and crushing of the cabin area without an ensuing fire. Flying projectiles resulting from the accident sequence or from explosions may produce injuries, or the impact deceleration may cause the occupants to be hurled forward and injured. Quite frequently several buildings are struck before the disabled aircraft comes to rest, and victims on the ground may be in need of immediate assistance. The large swath cut by the plane and the volume of smoke and fire normally attending such accidents increase panic and confusion.

When sizing up the crash scene, note the type of aircraft. A large airplane may have many passengers, which complicates the rescue problem. Military aircraft may be carrying bombs, guns, and

Rescue from Aircraft Accidents

Fig. 6-3. High-impact aircraft crashes offer few opportunities for crew and passenger rescue, but lifesaving responsibilities in the ground structures may be innumerable. (Courtesy of the New York City Fire Department.)

Fig. 6-4. Aircraft fuselages often remain relatively intact in low impact crashes. Unless there is a fire, the passengers are usually able to climb from the airplane with minor injuries or none at all. In this crash, the fuselage and surrounding ground was covered with foam to reduce the possibility of a fire resulting from spilled fuel. (Courtesy of the Denver Fire Department.)

other ordnance, which presents an additional hazard. Large planes will have a much larger flammable-liquid spill than small aircraft.

Extricating the Occupants

The rescue of survivors is the first action. Once the rescue has been made, or recognized as impossible, the best policy is to keep all surplus personnel clear of the crash. A perimeter of 1,500 feet (450 meters) can be assumed safe.

There is some danger that rescuers may overlook persons who have been thrown clear by the final impact, or who may be in part of the craft broken off in the crash landing. A thorough search must be performed. Do not presume that people in the wreckage are dead just because they are not moving. The large percentage of unconscious victims who survive such accidents indicates that many who at first glance appear dead need assistance urgently. If the airplane has crashed into a building, consideration must also be given to the possible casualties among its occupants.

Aircraft survivors who have escaped may be able to inform rescue personnel about the number and the possible locations of persons who are still inside the plane. Never take it for granted that all crew members and passengers have escaped. A complete inspection of the aircraft must be made.

Only one rescuer should make the first entry into an aircraft. The other members will stay at the opening until the first rescuer determines the existing conditions. The outside rescue members will be gathering hose lines and equipment while preparing to enter the fuselage; they can also warn the other personnel of fire or explosion dangers that may develop. This procedure constitutes part of the initial size-up to determine the extent of the rescue problem.

After entrance is gained and the problems evaluated, the most important duty is to locate and determine the condition of injured aircraft occupants and to start removing them. Inside, the aisle is often only 15 inches (38 centimeters) wide, and many emergency window exits measure about 19 by 26 inches (44 by 65 centimeters).

RESCUE FROM LAND AND AIR VEHICLES

This cramped space, in addition to the possible fuselage distortion or breakup, hinders rescue of passengers with broken bones. Rescue personnel must adjust themselves to these difficulties.

If the inside of the aircraft is on fire, call for hose lines to control the blaze while rescues are being performed. When immediate hazards are beyond control, occupants must be transported immediately. If instant evacuation is not possible because of entrapment or hindrance of wreckage, the rescuers must attempt to keep the fire away from the area where people are trapped.

Emergency care must be administered to all survivors. Even though no physical injuries are evident, the majority will be suffering from shock in one degree or another. Severe lacerations and fractures are common in this type of accident. However, watch for possible strokes and heart attacks that may be brought on by extreme stress and fear that these victims have just experienced. If the danger of fire or explosion exists, the victims should be removed to a safe area before extensive emergency care is attempted.

When time permits, turn all master electric switches off and disconnect the batteries. This will reduce the dangers of fuel vapors' catching fire from a spark of electricity. No part of an aircraft structure should be moved unless such motion must be made to make a rescue possible. When the aircraft is broken up, electrical cables may be severed or damaged. If the master switch and batteries have not been disconnected, the smallest movement of the aircraft wreckage may cause a damaged cable to spark and ignite fuel vapors.

Fatalities

Investigators of aircraft accidents consider the positions of all occupants in the wreckage vitally important, whether they are dead or alive. When fire is not a problem, all bodies of obviously dead casualties must be left in place until the coroner, governmental investigator, or other responsible official has given permission to remove them. As long as there is any indication of life, of course, every effort must be directed toward extricating the victim from the plane and administering lifesaving emergency care. Whenever possible, make a note of the seat location or other identifying position where the casualty was discovered.

Obviously dead casualties should not be moved from their locations, unless it is necessary to prevent their cremation in a fire. A knowledge of their seat positions and a check of adjacent luggage and other belongings may supply the only way to identify the victims.

To assist in marking the exact position of a dead or injured casualty, place corresponding tags on both the victims and their locations in the wreckage. When the casualties are mangled so that portions of corpses are discovered, each body part must be individually tagged and recorded. When bodies are discovered away from the wreckage, drive a stake into the ground and tie the tag to it.

Governmental Agencies

Crashes of all types involve legal aspects that everyone should be aware of. In the case of a civilian plane crash, the National Transportation Safety Board (NTSB) will be in charge, with the Federal Aviation Administration (FAA) investigating some light plane accidents. The nearest military installation will promptly dispatch personnel to handle crashes of military aircraft.

Except for the first priority of saving lives, including searching the wreckage for victims and doing what is necessary to fight fires, as little as possible should be disturbed at the scene without the direction of the governmental authority in charge. Until responsible officials arrive at the scene it will be the duty of the rescue personnel to make sure that none of the wreckage is moved, except for parts of the wreckage that constitute a hazard to life or property.

All spectators and souvenir hunters should be kept beyond the safe perimeter, and a no-smoking policy should be enforced. The hazard of volatile fuels is always present at the scene.

FORCIBLE ENTRY

Rescue personnel may have considerable difficulty at most aircraft crashes in gaining rapid access into all sections of the wreckage.

Normal Entry Procedures

All aircraft have doors used for the normal entry and exit of passengers and crew. Always try these entry points before attempting to force an entrance. Doors may be found on either or both sides of the fuselage, but they are normally on the left side. They open outward and are fastened on the inside by latches. Generally, the doors will have handles or other methods of opening both on the inside and the outside.

Emergency Exits

Emergency exits are particularly important because impact forces can jam the normal means of egress and, in the event of fire, evacuation must be extremely rapid. These emergency exits have become even more significant with the advent of jet aircraft, because a heavier fuselage skin and stronger structure is necessitated by the requirements of high-altitude flying and greater pressurization of the cabin atmosphere. These factors make forcible entry through the fuselage skin practically an impossibility, even with power tools. Passenger planes have one or more emergency exits in addition to the regular doors; the number is determined by the seating capacity. These exits must be easily located and operated, they must be designed to resist jamming, and they must be operable from both inside and outside of the aircraft. On some aircraft a wide band of contrasting color will indicate all doors, hatches, and windows that are externally operable. In spite of these regulations, emergency exits are often difficult to locate and open from the outside.

Emergency Entry

If a door or hatch is jammed from the impact of the crash, attempt to force it open by prying or cutting around the frame or at the hinges. Sections of the wreckage may be opened up by removing broken sections of windows. The sharp pick of a fire axe should be driven through the plastic at a point close to the corner or edge; this causes a series of long cracks that will weaken the entire section. Holes made in each corner of a window will allow the forced removal of a rectangular piece of plastic or Plexiglas. The portion can then be knocked out in one piece. The best method of removing large sections of plastic or Plexiglas is by using a power circular saw with a steel cutting blade; adjust the cutting depth so it just penetrates and cut around the window, staying close to the frame. Plexiglas is difficult to break when it is heated. Spraying the plastic with CO_2 will make it brittle and enable a hole to be chopped through the window with an axe or some other sharp instrument.

RESCUE FROM LAND AND AIR VEHICLES

Some aircraft have markings on the sides to indicate where entry may be safely made with cutting tools to enter the cabin; most do not. It is always hazardous to attempt forcing an entry through the sides of an aircraft because of the fuel, oxygen, hydraulic fluid, and other lines, together with electrical wiring, which may be under the skin. The rescuer who cuts along the window lines to achieve entry will usually meet with fewer obstructions.

Whenever power tools capable of emitting sparks are used around any wreckage, and the possiblity of flammable liquids or vapors exists, hose lines must be stretched and ready for use. Many rescue personnel have been killed and injured, and trapped victims cremated, when fuel vapors were ignited during extrication operations.

Access Markings

If entry cannot be made through doors, windows, hatches, or other openings, then as a last resort a hole must be cut through the metal fuselage. Military and some other aircraft are clearly marked so that openings can be cut through them if need arises. These standard access markings are in yellow with red letters that read "Cut here for emergency rescue." These emergency cut-in areas indicate where entry can be forced without encountering oxygen or flammable-liquid lines, electrical wiring, structural members, or other obstructions. To cut through the metal fuselage skin, use a metal-cutting power saw. The upper three sides must be fully cut; the section is then bent outward and downward to prevent leaving a sharp edge where it may cause injuries.

MILITARY AIRCRAFT

In addition to the problems and hazards presented by civilian aircraft crashes, military-plane crashes may possess peculiar dangers that should be anticipated. Some of these dangers are common, while others are rare, but rescue personnel must be constantly aware of the special hazards characteristic of military aircraft.

Danger Areas

Armament. Military airplanes may be carrying large fuel loads, external fuel tanks, and special weapons such as guns, cannons, bombs, rockets, and missiles. Flammables and high explosives are extremely vulnerable to heat; keep all flames and high temperatures away from these areas. Do not approach the place from the front in case a gun should fire accidentally.

Canopies. Canopies and hatches are jettisoned upward and aft of the cockpit area. Remain clear of the canopy jettison area until the canopy has been released or secured.

Engines. *If rescue is attempted before the jet engines are shut down, rescue personnel may be killed or injured.* The intake ducts of jet engines create enough suction during engine operation to ingest a 200-pound (90-kilogram) person from 25 feet (7.5 meters) away. Never stand in front of, to the side of, or directly to the rear of the intake ducts when the engines are running. Engine exhaust temperatures can vary from 3,000°F (1,649°C) at the exhaust nozzle to 100°F (38°C) at 150 feet (45 meters), so remain clear of the exhaust outlets at the rear of the plane when engines are operating.

Turbine blades will exit at a high velocity in the event of engine disintegration. Stay clear of the turbine plane of rotation at the sides of the fuselage until the engine has been shut down.

If the pilot is incapacitated and cannot perform the engine shutdown, a rope securely attached to rescue personnel will prevent their being sucked into the intake while they unlock and jettison the canopy or otherwise gain access into the cockpit. Gaining entry to the cockpit while the engines are running is hazardous because of the possibility of being drawn into an intake duct. To reduce this hazard, drive a crash truck up against the duct to block it. Personnel can then step from the top of the truck to the wing. Water or foam will probably stop the engine if large quantities are sprayed into the intakes.

Engine shutdown can be effected through one of the following actions: retard engine throttles into cutoff, turn the engine master switches off, or depress the engine fire-extinguisher buttons.

Laser

Some aircraft produce laser radiation that can cause instantaneous and permanent blindness, even at an extreme distance. Do not look at the laser beam. This beam is pointed downward, but reflection or refraction makes the beam travel in unpredictable directions.

Radar

Radar transmitters emit high-frequency energy directly ahead of and to the sides of the aircraft. The hazard of radiation burns is present when the radar is operating. Avoid unnecessary exposure in the zone of possible radiation, which is forward of the engine intakes.

Tires

The tires normally contain air under high pressure. They are extremely dangerous because they may burst if excessively heated. Stay clear of the wheels.

Egress Systems

Rescue crews should become familiar with the egress systems provided in many military aircraft. These systems permit crew members to catapult themselves (and the canopy and seat) clear of the plane when they are forced to bail out while the plane is in flight. Safety pins are used to prevent inadvertent jettisoning of the canopy and firing of the ejection seat catapult while the plane is on the ground. The pins are removed before each flight so the ejection equipment can be operated without delay. When making rescues, these safety pins must be inserted by rescue crews if the pilot is incapacitated, to avoid the real danger of accidental operation of the ejection system. An egress system is a combination of two arrangements: (1) the canopy jettison system and (2) the seat ejection system.

These mechanisms create some serious problems when rescuers must gain access to the cockpit or flight deck of the aircraft to remove a pilot or crew member without accidentally firing the seat or seats. No attempt will be made to cover all the military aircraft and their particular egress systems, but the basic facts will be offered.

Canopy System. Canopies are covering for the pilots and are most often found on fighter-type aircraft. On most jet bomber aircraft, crew positions are covered with hatches. When in flight, canopies and hatches are held firmly in place, but in an emergency they must

RESCUE FROM LAND AND AIR VEHICLES

be jettisoned before seats are ejected.

All aircraft egress systems will differ somewhat, depending on the type and manufacturer. There are various ways of removing canopies or hatches to gain access to the injured crew members. Most access openings can be forced from the outside by one of the following methods: opening them by hand, operating the egress system with electricity or air, or jettisoning the canopy or hatch. Because there is no room for error, all rescue personnel must be completely familiar with the particular egress systems on planes likely to crash in their area of operation. Once rescue personnel have gained entrance to an aircraft cockpit, it is possible to accidentally jettison the canopy by tripping the system from inside the airplane.

Do not jettison the canopy if there is fuel or fuel vapor in the area, because a fire or explosion may result. Manually open the canopy or use forcible entry to gain access to the cockpit. Entry may be gained by using a power saw with a metal cutting blade; adjust the blade to cut about 3/4-inch (18.75 millimeters) deep and saw around the periphery of the canopy section while staying close to the frame. Since the plastic will not shatter or break, a segment large enough for crew removal must be cut out completely.

Power saws generate a steady stream of sparks that are hot enough to ignite flammable vapors, especially if steel or iron is struck. Rescue personnel and victims have been killed and injured when a fire occurred during disentanglement operations. Any time there is the slightest chance that flammable liquids may be present, hose lines must be stretched and be ready for instant use before activities are started.

Spraying the canopy with CO_2 will remove the resilience from the plastic, enabling a hole to be chopped through the canopy material with an axe or some other sharp implement. Remember that a hole must actually be chopped through the canopy, since the plastic will not shatter and break. Exercise caution when chopping because a sharp tool may penetrate the plastic rather easily after a couple of blows, and injure the trapped crew member or jettison the seat.

Seat Catapult. Canopy and hatch jettison systems present some dangers to rescue personnel, but these hazards are small when compared to the threats of the seat ejection system. The force of a catapult will blow the seat and a fully equipped crew member as high as 150 feet (45 meters) into the air. If the canopy is off and the rescuer tries to pull the pilot out without making the seat firing mechanism safe, it may be launched and kill both the pilot and the rescuer.

When removing a crew member, the rescuer in most cases must work directly over the seat—picture what would happen if the ejection seat fired during the rescue attempt. Therefore, the first job is to deactivate the canopy jettison and seat ejection systems. Some seat catapults are deactivated by slipping safety pins or spikes into the proper holes in the seat armrests; others must be disarmed by cutting the tubes behind the pilot's head. Because working around a seat ejection system is extremely hazardous, it is essential to have as much knowledge as possible of the particular aircraft and the proper tools to work with.

Pilot Rescue

When rescuing a crew member, it is imperative that the person's oxygen mask be removed. This action precludes any possibility of suffocation from lack of oxygen during the rescue operation. The face mask is removed by pulling downward on the green cord beneath the chin and lifting the face piece free. This must be done before detaching the oxygen supply hose that is attached to the pilot's suit. Then release the pilot's seat belt and shoulder harness by unlocking the buckle at the person's waist or by cutting the belt. The victim may now be lifted out by the shoulders. The rescuer should stand on the fuselage just behind the victim to gain maximum hoisting power.

Armament Hazards

The armament systems of military aircraft can create a hazard to all personnel. Whenever an aircraft crashes, there is always the chance that ammunition or bombs may be aboard. The safest rule about machine guns and cannons is to stay away from the business end and out of the line of fire. Guns can and do fire at unexpected times. If a gun is loaded and has a round in the chamber, and if the gun barrel is hot, the round may "cook off" and the projectile will be ejected from the muzzle as though the gun had been fired manually. It may be one round or many. While a gun will not run wild in automatic fire after a round cooks off, it will spontaneously reload and other rounds may be fired individually.

Another gun hazard is the possibility of a short or other defect in the electrical firing circuit because of burned insulation or crash damage. Whenever possible, rescuers should reach into the cockpit and make certain that the gun switches are in the safe positions; this prevents the guns from firing if the trigger button on the control stick is pressed accidentally, but does not guarantee that the guns will not shoot when there has been fire damage.

Heat can also cause rockets to explode or fire. This can cause injury or death to personnel behind the rockets as well as to those in front of them. If a rocket warhead ignited, the blast effect would probably not only result in the destruction of the aircraft structure, but would also detonate any other warheads aboard. It is important to keep these weapons cooled below their ignition point with streams of foam or water. If ignition of the rocket package has occurred, it is best to avoid the use of water spray, since it adds to the intensity of burning rocket propellants.

Bombs may explode if exposed to fire. Ordnance experiments prove that certain types of bombs will detonate within a few minutes if subjected to intense heat. It is well, therefore, to attempt to keep bomb bays or wing racks cool whenever possible. The hazard is related more to high-explosive detonation than to any other factors, and this applies to nuclear weapon components as well as to conventional bombs. Nuclear weapons in aircraft present no greater danger than high explosives, except for minor chemical and radiation hazards.

The usual procedure followed by military crash crews when a fire occurs aboard an aircraft known to be armed with bombs is to advance immediately to take whatever rescue procedures may be expedient in the first few minutes, with the amount of time they spend working determined by the intensity of the fire. Then, if fire control has not been accomplished, all personnel should retire to a safe distance until the explosion hazard has passed.

QUESTIONS FOR REVIEW

1. The best method of handling emergency incidents arising from an aircraft accident is to
    a. adopt rigid standard operating procedures.

RESCUE FROM LAND AND AIR VEHICLES

    b. preplan and practice.
    c. follow the procedures adopted by a large metropolitan city.
    d. demand that the city purchase the larger fire apparatus required for structural fire fighting.

2. If an airplane crashes in a residential area, which of the following should receive the most consideration?
    a. Preventing ignition of the fuel spill
    b. Searching the residential wreckage for victims
    c. Removing the aircraft victims before the plane ignites
    d. Checking to be certain that military armament will not excessively jeopardize rescue personnel and bystanders

3. When possible, always approach the scene of an aircraft crash
    a. with the wind blowing on the backs of personnel.
    b. facing into the wind.
    c. with the wind blowing at an oblique angle.
    d. at right angles to the wind direction.

4. When advancing fire streams to combat an aircraft fire and rescue trapped occupants, which of the following should receive the least consideration?
    a. Wind direction
    b. Ground slope
    c. Fuel spills
    d. Protection of exposed adjacent structures

5. When Plexiglas becomes difficult to break because of the heat from a fire, entrance into the aircraft may be expedited by
    a. cutting an adjacent metal panel.
    b. checking for an alternate entrance.
    c. spraying the plastic with a carbon dioxide (CO_2) extinguisher.
    d. using a power saw with a steel cutting blade.

PART 3 METROPOLITAN RESCUE

7 Rescue from Collapsed Buildings
8 Rescue from Elevators
9 Rescue from Electrical Contact
10 Rescue of Trapped Victims
11 Rescue from High Places
12 Rescue from Burning Structures
13 Rescue Using Fire Apparatus

CHAPTER 7

Rescue from Collapsed Buildings

Most persons have the impression that large buildings are massive, durable places of refuge, places of safety in the event of disaster. Therefore, when a structure collapses—because of an earthquake, fire, explosion, tornado, hurricane, or other natural or human-caused catastrophe—panic and confusion will usually result. Rescue personnel responding to such a disaster incident will probably find that their operations must be conducted under difficult and complicated conditions. It is crucial that rescue operations be carried out systematically, and that a definite plan be followed (Figure 7-1).

SIZE-UP

Even the best rescue leaders tend to overestimate the difficulties when they first approach a large rescue operation, because the job appears so huge and complex. This reaction is quite natural; at such times, the officer in charge must exercise coolness, perseverance, and courage. The leader must make full use of the knowledge gained in previous experience and training.

The size-up is an orderly and methodical procedure for gathering facts when analyzing the rescue situation. It is thorough, accurate, rapid, and continuous throughout the entire emergency, until all operations are completed. Much can be done to locate and rescue trapped persons if the size-up is undertaken in a systematic and logical manner. The size-up forms a basis for decisions on personnel and equipment required and the techniques and methods to be employed. It eliminates wasted motion and promotes safety. A continuous survey of hazards should be made at the scene of the emergency to prevent additional injury or death to victims, bystanders, or rescue workers (Figure 7-2).

The time of the day should be taken into consideration when sizing up the rescue problem. An incident at a school will be treated differently during school hours than in the evening; the rescue problems at a hotel or apartment house will be more complex at night than during the working day.

Rescue from Collapsed Buildings

RESCUE OPERATIONS

While specific rules cannot be formulated for each rescue problem, standard procedures that apply to every rescue operation involving collapsed buildings can be established. The officer in charge must approach all decisions logically; work must be planned systematically. When rescue operations are prepared and performed in stages, there is less possibility of overlooking important points and the casualties will be more accurately located and promptly extricated.

Rescue personnel and fire fighters approaching the rescue scene should be wearing a full complement of safety clothing, including helmet, heavy coat and pants, boots or safety shoes, and gloves. If there is any chance of toxic or unbreathable gases or vapors, self-contained breathing apparatus must be worn.

Trapped Personnel

Because there is always the possibility that rescue or fire fighting personnel may be killed, injured, or trapped if engulfed in a fire or explosion or caught beneath a collapsing structure, everyone entering or leaving the collapsed building must be constantly accounted for. Bodies of emergency service personnel whose absence was never noticed have later been discovered during cleanup operations.

Fig. 7-1. When a tank ship crushed a structure on shore, extrication of the victims was complicated and hazardous because of the ensuing fire. In this incident, the explosion blew the superstructure shown in the background completely over the office building. A large volume of heat and smoke from the burning ship's hull, which is behind the photographer, and the ship's superstructure in the background forced fire fighters to blast a path through the flames with hose lines to check the building for trapped and injured victims. The force of the explosion sprayed the area with chain, masts, anchors, cable, and other debris, further complicating rescue and fire fighting operations.

Fig. 7-2. Extricating trapped victims from collapsed residences during a massive landslide is hazardous because the structure is usually very unstable while it is still in motion. Even though the rescued residents may vehemently assert that all occupants have escaped, the structures still must be systematically searched for trapped or injured intruders or forgotten visitors; this situation calls for procedures like those used by fire fighters at burning buildings. If the structure is somewhat stabilized, a thorough room-by-room search must be started and kept up until all sections have been checked. If the building is unstable and further slippage is a distinct possibility, then it is necessary to calculate whether the possibility that a person may be in the structure will warrant jeopardizing the lives of rescue personnel.

Personnel should be organized into squads or units, with an officer or designated leader in charge. No member should be allowed to wander around or work individually. If convenient, companies or squads that arrive in a group will stay together. A roster that accurately lists the members of the unit actually responding to the incident should be maintained and constantly revised. This roster is presented to the officer in charge at the command post or kept in a prominent place in the vehicle cab. The list must be readily available in the event that the squad leader becomes trapped or otherwise incapacitated.

Whenever a rescue squad is working at an accident site where emergency personnel may become trapped—a collapsing building, landslide, or avalanche—a roll call can be conducted to determine whether any crew members are not accounted for. The roster will permit an immediate assessment of the number and identification, by name and unit, of all trapped crew members. This information is necessary so a determination can be made of the extent of the required rescue operation, the amount of assistance needed, and the area where the victims might be located. The roster will also make it possible for search procedures to be discontinued as soon as all known casualties have been located and rescued. With the possibility of further collapse making all operations hazardous, an exact knowledge of the trapped crew members is essential.

One officer or squad should be given the responsibility of keeping track of all victims removed from the structure and transported from the scene, both civilian occupants and emergency personnel. Unless perfect records are maintained, it is difficult to know when the urgent and hazardous rescue operations can be terminated and the more orderly and safer routine procedures begun.

Team Effort

Officers must supervise their rescue units, not engage in actual manual operations. This will allow them to devote their complete attention to coordinating the efforts of their personnel, observing the surrounding area for unsafe conditions, and rotating members to avoid excessive fatigue. Managing and controlling rescue operations will enable the officer to provide far more assistance than could be offered by adding one more pair of laboring hands to the work force.

Only the number of members actually needed for the job should be permitted in the danger area. When the particular assignment can be accomplished by one, two, or three crew members, the remaining personnel of the unit should be ordered to remain outside the danger area and used for relief on a rotating basis. If the undertaking requires the entire squad, a relief unit should be assigned to the same task; these two squads should rotate as units, alternately working on the assigned task. When the operations are lengthy, rotation of personnel is essential. Although perseverance and courage are important, this type of rescue operation also requires exceptional physical and mental efforts. Fatigue develops quickly, resulting in accidents, injuries, and less productivity.

Safety Precautions

Rescue work consists of individual efforts welded by teamwork. Often only one or two members can work in the danger area, but they depend on other personnel to watch for developing hazardous conditions, to keep them supplied with tools and equipment, and to stabilize the structure or debris to create a safe working environment. The physical safety of personnel engaged in rescue operations must be of prime concern—not only to the officer in charge, but to every crew member. The chance of a second collapse is often very real, resulting in a more tragic occurrence than the original destruction. When there is the danger of another immediate collapse, an organized operation will produce the fastest, safest, and most productive results.

Efforts to remove occupants from a collapsed building may expose rescuers to more dangers than are faced by the victims. Rescue personnel must constantly observe all safety precautions to protect themselves from injury. Each person on the team must be alert to any change in conditions that could raise an additional threat to safety; equally important, whoever notices a new peril must quickly sound the alarm. If possible, take immediate action to avert the threat, or suggest action requiring efforts by others. Informed of the changed conditions so the size-up can be altered, the officer in charge can make a decision about strategy and tactics. Rescue personnel witnessing a structural collapse that engulfs workers and bystanders often have a tendency to rush headlong into the danger area and attempt to forcefully dislodge rubble and debris without a plan or tools. At times this is necessary and productive, especially if the victim is on top of the debris and can be removed easily. On most occasions, however, it is best to quickly study the situation, determine if there is an imminent hazard of further collapse, and decide upon the safest and fastest rescue procedures.

A shifting mass of debris and structural components must be stabilized, to forestall the danger of further movement or collapse while rescuers are working to extricate trapped victims. Shoring may be the necessary first step before actual rescue work can be conducted. Whenever jacks, levers, or other equipment is used to raise heavy objects, insert wooden blocks or cribbing between the object and the ground or floor, because there is always danger that the jack or object will slip and fall. As the jack is raised, a block or wedge-shaped piece of wood can be pushed into the increasing space between the raised object and the ground. Correctly placed cribbing will restrict accidental slipping and falling to a minimum and ensure safer operations.

Whenever personnel enter a building, for rescue or for fire fighting, an exit should always be kept clear. There are two very good reasons for this rule: (1) a clear passage to safety must be maintained so victims can be dragged or carried to clean air by the shortest route, and (2) a clear course is required to provide a flow of fresh air. The air in a building, even when a small fire is burning, is never completely clear, and the flow of fresh air through the opening will allow a more breathable atmosphere.

SIGNS OF BUILDING COLLAPSE

Fire fighters engaged in combatting structure fires are often caught in a collapsing building. The possibility of structural collapse should be of primary consideration in the development of fire fighting strategy and tactics by the officer in charge of a serious fire situation. Were the factors leading to a collapse simple, always evident, and easily recognizable, the problem would be greatly minimized. There are few instances in which the probability of destruction is clearly evident.

Structural failures often occur in mercantile and industrial buildings that have large open areas without supporting columns.

METROPOLITAN RESCUE

Collapse of residential structures severe enough to endanger rescue personnel or fire fighters seldom occurs. Perhaps the greatest danger to emergency personnel in such situations is the chance of falling through a weakened roof or floor into the fire. For this reason, rescue personnel who enter burning buildings usually try to stay close to the walls, especially when the floor has a somewhat spongy feeling (Figure 7-3).

The entire building or major parts of it may be unsafe. The walls may be cracked or leaning, floors or roofs may lack proper support, or the entire building may be hazardous because of age, vandalism, or previous fires. The cumulative effect of heat from preceding fires may have weakened floor beams, columns, bearing partitions, and walls. Because these conditions may not be apparent on arrival, all personnel must be alert for spongy roofs or floors, leaning walls, beams pulled out of walls, and missing interior partitions and columns. Any of these structural defects should be reported to the officer in charge immediately. Following are some of the situations that could warn alert personnel that the dangers of a building collapse are greater than normal (Figure 7-4).

Arson

Incendiary fires offer conditions more hazardous than normal. Accelerants, such as gasoline, are commonly employed to increase the burning rate of a structure and its contents. Arson is often intended

Fig. 7-3. When a residence does collapse, disentangling the casualties from the wreckage may offer problems. Rescuers may require all the skill and expertise that they have gained from extricating injured victims from vehicle accidents. When crushed or blocked doors and windows prevent entrance and egress, holes cut through the roof and wall sections with gasoline-powered chain or rescue saws will usually solve the dilemma. Before starting any operations, it is necessary to shut off or disconnect all electrical power wires and gas and water piping into the structure, for the protection of rescue personnel.

Fig. 7-4. Collapsing walls and floors in derelict buildings, whether in a large city slum or the middle of Death Valley, kill and injure many people every year. Structures constructed many years ago did not benefit from building techniques taken for granted today; consequently, collapse is a constant hazard. Explorers, visitors, looters, and other people entering derelict buildings must be constantly alert for any signs of structural weakness. Warning signs should be posted on derelict structures, and accidents involving old buildings should be prominently covered by the media, to educate the public on the hazards old buildings present.

Rescue from Collapsed Buildings

to cover up and destroy any evidence that a crime has been committed. The increased concentrations of flames and heat in this type of fire quickly weaken structural components, so that destruction is inevitable. The weight of water from attacking fire streams is then often sufficient to induce collapse.

Looting

Recently vacated and incomplete buildings are frequently targets of thieves who remove anything salvageable. Looters regularly cut floor beams, damage fire escapes, and cut large holes in the floors and roof to expedite lowering of salvageable material from upper floors. All of these actions affect the stability of the building.

Demolition

When a building is being torn down, its structural integrity is usually severely weakened. Caution must be observed when responding to any incident where demolition operations are being conducted.

Repeated Fires

Repeated fires in a building will have a cumulative effect, progressively weakening a structure.

Heavy Floor Loading

Businesses that usually require a heavy concentration of weight on the upper floors of their buildings, such as printers, plumbing supply firms, and warehouses, may find that so much weight has jeopardized the integrity of their structures. When a fire weakens the components or fire streams add to the floor loading, collapse of such buildings is almost inevitable.

Large Fire

When two or more floors are engulfed in a hot, roaring fire, the structural integrity of a building is quickly reduced to the point that collapse may be imminent. The fire resistance of the building will largely depend on its construction.

Unprotected Steel

Temperatures over 1,000°F (538°C) can soften steel supporting members and bring about the partial collapse of a structure. When you can see unprotected steel structural components, expect trouble if the steel is vivid red or white; at that point the metal is sufficiently heated to lose part of its strength.

One of the reasons single-story buildings with flat roofs collapse early in a fire is that unprotected steel bar joists are often used for support in such buildings when it is desirable to have a wide span free of columns at a minimum cost. Anticipate the collapse of any roof supported by unprotected steel, because the metal will either be heated to the point of distortion or will expand under high heat to the point where the steel will push out a wall far enough to cause collapse. The intensity of a blaze serves as a warning to fire fighters entering a building. It has been estimated that once a roof supported by unprotected steel has been subjected to intense fire, it can be expected to collapse in as little as fifteen minutes.

Sagging Floors

An accumulation of water on floors may indicate bowing of the floor beams or sagging of supporting columns due to the fire on a lower floor, or that the accumulated weight of the water is overloading the floor structure. When floors sag from excess weight, beam ends may pull out of the supporting walls.

Falling Plaster

If plaster and other building materials are sloughing off and falling from walls and ceilings, a structural shifting of the building may be in progress. These signs could precede a structural collapse of some consequence.

LOCATING TRAPPED CASUALTIES

In an emergency, accounting for all persons who might be victims is difficult. To be effective, this action must be methodically planned and carried out. Although the appearance of most collapsed structures would lead one to believe that no one could possibly be alive in the rubble, experience has shown that there is a high rate of survival in such disasters. This is particularly true when residential buildings collapse in a pancake fashion. Searching for victims and determining their location can be facilitated considerably if a person who is familiar with the layout of the structure is available for interrogation (Figure 7-5).

Information from family members, survivors, and rescued occupants will probably offer many valuable leads about trapped casualties, including the number of persons trapped and the best area to search. Other sources of information are the police or fire services already on the scene and bystanders. The assertions of residents who positively state that everyone is out of the building cannot relieve the rescue crew of the responsibility for conducting a search; there may have been visitors or unauthorized trespassers on the premises. Perhaps the most valuable technique in such cases is observation by trained rescue personnel, who, through their experience and training, can best analyze the situation.

When casualties are trapped in large piles of debris, a silent listening period is a good way to begin operations. Members are stationed around and on the rubble or collapsed building. The leader asks for silence and then calls loudly into holes in the piles. A period of silence follows each call. If there is an answer, the personnel stationed around the debris will be able to locate or approximate its origin.

THE FOUR STAGES OF RESCUE

Stage I, Immediate Rescue. Immediate rescue takes place after the initial size-up and before the completion of the general reconnaissance. It includes persons who can be seen or heard and those trapped in voids and other areas where their exact location is known.

Stage II, Exploration. Exploration consists of searching places where trapped persons are likely to be found, such as strong and sheltered parts of buildings; specially constructed shelters; spaces under stairways; spaces in basements; locations near chimneys; spaces under partially collapsed floors; rooms with exits blocked by debris; and the fringe areas around explosions (buildings where there may be partially trapped persons suffering from shock, cuts, or fractured bones). Exploration should never be omitted. It may be carried on at the same time as Stage III.

Stage III, Selected Debris Removal. Although some debris removal is necessary in Stages I and II, Stage III covers more thorough debris removal. It takes place when persons are still known to be missing, but can neither be seen nor heard. This stage is carried out according to a plan based on one of or combinations of the following facts:

1. Last known location of the person
2. Location and condition of the debris
3. Direction the victim may have fallen during the collapse of the structure
4. Location of voids formed by the collapse
5. The "calling and listening" method
6. Removal of debris up to the probable location of the trapped persons

The squad leader then decides on the method of rescue and the use of crew members and equipment.

Stage IV, General Rubble Removal. General rubble clearance is the method used after all other methods have been used and persons are still missing. This consists of stripping the area systematically.

This work must be done rapidly and with extreme care. Unless every precaution is taken, victims may be further injured or killed. Rescuers must be especially careful when using any type of power or hand tool. Heavy equipment may be requested from contractors, road departments, or other companies or agencies having skip loaders, bulldozers, cranes, and trucks.

General rubble clearance should not be confused with the removal of rubble to clear a site as a cleanup job; general rubble removal is an important part of the rescue operation.

DAMAGE TO BUILDINGS

The construction of a building gives some indication of the way it may collapse as the result of a blast, earthquake, cyclone, or other disaster (Figure 7-6). Buildings of the same class and type of construction collapse in much the same way, and common factors are present. Rescuers should study these factors, since this knowledge will prove helpful when extricating casualties.

Almost all types of damaged buildings will contain voids or spaces in which trapped persons may remain alive for comparatively long periods of time. To know where these safe places may be, it is necessary to know the characteristics of various types of construction.

Framed structures are of two types: (1) large multistoried framed buildings, such as hotels, office buildings, schools, hospitals, and loft buildings; and (2) smaller residential houses.

The larger multistoried framed structures are those having columns, beams, floor joists, and roof beams or trusses that support the weight of walls, partitions, and roof. These may be built of wood, steel, or concrete.

The smaller residential framed structures are usually limited to a height of two or three stories. The framework consists of posts or columns of wood, masonry, or steel in the basement, with wood or steel basement ceiling beams which, together with the exterior walls, support the upper floors and roof. The exterior walls are built with wood wall studs, the exterior siding with wood sheathing, or

METROPOLITAN RESCUE

Fig. 7-5. There is usually a high survival rate when buildings collapse in a pancake fashion. This type of collapse creates many large voids in the structure where victims can remain in a relatively comfortable environment until they can be extricated. Doors, windows, and other means of entrance and egress are normally crushed or blocked, but gasoline-powered chain or rescue saws will quickly provide a hole through roofs or walls with relative ease. Care must be taken to shut off or disconnect all electrical power, gas, and waterlines in the vicinity.

Fig. 7-6. Every year many fire fighters are killed or injured by collapsing walls and floors during the suppression of structure fires and the rescue of building occupants. Everyone on the fire ground must be constantly alert and watchful for any signs of an impending collapse, such as cracks, bulges, leaning walls, spongy roofs or floors, or other structural defects. (Courtesy of the Los Angeles City Fire Department.)

masonry, and the interior finish with lath and plaster. The floor joists and roof rafters are wood.

Unframed or wall-bearing buildings are those in which the foundations and walls support the weight of floors, roof, and interior partitions which are not bearing walls. These structures are usually built of masonry walls with wood or steel floor joists and roof rafters.

Combinations of both framed and unframed construction may be found in industrial buildings or those having additions built to the original structure.

Damage Effects

Framed structures generally withstand blast and other damages better than unframed buildings because of the tendency of the force to be distributed throughout the framing. Wall panels may be blown in or out without demolishing the frame and floors may remain intact or only partially collapsed. The collapse of a wall of a framed building does not cause the collapse of the building. Debris and rubble will result from the effects of a blast or earthquake on a framed building, but not to the degree likely to result when an unframed building is involved.

Reinforcing rods and fire-distorted structural steel frames may create difficult and hazardous rescue problems. However, these materials will create many safe places from which people may be rescued. Rescue from framed structures may not be as difficult as from unframed buildings, except that framed buildings are usually large and multistoried and contain many occupants, and rescue of more people from greater heights may present additional problems.

Extensive collapse may result from damage to unframed buildings. When bearing walls are destroyed or damaged, the floors are likely to collapse completely or become extremely dangerous. If bearing walls are damaged near their foundations, the remaining parts of the upper walls are likely to be unsafe. The large amounts of debris and rubble which usually result from damaged masonry buildings often cause rescue operations to be complicated and time consuming.

Types of Collapse and Formation of Voids

When floor supports fail in any type of building, the floors and roof may drop in large sections. These sections may form voids. If these sections remain in one piece and are supported on one side and collapsed on the other, they form lean-to collapses (Figure 7-7). The hanging collapse is a variation of the lean-to collapse, in that its lower edge is not at rest, but is hanging free. In some cases a floor or roof section may be suspended from one of its four corners, rather than along an entire edge. Persons may be trapped in the voids thus created.

The weakening or destruction of bearing walls may cause the floors and possibly the roof to collapse, one on top of the other, into a lower floor or basement. This is referred to as a "pancake" collapse (Figure 7-8). Persons may be trapped between the layers of these "pancakes."

When, as a result of collapse, the weight of heavy loads, such as furniture, equipment, or rubble and debris, is concentrated near the center of a floor, a V-type collapse may occur (Figure 7-9). Heavy furniture may support a collapsed section of floor or wall, creating voids where casualties may be located.

Wood-framed dwellings generally collapse in very large panel sections. Walls, for example, come apart in panels composed of studs

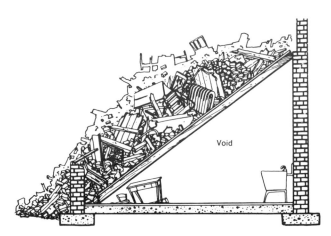

Fig. 7-7. When floor supports fail in any type of building, the floors and roof may drop in large sections. In the lean-to type of collapse, a large void that may shelter victims is formed when one side of the building remains supported. Victim extrication is best accomplished by entering the void along the wall where the wreckage is supported. It is essential that the debris be disturbed as little as possible, to reduce the chances of further collapse. If further movement seems possible, stabilize the wreckage with jacks or timber braces.

Fig. 7-8. The weakening or destruction of bearing walls may cause the floors or roof to collapse, one on top of the other, into a lower floor or basement. This is referred to as a pancake collapse, and victims may be trapped between the layers of these "pancakes." The voids formed by a pancake collapse are usually small, and their location cannot be predicted with any certainty. It is best to use the "call and listen" method of discovering victim location before attempting any extensive movement of debris.

and siding, and floors break up in sections of joists and flooring that are still somewhat intact. Many protective voids are generally formed in this type of building collapse.

HAZARDS FROM DAMAGED UTILITIES

The destruction of buildings and industrial facilities by any catastrophe will invariably result in ruptured electrical, water, gas, and sewer lines. Other hazards will be escaping gases and chemicals used in refrigeration units and in certain industrial operations. These utilities create serious problems for casualties and rescue personnel. To ensure maximum safety to both, rescue personnel must have a knowledge of the hazards, as well as an understanding of the utility supply-line patterns.

Repair of utilities, maintenance of service, and capping of lines during emergencies are responsibilities of the utility companies, in cooperation with the local government. However, rescue personnel may have to take certain emergency measures necessary to save a life. Information concerning utilities and emergency methods for shutting them off should be available from the utilities or the appropriate department of the local government. Rescue personnel should be instructed in the proper method of shutting off water, gas, and electricity and informed of the probable locations of shutoff valves and master switches. These locations may vary from one city to another, but they are reasonably consistent within a city. Information regarding the types and locations of dangerous chemicals and gases used by each industrial plant in the area should be obtained from the plant protection officials.

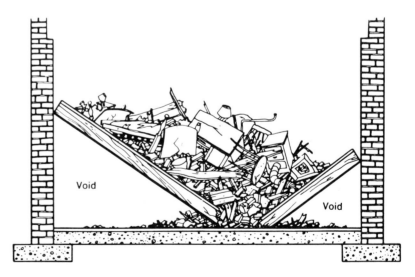

Fig. 7-9. When the weight of heavy loads—such as furniture, machinery, rubble, or debris—is concentrated near the center of a floor, a V-type collapse may occur. Walls, heavy furniture, or other large objects may support a collapsed section of the floor or wall, creating a void where casualties may be located. Entrance may be made into the voids along the walls; it will be necessary to stabilize the wreckage with large jacks or braces before attempting to move the debris.

Water

Water from broken pipes may flood basements and other low areas, endangering trapped persons. Rescue personnel may shut off the flow of water at the water meter or at the street shutoff valve. This valve requires a special water key which should be procured locally as part of the squad equipment. Shutting down the larger water mains must be done by utility or public works engineers.

Pumping equipment for use in flooded areas may be provided by the utility concerned, or by the public works or the fire department when requested. Sandbags should be carried on the rescue vehicle; when filled, they may be used to divert the flow of water from a rescue area or to prevent flooding.

Domestic Gas

Escaping gas in basements and other confined areas of damaged buildings creates the danger of explosion. Rescue personnel should observe the following safety rules:

1. Never look for a suspected gas leak with a match or other open flame.
2. Never attempt to ignite a gas leak; this should be done only by experienced maintenance personnel.
3. Never smoke in areas where gas may be present or suspected.
4. Do not attempt to extinguish gas flames except by shutting off the gas.
5. Use a self-contained breathing apparatus in areas containing gas.
6. Take every possible precaution to prevent igniting gas. Sparks caused by the use of tools and power equipment can cause ignition.
7. Do not shut off large street gas mains. This should be done only by utility or public works maintenance crews.

Local policies and procedures for shutting off gas under emergency conditions should be followed by the emergency services. Information relative to the type of gas (natural or manufactured), special hazards involved, and the system of gas distribution should be obtained. If special keys are necessary to shut off gas at the street, they should be available to each rescue squad. The valve for each service connection is usually located at the meter and can be shut off with a pipe or monkey wrench.

Rescue personnel should locate gas shutoff valves for buildings in each block. Building attendants and plant protection engineers should be able to assist in locating shutoff valves in apartments, industrial plants, and other large buildings.

Electricity

Live wires present a serious hazard to trapped casualties and to rescue personnel. Rescue personnel should observe the following precautions:

1. Assume that all electric wires are energized unless they are known to be dead. The fact that the wires do not sputter or spark is no indication that they are dead.
2. Live wires should be handled only by persons trained in the proper procedures.
3. Never attempt to move wires on the ground, dangling from poles or trees, or hanging slack between poles, except when a life is at stake; then wires should be moved with dry objects that do not conduct electricity, such as poles, boards, or ropes.

4. Avoid pools of water close to live wires; they may be just as dangerous as live wires. Avoid all other conductors such as metal doors and wire fences that may be in contact with high-voltage wires.
5. Never attempt to cut high-voltage wires. This should be done only by experienced maintenance crews.
6. Insulated wire cutters provided in rescue equipment kits should be used to cut wires carrying ordinary house current.
7. The electrical supply to a damaged building should be shut off at the master switch, usually located near the meter or fuse box.
8. Keep the rescue truck and other vehicles away from areas where wires are down.
9. Be especially cautious at night when it is difficult to see wires.

Sewers

Broken sewers may create problems due to flooding and escaping gas. Dams can be improvised to divert the flow of broken sewers away from trapped casualties and working areas of rescue personnel. Open flames should be avoided in the presence of sewer gas, since it can be explosive as well as toxic. Rescue personnel should wear self-contained breathing apparatus when working in areas contaminated with sewer gas.

RESCUE TECHNIQUES

Knowledge of specific rescue techniques is necessary to carry out a rescue operation involving a collapsed buiding. All rescue operations normally require the use of several techniques. Selection of the most effective techniques under the varying conditions of rescue work often means the difference between the survival or death of a casualty.

Debris Tunneling

Tunneling is used to reach casualties, usually when their location is known. It is slow, dangerous work, and should be undertaken only after all other methods have been exploited. It is used primarily for connecting existing voids. Tunneling should be carried out from the lowest possible level, should not be used for general search, and must not be aimless. Occasionally, however, tunneling may be used to reach a point, such as a void under a floor, where further searching may be conducted (Figure 7-10).

A tunnel must be of sufficient size to permit rescuers to bring out casualties. It should not be constructed with abrupt turns. Tunnels as small as 30 inches (75 centimeters) wide and 36 inches (90 centimeters) high have proved to be satisfactory for rescue work. Whenever possible, tunnels should be constructed along a wall, or between a wall and a concrete floor, in order to simplify the framing required.

Constructing a vertical shaft may be considered a form of tunneling for vertical or diagonal access. Usually these shafts are made through the earth after debris has been cleared from the surface. They are often made to reach a point where a basement wall must be breached. Shafts should not be sunk where water or gas service lines enter buildings. Layers of soil and gravel carrying water should be avoided.

Service gas pipes, water lines, and buried electric conduits may be encountered. Avoid cutting these lines. If cutting cannot be avoided, the cut ends must be sealed. Water pipes may be sealed with wooden plugs. Pressure in gas lines usually is lower and may be stopped with moist clay or a wad of rags. If workers from utility companies are available, they should work on these service lines. On gas or water lines three inches or more in diameter, pressure may be so great that they should not be cut, since it may not be possible to plug them unless the mains are shut off. It may not be advisable to shut off a water main, since it may interrupt the water supply for fire fighting. Heavy underground electric lines should be cut by a utility crew. All electric lines encountered should be regarded as energized until proven otherwise.

Debris tunneling is very different from tunneling through undisturbed earth, although strutting and bracing are necessary in both cases. The speed at which a debris tunnel can be constructed varies with the nature of the debris and the size and shape of tunnel required. Because debris is unstable and key beams have to be left in place, the shape and path of a tunnel through debris is often irregular. For this reason, a definite pattern of timbering, as in a tunnel through earth, may not be possible.

The size of timbers used for bracing is determined by the nature of the job and the equipment and material available. It is always better to use timbers that are too heavy than those that are too light, because of the uncertain weight they must support.

In debris tunneling, rescuers must be constantly alert for key timbers, beams, and girders, the disturbance of which could cause movement of the pile and collapse of the tunnel. To avoid any acciden-

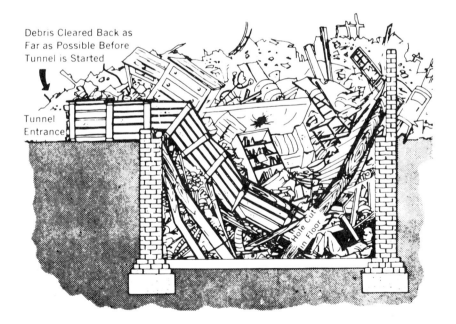

Fig. 7-10. Tunneling is a slow, dangerous method of reaching trapped casualties; it should be undertaken only after all other methods have been tried. A tunnel should be of sufficient size to permit rescuers to extricate victims. Whenever possible, tunnels should be constructed along a wall, or between a wall and a concrete floor, in order to minimize the framing required.

tal movement, horizontal pieces should be secured by a prop placed under them, with room for passage of both crew and stretchers. Recognizing these key pieces may be difficult; bracing everything in the tunnel as the work proceeds will help prevent accidents. Time spent in careful bracing is not wasted, compared with the time necessary to reconstruct a collapsed tunnel.

When piles of debris are large, shafts may be useful. It may be advantageous to sink a shaft in order to reach a basement level or a basement opening, and then tunnel horizontally to reach a victim (Figure 7-11). It is important to remember that even though the materials in the area appear to be solid, the sides of the shaft must always be braced and timbered, and the timbers wedged securely in place.

Timbering and Lining Tunnels

The recommended method for constructing a debris tunnel is by using frames and forepoling. Frames are the primary supporting elements of the tunnel. They should be prefabricated outside the tunnel and assembled in position as the work progresses. Forepoling is the use of planks or boards driven between the collar and crown bar of one frame and extending beyond the next frame into the debris. The crown bar is the horizontal piece connecting the tops of the two vertical struts of each frame. Two-inch (5-centimeter) spacer blocks are attached at each end of the crown bar; a horizontal piece (the collar), parallel with the crown bar, is attached to the spacer blocks, thus leaving a 2-inch (5-centimeter) space between collar and crown bar. The first three frames of a debris tunnel are permanent frames; thereafter, as debris is cleared, temporary frames support the forepoles until permanent frames are put into place. Material for timbering and lining debris tunnels can usually be found in the wreckage.

To start a tunnel using the forepole method (Figure 7-12), three frames are constructed. The first frame does not require a collar or spacer blocks at the top, nor do any of the temporary frames. The second and third frames, and all other permanent frames in the tunnel, require spacer blocks and a collar. Frame 3 is set first against a cleared vertical face of debris and then frames 2 and 1 are placed next at approximately 3-foot (90-centimeter) intervals back from the cleared vertical face of debris, and are solidly braced. Frame 1 should be diagonally braced to stakes driven solidly into the ground, about 2 to 3 feet (60 to 90 centimeters) in front of each strut. After the frames are in place, the top from frame 1 to frame 3 is covered with long pieces of lumber, such as floor joists, roofing, or flooring. Beyond frame 3, forepoles need to be long enough to overlap only from one frame to the next.

The sides are lined in the same manner as the roof of the tunnel, driving boards between the frame struts and the rubble. To insure stability of the tunnel thus far completed, debris is piled against the sides and over the top. When completed, the frames should be completely covered, with the exception of the first frame and the diagonal braces.

When debris is removed about 2 feet (60 centimeters) beyond the third frame, the load on the forepoles may make it necessary to construct a temporary frame firmly wedged under them until enough debris is removed to permit construction of a permanent frame. The temporary frame should be removed after the permanent frame is

Rescue from Collapsed Buildings

properly braced and lined. This procedure is repeated until the tunnel is completed.

Usually, the debris of a demolished structure includes small rubble and dust, which will tend to trickle through the timbering. At first this may not seem important, but the escape of this material in quantity may disturb the mass of debris, causing internal movement. Therefore, a tunnel through small loose debris should be boarded as closely as possible.

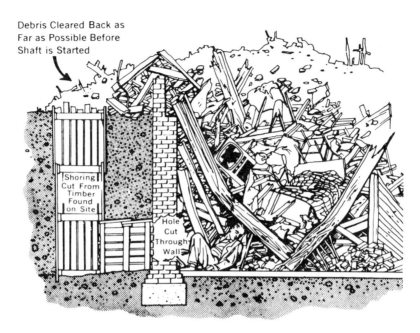

Fig. 7-11. To simplify tunneling to reach a trapped casualty through a jumbled mass of debris, it may be advantageous to construct a vertical shaft through the earth after rubble has been cleared from the surface. When the desired depth has been attained, a horizontal shaft is dug and the wall breached to reach the entrapping void. Constant vigilance is necessary to avoid encountering buried water and gas service lines, sewer pipes, and electrical conduits.

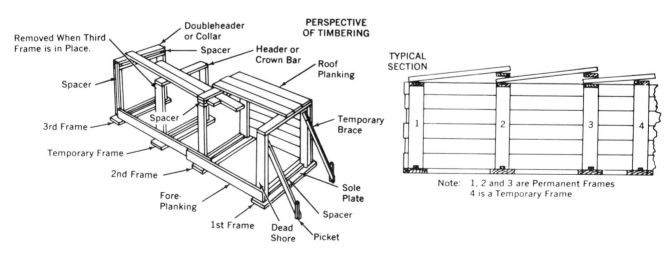

Fig. 7-12. The recommended method of constructing a tunnel through debris is by using frames and forepoling. Frames are the primary supporting elements of the tunnel and they should be prefabricated outside the tunnel and assembled in position as the work progresses. As rubble and debris are cleared, frames are placed in position to support the planking along the roof and sides of the tunnel. The planking should be boarded as closely as possible to prevent small loose debris from trickling through the timbering.

Rectangular framing has certain disadvantages in debris tunneling. Since frames are not rigid, unbalanced side pressures may cause them to collapse. In some instances, short debris tunnels with small cross sections may be constructed in the form of a closed triangle, using heavy planks keyed together at the ends (Figure 7-13). Regardless of the method used, the strutting or lining in a debris tunnel must be as rigid and tightly wedged as possible. Rigidity and wedging will keep the lining in position and prevent it from being broken by the impact of shifting or moving debris.

When there is a doubt regarding the quickest means of access, two or more methods may be tried simultaneously. For example, a basement may be reached by one or more tunnels, or by a shaft from the outside; all these methods are attempted at the same time.

Rescue personnel may have to remove persons from under collapsed basement walls or from basements still intact but with exits closed by debris (Figure 7-14). The squad leader may consider several different approaches; for example, breaking through the wall from an adjacent basement to reach lean-to spaces. Manholes or coal chutes may be cleared of debris to provide an entrance. When ground floors have not collapsed, a small area may be cleared either by tunneling along the floor (or otherwise removing debris) and cutting a hole in the floor to gain entrance to the basement (Figure 7-15). When floors have fallen and the basement is completely collapsed, a sloping tunnel may be constructed from the edge of the debris downward to the floor of the basement (see Figure 7-10). A shaft may be sunk next to the building and into the ground along the basement wall, through which a hole into the basement may be made (see Figure 7-11). If the basement ceiling has collapsed, a solid mass of debris may be revealed when a hole is cut through the outer wall. In this case, a trench or tunnel may be constructed along the outer face of the wall, and a hole driven into the wall at another point.

Fig. 7-13. Unbalanced side pressure may cause a rectangular tunnel to shift or collapse. In some instances it may be advantageous to construct tunnels in the shape of a triangle, using heavy planks keyed together at the ends. Rigidity and wedging will keep the tunnel lining in position and prevent it from being broken by the impact of shifting or moving debris.

Fig. 7-14. Occupants were trapped when a structure collapsed in a pancake fashion; they survived because they were contained in a large void. Rescuers heard the people caught in the rubble, used an electric saw to cut through the planking, and removed the victims, who were relatively unharmed.

Rescue from Collapsed Buildings

If a floor has collapsed forming a void against one wall, and there appears to be a void against the opposite wall, a tunnel may be driven through the debris from the first void toward the opposite wall to reach the second void. However, it should be remembered that debris tunneling is one of the hardest jobs in rescue work and should be undertaken only when other means of gaining access are impractical.

Tunnel atmospheres known to be contaminated with toxic gases or deficient in oxygen will require that workers wear oxygen or self-contained breathing apparatus. Whenever a rescue worker is using a mask inside a tunnel, some means of communication should be provided to the outside. Although ordinary portable telephone equipment is unsafe for use in atmospheres containing explosive mixtures, self-energizing telephones are safe. A lifeline should be provided, not only as a means of locating a worker should the tunnel collapse, but also as a means of communication. Persons who collapse in a toxic atmosphere only a short distance into a tunnel can be pulled out by their lifeline. They should not be pulled a long distance or around corners, since their face pieces may be dislodged, leaving them without mask protection. A worker who becomes unconscious should be carried out by fellow workers who have been standing by with proper equipment.

Trenching

Frequently an open trench can be completed more quickly than a tunnel if debris is not piled too high. Trenching and tunneling operations may sometimes be combined, with a trench extended into the debris until a tunnel becomes more practical (Figure 7-16).

Fig. 7-15. When persons are trapped under collapsed basements or in basements still intact that have exits closed by debris, entrance may be made by tunneling along the floor and then cutting a hole to gain entrance into the voids. Care must be taken to stabilize the debris sufficiently with planking and bracing to prevent any shifting of the wreckage.

METROPOLITAN RESCUE

To trench through debris, start by removing the larger pieces of timber, stones, or other objects from the face of the pile nearest the objective. Then clear a way through the debris by shoveling and other hand methods, removing the minimum amount of material necessary to provide a safe passageway. Progress is governed by the type of debris through which the trench is made. Trenching may be dangerous. If a trench collapses, the worker has little chance of avoiding injury. To avoid collapse or dangerous movement of the sides of a trench, bracing or some other method of retaining the sides may be required.

One satisfactory method of trenching is to drive sufficient sheet piling (usually lumber found at the site) into the ground. Additional support may be provided by horizontal bracing of the sheet piling, using screw or building jacks if available, or wood struts between the two retaining walls as necessary.

Material removed from a trench should be piled at some distance from the edge, so it will not fall back into the trench or have to be moved again. The size of the trench will be governed by its purpose and the nature of the debris.

Trenching is used to reach a specific point, not for general clearance. A crew may decide to start two or more trenches to a given point simultaneously, since it is not always possible to determine the fastest route.

Breaching Walls

Many different types of construction will be encountered in rescue operations. These include walls made of brick with lime mortar; brick with cement mortar; stone; concrete; or concrete block.

When cutting through walls or floors of large buildings, locate sections of the structures in which cutting can be done most quickly and safely. When cutting through walls, rescuers must be sure that support beams and columns are not weakened. After a building has been subjected to damage, the parts left standing may appear sound, although badly shaken and cracked. Therefore, when cutting away wall sections, especially with air hammers, care must be taken to prevent further collapse.

Openings large enough for rescue purposes usually can be made in brick walls without danger of the masonry's falling. The bricks should be removed so that the opening is arch-shaped.

Concrete walls and floors, especially when they are reinforced, are difficult to cut through. Pavement breakers or other power tools will be helpful. Crews should request such equipment from the city department responsible for street maintenance or from other sources. In all walls and floors except concrete, the best method is to cut a small hole and then enlarge it. With concrete, however, it is better to cut around the edge of the section to be removed. If the concrete is reinforced, the reinforcing bars can then be cut by a hacksaw or torch, and the material removed in one piece. If a torch is used, be sure explosive gases are not present, and that flammable materials are not ignited. A fire extinguisher should be kept nearby.

Ventilation

In constructing a tunnel or gaining access to a basement, dust and the possibility of gases may make work difficult. Each person must

Fig. 7-16. Trenching is used to reach a specific point, not for general clearance of wreckage and debris; an open trench can be completed more quickly than a tunnel if debris is not piled too high. Trenching is accomplished by removing the larger pieces of timber, stones, bricks, and other objects to clear a path. Bracing and timbering is normally necessary to support the debris and to prevent it from falling or shifting into the open trench.

have an adequate supply of fresh air. To work in areas with dust or gases a filtertype mask can be used; limitations of this type of mask must be understood—at least 16 percent of the atmosphere must be oxygen for it to function. In many cases, even when a mask is used, fresh air and ventilation must be provided. If air compressors and air movers are used, care must be exercised not to recirculate the air removed from the tunnel. When using gasoline-driven compressors, be sure that exhaust fumes do not enter the compressor's intake. Use of an acetylene torch in a confined space may cause the air to become foul, and trapped persons and workers may suffer if proper ventilation is not provided. An acetylene torch should never be used where there is danger from explosive gases.

Induced ventilation is the most effective means of dealing with gas hazards in confined spaces and may be the only way to save persons trapped under debris in the presence of escaping gas. An air compressor is the simplest means of inducing ventilation. However, high-velocity air from an open nozzle or hose will stir up dust; the force of the air must be modified in some way. Placing the end of the hose into a tin can and tying this assembly in an empty bag will reduce the force of the air and eliminate much of the dust hazard.

RAISING AND SUPPORTING STRUCTURAL ELEMENTS

Raising and supporting structural elements of collapsed and damaged buildings may be necessary before rescue workers can gain safe entrance into voids.

When floors collapse they usually tend to hold together. Walls frequently fall over in large sections. By holding together and falling in large sections, floors and walls often create voids which, to some extent, protect people under them. Jacks, levers, and blocks and tackle must be used to lift such heavy sections. In this type of operation safety cannot be overemphasized. The squad must make sure that in raising beams or sections of floors and walls, the stability of the rest of the building is not disturbed, causing further collapse. Be sure that any raised section is firmly supported by struts or cribbing before allowing anyone to crawl under it.

Shoring

Shoring is the erection of a series of timbers to stabilize a wall or prevent further collapse of a damaged structure that endangers the conduct of rescue operations. Only temporary shoring will be constructed by rescue squads. Shoring should not be used to restore structural elements to their original position. Any attempt to force beams, sections of floors, or walls back into place may cause further collapse and damage. It is important, however, to secure all shoring in position. This must be done gradually and without shock to the structure, using bars and wedges or jacks.

Frequently a weakened foundation or damage to the lower portion of a wall makes it unstable; causes of instability in a damaged structure may vary. Since there is no standardized method of approach, situations requiring shoring and the support of walls and floors call for careful planning and good workmanship. The lower part of the wall and its footing or foundation must carry the entire weight of the structure above it. If it is damaged, or is weakened by the removal of an adjacent supporting structure, it may buckle and crumble. Therefore, bracing or shoring on lower parts of the wall should be stronger than corresponding work on the upper portions.

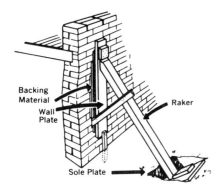

Fig. 7-17. A raking shore is used to stabilize a bulging or leaning wall to prevent it from collapsing, particularly if excavation or tunneling is being conducted next to it. The number of rakers required depends upon the height of the wall to be supported and the number of floors carried by the wall. Normally, the braces should be placed at intervals of eight to twelve feet.

Raking Shore

If a wall is bulging or out of plumb, shoring may be used to brace the wall or hold it in position, especially if excavating or tunneling is being conducted next to it. This type of shoring is called bracing, pushing, or raking shore. The principal parts include wall plate, raker, and sole plate (Figure 7-17).

If possible, the wall plate should be of one solid piece throughout its length. When used against a bulging wall, it should be backed with timbers to provide uniform pressure. Rakers are best formed with square timbers. The number of rakers varies, depending upon the height of the wall to be supported and the number of floors carried by the wall. There should be one raker for each floor carried by the wall and it should be set so that its foot forms a sixty- to seventy-degree angle with the ground. A cleat should be nailed to the wall plate where it meets the head of each raker. The wall plate must be secured to prevent it from sliding upward as the rakers are tightened into place.

Soleplates should be placed to take the thrust of the raker at an angle exceeding a right angle, so that it will become a right angle when the raker is tightened. The raker should not be tightened with a hammer. Instead, a small notch should be cut in the foot of the raker, and a pry bar should be used in the notch to tighten the raker. Struts or braces may be fixed to prevent any movement of the foot of the wall plate, and to prevent it from sliding up under the pressure of the raker. Struts should be spiked to the raker and wall plate. In supporting a wall, wall plates and rakers should usually be placed at intervals of from 8 to 12 feet (2.4 to 3.6 meters), depending on the circumstances, the type of wall, and the degree of damage.

Flying Shore

A flying shore is used to brace a damaged wall when a sound adjacent wall can be used as a means of support (Figure 7-18). The principal parts are horizontal beams, wall plates, and struts. Other items are cleats, wedges, and straining pieces.

To erect a flying shore, the cleats should be nailed on the wall plate, one pair to support the horizontal beam or shore, and the others to support the struts. The struts should be set at an angle not greater than 45° to the horizontal beam and kept apart with straining pieces. The length of the straining pieces is determined by the length of the horizontal beam. Proper measurements and angles can best be achieved by laying out the job on the ground prior to erection. While personnel hold the wall plates in position, a horizontal beam is placed on the center cleats, and tightened with wedges and shims inserted between the shore and wall plates. Next, the struts and straining pieces are placed into position. Cleats may be used to brace the shore more rigidly. The wall plate should be of one continuous length, with packing between the wall and wall plate, if necessary, in order to provide uniform pressure.

Flying shores should be placed along a wall at intervals of eight to twelve feet, depending on the situation, the type of wall, and the degree of damage. They are not recommended for use between walls separated by more than 25 feet (7.5 meters).

Dead or Vertical Shore

A dead or vertical shore is used to carry the vertical dead load of a wall or floor (Figure 7-19). The principal parts are the strut, the soleplate, and the headpiece.

Struts should be made of square timber, if possible, and should be heavy enough to carry the maximum expected load. It is difficult to estimate what load a strut must carry and to gauge what load the strut timber can support. However, in strutting a damaged building, the following principles apply: for a given size of timber, the shorter

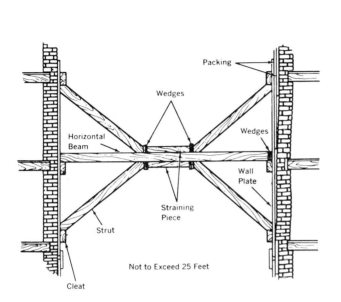

Fig. 7-18. A flying shore is used to brace a damaged wall when a sound adjacent wall can be used as a means of support. Proper measurements and angles can best be achieved by laying out the job on the ground prior to erection. Flying shores should be placed

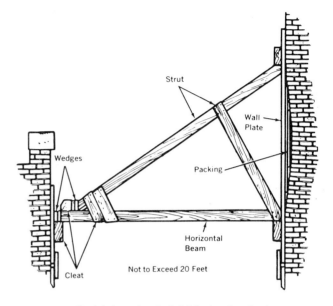

along a wall at intervals of eight to twelve feet, depending upon the situation, the type of wall, and the degree of damage. They are not recommended for use between walls separated by more than twenty-five feet.

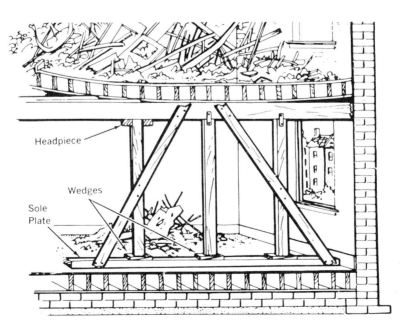

Fig. 7-19. Dead or vertical shores are used to support and stabilize a floor or roof that is so damaged and weakened that it is in imminent danger of collapse. When struts are used on the upper floors of a building, the strutting should be repeated on all the lower floors so the load will have a solid foundation. An exception to this is when a strut can be supported on a heavy beam in a part of the building that has not suffered much damage.

Fig. 7-20. Strutting is used to strengthen window and door frames when they are unsafe due to cracked or damaged walls. Whenever a window or door is stabilized, every effort should be made to avoid blocking the opening; sufficient room should be left between the struts for rescue access.

the strut, the greater the load it can carry; a square strut is stronger than a rectangular one with the same cross-section area; a strut will be much stronger in service if its ends are cut cleanly so that they fit squarely into the soleplate and headpiece; struts should always be made a little heavier than appears necessary—the size used will be determined by the weight of the wall or floor to be supported, and by its height.

Wedges are set under the strut and the strut is driven into position with the wedges until it just takes the weight of the structure and no more. Wedges should not be driven tighter, since this would lift the wall or floor being supported and might cause more damage to the building.

The soleplate should be made as long and as wide as practicable in order to spread the load over a large area. The soleplates should not be placed over a cellar arch or timber floor if there is any doubt that the arch or floor can carry the load. In such cases, the soleplate must be supported from below. Where struts are used on the upper floors of buildings, the strutting should be repeated on all the floors so that the load will have a solid foundation. An exception to this is when a strut can be supported on a heavy beam in a part of a building that has not suffered much damage.

The headpiece should have approximately the same cross section as the strut. However, the load being carried will be a determining factor here, as will the span between struts when two struts are used. This span should be kept as small as possible, because the smaller the span, the greater the load the headpiece can carry.

Strutting

Strutting is employed to strengthen window and door frames when they are unsafe due to cracked or damaged walls (Figure 7-20). Two methods of strengthening such openings are shown; many methods may be employed, but in any case, sufficient room should be left between the struts for rescue access.

Removing Walls

When it is necessary to remove entire walls or parts of them to reach a specific location, the safety of trapped persons must be considered. The crew should make a careful study before a wall is removed to determine whether or not the removal will further weaken the structure and add to the danger. Shoring or bracing of walls of adjacent buildings may be necessary so that part of a structure can be taken down in safety.

Working from the top down, walls may be removed, piece by piece, with picks, hammers, crowbars, and other tools and equipment. An entire wall or section may be toppled by a cable attached to a vehicle or winch, if this can be accomplished without endangering lives or causing other structures to collapse.

When dismantling a building from the top down, the work should proceed systematically, story by story, and the work on the upper part of the wall should be completed before the lower section is disturbed. If the wall is to be toppled, its direction of fall must be controlled. To control the fall, the line of weakness of the wall must be determined. If there is no obvious line of weakness, it may be necessary to create one by cutting away sufficient material at a suitable point so that as much of the wall as possible can be pulled down in the desired direction in one operation.

If a cable or rope is used to topple a wall, it must be securely fastened, so that the force exerted will not pull out a hole but will topple the entire section of the wall with it. The longer the cable, the more horizontal the pull and the greater its effect. The cable must be long enough to prevent the wall from falling on the men or machines doing the pulling.

Leaning a ladder against a wall may not be safe when attaching a cable to a severely damaged wall. An aerial truck ladder can be used or the boom of a crane, if one can be obtained. The cable can be placed in position by using a throw line to raise it to the top of the wall. If winches are used, the drum should be allowed to run free as soon as the wall begins to topple, to relieve the cable and drum from any shock.

DEBRIS HANDLING

Shovels, picks, and other standard hand tools are used in debris removal only when the location of the casualty is definitely known and all other casualties have been accounted for. Recognizing a body in debris is sometimes difficult. Therefore, tools, especially picks (if used at all), should be used with great care to avoid further injury to the victim. Debris close to a casualty's known or expected location should be removed only by hand. Rescue personnel handling debris should wear gloves to prevent minor injuries. Debris should be removed to areas clear of the damaged building in baskets, buckets, and wheelbarrows. Only when it is reasonably certain that rubble or portions of buildings to be removed do not conceal other casualties should cranes, power shovels, and bulldozers be used for debris clearance to gain access to casualty locations, or to prevent further damage or collapse which may hinder rescue operations. Such heavy equipment should be under the operational direction of the rescue personnel. If it proves necessary to pile debris in the streets, rescuers should remember to avoid blocking traffic. All debris that has been moved should be marked.

Rescue personnel, with the help of the police, must exercise constant vigilance to prevent disorganized and unsupervised groups from digging at random in the area. Workers should not climb over piles of debris unnecessarily; the disturbance may cause further collapse, making the rescue problems more difficult and decreasing a casualty's chance of survival. As debris is removed, rescuers should provide necessary support or stabilize the debris to prevent dangerous movement and further collapse.

QUESTIONS FOR REVIEW

1. When confronted with a collapsed building that possibly has many trapped victims, it is most important that
    a. sufficient personnel be summoned to extricate the victims.
    b. adequate equipment be available on the scene.
    c. rescue operations be carried out systematically.
    d. the officer in charge supervise all operations.

2. At 1:30 P.M. on a Wednesday, the rescue squad receives a report that a large structure has just partially collapsed. Which one of the following occupancies would probably create the greatest number of casualties?
    a. Elementary school
    b. Three-story hotel

c. One-story apartment house
d. Chemical warehouse

3. Which of the following is the most important duty for the individual members of a rescue operation to observe?
   a. Listen and check for trapped victims.
   b. Keep the commanding officer apprised of the progress of rescue operations.
   c. Conserve energy in case a new emergency arises.
   d. Constantly watch for any change in conditions that could jeopardize rescue personnel.

4. In which of the following collapsed structures is the chance of discovering live trapped victims greatest?
   a. A lean-to floor collapse of a factory
   b. A pancake collapse of a residence
   c. A V-shape floor collapse of a hotel
   d. A hanging collapse of a school

5. When a large amount of debris must be moved so rescue personnel can check for trapped casualties, the safest method is to use
   a. skip loaders.
   b. power shovels.
   c. picks and hand shovels.
   d. hands only.

CHAPTER 8

Rescue from Elevators

Elevators are provided to move passengers and freight with the least effort and inconvenience. These units vary from simple single-car installations to highly complex, computerized, multicar groups of elevators. In earlier times, elevators were merely a step-saving convenience; now, tall structures make them a necessary part of daily existence.

Elevator equipment is designed to protect the passengers by preventing movement of the car if the correct conditions for operation are not satisfied, and by stopping the car if any unsafe situation occurs while the elevator is moving. Any make or model of elevator can unexpectedly stop from time to time (Figure 8-1). Cars may cease to function for any number of reasons: loss of power, functioning or malfunctioning of a safety device, or failure of component parts. Loss of power may be due to a general blackout, building power failure, or trouble with the electrical supply for a single elevator. The older the installation, the more likely it is that the components with electrical contacts will malfunction. There are over a dozen safety devices provided to protect elevator passengers; unfortunately, these same devices may cause a car to stall between floors (Figure 8-2).

Occasionally, even a perfectly operating elevator may stop between floors when a passenger maliciously or inadvertently operates one of the control switches. When there are children in the stalled car, the possibility of this type of malfunction should be considered first.

No matter why the elevator has stalled, the principle consideration must be for the safety of the passengers. Handling of an incident of this type correctly requires a planned procedure and a direct plan of operation. The inexperienced, untrained rescuer will create problems rather than eliminate them.

PSYCHOLOGICAL CARE OF PASSENGERS

It is imperative that emergency personnel reach the scene of a stalled elevator quickly; panicky passengers or helpful bystanders have complicated simple emergencies by taking unnecessary actions, sometimes with tragic results.

While the passengers may be physically safe, their mental state must be given serious consideration. In any rescue situation, the first responsibility of emergency personnel is to reassure the passengers that they are not in any danger and convince them that they will be freed as soon as possible. People trapped in a stalled elevator react differently: some remain calm and stoical, others become frightened and irrational, a few display anger and pound on the door or controls, while others consider the whole situation comical. If the lights fail or a person is troubled by the fear of being confined in a small place (claustrophobia), panic or hysteria may affect the passengers' composure.

Prevent panic by reassuring the passengers that they are in no danger; let them know that their troubles are known and understood, and that everything possible is being done to speed their release. Talk to them over the car telephone or intercommunication system; if these are not convenient, converse with the passengers through the car doors. A few well-chosen words in a firm and positive voice will reassure them that competent help has arrived and that they are in no danger.

Ask whether there are any sick or injured persons in the car; the presence of an unconscious heart-attack victim will justify taking different procedures than if the passengers were all well and comfortable. Suggest that the passengers sit on the floor, so they will be more relaxed and complacent. Inform them that the car may be moved, but that they should not worry. Advise the occupants not to touch any of the controls or to attempt to force open the doors unless they

Fig. 8-1. Rescuers labor to release victim's leg, caught between an elevator car and the shaft. Note the wedge in the upper right of the photo, used to provide as much space as possible. (Courtesy of the New York City Fire Department.)

Rescue from Elevators

receive specific instructions. Avoid mentioning falling or danger, since this may bring on misgivings, fright, or terror.

QUALIFIED ELEVATOR MECHANICS

Urban areas usually have elevator companies that supply 24-hour emergency service and that can provide a repair person on the premises within an hour after notification. In rural areas there may be a longer delay before a qualified service person can be obtained. The more isolated a region is, the more people will rely on rescue personnel to provide emergency service.

If a rescuer is given instructions by a mechanic, they should be followed precisely. Request sufficient supplemental information from the mechanic (if it is not volunteered) so the rescue crew will not only know what to do, but why it should be done. This will ensure more intelligent cooperation between the rescue squad and the mechanic, and it will help train them for other incidents. A rescue squad calls on a mechanic not to relinquish command or to relieve themselves of responsibility, but to obtain reliable expert advice and to expedite the rescue. Remember, elevator mechanics and companies have legal responsibilities to fulfill; their advice will usually be well founded and consistent with good safety practices.

ELEVATOR CONSTRUCTION

An elevator consists of a car traveling in a vertical shaft called a hoistway. Steel guide rails, usually two in number, are fastened to the sides of the shaft. Guide shoes keep the car in sliding contact with the guide rails to prevent sway or lateral movement.

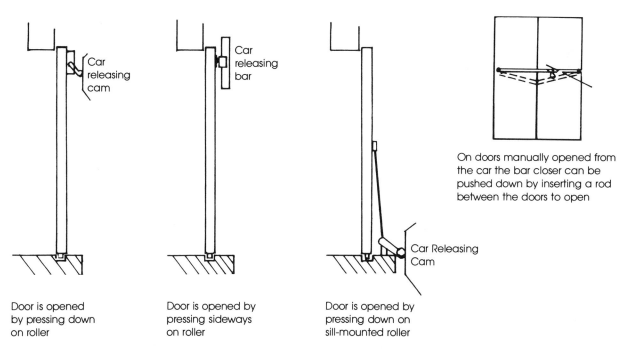

Fig. 8-2. Most modern automatic elevators are equipped with hoistway door locks which are automatically released by a cam on the car when the car is at (or within a few inches of) floor level. The most important thing to remember is that all hatchway doors are unlocked on the inside by a cam or roller which projects away from the door, toward the elevator car; by pressing the cam down or by pushing the roller, the door can be opened. Often the projecting cam or roller can be reached and moved with a long stiff wire, thin rod, or pole through the door cracks.

The car generally consists of a cage of light metal or metal-supported plywood panels encircled by a heavy steel vertical sling, or upright frame. The steel double beam that forms the top cross member of the sling is called the crosshead. To the crosshead is fastened the hoisting cables and the upper guide shoes.

Safeties are found on cars supported by hoisting cables. These brakes are designed to stop the car by clamping it to the guide rails, should it descend too fast or fall because of broken cables. The safeties are controlled by the elevator governor.

All electric and many hydraulic elevators are counterbalanced. The weights for counterbalancing are generally on the back of the hoistway; however, they may be on the sides.

Types of Elevators

There are two different types of elevators: electric and hydraulic.

Electric. Electric elevators are powered by an electric motor driving a drum or sheave to raise and lower the hoisting cables.

Hydraulic. Hydraulic elevators depend on an electric-powered pump to furnish fluid pressure for operation. The plunger-type of pump works like the grease rack hoist in a service station. The car rides on top of a tall piston; the piston cylinder extends deep into the ground, and water or oil pressure is used to raise and lower the piston and car. The car is also partly supported by steel wire cables that pass over a sheave and then down to a set of counterweights.

Communications

Communicating with the car may not be as simple as it seems. Two methods can be used: telephones or voice communication through the hoistway doors. Some installations are equipped with a two-way intercommunication system.

Telephones are the better way to contact the passengers if there is a direct line available so a telephone operator is not needed to relay messages. Many elevator codes require that there be phones in the cars, but they do not stipulate that a telephone be available within the building to provide direct communication with the car. Often the telephones inside elevators are connected to answering services far removed from the scene, so there is no way for personnel on the scene to communicate by phone with the passengers. Beware of relaying directions through operators, because omitted information can easily compound the problem.

The only other method of communicating with the passengers is to yell through the hoistway doors. When outside distractions, bystanders, and other elevator cars are present, the noise they make may have to be controlled before any type of conversation can take place.

Car Lights

Another important point to remember is that the lights in the car are no indication of whether or not the car's power supply is operational. Elevators are now required to have a separate circuit for lighting; the lights will remain on even when the circuit that services the operation of the car is interrupted. Conversely, do not assume that the lights in the car will be inoperative if there is an elevator power failure. Remember, the lighting system is on a separate circuit and has no relationship to the actual operation of the car. Some very old installations may not have this feature, but they are few in number.

This lighting feature is extremely beneficial for easing the occupants' fears and other emotions; it may be an important factor in avoiding panic. There are times when the gleam from the interior car lights will help rescuers locate the position of the car in the shaft.

Of course, if there is a general power blackout or a loss of electricity in the building, the lights cannot function. Sometimes this event is provided for by equipping the car with a battery-powered lantern that automatically lights when the power is disconnected.

Call Buttons

The call buttons on elevator landings are of interest to emergency personnel because they can result in loss of control of an elevator. Call buttons have been blamed for causing deaths in fires because under some conditions heat or flames can order the car to respond to a particular floor.

There are two different types of elevator call buttons: the ordinary push-button switch that requires a slight movement to close two contact points and touch buttons that contain a vacuum tube and require no operational pressure. Tests to determine the reliability of these call buttons under fire conditions have given a variety of results. Most of the units registered a signal at between 425 and 500°F (218 and 260°C) when the internal insulation material was burned off and a short circuit resulted. Some of the buttons were activated when subjected to the flame of a burning wooden match.

The tests proved that all the types of call buttons currently in use can be destroyed by some degree of heat and can, at the time of disintegration, create a false call within their systems. Electronic-touch buttons were not activitated by any amount of heat until the parts collapsed, but all the call buttons could be activated by the flame from a wooden match.

Emergency Unlocking Devices

Many cities and states require that hoistway doors be equipped with emergency unlocking devices operated by a special key-like tool. These devices will trip the door interlock mechanism when they are inserted through a hole in the hoistway door.

Some manufacturers supply these emergency unlocking devices on the hoistway doors on the basement and first-floor lobby landings, some on the bottom and top landings, and others on all the floors. There are various types of emergency keys for the unlocking devices of different manufacturers; the correct one is generally kept by the maintenance personnel of a building.

Although there are many types of emergency unlocking devices, they all operate by applying pressure on the pickup roller, which releases the hoistway door locking device. Rescuers operating in areas where these devices are installed should preplan by inspecting the elevators to locate the locks and by having prior training in the use of the tools needed to trip these emergency unlocking devices.

When keys are not available, interlocking mechanisms in most cases can be tripped by improvising tools made from a coat hanger or welding rod. These unlocking tools normally have to be made before the emergency or manufactured on the scene by looking at an adjacent car. The disadvantage of attempting to carry the tools is that every building may require a different tool design. Excessive pressure is not necessary when using an interlock tool; the important thing is knowing how and where to insert the tool.

ELECTRICAL POWER

Although a complete power failure rarely happens today, it can immobilize an entire city. When responding to a stalled elevator incident, see whether there is a general power failure. After entering the building, see if the structure has a power failure; it may be that fuses have blown or circuit breakers have tripped in the building's main power panel. Larger buildings usually are equipped with auxiliary generators capable of supplying enough current to provide emergency lighting and to return one car at a time to the ground floor. If the power failure is extensive and no emergency power is available, a serious problem exists. There will be many stalled elevators with trapped passengers who have no interior lights or ventilation to ease their minds and provide comfort.

Main Electrical Panel

If the building appears to have electrical power and lights but none of the elevators are operating, check the structure's main electrical panel to be sure that the elevator circuit breakers have not tripped or that fuses have not blown. Be certain that the switch, which should be clearly marked ELEVATORS, is in the ON position.

Elevator Switches

If only one elevator is stalled, the main power source for the elevator in question should be located. The main switch for a specific elevator will generally be found in the machinery room for that elevator. Machinery or motor rooms are normally located at roof level in a penthouse, but could be placed in the basement. Elevators in high-rise structures seldom traverse the entire height of the building; the machinery rooms for an elevator serving the lower or intermediate floors will be found on the machinery floor located above the highest level served by that particular car.

Generally, the machine-room switches are in direct sight of the selector and hoisting equipment. In buildings with more than one elevator, the switches should be numbered; it should not be necessary to check or shut down all of the units. Should there be any doubt about which is the correct switch, check them all to see if fuses have blown or circuit breakers have tripped. Unless there is a competent, trained person on the premises, it is best not to try to remove and replace fuses; because of the current involved, an untrained repair person could easily be injured. If a circuit breaker has tripped, it can easily be reset.

Before rescue personnel start any operations to remove trapped occupants from a stalled elevator, one person should be sent to the machinery room to place the switch in the OFF position. Under no circumstances should the person assigned to this duty leave the assigned post nor allow anyone to tamper with the switches until ordered to do so by the officer in charge.

Safety Switches

There are usually two safety switches controlling the electric current to the elevator; these are for the protection of maintenance personnel working on the machinery. One switch will probably be located on the top of the car, the other at the bottom of the shaft. These may be operated to deactivate the elevator.

Stop Switch

A stop switch on the interior elevator operating panel controls the current to the operating machinery; it must be ON in order for the car to operate. It may be labeled EMERGENCY SWITCH or EMERGENCY. Some elevators are equipped with a toggle switch that is clearly labeled ON and OFF. Other control panels have a button that is usually colored red; when this button is pushed in, it is in the OFF position and it must be grasped and pulled out to restore power.

Occupants of a stalled elevator should be asked to check the control panel to see if someone has accidentally or mischievously touched the stop switch. Before rescuers attempt forcible entry, passengers should place this switch in the OFF position to keep the car from moving.

BASIC ELEVATOR RESCUE

Several basic elevator rescue techniques can be adopted by any emergency unit called to a building where passengers are trapped in a stalled elevator. While these procedures appear simple, they should be preplanned by inspecting the elevators in the structures to which response may be made and by then deciding which are the best methods for handling possible emergency calls. This preplanning includes learning where the emergency keys are kept, how to take emergency control of the cars, what type of interlock devices are on the doors, the best method of releasing the interlock devices, where the electrical power switches are located, and many other details important in case of an emergency.

Locating the Car

Before any rescue attempt can be made, the stalled car must be located. There are six basic ways to find it:
1. Hoistway doors in the older models are usually equipped with glass windows. Visually checking the elevator shaft with a flashlight will often locate the car.
2. The first-floor lobby position indicator (if provided) will indicate the location of the car.
3. If the car telephone or intercom system is operable and a complete power outage has not occurred, the interior car position indicator will give the correct information to the passengers, who can then relay it to the rescuers.
4. In cases involving a multicar hoistway, if elevators adjacent to the stalled car are operating, use one to ride up; stop frequently to look across and upward between the car and the hoistway door to locate the stalled car.
5. Locate the machinery room and check the selector equipment for the exact location of the stalled car. With a little practice, one can tell the location of a car with this equipment, even if the car is between floors.
6. In a single-car hoistway, the best method for locating the stalled car may be to look upward from the bottom pit or downward from the top machinery room.

Checking the Cables

On electric elevators, check the hoisting cables to be sure they are not slack before attempting to remove the passengers. Slackness in the cables indicates that they may be damaged or broken, and that the car is being held in place by safeties—in these conditions the car may move at any time. The car should be supported or blocked into position before attempting to remove the passengers. Remember that the car should never be moved upward when hoisting cables are broken or damaged. The car is being held by the safeties and the slightest upward movement could cause the brakes to dislodge; the car could drop if the safeties failed.

Cars are best secured by gaining entry to the area above the car by opening the next higher hoistway door. Then secure the crosshead of the car with cables or ropes to the shaft cross beams, building columns, or any other secure building member.

After the car crosshead is secured, the car should be further secured by props or jacks from below. It must be remembered that this bracing and lashing is to provide extra security and safety, not to raise the car.

Moving Hydraulic Elevators

Some hydraulic elevators may be moved to a lower level by bleeding off the pressure in the machinery room. When the floor landing level is reached, the doors should open automatically, if the car still has electrical power. This method should not be attempted unless someone who is familiar with the system is present.

Rescuing Passengers Who Can Cooperate

Establish contact with the car occupants and have them check the interior operating panel to be certain that the emergency switch is in the ON position. Ask them to turn the switch to the OFF position and then back to the ON position to be certain that a good contact is made.

Instruct a passenger to push the DOOR OPEN button on the car panel; a rescuer should simultaneously press and hold the landing call button. This action should reinforce a weak electrical contact, if this is the problem. If the car is within floor range, the doors may open; if the car is out of the floor range, it may move slightly until it is in position.

One of the most common causes of a car's stalling between floors is a poor contact in the electrical switch at the hoistway interlock. Often the interlocks become weak or dirty and the movement of the car will cause the points to lose contact, thus stopping the car immediately. With the power ON, have a passenger shake the inside car door; emergency personnel should simultaneously shake the exterior hoistway doors nearest to the car. If the car does not resume operation, then shake the hoistway door where the passengers entered and the doors in between. Next, shake each hoistway door on all landings; if this is not practical because of the building height, shake as many as possible. Should a car be at or near a landing, and thus be within floor range, the door may open. In other cases, the car may proceed to the next programmed floor and the door may open as usual.

When the doors are equipped with a photoelectric beam to prevent them from closing on entering passengers, and if the car is within floor range, the doors may be opened by interrupting the light beam. These photoelectric eyes are usually placed in the doors at calf level and hand level. The beams may be interrupted by sliding a piece of stiff thin paper or cardboard between the doors or between the door and the jamb.

Have a passenger turn the emergency switch to OFF and send someone to the machinery room to turn the main power switch to OFF. If the car is within floor range, the door locks may have already been released. In these cases, merely pulling with hand pressure from the landing side may open the doors. Instruct the passengers to push or pull open the interior car door; this may open both doors. If the car is beyond floor range, only the inside door will open. It will then be necessary to instruct the passengers to move the pickup roller or link in such a manner that the hoistway door interlock is released; this requires little pressure, and the exterior hand pressure from the landing side will help to open the door.

ADVANCED ELEVATOR RESCUE

If the preceding basic rescue procedures are not successful, it may be advisable to wait for an elevator mechanic to free the passengers. Under most conditions, removing the occupants of a stalled elevator is serious, exacting work. It takes a lot of experience to examine an elevator and make sure that it is safe; it requires skill and good judgment to release trapped passengers.

Unless there is a fire in the building, an injured or sick passenger in the car, or imminent danger of a building collapse, explosion, or earthquake, the safest course of action is for the occupants to remain in the car. Check to be certain that a qualified elevator mechanic has been called and is responding to the call. Soothe the passengers and make them aware of what is happening; this should help avoid a state of hysteria or panic. It is better to leave the car occupants uncomfortable and safe than to jeopardize them by employing inexperienced rescuers with improper equipment.

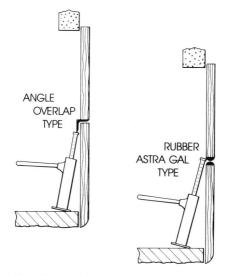

Fig. 8-3. Freight elevator doors that move vertically can usually be forced open with a hydraulic jack, hydraulic rescue tool, pry bars, or other spreading tool.

If there are a great number of stalled elevators because of a general power failure, if a medical problem is complicating the situation, or if there will be a lengthy delay before a mechanic can arrive, then more advanced techniques are justified.

Using pry bars and brute force to open hatchway doors may cause hundreds of dollars' worth of damage and keep the elevator out of operation for many days while repairs are being made. Rescuers should take time to study the alternative means of opening the elevator doors, since this will save time, effort, and expense in the long run. The automatic door locks or door-opening devices cover a wide range of variable designs, are electrically operated, and can function to open both the car doors and the hoistway doors at the same time.

Most automatic elevators are equipped with hoistway door locks that are automatically released by a cam on the car when the car is at, or within, a few inches of floor level (Figure 8-3). If the elevator is this close to the floor, opening the hoistway doors can be accomplished by merely pulling on the doors. When power is available, and the car is at the landing, the doors may be opened by pressing the hall button or by instructing the occupants in the car to press the DOOR OPEN button on the car operating panel. There are times, however, when the car will be stuck near a floor but will not be within the range in which these methods are successful.

There are many types and designs of elevators with locking devices mounted on the inside of the doors in different locations. It is impossible to describe them all in detail. Possibly the most important thing to remember is that all hatchway doors are unlocked on the inside by a cam or roller that projects away from the door toward the elevator car; by pressing the cam down or by pushing the roller aside, the door can be opened. The first step in releasing a door lock is to obtain access to the elevator hatchway. Sometimes the releasing mechanism can be tripped with a rod, pole, or hook after opening the door on the floor above the stuck elevator, or by positioning an adjacent car that uses the same hoistway alongside, or by forcing the door open slightly and working through a crack (Figure 8-4). If all else fails, a hole in the door or adjacent wall may have to be breached.

If the decision is made to continue attempts to remove the passengers, reassure them. Be certain that both the emergency power switch in the car and the main power switch are turned off, and that someone is stationed in the machinery room to be sure that no one activates the power unless directly ordered by the officer in charge. Serious injury could result if the power was not disconnected, or if some of it was restored while rescuers and passengers were attempting to enter or leave the car. The lighting and ventilation systems will probably remain operational, since they are normally connected to independent electrical circuits.

Securing the elevator by shutting off the power will prevent the motor from operating; this action will also cause the brake safeties to be applied. This should stabilize the car and prevent any movement. On older installations the condition of the brakes may be doubtful; under these conditions it is advisable to place a crowbar or piece of lumber between the spokes of the hoisting sheave to prevent rotation. If there are no spokes, attempt to wedge the sheave so as to hinder movement.

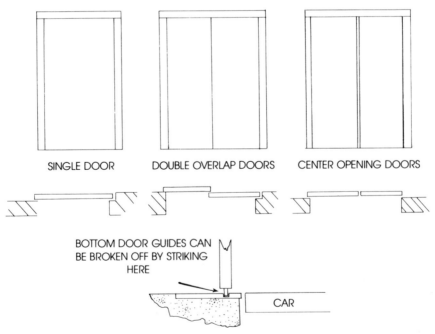

Fig. 8-4. Elevator doors are of three types: single, double overlap, and center opening. Each type has its peculiarities that must be considered when forcing the doors open. They do have one common denominator, however: all passenger elevator doors have bottom door guides that can be cut off with a saw or broken off by striking with a sledgehammer. Once the bottom guides are severed, be careful not to swing the doors in so far that they are released from the upper tracks and fall into the elevator shaft. Remember, too, that the severed doors will not prevent anyone who leans or falls against them from falling into the shaft.

Operating Door Interlocks

If a multicar hoistway is encountered, bring an adjacent car level with the stalled car. A crew member in the rescue car can insert a 6-foot (1.8-meter) pole or stick between the striking post and the door buck of the car to trip the door lock on the stalled car. Keep in mind that regardless of make or model of the elevator, the rescuer should always push the roller or cam away from the leading edge of the door to disengage the lock.

When a single-car hoistway is encountered, the lock can sometimes be tripped by forcing the door open slightly and then using a long hook fastened from a coat hanger or other stiff wire to trip the lock.

If the hoistway door on the landing above can be forced open, the door lock can probably be tripped with a pole or stick.

Removing Passengers

Frequently an elevator will stall between floors and the usual methods of car movement and relocation will be unsuccessful. There are fewer safety problems and chances for injury if all rescue work on the stalled car is performed from the floor above the car, rather than from the floor below. If the car is located closer to the lower floor, it should be used as the base for rescue operations, rather than attempt to pull passengers upward through a tight narrow space. Occupants should not be squeezed out through the doors if there is less that 3 feet (0.9 meter) of space between the car opening and the landing floor.

When working from the floor above the car, place a short ladder into the car and lower it to the passenger level. One rescuer should enter to assist the occupants. Other crew members should then place themselves at the top of the ladder to help the passengers as they reach the floor landing. Should fear, panic, or hysteria exist among the occupants, fitting a life belt and line to the troubled persons for support may give them confidence.

Working from the floor below the car exposes the open shaft, multiplying the safety hazards of the rescue operation many times; it is imperative that the open hatchway be barricaded and securely protected to prevent anyone from falling down the open shaft. Place a ladder between the car floor and the landing. Passengers may have to lie face down and slide out backwards until their feet are on the ladder.

Forcible Entry

If all else fails, prying or cutting can be undertaken. The best policy is to wait for a qualified mechanic to arrive, rather than to undertake this type of operation without competent help. Unless the hoistway doors are provided with an emergency unlocking device operated by a special key to open the door on the landing side, the hoistway doors may have to be forced open by an axe or similar tool to permit entry for removal of the passengers. Whichever method is decided upon, every effort must be made to keep the door on the track and hanger from which it is suspended. To remove a door completely from the hall side and try to keep possession of it is an almost impossible task. Trying to do this might result in the door's falling and striking the top of the car, doubtless causing a further deterioration in the composure of the trapped passengers.

The interlocks on automatic elevators are usually mounted near the top of the door, on the jamb side of one-speed or two-speed sliding doors. For center-opening doors the interlocks are near the top where the two doors close together. Forcing the doors is only possible if the power has been turned off. On side-opening doors, pry at the top on the side where the door closes. On center-opening doors, pry between the doors at the top. Prying the door open should break the small screws that hold the latch in place. When the exterior door is open and the current is off, the inside door can be opened manually from the inside or outside.

In some cases, cutting a hole in the hoistway door is the course of action that will cause the least damage. With an air chisel or rescue saw, cut a hole large enough so that a hand can be inserted to trip the interlock.

The bottoms of the hoistway doors of newer elevator installations are equipped with track guides called *gibbs*. These guides are fiber or metal L-shaped blocks, which are generally held in place by two small machine screws. There is usually one set of gibbs for each door; on very large doors there may be more. Their purpose is to stabilize the bottom of the hoistway door. The most efficient method of cutting these guides is by inserting a sheet metal saw at the bottom center of the door and cutting each guide along the door access. A single hacksaw blade may be required if a thick carpet or kick plate in front of the doors prevents insertion of a sheet metal saw. If the saw binds while cutting, a slight upward pressure with a crowbar or claw tool may be applied beneath the door.

If a sheet-metal saw is not readily available, the door can be forced by breaking the gibbs with a sledgehammer; little or no damage will result from this method. Place a block of wood at the gibbs location at the center bottom of the door and strike it several times with a sledgehammer.

After the gibbs have been cut, remove the saw and carefully apply pressure at the lower center of each door; the doors will push inward into the hoistway and offer a good view of the shaft. Extreme caution should be used, since the doors can be pushed from their tracks and fall into the shaft. Distances of over 18 inches (45 centimeters) may dislodge the door from the overhead tracks. When released, the doors will return to their normal position, and negligible damage has been done. Once the gibbs have been cut, door operation will cease and the unit will be out of service. Place a barrier across the doors to prevent people from entering the elevator.

The purpose of cutting the gibbs is not to permit the physical entry of rescuers into the shaft, but to permit observation and allow the rescuer to trip the hoistway door interlocks on that landing for the stalled car.

If entry cannot be made into the shaft by any other method, it may be necessary to breach a wall. Make the hole where the least amount of building material will be encountered and so the hoistway door lock will be within arm's reach. If the doors cannot be disengaged by pushing the roller or cam because of some obstacle in the hoistway, sometimes the excess aircord or chain of two-speed sliding doors, (which runs along at the tops of the doors) can be pulled down far enough so that it can be cut with bolt cutters. Accomplishing this will disconnect the slow-speed door from the fast-speed door. Since the lock is always fastened to the fast-speed, or leading, door, this will permit entrance into the shaft through the slow-speed door opening. This method could also be applied to some single-speed, center-opening doors because the cutting of the aircord or chain would release one door. Cables should not be severed if the lock can be reached and released.

EMERGENCY EXITS

If the preceding solutions have not proved successful and the passengers have not been rescued, the car emergency exits may be the answer. There are two types of these exits: they are located at the sides and the top of the car. Before using either, the electrical power should be checked to be sure that it is still in the OFF position.

Top Emergency Exit

The removal of passengers through the emergency exit in the car top should be used only as a last resort, because it is difficult for aged or infirm occupants to climb ladders, and because the mechanism on the top of the car leaves very little space for a person to stand.

The rescue squad will reach the top of the car by opening the hoistway door at the floor above the car. The top exit panels are held in place by thumbscrews with screwdriver slots on the outside end; they are so arranged that the exit cover can be opened from either the inside or the outside of the car. The cover is either hinged or secured to the car by a chain; it will open outward and the space above the cover is unobstructed by equipment. Frequently, a rope or chain ladder is provided in a metal box on top of the car, with one end of the ladder fastened to the car structure. If no ladder is provided, or if the passengers will have trouble climbing a limber ladder, a short ladder may be lowered through the access hole. Pulling passengers up by their arms should be avoided, since this can be painful and lead to injury.

One rescuer should descend into the car to steady the ladder and assist the passengers, and a second should be stationed on the car top to assist passengers onto the ladder leading to the floor above. Although the space above the top exit is supposed to be unobstructed, it may be necessary to step over the car crosshead or other equipment. Care should be taken to prevent stumbling or injury. A third rescuer can then assist the passengers off the ladder onto the floor above. It may help to place a safety belt with a line attached on the occupants before they start the ladder ascent, to steady them and to prevent hysteria or injury.

Side Emergency Exit

When there is a bank of elevators with two or more cars traveling in the same hoistway, passengers may be removed through the side emergency exit if the adjacent elevator car in the hoistway is still operating.

Side emergency exits are either hinged or removable; the exits open inward. The removable exit is held in place by at least four fastenings, so arranged that they can be operated by hand from both the inside and outside of the car; fastenings are designed so they cannot readily be removed from the panel. The hinged panel is provided with a lock arranged so that the catch can be operated from the inside of the car by a removable key, and from the outside of the car by a nonremovable handle. The key is kept on the premises by the person responsible for the maintenance and operation of the elevators and in a location where it will be readily available to qualified personnel in case of an emergency, but not accessible to the general public. Side exits are usually at least 16 inches (40 centimeters) wide and at least 5 feet (1.5 meters) high; they are so located that there is free access to the side exit of the adjacent car.

If one car is stalled between floors, the adjacent car can be run up to the same level, the side exit in the rescue car opened with the key, and the side exit in the adjacent stalled car opened from the hoistway side by using the nonremovable key or by operating the fastenings of a removable hatch.

The passengers may then be assisted from one car platform to the other across a short plank. After the occupants have been transferred, the side exit doors should be closed and fastened, the plank or any accessory equipment removed from the hoistway, and the elevator run down to the main floor.

EMERGENCY SERVICE PROVISIONS

The first concern with elevators in an emergency is to stop normal operation. When an EMERGENCY SERVICE key switch at the main floor lobby panel is turned to the ON position, elevators immediately close their doors and travel nonstop to the main floor if stopped at a floor with doors open. They will continue nonstop to the main floor if already moving in that direction or they will stop and reverse without opening doors if moving away from the main floor. When an elevator is on ATTENDANT OPERATION, a signal in the car tells the operator to return to the main floor. The same sequence is initiated if one of the heat- or smoke-sensing devices at an elevator landing is activated.

When an elevator has been taken from normal service and brought to the main floor it is no longer hazardous, but it is not very useful either. The key used to park elevators at the main floor can also be used in the elevator car to operate a key switch that permits the elevator to be operated from the car only. In this emergency operating mode the car will not respond to hall button calls or operate doors automatically; it is completely controlled by the operator in the car. The operator can make on-the-spot decisions about what can be done safely with the elevator, and has the knowledge and equipment to escape from the car if the elevator fails.

If an elevator starter or maintenance person is present, this person should be instructed to place a passenger car on independent service. This can be accomplished in a matter of seconds at the lobby floor starting panel or at the station panel in the car itself. This procedure will place the operation of the car solely in the hands of the operator and reduce the chances of the car's going to the wrong floor (such as the fire floor). This procedure also permits operating personnel to leave the car unattended and ready for their immediate use.

Freight Elevator

The use and selection of elevators by emergency personnel should receive thorough consideration. Because of its independent service features, the freight or service elevator should be the first choice.

Most large buildings contain freight elevators; many of these have direct access to outside service areas. Freight elevators have the following advantages over passenger elevators: they bypass the main lobby and other congested areas; they often originate near outside loading docks where rescuers and apparatus can most easily respond to orders; they are not controlled by computers; their operating controls are simpler than the controls of passenger elevators; complete manual control is easy; building occupants are not competing for their use; and they service the entire height of the structure. Some fire departments have a standard operating agreement with the management of structures equipped with freight elevators, that a freight elevator will be attended and waiting at the ground floor any time a fire or other emergency incident is reported.

Fire

Occupants and visitors of a structure tend to leave it the same way they entered. This could prove deadly if too much dependence is placed on the use of elevators during a fire. Today most elevators are automatic, with no attendant on hand to police emergency evacuation of a building via the elevators. If people press the call button at the floor where there is a fire, an elevator will stop there automatically, exposing its occupants to fire, heat, and smoke. If a crowd of people trying to enter a car blocks its doors, automatic safety devices may keep the doors from closing and the elevator from starting; this will delay or prevent movement of the car. Electric-eye devices are often used to prevent doors from closing on entering passengers; smoke will intercept the light beam just as a human body does, and hold the doors open.

Most fire departments recommend that the elevators in buildings where there is a fire or the threat of a possible fire should not be used until the cars and shafts have been inspected and deemed safe. Until the elevators and shafts have been inspected and determined to be safe, all cars should be summoned to the ground floor and inactivated with the STOP switch. A fire fighter should be stationed in the lobby to prevent their use.

Before entering the elevator, check to see whether there is smoke in the shaft or whether the car shows any indication of having been exposed to excessive heat. Failure to detect smoke in the shaft by shining a flashlight up between the car and the wall does not conclusively prove that there are no hazards. If the fire appears to be extensive, or if the cars and shafts may be involved, the elevators should not be used until they have been checked for safe operating conditions; it will usually be necessary for the initial fire attack company to climb the stairs to the fire floor and make a close inspection. After the elevators have been checked, the command post should be informed about their availability.

Taller structures are provided with two or three banks of elevators; they seldom have cars that service the entire height of the building. The low-rise elevators service only the lower floors and the high-rise are restricted to the upper floors. The hoisting machinery is generally located on the floor above the highest story serviced by a particular bank of elevators. When a fire involves the upper floors, and caution in using the elevators is warranted, the low-rise bank may safely be used to its limit. There are usually no openings from the high-rise shaft into the lower intermediate floors. When the fire is in the lower stories, and the shaft is not involved, it may be safe to use the high-rise bank of elevators to get above the fire.

Fire fighters have been killed by using an elevator to check for a reported fire; if the doors open and excessive heat and smoke are present, it may be impossible to move the car or escape from the floor. If the decision is made to use the elevators on the initial size-up, the fire fighters should be taken two floors below the reported fire floor and the stairways used to proceed further. It is good practice to use the STOP switch to halt the car before the doors open automatically; the inside doors can then be forced open slightly so the outside doors can be checked for excessive heat.

When there is a fire or other emergency, control of all elevators should be taken as soon as available personnel permits; place the elevators on manual control and provide a fire fighter operator for each one. Elevators being used to evacuate occupants should be operated by emergency personnel to allow the maximum number of people to be transported in the shortest period of time.

FIRE ONLY Button

The FIRE ONLY button will override the photoelectric cell and will automatically close the door even when the beam of light is broken. The button must be depressed until the car door is completely closed. Some doors will close completely if the doors are within 1.5 to 3 inches (3.75 to 7.5 centimeters) of the jamb, even if the button is released; others must be closed completely before the rescuer can safely release the button. If the pressure is removed from the button before these minimum distances are reached, the doors will automatically reopen.

Occupant Evacuation

When there is a fire alarm, the building occupants should be directed to leave the building by the stairways. The basic objections to the use of elevators as emergency exits are:

1. Persons seeking to escape from a fire by an elevator may have to wait on the landing for some time, during which time they may be exposed to fire or smoke, or panic may develop.
2. Automatic elevators respond to the pressing of buttons in such a manner that it is possible that an elevator being used to descend from floors above a fire might stop automatically at the floor involved in the fire, and that the doors might open automatically; this would expose the occupants to fire and smoke. Once a car stops at a floor engulfed in flames and smoke it is not likely to leave, because the heat will probably cause the elevator to malfunction after the doors open. Following a fire, it is not unusual to find several inoperative elevators at the fire floor.
3. Modern elevators cannot start until the doors are fully closed. A large number of people trying to crowd into an elevator in an emergency might make it impossible for the elevator to start. In most cases, heavy smoke will delay the closing of elevator doors equipped with an electric eye. This is also true of car doors equipped with electronic door detectors. These electronic detectors, used instead of electric eyes, are also subject to failure caused by heat and heavy smoke.
4. Any power failure, such as the burning out of electric supply cables during a fire, may render elevators inoperative or may trap persons in elevators stopped between floors. Fire conditions may not allow time to rescue trapped occupants through emergency escape hatches or doors.
5. Seconds count in some instances. Even if a person remains totally rational while waiting for an elevator during a fire,

alternate escape routes may be cut off by toxic products of combustion accumulating in passageways, by doors protecting passageways being opened to fire and smoke from nearby areas, or by any of a number of other events that a totally rational, alert person has no way of knowing about. The elevator may never arrive.

Fire in the Elevator Hoistway

If a fire should develop or spread into an elevator hoistway, remember that there is very little in a modern shaft that can burn. Therefore, a fire in the hoistway will usually be of limited size, and smoke will be the principal hazard. However, heat and smoke from burning materials on the lower floors will readily rise in the shafts after the hoistway doors fail. All cars in a common hoistway exposed to fire or smoke should be brought to the ground floor and evacuated.

If the fire is in the elevator pit or under the car, the cars in the common hoistway should all be brought to the next to the lowest landing and secured. The fire department can then enter the shaft at the lowest landing by using the hoistway unlocking device or emergency key, or by forcing the doors.

Access may be obtained in the same manner at the top landing, if the fire is in the upper part of the hoistway. When opening any hoistway door, remember the hazard of an open shaft.

Fire in the Elevator Machine Room

If a fire should develop in an elevator machinery room, pull all main line disconnect switches first; these switches should be near the machine room door. Do not use water on the fire, because of the hazards of electrical shock; use carbon dioxide, a dry chemical extinguishing agent, or some other nonconducting extinguishing agent approved for electrical fires.

Earthquake Damage

Earthquakes usually have recurrent shocks and can cause the following problems: elevators may stop between floors and trap the occupants; counterweight guide shoes and rails can become loosened; machinery room equipment can become damaged; failure of the suspension system may occur; beams may be loosened, deflector sheaves can become released, or ropes and cables can become broken or slackened; hoistway doors may bind so they cannot be opened; or the door locks may be broken, which would allow the doors to sag and leave an open shaft.

If an earthquake should occur, proceed as follows: stop all elevators, with a single disconnect switch if possible; inspect all elevator mechanisms and equipment thoroughly after the probability of recurrent aftershocks has passed; do not operate the cars through the shaft at fast speeds until the hoistway and machine room equipment has been thoroughly inspected by competent elevator personnel; do not carry passengers on elevators until the inspection has been completed and all necessary repairs have been made.

In case of an air raid or enemy attack, follow the same procedure as for an earthquake. If a short warning is given, bring all elevators immediately to the ground floor and disconnect power to all cars. If the building suffers any significant shock or damage from a blast or concussion, secure all elevators until a complete inspection is made.

During the mass evacuation of a structure, the elevators should be operated by an assigned trained attendant. A qualified building official should decide whether any elevators shall be left running for the use of people who stay in the structure.

Pinned Victim

The method used to remove a person caught between the elevator and the hoistway may save the victim's life or minimize the extent of injuries. Before taking any action, analyze the problem carefully. The following questions should be asked: Are the ropes and cables slack? Should the car be supported to make sure it will not move while the rescue is taking place? Where and how is the person caught? In what direction was the car moving when the accident occurred? Can the injured person be removed by moving the car up or down?

Geared machines are provided with a crank in the machine room that fits over the motor shaft after the guard over the end of the motor is removed. By turning the crank by hand the elevator can be slowly moved up or down. A hydraulic rescue kit with the spreader attachment would probably be the best means of forcing the car far enough to allow the victim to be removed. If no other means is available, sometimes the car can be moved far enough by prying with a lever or heavy timber. The car can also be moved electrically from the machinery room by trained and qualified elevator mechanics who are thoroughly familiar with the controlling equipment.

Some additional car movement from front to back may be obtained by removing the top guide shoes; these are usually fastened to the crosshead by four bolts.

If none of the foregoing methods can be used, the rescue squad will probably have to resort to a cutting torch. The prime consideration is the quick and painless removal of the victim.

QUESTIONS FOR REVIEW

1. When confronted with a stalled elevator, rescuers should give first consideration to the possibility that
   a. a fuse has blown.
   b. one of the passengers has inadvertently or maliciously operated one of the controls.
   c. electrical power to the building may be off.
   d. the motor may have burned out.

2. Before rescue personnel start any operations to remove trapped elevator occupants, which of the following procedures should they follow first?
   a. Block the car to prevent its falling
   b. Call the power company to check the electrical circuits
   c. Force open the doors to check the physical condition of the passengers
   d. Place the power switches in the OFF position

3. If an elevator car is stalled when it is level with the landing floor, which of the following procedures is most likely to open the door without any damage?
   a. Ask a passenger to press the DOOR OPEN button.
   b. Use a pry bar to force the door interlocks.

c. Break the gibb guides at the door bottoms.
   d. Remove the power fuses.

4. Elevator door interlocks can sometimes be tripped with a
   a. pry bar.
   b. battering ram.
   c. long stiff wire.
   d. air jet.

5. The FIRE ONLY button in modern passenger elevators is designed to
   a. allow the car to progress more rapidly under emergency conditions.
   b. allow the doors to close automatically under smoky conditions.
   c. restrict elevator car operation to the exclusive use of emergency personnel.
   d. stop elevator car operation when the shafts may be on fire.

CHAPTER 9

Rescue From Electrical Contact

Everyone has seen lightning and is aware of the injuries and deaths that it can cause. The most dramatic and feared effects of electricity are those caused by lightning. Less exciting, but more frequent and responsible for a greater number of serious injuries and fatalities, are the smaller electrical shocks related to inadequate shielding or insulation, old equipment, improper grounding, accidental contact with energized wires or equipment, or general ignorance of how to work with electricity and electrical equipment. In one recent year 1,157 fatalities occurred in this country from the misuse of electric current. More than two million people suffer serious burns from accidents involving electricity in this country every year.

Emergency medical technicians (EMTs) must be thoroughly instructed in basic electrical theory and the hazards of electricity before they can accomplish a safe rescue without endangering themselves or others participating in the operation. If it is necessary to clear an energized wire from a victim, or to release the victim from live equipment, techniques have been established that provide relative safety to the rescuer who performs them carefully.

The electrical shock or burn victim requires prompt action because in a matter of moments tremendous damage can be done to the skin, vital organs, and underlying tissues from contact, explosion, blazing arc, or fiery flash. While paramedics should act quickly, they should also act with caution to prevent themselves or others from being injured or killed during their efforts to administer emergency care to the victims of an accident involving electricity (Figure 9-1).

ELECTRICAL PRINCIPLES

All emergency personnel must have a good general knowledge of the principles and hazards associated with electricity.

Electrical Currents

Electricity consists of a flow of energy through conducting substances. Most metals, water, moist ground, and the bodies of people will conduct a flow of electricity with varying degrees of efficiency. Nonconductors, such as dry air, ceramics, glass, and other such nonmetallic substances, are used as insulators. Generally, the colder a conductor is kept, the less resistance is present; however, sufficient heat can change an excellent insulator into a conductor. Personnel must be cautious at fires because the dielectric properties of an insulating material can be lowered sufficiently by excessive heat to allow dangerous quantities of current to flow where it will be least expected.

Current flow can be continuous and in one direction, or can reverse its direction frequently. The former is direct current (DC), and the latter, alternating current (AC). The effects of alternating current are of most concern to paramedical personnel because alternating current is by far the current most often used to power electrical devices. Furthermore, AC current is more dangerous than DC because of the damaging effects of prolonged exposure to rapid reversals of direction of flow and polarity across biological membranes. In the United States, 60-cycle AC, which is now referred to as 60 hertz or Hz, is used. This means that the current flow is reversed 120 times per second.

Electrical Terms

Ampere: a measure of current flow. Extending the analogy to water, measurement of electricity in amperes may be likened to the measurement of volume of water in gallons per minute.

Impedance: relates to an AC circuit. It is analogous to resistance in a DC circuit.

Fig. 9-1. An extremely hazardous condition was created when a truck knocked down a power pole, resulting in high tension electrical wires and a transformer's falling onto a residential street. In such cases, rescue personnel should not attempt to cut or move energized electrical wires, but should immediately call for utility line workers, and keep people from entering the hazardous area until the current has been turned off. Photo shows a trained utility line worker cutting the wires to create a safe condition. If victims are trapped in a vehicle by fallen wires, they should be told to remain in the car until it has been declared safe. (Courtesy of the Los Angeles City Fire Department.)

Ohm: the measure of resistance or impedance. One ohm is the resistance through which a potential of one volt produces a current of one ampere.

Resistance: the property of a conductor to impede or oppose current flow. It may be compared to friction.

Voltage: the force driving electrons through a conductor, the electromotive force. Measured in volts, it may be compared to water pressure expressed in pounds per square inch to describe the force driving the movement of water.

Ohm's Law

The amount of current flowing between two points of different electrical potential varies inversely with the intervening distance. The greater the distance between two points, the greater the resistance. Thus, large amounts of current flow when the resistance is low, and only small amounts flow when the resistance is high and the voltage remains constant. These relationships have been summarized in Ohm's law, the formula for which is $A = V/R$ or $V = A \times R$. In this expression, current in amperes (A), is directly proportional to voltage or electromotive force (V), and inversely proportional to the resistance (R) measured in ohms.

Ordinary line voltage fluctuates within a fairly narrow range around 120 volts. Therefore, the most important variable in applying Ohm's law to clinical medicine is skin resistance, since it will determine current or amperage, the primary cause of the biological effects of electricity.

An electric current consists of a flow of electrons through a conducting pathway or circuit. For current to flow from one point to another along this conductor, the circuit must be complete (closed) with a difference in potential (voltage) existing between the two points. Completion of a circuit from the energy source to the ground or to another conductor will allow a continuous passage of current through the wires. Because a complete circuit from the source through wires to the electrical equipment, and then back to the source is necessary before electrons can flow, a break in the circuit will halt the current flow; this is what happens when a switch is opened. For example, if a high-tension cable breaks, current will not flow until the circuit is restored again through some conductor. In this case, the circuit may be completed again when the broken end of the wire touches the ground or anything else through which the current can flow to the ground. If a person touches an energized wire, a circuit may be completed and the current will flow through the person to the ground.

Since electricity readily flows through water, wet ground, puddles, and damp insulating materials, extremely hazardous situations can occur at rainy or wet rescue scenes. Many persons have been electrocuted because they were standing on damp ground or in water while working around charged electrical equipment.

Electrical Utilities

Circuits carrying electricity from power companies to consumers are divided into transmission, distribution, and service circuits. Transmission and distribution circuits carry high voltages; service circuits carry low voltages.

Transmission circuits are the wires from the generating plants to the distribution points. Line voltages are usually above 66,000 volts.

Distribution circuits refer to the wiring from the distribution points to the transformers which are servicing the consumers. Line voltages vary from 600 to 66,000 volts.

Service circuits contain the wiring from the low voltage side of the transformers to the consumers' homes or places of business. Line voltages are 600 or less. These low voltages should not be regarded as safe because current with a voltage as low as 110 can be lethal to a well-grounded person.

HAZARDS OF LIVE WIRES

Emergency personnel are often dispatched to locations where high-tension electrical wires that should be suspended on poles and crossarms are sagging or have fallen. Even when no one is injured or immediately jeopardized, this is an extremely hazardous situation that constitutes a menace to the public.

A fallen high-tension wire may charge the ground surrounding the wire and make it possible for a person to be electrocuted without touching anything. Higher voltages and a high moisture content in the soil increase this hazard, and pools of water are as dangerous as the wire itself. Metal doors, wire fences, and vehicles are similarly dangerous. A live wire may whip around in much the same way as a garden hose discharging water through a nozzle. However, the fact that a wire is not whipping, sputtering, or sparking should not be taken to indicate that it is dead. *Rescuers should assume a wire is live unless there is definite reason to believe otherwise.*

When called to the scene of an accident involving electricity, rescuers should notify the power company immediately. Unless a victim who is still alive must be rescued, no attempt should be made by rescue personnel to touch, move, or cut any wires. Assume that every wire is energized and dangerous. Never cut any energized electrical wires if this can be avoided. Anyone attempting to cut the wire will be too close to the electrical charge and may receive severe burns from the flash or be wrapped into the ends of the wire when it whips free. Whenever any wire must be cut, use the proper tools, wear line workers' rubber gloves, and observe all safety precautions. Reach up as high as possible to cut the wires so the dangling ends will not remain a hazard. Cut the wires one at a time and then bend the live ends away from each other so there will be no danger of their touching. Clip the wires on both ends, and then coil and remove the section of wire so it will not continue to constitute a hazard.

Wires may fall on fences, automobiles, railroad tracks, metal buildings, or other conductive materials. If this happens, a lethal charge of electricity could cause death or injury at a considerable distance from the downed wire. In Southern California a traffic accident caused a transmission line carrying an extremely high-voltage cable to drop across a freeway. Since the metal chainlink fence in the center of the freeway was supporting the energized wire, a radio warning was broadcast to alert everyone within several miles of the accident to avoid touching the fence, because it could transmit a lethal charge of current.

A fallen wire lying in a puddle can energize the entire ground area. Electric arcs and flashes resulting from such a situation can produce skin burns and eye injuries, even when the victim has no direct contact with the wire.

Broken wires may work loose from the poles and crossarms, causing the next span to drop; therefore, all rescue personnel not actively engaged in rescue operations and all spectators should remain at least one full span away from either side of the break. Vehicles and equipment should also be moved to a safe location. At night rescuers must use additional caution, because it is difficult to see the wires and because electrical arcs and flashes may damage the eyes.

It is extremely hazardous to approach any fallen high-tension wire, since the earth may be energized for a considerable distance around the area where the wire touches the ground. It is impossible to say what will usually be a safe distance, because the size of the hazardous area varies according to the voltages involved and the ground conditions. The dangers are greater during damp or rainy weather.

Circuit breakers are used to de-energize the wires if the current flow exceeds a predetermined quantity, as would happen if there were a short circuit caused by a broken and downed line. Most of these devices will automatically reestablish the circuit if the short was caused by a momentary overload. De-energized wires lying quietly on the ground may suddenly become energized and start to whip around without warning; therefore, it is safest to consider all wires as potentially live and dangerous.

Many persons are killed or injured when they inadvertently contact high-tension electrical wires. Among the recent victims accidentally touched the overhead wire that provides power. Other youths were killed or injured by a subway third rail or while retrieving kites or bird nests in the vicinity of high-tension wires. There have also been numerous casualties among persons installing television or amateur radio antennas, who allowed the equipment to touch a power line.

When a rescue company is nearing the scene of an electrical emergency, they should approach the area with caution. Stop the vehicle well away from the fallen wires, since they are difficult to see, particularly during stormy weather. Be very careful when descending from the vehicle, because a wire lying in a pool of water may have rendered the entire area lethal. If any portion of a vehicle is touched by a live wire, the entire vehicle will probably be energized. It is improbable that the contact between the vehicle and the earth will be efficient enough to cause the electrical charge to form a circuit through to ground. No one should touch any vehicle, fence, or structure that could be energized.

Electrified Booby Traps

With the increase in burglaries during recent years, honest citizens have started devising new methods of protecting their belongings. A new type of protection consists of improvised devices rigged to electrocute uninvited guests and attached to windows, doors, gates, and other openings. Although effective, these illegal innovations unfortunately create a hazard for rescue and fire fighting personnel who may be forced to use these booby-trapped entrances for emergency access.

Because of the possibility of encountering these electrocution devices, all windows and other openings should be carefully inspected before rescuers attempt to enter a structure. If any suspicion exists, check for wires, sparks, or out-of-place pieces of metal. There may be wires strung across the window and a ground plate on the floor; it may be wise to probe first with a short pike pole. If any indications of a booby trap are discovered, turn off the power in the building at the main power panel and leave it off until the device has been inactivated.

Electrical Generating Stations and Substations

Safe and proficient action at fires and rescue incidents, whether in generating stations, substations, or in the field, depends upon knowledge and understanding of the potentials of electricity and the proper procedures to be followed. Rescuers must realize that extremely high voltages generate an electromagnetic field around the conductor. Since this force extends away from the conductor, death or injury can occur to rescuers without their actually touching the electrified wires and equipment.

Generating stations supply bulk electrical power (voltages in some areas exceed 345,000 volts) to substations, where it is then reduced to a lower voltage and distributed for local use. The power is conducted through a pipe or bar-type conductor (Buss bar), which can easily be mistaken for a handrail or guardrail. Pipes which are supported or suspended by porcelain insulators are electrical conductors, regardless of the angle at which they are mounted.

Rescuers confronted with a fenced and locked generating station or substation should not enter, because fenced-off sections designate extremely hazardous areas. Call for a power company representative. Entry or operations must not be undertaken without the specific guidance of a knowledgeable person. It may not be possible to de-energize the equipment.

Under no condition should metal ladders or metal equipment such as claw tools, lockbreakers, or other forcible-entry tools be carried inside the gates of stations because of the extreme hazard of accidentally touching an energized component.

Many switch and transformer rooms in large buildings and power utilities are equipped with built-in carbon dioxide extinguishing systems. If the door is kept closed, the gas will put out the fire. If the door is opened before the heat has had an opportunity to dissipate, the flammable gases may reignite when a fresh supply of oxygen is admitted.

Water in any form should not be directed against electrical equipment without the advice of the utility representative. While the use of a spray or fog stream may be relatively safe around energized electrical equipment so far as personal injury is concerned, the water that drains away from the location may cause short circuits in other locations.

Shutting Off Power

Whenever a building has been sufficiently damaged by a fire, explosion, or other incident that the electrical wiring and other devices are unsafe, the power should be shut off. Leave the power turned on as long as possible so that lights, elevators, and other electrical conveniences can be used. When it becomes necessary to interrupt the electrical service, it is best to shut off the power only in the section of the structure that is hazardous.

The best method of disconnecting the electrical service is to open the main switch. If the switch panel or box has been damaged or is wet, or if it is necessary to stand in water while opening the switch, under no condition should the switch handle be grasped manually. Use a pike pole, piece of rope, or some other nonconducting object or wooden-handled tool to operate the switch. Use tested and

METROPOLITAN RESCUE

Fig. 9-2. Fire fighter insulated from the ground and the wires by a dry ladder, cutters, and gloves is cutting the low voltage electrical wires so the fire fighters combatting a structure fire will not run the risk of electrocution. Cutting the wires at the drip loops where they enter the structure will deenergize the electrical circuits without causing the wires to drop into the street.

Fig. 9-3. The best place to cut electrical wires is at the drip loops, where current enters the building.

approved rubber safety gloves (if at all possible) when working around energized electrical equipment.

Another method of disconnecting the electrical service is to cut the wires on the outside of the building. It is best to adopt the rule that rescue personnel should never cut electrical wires unless human life is endangered. Trained power company personnel are usually available for a quick response to the scene at any time of the day or night. It is a bad practice to unnecessarily subject anyone to the dangers of possible electrocution (Figure 9-2).

Most departments do allow their personnel to cut the low-voltage (600 volts or less) service wires into a structure if utility workers are not readily available. The best place to cut the electrical service to a building is at the drip loops, which are the portions of the wires between the anchor insulators and the point where the wires either enter the building or descend down the wall (Figure 9-3). If the wires are cut at the drip loops, the wires between the pole and the building will not drop to the ground and jeopardize other persons.

MEDICAL EFFECTS OF ELECTRIC SHOCK

Electric shock refers to the effects and reactions produced by passage of an electric current through the body. There are three ways in which shock can kill or seriously injure a person who accidentally becomes part of an electrical circuit. The tissues of the body surface may be burned extensively; there may be severe deep burns with apparently minor surface damage; or vital internal organs of the body may be injured. The severity of shock and the consequent damage a person who becomes part of an electrical circuit can receive depend on several factors:

1. Quantity of current flow measured in amperes.
2. Tissues and organs lying in the pathway of current through the body.
3. Length of time the body is in the circuit.
4. Area of exposure.

The pathway of current through the body is important because of the type of damage that might result. Current flowing from one finger to another on the same hand would not pass through vital organs, whereas current flowing from one hand to the other or from one hand to the feet would pass through the heart and lungs. The electrical contact breaks down the resistance of the skin; the more extensive the burn, the greater the flow of current.

When an energized electrical line, even one of low voltage, touches the palm of the hand, the muscles tend to contract. This reaction causes a person to freeze to the line. Persons feeling their way through the dark in areas where they may accidentally contact a charged wire, fence, metal door, or other electrified object should explore with the backs of their hands rather than their palms. An electric current on the back of the hand will cause the elbow muscles to tighten, and this will pull the hand away from the contact.

You might think that a shock of 10,000 volts would be more deadly than one of 100 volts, but this is not necessarily so. Individuals have been electrocuted by appliances using ordinary house currents of 110 volts alternating current, and by electrical equipment in industry using as little as 42 volts direct current. Many people have accidentally touched an automobile engine spark plug and received a shock that was jolting, but not damaging. Spark plugs require between 25,000 and 40,000 volts to fire, but the amperage involved is extremely low. The real measure of a shock's intensity lies in the amount of current (amperes) forced through the body. Any electrical device used on a house wiring circuit can, under certain conditions, transmit a fatal current. Note that voltage is not a consideration, although it requires voltage pressure to make the current flow; the amount of amperage will vary, depending on the body resistance between the points of contact.

Skin Resistance

Ordinary line voltage fluctuates within a fairly narrow range around 120 volts. Therefore, the most important variable in applying Ohm's law to clinical medicine is skin resistance, since the resistance will determine the current (or amperage), the primary cause of the biological effects of electricity.

Skin provides a high-resistance barrier to the transit through the body of externally applied electric current. Dry skin has a higher resistance than wet skin and, therefore, offers less shock hazard. The danger is greatly increased when the skin is wet because moisture lowers electrical resistance and facilitates electrical contact.

Thick skin on the palms of the hands and bottoms of the feet offers a higher resistance to passage of electric current than thinner layers. Moistening of the skin with sweat or water reduces this resistance. Other body tissues also provide resistance or otherwise act to divert or diffuse a current that penetrates the skin barrier, but they offer considerably less impedance to current flow than does the skin.

Effects of Current

It is the amperage, rather than the voltage, that is the real criterion of shock intensity. One-tenth ampere is usually fatal if it lasts one second or more. If a person can break free from the current instantly, the shock duration may be too short to do much damage.

The smallest alternating current shock that a person can feel through intact skin is 1 milliampere (0.001 ampere). This amount produces only a slight tingling reaction. Applied to mucous membrane or to a break in the skin, both of which offer less resistance, the same amperage creates a more intense sensation. Increasingly uncomfortable sensations are produced by increments of current, up to about 16 milliamperes. Beyond that level, a current applied to skin will produce pain.

A person grasping a bare (uninsulated) wire carrying current, with 16 milliamperes traversing the limb, will be unable to relax the grasp because of persistent muscular stimulation. This is referred to as the "let go" current, although it might be more accurate to call it the "can't let go" current. In this situation, all the muscles are stimulated, but flexor action is dominant, since these muscles are stronger than the extensor muscles.

Amperages above 20 milliamperes may cause labored breathing and increased pain; unconsciousness may result. At 30 milliamperes paralysis of the phrenic nerves or of the nerve centers in the medulla part of the brain may halt respiration. Artificial respiration must be promptly started in an attempt to restore breathing.

At 100 milliamperes there may be sufficient heart stimulation to cause ventricular fibrillation. Larger quantities of electricity sometimes tend to become diffused and to follow numerous routes through or across the body. They are capable of causing burns and local tissue damage, which may be severe. Greater shocks produce sustained contractions of various muscle groups and vital organs. There may be burning of internal organs, even if the injuries at the points of electrical contact are small and insignificant.

There may be no pulse and a condition similar to rigor mortis may be present, because, at times, muscle spasms will occur to such a degree that the casualty is extremely rigid. These are manifestations of shock, and are not indications that the victim has succumbed. Manual types of artificial respiration and massage will aid in relaxing the muscles.

Damage to the Heart

Since the heart muscle is particularly sensitive to electrical stimulation, the cause of death from electrocution is most often ventricular fibrillation. For a current to do this damage, however, the victim must be grounded, there must be contact with a source of current, and the heart must be in the pathway the electricity subsequently follows to ground. The combinations of circumstances producing this catastrophic arrhythmia are numerous, and can develop any time the heart is included as a component in an unplanned electrical circuit.

One-tenth of an ampere (100 milliamperes) of electricity flowing through the body in a path that includes the heart will probably cause ventricular fibrillation. When this happens, the heart fails to pump blood and death quickly results. After the electrical hazard has been eliminated or access to the victim can be safely accomplished, EMTs should initiate emergency medical care; cardiopulmonary resuscitation will probably be necessary until a defibrillator can be utilized to stop the uncoordinated fibrillations. If the blood has been kept adequately oxygenated and there is no other massive trauma, the heart will resume a normal rhythm after defibrillation. It must be remembered that resuscitation has priority over other emergency procedures.

At higher amperages, a sustained myocardial contraction is evoked that lasts as long as the current is applied. This muscular contraction is so severe that the heart is forcibly clamped so that no beat occurs during the shock duration; this action protects the heart from going into ventricular fibrillation and the victim's chances for survival are good. However, once the current is discontinued, the myocardium may relax and resume its normal rhythm. This principle is the basis for the use of electric shock with a defibrillator to terminate ventricular fibrillation.

Fractures as a result of falls are frequently suffered by electrical shock victims. The fracture may have resulted from a muscular spasm of sufficient intensity to break a bone, or the shock itself could have thrown the victim against some object. Fractures should be suspected in all electrical shock patients and the broken bones should be immobilized after the victim's circulation and respiration have been restored and stabilized.

ELECTRICAL BURNS

The chances of recovering from an electrical burn are related to many factors: the extent of the burn, the nature of the trauma, the victim's age and general health, other injury complications, and the availability of the correct medical care, such as specialized burn treatment centers.

Electrical burns range from the simple to the extremely complex. A patient's condition may include various degrees of burns. It is important to make a correct and early assessment and treat the trauma accordingly. The initial treatment of electrical burns is most important and should follow established basic principles of burn management.

Emergency Care

Electrical burns should be covered with a dry, preferably sterile, dressing. These burns usually extend deep into the body and are found in two places, the point of entrance and the place of exit. Look for the second wound and cover it also with a dressing, if necessary.

To properly provide initial emergency medical care to the victim of a high-voltage electrical burn, one must appreciate the causes and manifestations of its occurrence. Prognosis and degree of electrical injuries are difficult to determine at the scene; the extent and depth of the trauma are deceptive and difficult to diagnose in the early stages. These injuries also appear to progress for several days due to related damage to the vascular and nervous systems. Of great consequence are injuries to the hands, feet, and face; deep, full-depth burns in these areas usually result in some permanent loss of function.

Prolonged flame burns, such as those produced by a sustained electrical arc or flash that ignites clothing and remains in contact with skin surfaces, usually result in full-thickness skin injury. Damage to the respiratory tract is a distinct possibility because the victim may have inhaled flames and superheated toxic gases, especially if the accident occurred in a confined space. Acute inflammation and edema of the epiglottis, vocal cords, and upper trachea are the most common initial respiratory injuries.

In a high-voltage, high-amperage accident, tissue damage may be massive. Large areas of skin and flesh may be burned away, leaving a gaping wound that extends inward to the bone. The current may travel a path that follows the arteries and veins of the circulatory system or the internal organs.

Electrical injuries are known as "iceberg" injuries because often the only outward signs of such injuries are small entry and exit wounds. These mask substantial internal injuries where the current followed the path of least resistance through muscle and bone tissue, beneath seemingly undamaged skin. Severe electrical trauma may often be accompanied by a serious flame burn, resulting from the ignition of clothing by the electrical spark or arc.

Effects of Lightning

Among the many effects of being struck by lightning are loss of consciousness, burns, muscle weakness, loss of hearing, amnesia, myocardial damage, minor electrocardiographic changes, hypertension, and damage to the lens and retina of the eye. Death results from respiratory paralysis and ventricular fibrillation. All injuries, except ocular damage and hearing loss, are reversible in survivors. In managing the victim struck by lightning, first attention must be directed to restoration of heart action and ventilation. Later, fractures and burns should be treated and supportive measures provided. The unconscious victim must receive medical attention before the conscious patient; the latter is alive, whereas the former must be resuscitated.

METROPOLITAN RESCUE

RESCUE FROM ELECTRICAL CONTACT

Everyone must be prevented from approaching too close to wires or other electrical equipment that might still be energized. It is important that enough guards be placed around the scene to restrict unauthorized persons from entering until the utility lineworkers declare the area safe.

A victim in an automobile charged through contact with a fallen wire should remain inside the body of the car until the power has been shut off. Even though the automobile itself is charged, there will be no flow of current to the ground and the passengers will be safe in an upright car if they remain inside it, because the rubber tires will insulate the vehicle from the ground. Persons attempting to escape from the vehicle or to help the passengers will be subjected to current flowing through their bodies if they contact both the vehicle and the ground at the same time.

There is a common belief that a person who is trapped in an automobile with live wires on top of it can jump to safety. No such directions should be given, as this act can have deadly results. The person could slip in attempting the leap and complete the electrical contact between the energized car and the earth. The same results would be obtained if a hand or foot remained in even the most fleeting contact with the automobile when the other foot touched the ground. Insist that the victims stay inside the car, because any other procedure would be dangerous. The EMT should instruct the occupants about administering emergency care to themselves and other victims until the lineworkers arrive.

In some accidents it is impossible for the car occupants to remain in the vehicle: for example, if the auto is on fire. In this case, the occupants should not step out of their vehicle, but should jump as far as possible. When descending, they must make sure that hands, feet, and body clear the vehicle completely before touching the ground.

In one instance, an automobile with two persons in it struck a power pole carrying high-tension wires. One wire broke and dropped on the car. The two victims, dazed and confused, were still in the automobile when responding paramedics arrived. The paramedics recognized the danger and advised the two persons to remain where they were until the utility company could respond. The driver panicked, opened the door, placed one foot on the ground, and immediately received a lethal charge of current. The passenger remained in the car and was later removed relatively unharmed.

The greater the number of vital organs through which the current passes, the greater the chance of death. This fact should be remembered when removing a victim from electrical contact. The wire or the person must be moved so that there will be no further damage to the vital organs and so that any unavoidable additional damage will be to the limbs or lower parts of the body instead.

A long dry stick, pole, rope, or other insulated material should be used to drag away a casualty or to remove a wire (Figure 9-4). The wire should be pulled away by walking backwards; if it gets loose, it may spring away. It is important not to remove the wire by pushing away because if it gets away, it will spring back towards the rescuer.

Rope may be used to remove wires from the victim or casualties from the energized equipment. Dry manila rope is reasonably safe if the distance is not too short; wet rope may conduct a fatal shock.

Rescue from Electrical Contact

Most synthetic fiber ropes are nonconductive, will not absorb moisture, and are lightweight and strong. One of these may be used to loop around the wire or the victim, so that a safe rescue can be accomplished.

Some rescue squads carry a 100-foot (30-meter) length of 1/4-inch (6.25 millimeter) polypropylene rope with a 1/2-pound (0.23-kilogram) weight attached to each end. This weighted rope is used only to handle energized wires; any other use could soil the line so it would have less insulating value. To use it, an EMT would throw one weight under the downed wire and about 40 feet (12 meters) beyond. The other weight would be hurled over the wire so it would land close to the first weight. One squad member would then grasp both weights and drag the wire to a safe place. Tested and approved lineworkers' rubber gloves and protectors should be worn when handling any rope or other equipment that might contact energized wires or other equipment.

"Hot sticks" are especially designed nonconductive wooden poles employed by utility lineworkers to handle energized wires. These hot sticks may be carried by some rescue squads and should only be used by personnel wearing tested lineworkers' gloves and protectors. Hot sticks are used to push or pull a victim from a wire or the conductor from a victim.

FIRE FIGHTING HAZARDS

Fire fighting operations often jeopardize fire fighters by exposing them to electrical hazards. These dangers may be present anywhere in the country, because the rural areas are becoming as electrified as the cities. Fire fighters should study the problems and know how to protect themselves when they could become exposed to a possible electrical contact while combating a blaze. A standard operating

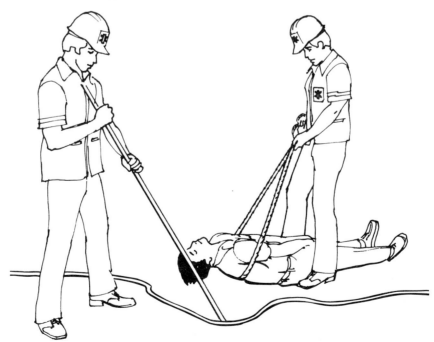

Fig. 9-4. Electric shock. Use a dry wooden pole, dry rope, or other dry material which will not conduct electricity to remove the wire from the person, or the person from the wire.

procedure that has been approved by the local power utility should be adopted.

Fire Streams

Fire fighters on hose lines or directing heavy stream appliances could be injured or killed if they should inadvertently direct a fire stream into or at energized electrical wires or equipment. The amount of current that may be conducted back to the nozzle will depend on the following factors: the voltage involved; the distance from the nozzle to the electrically charged line or equipment; the purity of the water in the fire stream; the size of the fire stream; and whether the fire stream is solid or broken.

The application of hose streams to charged wires or equipment should be avoided whenever possible. However, fire streams often must be utilized during fire fighting operations in close proximity to charged circuits, so all hose operators must be aware of the possible dangers involved, as well as the permissible procedures. Remember, the only nonconducting extinguishing agents considered safe are carbon dioxide and dry powder extinguishers that have a nonconducting applicator. Salty or dirty water, or the discharge from a soda acid or foam extinguisher, may have such high conductivity that no safe rule can be adopted. If it becomes necessary to use water that is known to have a high conductivity, such as ocean water, a spray or fog stream should be employed.

Certain procedures are suggested to minimize fire fighting hazards. For ordinary household circuits of 110 volts to ground, any type of nozzle can be held within a few inches of the charged wire without danger. The distance at which a straight stream nozzle can be positioned without discomfort is 3 to 4 feet (7.5 to 10 centimeters) from a conductor carrying 550 volts. It is safest for all personnel to consider every wire as one carrying high voltage because there is always the possibility of some remote contact with high-voltage wiring.

Spray and fog streams are the safest for all fire fighting activities in the vicinity of electrical currents. There is very little chance of a hazardous amount of current being conducted back to the nozzle because of the air spaces between the water droplets. Therefore, whenever possible spray and fog should be employed to cool down the equipment and extinguish the fire.

Be sure that the fog nozzle is not equipped with a long metal applicator that might touch the energized equipment. Also, the spray nozzle should not be adjusted while it is being directed on the electrical equipment because it might accidentally slip into the straight-stream setting when the nozzle is too close to the fire for safe use with a solid stream.

When directing a heavy stream through high-tension wires, a fog pattern should be used because the force of a solid stream could push the wires together and cause a short circuit. A spray stream is less likely to do this.

When handling ladders or long tools around electric wires, particularly if they are constructed of metal, extreme caution must be observed. Also, when climbing or descending a ladder, death or injury can result if a fire fighter barely brushes against or pushes live wires together.

When dropping a rope from the upper floors or roof of a building to use as a lifeline or to pull up hose or equipment, make certain that the rope falls close to the face of the structure. Throwing out a coil of rope could cause it to pass over or through high-tension wires; this would probably drag them together, causing them to arc, spark, and possibly fall to the ground.

Ladder and Boom Hazards

Because all cranes, aerial ladders, elevating platforms, and water towers are constructed of metal, they can be considered excellent conductors of electricity. However, vehicle tires are not very good conductors of current; therefore, if a ladder or boom contacts a charged wire, it will not necessarily become grounded. Electricity follows the path of least resistance. If persons standing on the ground touch the vehicle, the current will probably flow through them to the ground.

Exercise extreme care whenever operating any ladder or boom around electrical wires. Treat all the wires as though they were charged, even if the utility company claims that they have been de-energized; fallen wires in the vicinity may have re-energized them. The essential point is never to provide a path to ground for an electrical circuit. If the ladder or boom actually makes electrical contact, the safest course for the operator to follow is to get away from the controls and touch nothing until the power is turned off. Personnel on the ladder should keep their hands and feet on rubber-covered rungs until the circuit is cut. It is possible that a ladder or boom could swing into a high-voltage wire with enough force to cause the wires to come together; the covering on these wires is not of sufficient insulating strength to prevent a short circuit. The temperature of an electric arc is about 3,300°F (1,816°C). At this temperature, the wires burn through, break, and drop to the ground; this causes additional hazards. Also, the apparatus and personnel are in danger of damage or injury from the arcs and sparks.

If it becomes necessary to get off the apparatus while it is electrically charged, jump, so that contact with the truck and earth is not made simultaneously. Members on the ground must keep clear of the charged apparatus.

It makes very little difference from the standpoint of electrical conductivity whether or not ground jacks are in place. In dry weather they make a poor grounding contact. An exception would be if the jacks were resting on a rail or in a puddle of water.

Another fire fighting hazard is the possibility of contacting a live wire while cutting holes in roofs or walls during ventilating or overhauling operations. A chainsaw could easily cut through conduits to expose fire fighters to an electrical shock. Caution must be exercised at all times when working in a structure while the electrical circuits are still energized.

QUESTIONS FOR REVIEW

1. When comparing the principles of electricity to those of water flow, electrical voltage is best compared to
   a. pressure.
   b. volume.
   c. resistance.
   d. tendency to arc.

2. A wire has fallen from a pole; which of the following statements is the most inaccurate?
   a. The wire may not give any indication that it is live.
   b. Unless the wire arcs and sparks, it may be considered safe.
   c. Adjacent fences, cars, and other structures may become lethally charged if contacted by the conductor.
   d. Pools of water may be as dangerous as the wire itself.

3. If a fire must be combated where energized electrical equipment may be present, which one of the following is the safest procedure?
   a. Using straight streams at high pressures
   b. Using straight streams at low pressures
   c. Using fog or spray streams
   d. Using nozzles with long metal applicators

4. A 100-watt light bulb in a 110-volt circuit will require less than one ampere of current to be fully illuminated. About how much current is considered the minimum amount required to cause the heart ventricles to fibrillate?
   a. 100 amperes
   b. 10 amperes
   c. 10 milliamperes
   d. 100 milliamperes

5. An automobile has struck a power pole and wires are lying across the car body. Injured victims are still in the car. Which of these is the last thing that rescue personnel should do?
   a. Call the utility company to disconnect the current.
   b. Tell the conscious victims to remain in the car.
   c. Remove the most seriously injured victims from the car.
   d. Post guards around the area to prevent bystanders from being jeopardized.

CHAPTER 10

Rescue of Trapped Victims

Occasionally rescue personnel find a person caught or trapped in such a way that movement is restricted. It is particularly difficult for rescuers who arrive upon the scene to find a victim who is already mutilated, dismembered, or disfigured. A situation like this requires all the professional attitude, mechanical ability, experience, and ingenuity that the rescue crew possesses.

CONTROL AT THE SCENE

To expedite operations and reduce confusion, one member of the rescue crew must direct the entire operation and coordinate all efforts. It is a good idea to listen to suggestions from everyone and request advice from experts, but one person should control the situation and make all decisions.

The prime concern must be to prevent the victim's further injury or death; this may require bold, drastic initiative on the part of the crew. Consider the casualty as a person who is in real trouble. Remember, the unfortunate victim is in need of release from entrapment, relief from pain, and reassurance. In serious cases, the presence of a religious leader from the patient's faith is a great source of comfort.

In most cases, speed is important. Prolonged pinning of any portion of the body will cause serious damage. If there will be a considerable delay in disentangling the victim, send for a doctor immediately. Cutting or severing any part of a person's body, whether the victim is dead or alive, must never be attempted by anyone but a physician.

Remove all bystanders and curiosity seekers from the vicinity. Allow only necessary personnel to approach the scene. The horrified exclamations of onlookers do little to bolster the victim's courage.

MACHINERY ACCIDENTS

Because of the large number of heavy manufacturing machines of all types in this country, many persons become entangled in them or mutilated by them. Very often hands, feet, arms, or legs are trapped; about 90 percent of machinery accidents involve the hands and arms. Almost all the accidents around mechanical contrivances result from misplaced hands or from clothing that is caught in or on some moving part of the machinery.

Turning off Electrical Switches

The first thing to do is to shut off the power to the machine at the main switch. Station one member there continuously to guard the switch and prevent anyone from turning it on inadvertently. This is vitally important because rescuers have been maimed and killed when a machine started operating during an extrication procedure.

Disentanglement

Whenever there is a choice between breaking a machine and quickly removing a casualty with major trauma, disregard the machine. If disassembling the machinery to release the victim may take an extraordinary length of time, rapidly prying and forcing the parts with wedges, pry bars, sledgehammers, and hydraulic rescue tools may separate or spring the machinery components apart far enough to release the trapped person. It may be possible to manually reverse the machinery so the victim's limb is backed out; this is particularly true if the person is caught in a gear or roller. If there is an adjustment, release as much pressure as possible. While this maneuver will subject the body part to the damaging force a second time, probably inflicting an increased amount of trauma, this technique may be the only practical option.

When the victim is trapped in a small device, it may be transported with the person to the hospital. In other cases, the entrapping device can be disassembled from the larger machine so both the victim and the device can be transported. Medical personnel can then extricate the victim or disassemble the device in a more controlled environment.

Occasionally, someone sticks a head, hand, foot, or finger into a small hole and then cannot withdraw it. It may be that the body part became swollen because of the irritation or constriction, and perhaps the person just cannot find the correct combination of positions to allow withdrawal. Liberal application of grease or heavy oil, while messy, may permit the part to be withdrawn. Cooling the body part with ice may reduce swelling sufficiently to release the object. *Caution:* Do not use salt on the ice to speed up the process or frostbite could occur, even on a hot summer day. In other cases, a careful analysis of the problem and calculated manipulation of the trapped part may solve the problem.

When the entire body is caught, it is urgently important to provide sufficient space so the heart, lungs, and other vital organs can function until disentanglement can be performed. Even if serious injury has already occurred, block and wedge the machinery to prevent it from settling and causing further trauma. Victims with great pressure on their bodies may still be alive, conscious, and in very little pain. However, when pressure is released the massive damage and severe internal hemorrhaging will usually cause instantaneous death. When the victim is secured or supported by the entrapping machinery, the person must be suspended and braced so that release will not cause a fall.

Victims who fall into a hopper or pit and then are buried in material within it may be released faster by opening the bottom of the hopper or cutting a hole in the side than by attempting to dig down through the substance from the top.

IMPALED VICTIMS

Rescuers often discover victims who have been impaled when a pipe, rod, pole, or other such object has penetrated and passed through a part of the person's body. The casualties seldom survive this type of accident when the article has punctured the head, neck, or vital organs. In many cases, the person is impaled by an object penetrating an extremity or a section of the body. This kind of injury does not usually prove immediately lethal.

The impaling article must never be withdrawn from the casualty, nor should the victim be pulled loose from the object, as incalculable and unpredictable damage can result. Release of the pressure may start uncontrollable bleeding.

When the object is not so long and unwieldy that it would offer transportation problems, the article must be supported and steadied to minimize movement and the victim rushed to a hospital. If the object cannot be transported in its entirety with the casualty because of its length or weight, then firmly support the victim and the penetrating object and cut the article off about one foot from the patient's body.

When a person has been impaled by objects that are too massive to be cut or transported, a physician should respond to the scene so immediate medical care can be administered upon release. An example would be a railroad switch operator who had been impaled by car couplings joining through the abdomen, as happened in one incident. The victim may stay alive until the pressure is relieved, although death is usually instantaneous upon uncoupling. All efforts must nevertheless be made to preserve the person's life.

TRAPPED CHILD

Children may be trapped in many types of situations, ranging from locked rooms to garbage cans. In most cases they are uninjured, but frightened. Usually children will behave better if their parents are not present, because the relative's sympathy and concern often cause the child to be more frightened and unmanageable. In the absence of a parent or other relative, an experienced paramedic can usually calm the child and obtain cooperation. Any time a child, regardless of age, is treated more as an adult than as a little child, response will also usually be more adult.

Children with their feet, legs, or bodies caught in a barrel, drum, can, bucket, or other type of restricting container can usually be freed by reversing the sequence of operations that originally trapped them. Take time to analyze the situation; often the knee is at the bottom of the barrel with the foot higher, and attempting to pull the child out will only force the person in tighter. Reaching down inside and pulling the child's foot up may effect a complete release.

Children locked in rooms may be calmed and talked into unlocking the door. Often the child may be lured into opening the door by referring to the child as belonging to the opposite sex; calling a boy a little girl, or referring to a girl as a boy, will sometimes cause sufficient anger that the child opens the door to prove masculinity or femininity. Many bathroom door locks may be unlatched from the outside by inserting a thin screwdriver into a hole in the center of the knob. If the door cannot otherwise be opened, entry may be made through a window.

Fingers caught in rings, bottles, and other objects usually become swollen. Occasionally soap suds or oil will free them, or the swelling may be reduced with ice packs. A useful tool to carry on a rescue vehicle is a ring cutter.

Children may have their heads trapped between bars. Hydraulic rescue tools or an air bag may spread the bars, or saws and other cutting tools may be used to extricate the child. Every case will probably be different, but calm analysis should solve the problem.

GARBAGE DISPOSALS

Often a victim traps one or both hands in a garbage disposal unit while cleaning it. Removing the hand is usually simple if the victim is grasping a piece of silverware or other object, as the victim need only release the item. This reduces the size of the hand and the victim may pull it out easily. In other cases, the piping and disposal unit must be removed from below the sink. This leaves only the sink ring around the wrist, and the hand may then be worked out easily. To reduce friction, a liberal application of liquid soap and cold water may facilitate the release.

QUESTIONS FOR REVIEW

1. When a person is trapped in a machine so tightly that freeing the victim will be a lengthy process and the limb is obviously mangled beyond medical repair, cutting the victim's body to accomplish disentanglement and thereby save a life may be done by
    a. only a physician.
    b. any trained paramedic if a life is jeopardized.
    c. only trained medical personnel such as doctors, nurses, and technicians.
    d. any prudent person, when life is at stake.

2. Before starting to disengage a victim from an entrapping machine, the first operation is to
    a. administer a sedative to the victim to relieve shock and pain.
    b. obtain permission from the victim to relieve the rescuers of possible liability from causing further injury.
    c. shut off the power.
    d. obtain permission from the owner of the machine before causing any damage.

3. When a victim is entrapped in a small, portable device, the best remedy would probably be to
    a. dismember the device with a ring cutter.
    b. transport the victim and the device to a hospital.
    c. wait until body swelling subsides so the device will slip off.
    d. tell the victim to go to his or her personal physician.

4. Which of the following is least recommended when rescuers are confronted with a small boy who cannot withdraw his finger from a small hole?

a. Cool the finger with ice.
b. Cool the finger with ice that has been liberally sprinkled with salt.
c. Liberally apply grease to the finger.
d. Pour oil over the finger.

5. A small child that is trapped in a locked room will be most cooperative if
    a. ordered to open the door in a loud, authoritative voice.
    b. requested by a parent or friend to open the door.
    c. threatened with a beating for noncompliance.
    d. treated as a friendly adult.

CHAPTER 11

Rescue from High Places

Every day a vast number of people work, live, and play in high places. Some of these places are practically inaccessible. When an accident occurs, emergency personnel are called to the scene. Rescue operations are usually dangerous and difficult; there is room for only a slight margin of error.

ACCESSIBILITY OF VICTIMS

Rescue personnel become involved in many incidents where reaching the sick or injured victim is a major obstacle. Rapid treatment is vitally important, and often the mechanics of reaching the casualty may greatly lessen the chances for survival. Getting to the victim and bringing the necessary equipment to render adequate emergency care often requires imaginative and carefully coordinated steps.

Paramedics may be called to almost unapproachable locations to help victims suffering from heart attacks; strokes; electrical shock and burns; poisoning from toxic gases, vapors, and chemicals; falls from elevated structures, such as scaffolds and bridge structures; and every other type of illness or trauma. Such victims require immediate emergency medical care to help stabilize their condition before extrication, lowering, and transportation to a hospital can be attempted.

POLE TOP RESCUE

With a little imaginative adaptation, the techniques used for rescuing electrical workers can be employed when paramedics respond to an incident and discover a sick or injured casualty situated in some elevated, almost inaccessible location. Utility companies routinely train their personnel so that injured or incapacitated workers on power poles and towers can be efficiently and effectively cared for. Because electrical shock is a continuous hazard, procedures have been devised and adopted that enable rescue within the four-minute period during which the heartbeat must be restored if a person's heart has stopped or is in fibrillation. If an elevated rescue is not accomplished and emergency cardiopulmonary resuscitation (CPR) begun within this amount of time, the person will suffer permanent brain damage or even death. Since CPR requires that the victim be lying flat and face upward on a hard surface, the patient must be lowered to the ground as quickly as possible.

Preparation

When an injured worker needs help, the rest of the crew must be rapidly mobilized into a rescue operation. Before starting, the situation must be clearly analyzed. The condition of the pole, crossarms, insulators, and conductors must be sufficiently strong to bear the weight of another person. When working around extremely high voltages, it may be necessary to de-energize the circuits. The rescuer ascending the pole must be equipped with a strong rope, knife, rubber safety gloves, and other necessary equipment to disengage the victim from an energized wire. A rubber blanket may be used to get the rescuer or victim past an energized conductor. Do not skimp on protective equipment. The victim will not be helped if the rescuer rushes in and becomes a casualty as well.

Ascending the Pole

The rescuer will climb the pole and use the protective equipment to make a safe rescue. Keep the injured person's body from assuming a position that might be hazardous; the victim's body could be just as dangerous as an energized conductor. If the victim is in contact with an energized wire, the rescuer may have to stop well below the casualty until a safe clearance has been effected.

The best position for the rescuer, once the victim is cleared from a hazardous position, is slightly above and to one side of the injured person. This enables the rescuer to see the victim's face clearly and to give mouth-to-mouth resuscitation easily if required; it also makes attaching the lowering rope easier.

Emergency Care

The casualty's condition must be determined as soon as possible. If the person is unconscious and not breathing, time is of the essence. If the victim is conscious or unconscious but still breathing, time will not be as critical.

Conscious and Breathing. If the victim is conscious and breathing, offer reassurance. Most persons are in a state of traumatic shock after an accident, often unable to evaluate their own condition. Usually a few words of reassurance will relieve their fears. The victim may feel capable of descending unaided. If there are any doubts about the person's ability to climb down safely, tie the lifeline under the person's arms and play out the slack as the descent is made. Be ready to snub up on the rope if any support or steadying is necessary.

Unconscious but Breathing. If the injured person is unconscious but breathing, it is necessary to assume a position alongside the victim and ascertain that ventilation has not stopped. If the person does not immediately regain consciousness, prepare to lower the victim to the ground.

Unconscious and Not Breathing. When a person is unconscious and not breathing, immediately provide an open airway by tilting the head back and clearing the mouth, if necessary. Sometimes this simple

METROPOLITAN RESCUE

Fig. 11-1. Proper position to perform mouth-to-mouth ventilation during pole-top respiration.

maneuver will restore breathing. If not, administer five or six quick breaths of air mouth-to-mouth. If the patient responds by showing signs of life, if the color improves, or if the person coughs or moves, continue artificial ventilation until the patient is able to breathe without assistance (Figure 11-1).

If artificial ventilation does not help in five or six breaths, determine pulselessness by quickly checking for the carotid pulse. A check for widely dilated pupils can also be made, though this is not necessary.

If a pulse is detected, continue mouth-to-mouth breathing on top of the pole. When there is no detectable heartbeat, immediately start procedures for lowering the victim to the ground; effective CPR cannot be administered unless a person is lying down. If less than one minute has elapsed since the accident and there is reason to suspect cardiac arrest, administer a precordial thump before proceeding.

Lowering the Victim

Place a rope over the crossarm or other substantial part of the structure, preferably away from the center of the pole. Make sure that the line will run freely. If the load line crosses the fall line, it will be impossible to lower the patient.

Pass the rope high under the victim's armpits and tie it into a slip knot with three half hitches and cinch it tightly. Tie the knot in front, but slightly to one side of center front; if the knot is kept near the center of the body, it may rub the victim's face. Experiments have proven that an unconscious person cannot raise the arms high enough to fall out of the loop when this method is used. The line places pressure on the large muscles under the victim's arms, inhibiting the movement of the limbs so that they cannot be raised far enough to allow any release. If the line is tied high enough, it will not cause any serious injury to the chest or internal organs.

To lower the victim, remove as much slack as possible from the line. Take a firm grip on the rope with one hand and cut or unsnap the victim's safety strap. Guide the victim to the ground.

When the casualty reaches the ground, other members of the crew will immediately start CPR and administer other emergency care.

HOISTING THE VICTIM

The rescue of a person trapped on the side of a cliff or in a canyon is difficult and dangerous. The situation is often further complicated by injuries that the victim has sustained in a fall or other accident. It is usually easier to work from the top, above the casualty, but rescuers must be careful to stay a little to one side of the victim's location instead of directly above, in case rocks are displaced. One or two members can be lowered on ropes to the level of the victim. A vehicle with a winch is very helpful in these cases; otherwise, a long rope and many personnel are necessary (Figure 11-2).

Occasionally a helicopter can accomplish the mission, but rescue personnel on the ground are necessary to assist those in the helicopter. An aerial ladder truck positioned at the base of the cliff can sometimes be extended high enough to reach a trapped person. An aerial ladder can be extended horizontally over the edge of the cliff or ravine to act as a crane. A rope can be placed on the tip of the

Rescue from High Places

ladder and passed over it and down the cliff to the victim; members will line up on the upper end of the rope and pull in unison. Under some conditions, ground ladders lowered down to the victim may be helpful.

RAPPELLING

Rescuing endangered persons from elevated structures and cliffs is a dangerous task. Along with treating the victim for illness or injury, the rescuer is faced with additional problems in reaching the injured person. Unfinished buildings, chemical and refinery storage units, power and radio towers, tanks, cranes, industrial chimneys, elevators and shafts, dams, and amusement park rides all present unique situations that complicate the rescue procedure.

Fire fighters and rescue personnel have for years been taught how to slide down a lifeline; in a mountaineering climate, this technique would be referred to as rappelling down a rope. Whether it is referred to as sliding or rappelling, the procedure involves being safely suspended in a belt or harness and then controlling your descent down a line by means of changing the friction on the rope with some device. Rappelling is a quick, reasonably safe way to get down from high places with a minimum of equipment.

Heavy, bulky manila rope used with a fire department pompier hook is the most commonly used equipment for rappelling. Friction is controlled by the size of the lifeline, the number of turns around the hook, and the amount of force applied to the lower part of the rope with a gloved hand. This equipment is very heavy and rescuers tire quickly when carrying it to extreme heights becomes necessary. Also,

Fig. 11-2. A fire department straight ladder that is supported and steadied by ropes and pike poles allows a fire fighter to descend down a cliff to rescue a trapped victim. Note that lifelines are attached to both the rescuer and the victim to prevent any further mishaps. (Courtesy of Photo One, the San Francisco Fire Department.)

heavy manila rope tends to untwist in response to weight on it; this has a tendency to spin the person suspended on the line, often quite violently.

Modern rappelling devices and ropes, devised and improved by mountaineering and rescue organizations, are lightweight and far easier to carry and use. Braided rope has less tendency to spin. Modern lifelines have a core consisting of a bundle of straight fibers held by an outer sheath of woven or braided strands. Rescue lines should be made of nylon, dacron, or manila. Polypropylene, polyethylene, or other ropes that have low melting points must not be used for rescue operations because the heat developed through friction may reduce the strength of the line. In most cases, a low-stretch, braided 1/2-inch (1.25-centimeter) nylon rope is adequate. If at all possible, all lines should be well protected with padding at points where they pass over windowsills, ledges, or other edges.

Lightweight, compact controlled-descent devices have been invented. They are designed to enable a person to slide down a rope and stop in midair with both hands free for rescue work. This characteristic makes these devices ideally suited for the rescue of injured or stranded persons from high places, for rapid and completely controlled descents from any structure or formation where a rope can be anchored or suspended, or for any type of work where operations must be performed in an elevated or precarious position.

The basic principle of this equipment is quite simple. The energy absorbed by the devices in the form of rope friction is the key to their function. The designs allow the amount of friction to be increased or decreased quite easily when the operator desires to do so. Some of the devices incorporate a brake bar, snubbing post, or other parts that allow the operator to tie off the free end of the descent rope and remain motionless with both hands free. The speed of descent depends on the size of the line, the number of turns of the rope around the hook, and the amount of tension applied to the free end of the rope.

LIFESAVING NETS

The first lifesaving nets were hand-held blankets; later, an interwoven rope net was developed. What was ultimately required was a device small enough to fold into the compartment of a rescue apparatus, yet large enough to efficiently and dependably save lives.

The circular lifesaving net of today is a very useful device. The pioneer rope net was adequate in coping with the requirements made on it during the early part of the century. Today's taller buildings require more dependable apparatus. The modern net consists of a fabric bottom supported within a 9 1/2-foot (2.85-meter) diameter steel frame by 32 dual shock absorbers that distribute the initial force of the impact. The nets are available in half- and quarter-fold models for ease of storage; when folded in quarters the net's dimensions are approximately 7 feet (2 meters) by 2 1/2 feet (0.75 meter) by 1 foot (0.3 meter).

The shock absorbers reduce the impact normally absorbed by the personnel spaced around the net and make it possible for fewer members to handle it under safer conditions. At a hotel fire in Dubuque, Iowa, at least 30 lives were saved in life nets. At other fires a 4-month-old child was safely dropped from a third floor and a 70-year-old woman was caught after jumping from the sixth floor of a Chicago hotel. A Newark, New Jersey, rescue squad made two rescues of suicide attempts within one year; one was a 140-pound woman jumping from the fifth floor and the other was a 160-pound man jumping from the eighth floor. Although the recommended landing position is on the small of the back, both of these jumpers landed feet first without injuring any of the handlers.

A 160-pound body falling through space from a height of 75 feet (22.5 meters) attains a speed of nearly 50 miles (80 kilometers) per hour and exerts an impact force of approximately 6 tons (5.4 metric tons). Table 11-1 shows kinetic energies of a falling body at different heights. A simple, though inexact, method of calculating kinetic energy is to multiply the height by the weight.

Table 11-1. KINETIC ENERGY CHART FOR A FALLING 160-POUND BODY

HEIGHT OF FALL, FEET	LENGTH OF TIME, SECONDS	VELOCITY AT END OF FALL, FEET PER SECOND	MILES PER HOUR	FORCE OF IMPACT POUNDS
10	0.80	25.60	17.45	1,638
20	1.10	35.20	24.00	3,097
30	1.40	44.80	30.55	5,017
40	1.60	51.20	34.91	6,553
50	1.78	56.96	38.84	8,410
60	1.95	62.40	42.55	9,732
70	2.10	67.20	45.82	11,287
80	2.25	72.00	49.09	12,960
90	2.38	76.14	51.91	14,477
100	2.50	80.00	54.55	16,000

[a] Kinetic energy.

Not more than 14 members or fewer than 10 should try to use the net. Personnel should be placed evenly around the net with hands on every other grip; the net should be raised to neck level. Palms are spread and elbows are extended forward and slightly outward from the body. This allows the arms to move with the impact of the jumper and permits the elbows to clear the crew's chests; thus, injury from an elbow being forced into the body is averted. Net holders stand with their left feet forward and knees slightly bent for a resilient stance. Knees must never be locked. It is important not to span the hinged joints and rescuers should not purposely allow the net to give when the jumper lands. They should look up immediately upon assuming position (Figure 11-3).

The officer should immediately start talking to the jumper, calmly reassuring the person and taking all steps necessary to prevent a premature jump. The officer will personally direct the final positioning by grasping the frame of the net with one hand, palm down, and issuing the necessary order with a gentle shove.

The person jumping should leap with arms outward and high overhead. The legs and feet should be kept together with legs extended forward and feet held as high as possible. It is best to land in the center of the net on the buttocks or small of the back. It is important to avoid a feet-first impact.

Rescue personnel on the side of the net away from the building must watch the jumper and be prepared to take vigorous and positive moves with the net in order to catch the person. Personnel on the inside, with their backs to the building, must watch the outside members and follow through immediately with appropriate action.

If a jumper should drop directly down the face of the building, the net must be forced against the wall and raised slightly higher at this point to assist in taking the impact. Serious injury to those members between the net and the wall may result if they do not take one of several courses afforded them. The most desirable one, if possible, is to drop out of position and remain clear of the net, the wall, and other members. Other alternatives are to drop to the ground beneath the net and attempt to keep the body clear of the center point of impact; or drop to the ground outside of the net and parallel to the wall.

Members must be alert to the possibility that jumpers may strike overhead wires or other obstructions and thus change their course of fall. In the event that a number of victims attempt to jump, those caught should be quickly and gently rolled from the net on the side away from the building. The net should be relocated immediately in anticipation of another jumper.

At the scene of an emergency, rescuers must never expose the net in the open position except when anticipating a jumper. If the net is loosely held or laid on the ground in an open position, a jumper may believe the net is ready and jump.

Fig. 11-3. Fire fighters are supporting a circular lifesaving net ready to catch a falling or jumping victim from a collapsing structure. Shock absorbers between the tubular ring and the net will reduce the amount of impact normally forced on the net holders. Rescue personnel on the side of the net away from the building must watch the jumper and be prepared to quickly move the net to catch the victim. (Courtesy of the Los Angeles City Fire Department.)

REMOVAL OF CASUALTIES

Since rescues will be conducted under almost every conceivable adverse condition, the method employed for casualty removal will vary according to the location of the victim and the type of injury that has been sustained. In some rescue operations patients will have to be lowered from upper floors of buildings; in others, they will have to be hoisted from below through a hole in the floor. Some casualties may be helped down a ladder, while others may have to be lowered in stretchers or litters. Whatever type of stretcher or improvisationis used, the victim should be kept in a horizontal position, if possible. In many cases, however, it is necessary to lower a casualty feet first through a narrow opening, such as a hole in the floor or down an elevator shaft.

After removal, many victims will have to be carried over piles of debris and uneven ground before arriving at a transportation vehicle. Some persons will be injured; some will be unconscious. Speed in removal is important, but it should be consistent with safety and proper handling to prevent further trauma. The method used will depend on the immediate situation, the victim's injury and condition, and the available equipment.

LADDERS

Ladders are used not only for climbing, but also as improvised bridges, derricks, and stretchers. All personnel should understand their care and handling, and should practice with them to become proficient in their use. Familiarity in working with ladders at heights can only be acquired through many hours of practice.

Maintenance

Ladders must be kept clean and in good working order. Metal ladders should be inspected for corrosion and dents or cracks. Wooden ladders are routinely varnished but never painted, since paint conceals defects. Ladders should be inspected frequently and handled carefully. They are to be used only for the purposes for which they are intended, and must never be overloaded.

Carrying Ladders

The safest method of carrying a roof ladder is to place it on either shoulder, passing one arm through the ladder at the middle of its length. If the hooks are not foldable, the ladder should be carried with them in front where they can be watched. Other ladders should be carried with the heel forward so the ladder can be set and raised in one operation.

When two members carry a ladder, they should both stand on the same side of it. Each passes an arm through the ladder and grasps the rail on the second rung forward. When carrying a ladder through a crowded place, the lead crew member takes a position well to the front, using the outside hand to prevent injury to persons in the line of travel.

Climbing Ladders

To establish the proper angle for safe climbing and to assure maximum strength under heavy loads, the following methods are used to determine the distance the heel should be placed from the building:
1. Divide the extended length of the ladder in feet by five and

METROPOLITAN RESCUE

then add two. For example, if a twenty-five-foot extended ladder is used, the proper distance of the heel from the building is seven feet, since one-fifth of twenty-five is five, and two added equals seven.

2. Take one-fourth of the length of the extended ladder and place the heel that distance from the building. The proper distance for the heel of a twenty-five-foot ladder would be approximately six feet from the building, since one-fourth of twenty-five is about six.

Ladders should be placed so that the climber's body is perpendicular to the ground at all times. When the climber's arms are extended for climbing and the body is perpendicular to the ground, climbing is easy and safe. If the heel of the ladder is too far from the building, the climber must lean forward. If it is too close, the person must hug the ladder to keep from falling backward (Figure 11-4).

Ladders should be placed against the sill to either the far side or opposite the working side of a window, not in the center.

When climbing ladders, perfect rhythm is essential. To acquire this rhythm, the climber should step on every rung and grasp alternate rungs. It is not advisable to climb with the hands on the ladder rails, unless carrying a heavy object. It is best to climb near the center of the rungs, on the balls of the feet, and to keep the upper part of the body at arm's length from the ladder. Climbers should always look toward the top, not the bottom, of the ladder and never run up or down; climbing is performed best when it is done briskly, steadily, and smoothly.

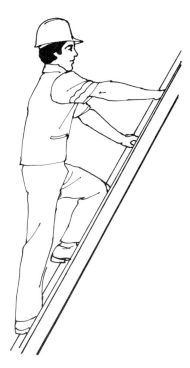

Fig. 11-4. Safe ladder climbing position.

When it is necessary to work from a ladder, a leg lock should be used for safety. To do this, pass the foot that is opposite the working side over the second rung above the one where the standing foot remains. Then pass the locking foot back so it hooks the rail or beam (Figure 11-5).

When extension ladders are not available, two ladders may be lashed together to reach the desired height. When lashing them together, join the beams, not the rungs. There should be two lashings on each side. Separate ropes should be used for each tie; make each lashing snug and tight (Figure 11-6).

Helping a Person Down a Ladder

Great care should be exercised when helping anyone down a ladder, even though the person being helped is conscious and uninjured; the basic techniques are the same whether the victim is conscious or unconscious. Rescue personnel should keep in mind that most people are unaccustomed to height. They may become frightened and either freeze into immobility or lose their hold. A rescuer who is not in the proper position runs the risk of being knocked off the ladder should the person who is being helped let go and fall. Keeping close to the person is essential for good control.

Fig. 11-5. Ladder leg-lock.

Victim Conscious (Figure 11-7a). When the victim is conscious, the best position for the rescuer is one rung below the victim, with the rescuer's arms encircling the person's body while the rescuer's hands grasp the rung. The rescuer should keep in step with the person being helped down the ladder and let that person set the pace. The rescuer should precede the person being helped, keeping the knees close together to ensure support in case the other person slips or becomes unconscious. The rescuer should also talk to the victim to help keep up morale and overcome fear. If the victim slips or becomes unconscious, the rescuer should allow the person to slide down until the victim's crotch rests on the rescuer's knee. By repeating this procedure for each step down the ladder, the casualty can be safely lowered to the ground.

Victim Unconscious (Figure 11-7b). When the victim is unconscious so that no cooperation in descending can be obtained, the rescuer may pass both arms around the person's body and grasp the ladder beams, instead of the rungs. By pressing the casualty's body against the rungs, the speed of descent can be safely controlled.

If the casualty is unconscious or unable to descend the ladder, a good method is for one rescuer to stand near the top of the ladder while grasping the beams. Other personnel then place the patient

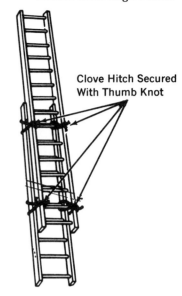

Fig. 11-6. If extension ladders are not available, lash two ladders together to reach the desired height.

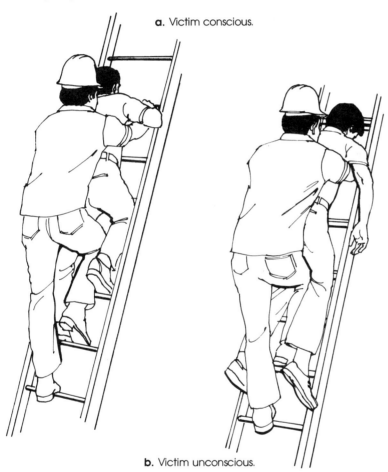

a. Victim conscious.

b. Victim unconscious.

Fig. 11-7. When helping a person down a ladder, keeping close to the person is essential for good control. The best position for the rescuer is to be one rung below the victim, with the rescuer's arms encircling the person's body and the hands grasping the rung. The rescuer should keep in step with the person being helped down the ladder, and let the victim set the pace. The rescuer should precede the person being helped, and should keep the knees close together to ensure support in case the victim loses a firm grip or loses consciousness. If either of these happen, the victim should be permitted to slip down until the crotch rests on the rescuer's knee. By repeating this procedure for each step down the ladder, the rescuer can lower the casualty to the ground.

across the rescuer's arms. The descent is made while the rescuer presses the victim against the ladder so the person's buttocks rest lightly on each rung as they descend. The rescuer's hands slide down the beams, instead of grasping the rungs.

Bridging Between Buildings

If rescuers are unable to ascend to the roof or upper floors of a structure, it may be possible to bridge the gap between the building where the emergency exists and an adjacent structure with a straight ladder. If the gap requires bridging an excessive distance, other ladders or lumber may be used to strengthen the ladder. Figure 11-8 illustrates bridging a gap between two rooftops. The same procedure may be used through windows.

QUESTIONS FOR REVIEW

1. A utility worker on a high power pole has received a shock through contact with a high-tension wire. Other personnel working in the vicinity determine that both the heart and respiration have stopped. Cardiopulmonary resuscitation must be started as soon as personnel can
    a. reach the victim, and must be continued without hesitation until medical help is obtained.
    b. reach the victim, but may be discontinued while the victim is being lowered to the ground.
    c. agree unanimously what the correct treatment should be.
    d. lower the patient to the ground.

2. When a person is trapped on the side of a steep cliff, it is usually easier for the rescuers to approach the victim from the
    a. top.
    b. bottom.
    c. side.
    d. top, unless the cliff is very steep.

3. Sliding down a lifeline from the top of a cliff or building is usually called
    a. descending.
    b. rappelling.
    c. lifelining
    d. down creeping.

4. When a person jumps from a tall building into a lifesaving net, the tremendous impact force is absorbed by the
    a. rescuers holding the net.
    b. air resistance, which prevents a falling body from accelerating to a high speed.
    c. shock absorbers supporting the net bottom.
    d. rigidity of the steel frame.

5. A thirty-foot-long ladder is raised so its tip rests against the wall of a building. To attain the correct climbing angle, the heel of the ladder should be placed out from the building about
    a. 12 feet.
    b. 10 feet.
    c. 8 feet.
    d. 6 feet.

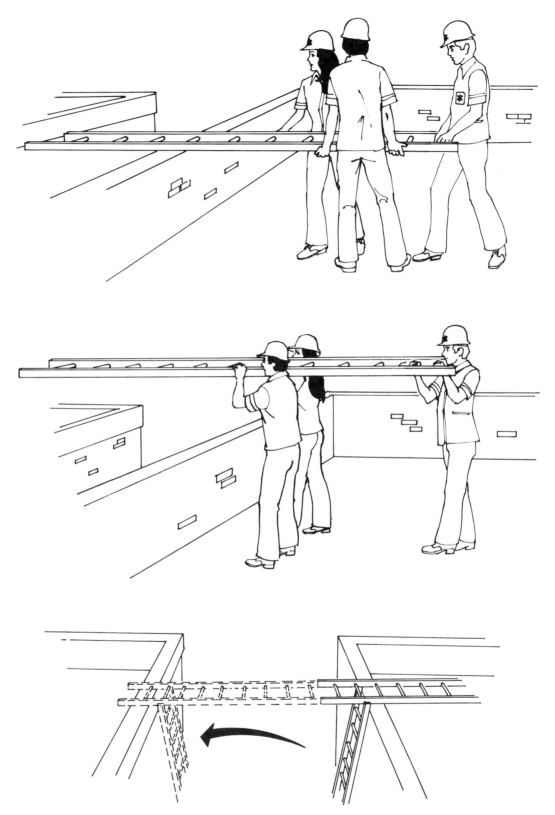

Fig. 11-8. A sturdy ladder can be used to bridge the gap between buildings, if victims must be removed from the upper floors of burning buildings or if rescuers must reach the roof of an adjacent structure. When the distance is narrow, the gap can be quickly bridged as shown in a and b. The bottom illustration shows the best procedure when the gap is wide. If the building is only one story high, use a long straight ladder to support the end of the bridging ladder while it is crossing the gap between the buildings. If a ladder is being used to bridge a gap between windows, a rope lowered from an upper floor will support the end of the ladder until it can be placed in the opposite window.

CHAPTER 12

Rescue from Burning Structures

The rescue of building occupants and their removal to a safe location must always receive first consideration at every fire and other emergency incident. Usually, a prompt, aggressive attack on the fire, quick extinguishment, and adequate ventilation procedures are the best methods of saving lives and preventing further extensions of the blaze.

A thorough search of the area should be conducted at every fire or other emergency, whenever there is any possibility that a human life could be in danger. Searching for trapped, injured, or unconscious victims is one of the fire fighter's more hazardous jobs. The search itself is usually done in heat and smoke, which compound the problems. The protection of a hose line might not always be available to searchers. Rescue operations at fires are often complicated by the fact that casualties must be located under the most hazardous and difficult conditions, and then removed to a place of safety before emergency medical care can be administered (Figure 12-1).

People who have never been in a structure under severe fire conditions cannot realize, or even imagine, what it is like. No amount of narration or portrayal could describe the situations and difficulties that might be encountered inside a burning building. Motion pictures give a false impression of fire behavior, because they seldom depict a blaze growing in size or intensity. Smoke and toxic gases are usually negligible in the films. The movie star picks up the beautiful victim with tender loving care and carries her with ease out of the building. In reality, fires often gain in intensity with almost explosive acceleration, heat and smoke usually force the rescuers to crawl along the floor with no visibility, and victims must be dragged out in any way possible.

As is true in all emergency operations, a calm and efficient demeanor on the part of rescue personnel will play a large part in reassuring and quieting the people involved in the incident. The

Rescue from Burning Structures

Fig. 12-1. Fire fighter equipped with a modern closed-circuit breathing apparatus rescues a small victim from a hot and smoky structure fire. (Courtesy of Biomarine Industries, Inc.)

prevention of panic at fires in hospitals, theaters, auditoriums, and schools is very important, because any large group of people can become excited and unruly. *Nothing will panic and arouse the public more than a loud, excited rescuer.* People will logically assume that because a professional emergency worker is excited and worried, conditions must be critical.

It must be constantly remembered that sometimes confining and extinguishing the fire can be the most effective action taken in the rescue operation. For example, a massive attack on a relatively small fire backstage in a theater might be a more practical lifesaving move

METROPOLITAN RESCUE

than dividing the available personnel between making an ineffective attack on the fire and simultaneously assisting patrons to leave the auditorium. Minimal assistance to calm people and direct them to the nearest exits would probably suffice. Under entirely different conditions, fire fighting activities would be left for late-arriving personnel, and all attention would be devoted to rescue operations. Only through an adequate and intelligent size-up can the most effective strategy and tactics be adopted and utilized.

FIRE BEHAVIOR

A typical structure fire usually begins with one small object in flames. If the blaze does not self-extinguish, the flames spread to other objects in the room and to the walls and ceiling, from which they radiate heat, thereby spreading the fire to other objects in the room. When sufficient heat has built up, the room reaches the point of flashover, at which time the whole room suddenly becomes engulfed in flames with high temperatures uniformly dispersed throughout the space. At this point the oxygen level in the room drops quickly and substantial quantities of carbon dioxide, carbon monoxide, and other fire gases are generated in an exceedingly short period of time. The time of flashover is the definitive point in fire spread. When this is reached there is no hope of human survival in the room of fire origin. Smoke, heat, and toxic and combustible gases are released; these impinge on neighboring sections of the building, and the fire starts to spread from the original ignition source.

Fire invariably travels upward. If its upward path is blocked, it will travel horizontally and, if both of these routes are hindered, the fire, heat, and smoke will progress downward. Personnel in a burning building should stay low where the heat and smoke conditions will be more tolerable. There is usually breathable air next to the floor. Improved visibility can also be expected.

Basic Fireground Operations

There are seven basic fireground operations: (1) rescue, (2) protection of exposures, (3) confinement of the fire, (4) extinguishment of the fire, (5) overhaul, (6) ventilation, and (7) salvage. With the exception of the last two items, these procedures should receive the attention of the fireground commander in the order in which they are listed. The last two procedures, ventilation and salvage, do not fit into any specific order. For instance, it may be necessary to ventilate a building to make a rescue possible. Or it may be feasible to make a rescue and protect the exposures before doing any ventilating. On the other hand, if a sizable section of the roof has burned off by the time the fire fighters arrive, the blaze itself may have done all the ventilating needed.

Boiled down to its essentials, fire fighting consists first of locating a fire, second of confining it, and third of extinguishing the blaze. It is that simple, but very often the simplicity of fire fighting is ignored or forgotten in the excitement, and sometimes the confusion, that prevails on the fireground.

Fire Fighting Strategy

In any fire fighting operation, minimizing life hazard comes first. This makes the stairwell, which is the main path of escape for any

Rescue from Burning Structures

occupants, the number one priority to protect from an extension of the fire, heat, and smoke. The stairway is also a pathway to the upper floors for fire fighters. Every effort must be made to defend it, no matter how fiercely the fire burns in other sections of the structure.

The second most important consideration is the rooms and apartments directly over the involved area. Fire could conceivably burn through the ceiling and then spread rapidly through the next higher floor.

If people are hanging out of the front windows when fire fighters arrive, they will naturally have to be taken down ladders or guided to other exits. A good point to remember here is that if a fire is severe enough to cause occupants to hang out of the front windows, there is a good possibility that they are also gathering at the back windows. This must be checked immediately.

Hose Lines

Personnel who are attempting to perform a rescue should have every kind of protection available, and foremost among protective equipment is a hose line with an adequate supply of water. The obvious advantages of a fire stream are its usefulness in controlling the fire in the rescue area and its cooling effect. The force of a water spray will also help ventilate the structure and provide cool fresh air, which will help the victims as well as the rescuers (Figure 12-2).

Another important advantage of taking in a hose line is that the hose automatically marks an escape route. If worsening conditions force a retreat and smoke has obliterated all visibility, the hose line will lead the search team out of the building and will also indicate a route for other personnel sent in to assist. Searches must be conducted quickly. For this reason, it may not be possible to use hose lines in all situations. However, as operations progress, additional lines should be advanced to protect searchers and trapped victims. Fire streams may have to be used to knock down the fire and to protect victims.

It may be necessary at times to delay searching an area until a charged hose line is ready to advance. The search team must then enter the structure behind the protection of the fire stream and, as the fire is controlled, search each room.

To expedite the search of the more distant rooms, the search team can occasionally leave the protection of the charged hose line. Before this is done, it is vital that the fire fighters on the charged line be told of this maneuver. If necessary, it may be possible to quickly reposition the fire stream to protect the search team's advance or withdrawal.

Fire fighting officers operating on the fire floor must keep in mind the presence of any personnel on the floors above. If it becomes apparent that fire streams are unable to hold the fire, or that there is a danger of building collapse, instant warning must be given to the crew above the fire. Steps should then be taken to provide escape by ladder or secondary exit, and an effort made to place adequate streams between the fire and the exposed rescue personnel.

Caution must be exercised when stretching hose lines to keep them from interfering with the evacuation of the occupants, unless a fire stream is necessary to protect their escape. When many persons have to get out of a building, rescue strategy must be considered

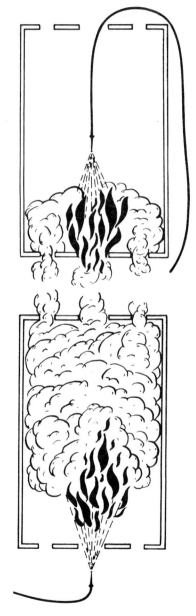

Fig. 12-2. The initial attack should be made by stretching hose lines into the uninvolved sections of the building to prevent further extension of the fire and to protect the occupants' means of exit.

even in the positioning of equipment and the stretching of hose lines. When many persons are using an enclosed stairway or a fire escape to evacuate a building, it is best to allow the people to get out of the structure before cluttering the stairs with hose, because the fire crew can be employed for only one thing at a time—either removal of occupants or stretching of hose lines.

When a fog or spray nozzle is advanced into a room, cellar, or other enclosed space, the water spray moves large volumes of hot fire gases ahead of it. If the building has been properly ventilated, the hot smoke and gases will be expelled to the outside by the velocity of the fire stream and by the force of steam generated by the heat and water. When ventilation has not been properly performed, the extremely hot gases will pile up against the far end of the enclosed area; then, as the fire continues to burn and to evolve additional heat and gases, the gases piled against the wall will rebound and progress back toward the nozzle. When the heated flammable gases reach a section of the room where there is an adequate concentration of oxygen in the atmosphere, they will burst into flame, sometimes with explosive force. Occasionally, this will occur near the fire fighters; death and serious injury have resulted. This is why any initial attack hose line in an enclosed space should be followed by a backup line for protection. If, despite ventilation, the worst happens, the backup line is immediately available to help recover control of the situation.

Ventilation

Ventilation, as applied to the fire and rescue service, refers to the planned and systematic clearing, from a structure, ship, or other area, of objectionable smoke, heat, and noxious gases. The purposes of ventilation are:
1. Rescue operations are facilitated by removing smoke and gases that are endangering occupants who may be trapped or unconscious.
2. The area is made more tenable for fire fighting operations. Fire fighters can enter the structure to search for victims and can approach close enough to the fire to extinguish it with a minimum of delay, smoke and water damage, and danger to themselves.
3. Backdraft conditions are reduced or eliminated.
4. Spread of the fire is controlled.
5. Toxic and explosive gases and vapors are removed from the structure.

Promptly clearing the heat, smoke, and toxic gases from a room or a structure could make the difference between a victim's dying and receiving sufficient air to survive. There are occasions when search teams and fire fighters have been unable to advance more than a few feet into a burning building until roof ventilation was accomplished. In such situations, the effects are often startlingly dramatic and the crews are able to proceed with a reasonable degree of safety and comfort that adds to their effectiveness. When roof ventilation is not practical, opening all windows as wide as possible and removing curtains and screens will usually suffice.

The way ventilation is accomplished to assist a rescue operation must be selected in relation to the overall effect it will have on the progress of the fire. No good is accomplished, and hazards are increased, if smoke and heat are relieved in rescue or escape areas by an action that speeds the advance of the fire into these very sections of the building. If the roof has a scuttle or a skylight, removing the cover is the quickest way to obtain at least a modest amount of ventilation. However, if the fire has attacked a large section of the structure, a large hole must be cut in the roof to achieve effective ventilation (Figures 12-3 and 12-4).

It would require an entire book just to comprehensively cover the subject of fire fighting ventilation. Officers should teach their crews and drill them in all aspects of this subject because effective ventilation is one of the most important factors leading to efficient structural fire fighting and rescue. In fact, often the success or failure of a fireground operation depends on the degree of attention devoted to ventilating a building. Because of the personnel hazards involved, fire fighters must be on the alert for backdrafts (smoke explosions).

Backdraft Explosions

Backdrafts, or smoke explosions, are a constant threat to emergency personnel. Every fire fighter and paramedic should understand their nature, causes, warning signs, and the measures that can be taken to prevent their occurrence and to protect personnel against them. They are not uncommon at fires; in many instances, it is only a matter of good fortune that more emergency personnel are not killed and injured by backdrafts. Whenever it is necessary to enter an atmosphere containing an oxygen deficiency (less than 16 percent oxygen), or when a building is heavily charged with smoke, gases, and suspended solids, a self-contained breathing apparatus should always be worn.

When common carbonaceous materials, such as wood and paper, are burned in an atmosphere that contains sufficient oxygen, carbon dioxide is generated. But when a fire occurs in a tightly closed structure, the fire gradually depletes the normal 21 percent of atmospheric oxygen. When the oxygen content is lowered below 16 percent, the combustion of a free-burning fire will slow down and smoulder. Incomplete combustion produces carbon monoxide, which is a highly toxic and flammable gas. A mixture of carbon monoxide and air has an explosive range of 12.5 to 74 percent by volume and an ignition temperature of 1,204°F (651°C). When these fuel gases are heated to their ignition temperature, they will burn when sufficient oxygen is present; they require only the admission of enough fresh air to cause rapid burning, which in turn may be enough to cause an explosion.

When a fire is confined within a building, conditions are comparable to a furnace burning with the doors closed and the damper shut. To ventilate at a lower floor before providing top ventilation would be like opening the door of the furnace while the damper is closed. Additional oxygen would be added with a corresponding increase in the intensity of the fire and in the generation of gases; this could turn the building into a raging inferno with flames engulfing all floors. Moreover, if sufficient oxygen is admitted into a lower floor before top ventilation is provided, there is always the possibility of instantaneous combustion of the confined and heated carbon monoxide mixture. Proper ventilation, achieved by opening the building at the highest point of the fire, will permit the heated gases and smoke to be expelled and will make entrance into the lower floors safer.

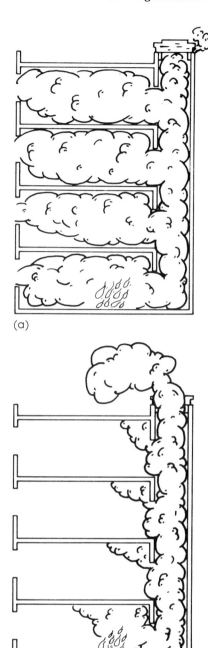

Fig. 12-3. (a) When the heat, smoke, and toxic gases cannot escape from a structure, they rise to the upper floors and then progressively fill the building downward, creating a mushrooming effect. (b) After the roof has been opened over the stairs or elevator shaft, the heated smoke and gases rise through the opening and the structure's atmosphere again becomes breathable.

METROPOLITAN RESCUE

Fig. 12-4. A prompt attack on the fire will be possible if strategic ventilation through an opening in the roof allows the heated smoke and gases to escape, so the atmosphere in the building is breathable (a). Fire fighters can be driven from the building, possibly with serious results, if a fire stream is projected down through a ventilation opening in the roof (b). This improper action cools the rising column of heated smoke and gases, thus reversing their orderly movement from the structure, and forcing the fire fighters inside the building to retreat. A hose line should be taken to the roof of a blazing structure only to protect personnel and adjacent buildings.

When the upper floors of a building are occupied, the mere presence of smoke in any quantity will cause apprehension, and may create a panic. For this reason alone, it is imperative that personnel should be sent to the involved floors at the earliest possible moment to comfort the occupants and ease their fears. In addition, of course, they will be concerned with their normal fire fighting and ventilation duties.

Because of the increase in intensity of the fire when it is supplied with additional oxygen, charged hose lines should always be in place and fully attended before extensive ventilation is attempted. To accomplish good, effective fire fighting and rescue, a great number of different operations must be conducted simultaneously; proper timing is the result of dedicated drilling and good teamwork.

Warning Signs of a Backdraft

A backdraft is a rapid, almost instantaneous, combustion of flammable gases, carbon particles, and tar balloons, which are emitted by burning materials when there is insufficient oxygen for complete combustion. Therefore, backdraft conditions do not exist except where a fire is burning in a confined space, but the dangers must be recognized and considered when ventilating and making forcible entry. Fortunately, backdrafts do not occur without forewarning; their very nature creates certain conditions that are usually discernible to knowledgeable personnel.

Everyone knows the appearance of a free-burning fire with sufficient oxygen for complete combustion; there are few persons who have never gazed into a bonfire, campfire, or fireplace. One indication of incomplete combustion is unusually thick, black smoke at an ordinary structure fire where common materials such as wood, paper, and cloth are burning. The smoke is thick and black due to the extreme lack of oxygen; this causes unburned carbon particles to be released into the atmosphere in great quantity. This is also an indication that generous amounts of flammable gases are being emitted.

Backdraft Indications from Outside the Structure

Whenever fire fighters are approaching a structure where there is a suspected fire, they should remain alert to the possibility of a smoldering fire. Feel the doors and windows before attempting to open them; backdraft conditions are possibly present if the window glass or door panels are hot and there is very little or no fire immediately adjacent. If visibility into the building is poor, and yet no fire is in evidence or there is just a faint glow through the smoke, it is reasonable to assume that a fire has been smoldering for long enough to produce the smoke causing such poor visibility. In such situations warm smoke may be emitted rather forcefully from around doors and windows, rather than oozing out. In these instances, the smoke is usually forced out intermittently in puffs.

When opening a door or window where a backdraft may be encountered, stand to one side of the opening, remain low, and have charged hose lines available and ready to attack the flames. Any time there is the chance of a backdraft, personnel should not stand in, or directly in front of, any door, window, or other opening. It is safest if no one remains within the V-shaped force pattern that will emanate from such openings. The gaseous products of the explosion

Rescue from Burning Structures

emanate from such openings. The gaseous products of the explosion will expand with some force as they are driven through the openings because of the lesser pressure of the atmosphere outside the building.

Backdraft Indications from Inside the Structure

There are some definite indications within a structure of an imminent backdraft, but unfortunately visibility inside a structure usually precludes such observations. However, there have been fire fighters who have lived through a smoke explosion and have shared their experiences. If the flames assume a pale or sickly yellow hue and seem to lose their liveliness, this is an indication that the fire is becoming oxygen starved and that flammable gases are being generated. The presence of a heavy volume of hot, dense smoke swirling around with great force near the ceiling, or at all levels of the room, is a sign of an imminent backdraft; the smoke is usually grayish-yellow in color and the swirling atmosphere is generally accompanied by a peculiar whistling sound as fresh air is drawn inward.

If fire fighting or rescue personnel are inside the building when backdraft conditions develop, they should leave as quickly as possible. Personnel who have been caught in an explosion state that they could see it coming; under hazardous conditions, crews should drop to the floor and crawl out. A backdraft starts with what appears to be a reddish rolling mass of fire that rapidly increases in size and spreads throughout the entire structure, overtaking the fire fighters in the course of travel. While the force of the backdraft is evident from floor to ceiling, the fire itself usually stays high and comes only to within a few feet of the floor. Personnel who drop prone are usually not burned, nor are they thrown about by the force of the explosion. However, those who remain standing are not only burned on exposed skin surfaces, but are thrown considerable distances.

Backdraft Prevention

Backdraft conditions are not always readily apparent upon arrival at a structure fire; moreover, they may develop after fire fighting operations are underway. Assuming that hazardous conditions are recognized upon arrival, every effort must be exerted to prevent the backdraft from occurring. Prompt and proper ventilation will be the correct precaution, but even this cannot be relied upon to absolutely prevent an explosion. Ventilate in accordance with prescribed methods and attempt to cool down the atmosphere and drive out the flammable gases and vapors through the use of fog streams.

An example of how unpredictable smoke explosions can be occurred when two engine companies and one truck company responded to a reported fire in a large single-story nightclub in the early morning hours. The doors and windows, both front and rear, were very hot, so there was no doubt that conditions were right for a backdraft. The first engine company laid hose lines to the front door, charged them, and had personnel ready to advance the fire streams. The second engine company laid lines to the rear door, charged them, and were ready to go, but stood by; it is seldom advisable to advance fire streams simultaneously from two opposite directions. All crews were equipped with self-contained breathing apparatus. When the hose lines were in position, the truck company used power saws to cut a large hole in the hottest portion of the roof; immediately, flames leaped twenty feet (6 meters) into the air,

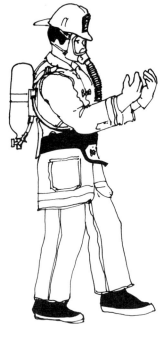

Fig. 12-5. When visibility is poor and hazards may be encountered, the body weight should be concentrated on the back foot as much as possible.

indicating that the hole had indeed been cut over the main body of the fire.

Upon being advised that the structure had been ventilated, the second engine company opened the rear door and windows, and took a position to one side of the doorway. The first engine company opened the front doors and windows and waited to see if an explosion would occur. A dense mass of thick black smoke poured from the front and back doors and windows; the roof opening still emitted flames very forcefully.

As the fire appeared to be confined to the rear half of the structure, it was decided to advance the search teams and hose lines from the front. This would prevent driving the flames through the uninvolved portions of the structure, which would happen if the rear fire streams were used. All the personnel except one were crawling on their hands and knees while working the fire streams to drive the heat and smoke ahead of them and out the rear of the structure; the one fire fighter standing was helping to advance a backup line in the rear. The smoke was so thick inside the building that visibility was absolutely zero. Suddenly a backdraft occurred; flames and smoke were propelled with tremendous force from the front and rear entrances of the structure. The personnel who were prone were not injured, but the one person who had been standing was very seriously burned. The crew at the rear door was still standing to one side, so they were not hurt. After the backdraft developed, the atmosphere rapidly cleared; the fire was quickly confined and extinguished.

Investigations showed that the correct fire fighting and ventilation procedures had been followed, and that all precautions had been taken. The nightclub had been redecorated many times and two separate and complete false ceilings had been installed under the original structure. The fire originated in a rear storeroom, spread up into the area between the false ceilings, and burned there until enough oxygen had been consumed to cause the blaze to subside into a mass of smoldering embers. The hole in the roof did ventilate the main body of the fire, but not the dense atmosphere of flammable gases inside the nightclub. Enough oxygen entered through the front and rear openings to create an atmosphere that was within the explosive range of the mixture; ignition of the mass occurred, and a backdraft explosion was inadvertently created.

This true experience shows that even when all precautions are taken, at the present time there is no method of being absolutely certain that a smoke explosion will not occur.

At a fire in a carpet warehouse, fire fighters were killed because the truck company personnel did not ventilate correctly. A large hole was cut in the roof before the nozzles were advanced; but the heat and flammable gases could not escape through the opening because no one poked a hole through the ceiling with a long pike pole. In this case, the ceiling was constructed several feet below the roof structure. Truck company personnel cut a large opening through the roof, felt down through the hole with an axe, could not detect a ceiling below, and moved on to other duties. The heat and flammable gases could not escape; they mushroomed down and filled the lower floors, and a violent backdraft explosion occurred after fire fighters were inside the structure.

Remember: When opening up a roof to ventilate a potentially explosive atmosphere in a structure, check to be certain that the heated smoke and gases are escaping through the hole. There may be an underlying ceiling to prevent the passage of smoke and gases; firedoors or an enclosed stairway may be interfering with the atmospheric circulation; or there could be a variety of other reasons that an opening in the roof is not allowing smoke and gases to escape. High-rise buildings often develop a condition that prevents the smoke from rising to the underside of the roof. The smoke from a fire in the basement or the lower floors will rise until it cools to the point that it is the same density as the surrounding atmosphere. This plug of cold smoke usually stratifies ten to fifteen floors above the fire, and it will effectively prevent the heat and smoke from rising any higher. Whenever vertical ventilation through the roof is prevented, fire fighters must depend on horizontal ventilation through windows and the use of power blowers.

Traveling With Poor Visibility

Emergency personnel often encounter conditions in which the visibility is either severely limited or nonexistent. The crews must be able to travel and still avoid whatever hazardous conditions might exist, such as open shafts and pits, energized electrical wires and equipment, or machinery.

A flashlight or other portable light should be used whenever possible, but very often such lights will offer little or no relief. When moving where visibility is poor, the only other option is to work by touch. Crawling or moving slowly in a crouched or kneeling position, while using the hands or a tool as probes by moving them from side to side, is safe, but slow. It is easy to become disoriented and lost while crawling, unless you are following a hose line or a guide rope.

When the amount of heat and smoke does not preclude adequate respiration (or if the rescuer is wearing breathing apparatus), a cautious method of traveling is by shuffling—not walking—while upright (Figure 12-5). The weight of the body should be poised on the rear foot until the advancing foot has tested the new ground and found that it is safe to move forward. The feet should not be lifted from the ground; they should slide forward, since this will make it easier to detect obstructions. Resting the weight of the body on one foot, the rescuer should slide the other foot forward to test the surface for clearance and stability. If the area is safe, the body weight should be gradually shifted to the other foot and the sequence repeated.

When any energized electrical wire or equipment is touched with the palm of the hand, even if the voltage is low, the muscles will tend to contract. The fingers will grip whatever they touch. This reaction will cause a person to freeze to the line. Persons feeling their way under circumstances where a charged wire, fence, metal door, or other electrified object may be accidentally contacted should proceed with arms outstretched and the palms of the hands towards the face. If the rescuer contacts an energized wire with the backs of the hands or fists rather than the palms, the current will cause the elbow muscles to tighten; this will pull the hand away from the contact and prevent it from grasping an exposed wire.

SURVIVAL IN A STRUCTURE FIRE

Anyone might someday find himself or herself the occupant of a burning building. Many of the casualties at residential fires could be avoided if a few simple precautions were observed. In the interests of public safety, fire service organizations should make a continuing effort to inform people in the community of ways to protect themselves in case of fire, as described in the following paragraphs.

When leaving an apartment or room during a fire, test the doors with the hands before opening them (Figure 12-6). If a door or doorknob is hot, or if smoke seeps out from around it, leave the door closed tightly and search for another exit. If the door feels cold, open it cautiously. Be prepared to slam it quickly if the hall is full of smoke or if heat or pressure is felt against the door. If the hallway is clear and free of excessive amounts of smoke and heat, proceed quickly to the nearest exit and leave the structure. Be sure to close all doors, since this will help prevent the fire from spreading. Doors must never be blocked in an open position.

A person trapped in an apartment or room who finds the exits blocked by heat, fire, or smoke should close the door tightly and seal the cracks around it. The person should also seal any vents to prevent them from admitting smoke. It is best to take refuge in a room with an outside window, unless the fire is extending upward on the outside of the structure by leaping from window to window. The more closed doors there are between the occupants and the fire, the greater the protection. If the exterior windows are free of heat and smoke, they should be opened from the top and bottom for ventilation. A sheet or other highly visible object hanging from a window is a signal that the occupant needs help. The person should wait for rescue aid to arrive, and not panic and jump prematurely from the window.

FIRE RESCUE BY POLICE OFFICERS

Evacuation of trapped and injured casualties from a structure fire often sounds relatively simple and uncomplicated during discussions of the correct techniques that trained fire fighting and rescue squad personnel should follow. However, law enforcement officers on routine patrol duty discover many fires or are the first emergency personnel on the scene. This is especially true for the early morning residential fires in which the hazards to life are especially great. These officers may have the responsibility for entering the burning building, alerting the occupants, and removing them from the structure without the safeguards that fire fighters and rescue squads routinely use, such as protective clothing, breathing apparatus, and experience at working in a hot, smoky environment.

If a police officer discovers a fire or other emergency incident that cannot be handled routinely, all pertinent information should immediately be transmitted to the dispatcher, so the fire department can be alerted. If the officer intends to enter the structure, this fact must also be conveyed to the dispatcher.

When there is more than one officer on the scene, a plan should be discussed before starting the operation. A prearranged place to meet after leaving the structure should be agreed upon before entering the building. Too often, officers have needlessly reentered burning structures in the mistaken belief that their partners had been trapped when, in reality, they were safe in an adjacent area.

If in plain clothes, officers should pin on their badges or other identification to provide recognition of their authority.

When approaching the structure, the officers should quickly question the escaped occupants and bystanders regarding inhabitants still in the building. Their responsibility should primarily be to alert the residents and assist small children, elderly persons, convalescents, and the blind.

Officers should walk, not run, to conserve their energy and air supply, since these assets will be very precious after they enter the building. It may be an advantage to take several deep breaths of fresh air just as if they were planning to dive into deep water.

It will seldom take as much time as most people think to traverse the halls and alert the occupants of a large hotel or apartment house about the fire, and to direct them to the nearest exit. In a simulated fire rescue, a police officer was able to walk from the squad car to a three-story apartment house, up the stairs, along the halls, and return to the car in under three minutes. Of course, more extensive structures or casualty complications will require more time.

Before entering a building or room, feel the wood panels of the door and the doorknob. If they are hot to the touch, that entrance must be avoided and another way to enter discovered. When deciding to enter a door or window, officers should remember that the dangers of a confined fire are very similar to those posed by a barricaded, armed criminal. Everyone must stay to one side of the entrance for protection from a possible blast of hot smoke and flames, a blast that could be just as deadly as a burst of bullets.

FIREGROUND SIZE-UP AND TACTICS

Before any search and rescue procedure can be started or any emergency operation performed, a size-up to detect the facts, probabilities, and possible complications of the situation must be made. Information must be gathered and evaluated before fire fighting strategy and tactics can be planned or search methods determined.

The size-up should begin before the company leaves their quarters and continue until the emergency situation is over. Information given at the time of dispatch may alert personnel to actions and tactics that may become necessary after arrival on the scene. While traveling to the location, it is considered good practice to mentally recall everything that is known about that particular section of the city or that precise building.

Size-Up Elements

After arriving on the fireground but before entering the structure, rescuers should note the width, depth, and height of the building, and the relationship of the fire to the building entrance. The volume of smoke and visible flame emanating from the structure will offer very graphic clues about the area of involvement. These signs will give some indication of the distance the rescue teams must travel before reaching the fire, and the most probable places where trapped or injured victims might be.

The type of construction is an important factor in rescue operations because it will determine whether the structure can contain the fire within the room of origin or whether it will allow the flames and heat to extend to other floors in the structure or to adjacent buildings. There are inherent dangers associated with a number of construction types. Some structures become unstable and are likely to collapse under fire conditions faster than others. Fire fighters and rescue personnel have been killed because they entered or remained in a building when the type of construction and the volume of fire

Fig. 12-6. Care must be used when opening a door. The door should not be opened if the door panels or doorknob is hot. Brace the door when opening it in case there are excessive pressure and heat on the other side.

indicated that there was imminent danger of collapse. It is everyone's responsibility to remain constantly on the alert for signs of structural danger and excessive fire spread, and to alert all personnel of the impending hazards.

A building heavily charged with smoke and high heat is ready for a flashover or a backdraft explosion. Many fire fighters and rescue personnel have been fatally or seriously burned because no one recognized the signs of the impending catastrophe. Every rescue operation must be planned and conducted so that personnel making the actual rescue effort will not themselves require rescue.

The type of occupancy and the time of day are extremely important factors to consider when sizing up a structure fire. Nighttime in a residential section suggests sleeping occupants; the same situation in a commercial area may mean empty buildings, except for possibly a security guard. Some businesses are naturally more susceptible to rescue problems than others. At two o'clock in the morning a hotel is more likely to have people in trouble than a warehouse. At two o'clock in the afternoon, a school fire would be handled differently than at two o'clock in the morning. A department store would undoubtedly contain more shoppers and other possible victims in the daytime than at night.

All personnel should carefully note and remember the locations of all doors, fire escapes, balconies, and other structural components and devices on and in the building. This information might later be critical to their survival or escape if they should become trapped by the fire or by the collapse of the building.

Rescue Priorities

Fire size-up and attack often become simultaneous activities on the fireground when it is instantly obvious that a rescue must be made or an exposure must be protected. Rescue should be the first concern upon arrival at the fireground, and the officer in charge must determine without doubt whether there is a possibility of anyone's still being in the burning building. The word of a hysterical bystander cannot be relied on, nor can it be ignored.

Rescuers should check with the members of the family, other employees, neighbors, and bystanders to determine whether there is anyone remaining in the building, and if so, where. A thorough search must be made, even if everyone has been accounted for. Occasionally, neighbors, friends, trespassers, or children will enter buildings without the knowledge of the occupants. A search must be made as rapidly and safely as possible; it should be systematic and thorough.

If a distraught man in nightclothes excitedly announces that his wife and children are still in the building, get as much information about their possible locations as necessary and launch an immediate rescue effort. If the same man says that his entire family escaped and are at a neighbor's house, the chances are that it is the truth. However, ask him if there was anyone else in the house; he may have forgotten about an overnight guest. Even if the residents of the burning structure assert that everyone is out of the building, a thorough search will have to be conducted in case there was an unauthorized guest or intruder in the house.

If a bystander or a neighbor claims that there are people still in the building, question him or her closely. Their information cannot be considered dependable, but it cannot be ignored.

When there is no one obviously trapped in the structure, it must be routinely searched anyway to make certain that no fire victim is overlooked.

If the need for a rescue is apparent upon arrival on the fireground, the efforts of all fire and rescue personnel may have to be concentrated on saving lives. This means that the main body of the fire may burn unchecked, and the exposures remain exposed, while the first hose lines are used to make the rescues possible. Very often, though, a prompt, massive attack on the fire and intelligent ventilation operations will save more lives than will assigning the available personnel to search and rescue activities.

Exposure Protection

Rescue work extends to the removal of persons from hazardous exposed structures. After rescue, the next action should be to protect the exposures, which could include the floor above the fire, adjacent rooms, and adjoining buildings. If a rescue is not too difficult and the initial response of personnel and equipment is sufficient, the officer in charge may be able to divide forces and accomplish both rescue work and the protection of exposures simultaneously. This, of course, is done whenever possible because it leads to confinement of the fire, which may facilitate the rescue.

The protection of exposures has the simple but vital virtue of reducing problems in the next fire attack step, which is to confine the fire. If the fire is in one room when the first companies arrive, the objective is to hold it to that room. If one floor is involved, then the goal is to keep the flames from spreading from that floor. If an entire building is blazing, then the fire should be confined to that structure.

When the fire is obviously confined, then the extinguishment phase of the attack starts. The pressure slacks off because the major problems have been taken care of.

As extinguishment progresses, some hose lines and nozzles are reduced in size; others are shut down. The extinguishment phase gradually progresses into the overhaul phase.

SEARCH AND RESCUE PROCEDURES

The search operation is divided into two basic categories: the primary search and the secondary search. The primary search is the immediate attempt to locate and remove trapped occupants. It is generally performed under the most adverse and hazardous conditions. The secondary, or follow-up, search is conducted as soon as fire conditions will allow a thorough inspection of the entire premises.

Primary Search

The crew arriving first will conduct the initial, or primary, search. This quest should be started on the fire floor, and should then move to the floor above the fire. The crews sent above the fire have the most dangerous position of any personnel at the fire. They are searching the area into which the fire is likely to extend, an area generally heavily charged with heat and smoke; because of this, it is necessary to assign personnel experienced in searching to the position above a fire. An inexperienced person should be teamed with an experienced member until the new person becomes proficient enough to be assigned to search alone above a fire.

METROPOLITAN RESCUE

A prerequisite of any search plan is that it be systematic. This will prevent missing victims or covering the same sections more than once. The primary search of a structure can combine several objectives, depending on fire and smoke conditions in the building. The major purpose of any search plan is to locate persons in need of rescue. At the same time, searchers can direct other occupants to exit paths from the structure. Searchers must use their judgment when large numbers of people must be evacuated. The decision must be made whether it is more important to assist the evacuation of the building or whether the rescue team should concentrate on continuing the search, and allow the fleeing occupants to escape any way they can. Varying conditions will lead to different actions. It is necessary that all search and rescue team personnel be sufficiently knowledgeable and experienced that they will be able to make rapid, intelligent decisions on the conduct of the lifesaving efforts.

Rescue personnel should be divided into two-person search teams whenever possible. Two or more teams may work together if the situation warrants such action. Full protective clothing, including gloves, helmets, and self-contained breathing apparatus, should be worn. Searchers must carry flashlights and forcible-entry tools. The two-person team operation will differ according to the structural conditions. In smaller buildings, such as homes, the team may split up as long as they stay on the same floor and maintain communication with each other. In larger structures, the two members should remain close together.

The search must be well defined and assigned to the teams so that the officer in charge will know where all personnel are working. This procedure is necessary in case the fire gains headway, structural collapse appears imminent, or some other hazard becomes apparent; it also prevents a duplication of effort.

When searching, it is essential that voice contact between members of the search team be maintained. Members can keep each other apprised of their progress and of conditions within the structure. Besides contributing to a better-coordinated search effort, this procedure will provide a degree of safety and encouragement for the search team. Portable radios should be used (if available) so searchers can maintain communication with each other. Communication enhances the safety of personnel and the coordination of their search effort, and enables them to obtain additional help if necessary.

Searchers for trapped or overcome victims may cause some structural damage. When forcible entry or ventilation is necessary, every effort must be made to keep damage to a minimum. If possible, master keys should be obtained from the management or employees to avoid the necessity of forcing doors. Damage will also be reduced by searching in a deliberate, systematic manner.

All personnel should remain alert for any evidence of arson, such as explosions, a strong odor of flammable liquids, descriptions of persons or vehicles that rapidly leave the scene, faces of people who have been observed at the scene of previous fires, spot fires burning in several locations not related to one point of origin, and any other suspicious occurrences.

When entering an occupancy to seek trapped or overcome casualties, check for the presence of fire in the search area. All movements within the fire building should be cautious and planned, especially if the searchers have entered sections above the fire. In general, it is best to avoid the center of the floor because it may have become weakened or burned away by the heat and flames below. It is usually safer and easier to follow along the walls to the door. Stairs must also be inspected to be certain that they will hold the rescuers' weight.

The floor above the fire is the most difficult and dangerous place to search. Occupants are seldom warned soon enough to escape, and they may be trapped or overcome by excessive heat, smoke, and toxic gases. Products of combustion can rapidly make the entire area untenable, and a sudden extension of the fire may cut off the escape route of the search team. These and other problems will probably occur and only a search team's training, knowledge, and experience will enable their survival.

There are, however, some basic precautions that should be observed. Before a team begins operations on the floor above the fire, an officer on the fire floor should be notified of that team's location. If there are enough personnel, a fire fighter should be left on the landing to check the progress of the hose line working below. If the intensity of the fire accelerates, an immediate warning should be sounded.

Before entering the room or apartment to be searched, the rescue team should gain access to a safe apartment on the same floor to use as a nearby place of safety should retreat become necessary. When searching the area above the fire, an effort should be made to ventilate not only that section, but also the room below. This will support the search team working below, as well as enable the hose line to advance.

Whenever fire is discovered above or at a location away from the main bulk of the blaze, it must be reported to the officer in charge and a call made for a charged hose line. When making a search above an area seriously involved in fire, a charged hose line of sufficient size to hold the blaze should be in position and operating. If this is not possible, rescuers must check to be sure that there is a secondary exit available to them.

If intense heat is encountered in a stairwell, it is generally the result of the normal upward movement of heated air and gases due to convection. There may be some tenable sections on the floors above the fire, and a concerted attempt should be made to penetrate this heat barrier or bypass it to continue the search. If a position on the stairs or in the hallway becomes untenable and entrance to an apartment or occupancy cannot be made, go down the stairs, not up. To be caught in a stairwell above a fire that gets out of control is exactly the same as being trapped in a stove flue.

Search Procedures

When entry at the front door is thwarted by heat and smoke, an attempt should be made to enter by another route. Consider using the fire escape, raising a ladder to a window, or going through a breached wall from an adjacent room. Entry through a breached partition is often the only way to bypass a blaze located near the only entrance into an area.

When moving about in a fire building, be especially careful when opening a door. Heat and flames may have accumulated on the other side of a partition; the rescuer who opens the door will be met with a blast of hot air and smoke. Always feel doors with the hands and open them cautiously. Often the fire can be isolated temporarily by closing a door, gaining more time for a thorough search of the remainder of the building. If an unusually long time is required to force doors above the fire floor, a check should be made from time to time to be sure that the fire is being contained by the fire streams below.

A search team must not allow any door to lock behind them because there is a possibility that they could become trapped. A long, wide strip of rubber, such as may be cut from a tire tube, with holes in the ends to slip over both doorknobs will allow a door to close but will prevent it from locking. The door may be kept open by placing the head of an axe or a wedge between the door and the jamb or floor. The use of any of these devices will also alert other search teams that the apartment has been entered and is being checked.

A trapped rescuer may be able to reach safety by breaching a wall into an adjoining room or occupancy. If trapped on an upper floor, enter a room, close the door, and open a window. Lean out, but stay low to remain out of the path of escaping hot gases. Safety is often as close as the nearest well-hung door. When trapped, with no other means of escape available, crews should find such a door and get behind its protection. When rescue personnel are cornered, they should throw a helmet outside to the ground, if that is possible. A helmet thrown to the ground around a fire building indicates that a person is trapped inside.

When intense heat or a heavy concentration of smoke is encountered, the searchers should stop and consider their situation. If the intense heat is evenly distributed throughout that section of the building, a decision must be made as to whether anyone without respiratory protection could possibly survive. The search team's responsibility is to find and save the living survivors. Time and effort must not be wasted on removing dead bodies during the primary search. Any delay in the operations caused by attempting to penetrate obviously intolerable conditions could increase the death toll.

Intense heat and heavy smoke at the ceiling level should not deter personnel from entering a room or apartment and making a quick search if it is known or suspected that an occupant may be present. However, if after the door is opened the environment is too hot to allow rescuers to proceed more than a few feet, the immediate area can be probed with the hands or a tool. Often a victim will lose consciousness before being able to unlock the door to escape. Rescuers can also use hands or a tool to probe the immediate area when entry through a window is denied because of heat.

As in many other rescue situations, searching is a race against time; aimless searching may cost the lives of occupants and rescuers alike. A thorough, efficient search method must be used. When searching an apartment or a hotel, two different systems may be used. One procedure would be to advance directly to the seat of the blaze and attempt to confine it by closing a door or using standpipe hose lines or extinguishers. The search would then be started at this point. This method is preferred when checking the section of the building where the fire started. When searching adjoining apartments and those on the upper floors, the search should start immediately upon entering the apartment or floor. Regardless of the starting point, every room, apartment, and other space must be entered and completely checked to be sure no victim is overlooked.

When the search is going to be made in a large area, thought should be given to placing a lifeline of small-diameter rope on the doorknob as a guideline. Once inside a room, the pattern that will

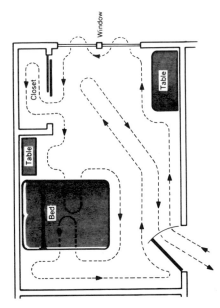

Fig. 12-7. Once inside a room, a searcher should make a complete circuit, keeping close to the wall, feeling under and on beds, and opening and feeling inside closets. As a final precaution, the room should be crossed diagonally to make sure that no one is lying in its center.

best ensure a quick and complete search should be followed. It is best always to work toward the second exit in case the entry point becomes blocked; this exit will sometimes be indicated by the direction of light or smoke travel.

Break a large picture window or plate glass store front from the top, and remove the hanging glass. Otherwise, anyone entering through the opening will be passing under a potential guillotine of glass.

Often after entry is made into an apartment, the search attempt is hampered by obstacles such as boxes, old furniture, and bicycles stored in the area near the entrance door. This situation usually indicates that the apartment has two entrance doors; one that is used, and one that is not being used. The area around the door that is not being used for entry purposes becomes a storage space. It might be easier and safer to find, and enter through, the door normally used for access.

A team of two searchers can check an average resident in a very short time by keeping on hands and knees, holding hands in order to not lose contact with each other, following the walls, and reaching out with outside hands and feet as far as possible. If heat and smoke prevent entrance to an area, that section must be probed with arms, legs, or tools to determine if anyone is lying close to a door or window. Victims are often found near an exit, particularly near the main stairs, fire escape windows, or other secondary means of departure that provide light, ventilation, or a possible escape route.

Smaller rooms may be searched by one person who follows a wall right from the entrance and continues to work around the room, feeling with the hands and feet. After one time around the room there will be a 4-foot (1.2-meters) perimeter that has been checked. The rescuer then crawls back and forth across the room, sweeping arms widely while traveling to cover the center of the room. One searcher can check a strip about 6-feet (1.8-meters) wide every trip across the room (Figure 12-7).

Rescuers should ventilate the structure as they move by opening the windows and doors, so long as the fire will not be extended by such ventilation. Because a fire increases in intensity when provided with a draft, the doors may have to be closed after entering. Fire conditions will dictate whether doors should be left open or closed.

When vision is obscured, the rescuer must feel ahead to check for open elevator shafts or other openings in the floor, and must remain alert to the possibility that fire or building collapse may cut off escape.

Rescuers must look under beds as well as on them and under the covers. An easy way to check under a bed is to insert a leg beneath the furniture and sweep it gently back and forth to locate any victims. When one has finished searching a bedroom, it is a common practice to flip the mattress into a U-position to indicate that the area has been checked. Since the mattress level of a crib or the upper level of a bunk bed is higher above the floor than the level of an ordinary bed, it is possible to overlook victims lying in bed when crawling through a dark, smoky room. The narrow, tapered wheeled legs of a crib and the extremely low mattress level of the lower bunk bed are characteristics that might help identify these critical search areas (Figure 12-8).

It is important to pause from time to time to listen for crying, coughing, moaning, or any other sounds indicating the presence of victims. It is necessary to check under beds, behind furniture, in closets, and in every other possible place a person may be concealed. Children and elderly people often hide from the fire in such places when they become frightened. Remember, basic instinct will drive adults in their panic and attempt to survive towards water. An especially careful check must be made in kitchens and bathrooms. A locked bathroom door is usually an indication that someone is inside. This means that bathtubs and shower stalls will have to be examined.

Every door, whether locked from the inside or the outside, must be opened so a search can be conducted. All doors within a given room must also be opened. Make certain that an opened door does not conceal either a victim or another door. Closets must be inspected. Do not pass up what might appear to be a pile of washing or clothing. It must be felt and inspected as it may be an unconscious victim. Small children often seek refuge in closets and toy chests (Figure 12-9).

The arrangement of furniture in a room sometimes creates spaces that are easily missed by searchers. For example, a sofa or

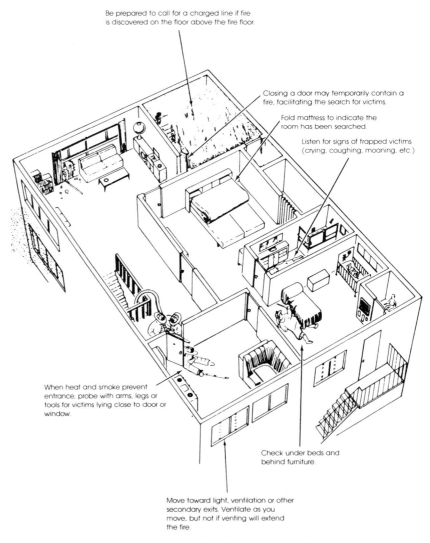

Fig. 12-8. Search methodically on the floor above a fire.

Rescue from Burning Structures

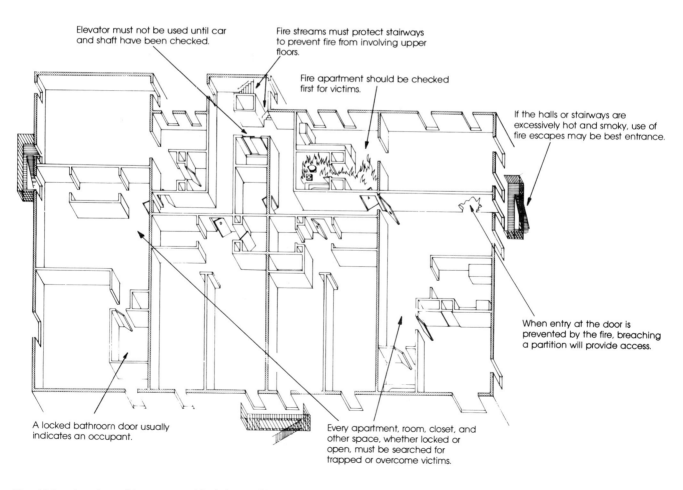

Fig. 12-9. Apartment houses and hotels must be carefully searched.

dresser placed diagonally across a corner of a room might conceal a body. When examining these spaces, avoid moving furniture, since this action may seal a door or even hide a victim. Dresser drawers must be checked. Experience shows that drawers are occasionally used as sleeping areas for small children and babies.

Areas near windows must be checked, since an occupant may have been trying to escape through the opening. Casualties are often discovered on the floor near windows and doors. When removing window draperies and curtains, care must be taken not to place them where they might conceal an unconscious victim.

Removing Victims

While searching, rescuers must constantly plan the best route for removing victims. A quick escape may suddenly become necessary if a developing hazardous situation should threaten the rescuers' lives. Everyone must keep in mind the heat and smoke conditions in the various sections of the structure, and how the environmental situation will affect a casualty without protective clothing or breathing apparatus. It is not always necessary, or perhaps possible, to escape the same way that an entry was made. It is well to consider using a fire escape, balcony window, or a fire department ladder, since these means of escape may be quicker and easier than leaving by the entry point.

Removing an unconscious victim is not always an easy task. If one person must remove the casualty without help, the important thing is to remain calm and quickly think out which removal route is best. Often a conscious victim on the verge of panic will grapple with a rescuer and cause both of them to lose their sense of direction. If a victim is in a state of panic or is highly excited, rescuers must be alert to the possibility that the victim might grab for the rescuer's breathing apparatus mask. It may be necessary for the rescuer and the victim to buddy breathe in a toxic atmosphere by alternately using a breathing apparatus mask. An attempt must be made to encourage the victim to have trust and to follow directions.

The discovery and removal of a casualty does not terminate the search efforts, even when all occupants have been accounted for. Any remaining unchecked sections must still be examined. Conscious victims who have been evacuated should be questioned immediately about the location of other possible occupants.

Upon discovery, a victim should be dragged to a place of safety. The fire fighter's carry should not be used because it requires considerable effort to lift an adult to the shoulders, and it also places both persons too high to have a source of breathable air in a smoky environment.

Marking System

A uniform system of marking the rooms, apartments, and sections of the structure that have been searched should be adopted and used by all search personnel. In a normal dwelling fire, no marking will be necessary because the teams will be in close contact with each other. However, in large hotels, apartment houses, office buildings, and other structures with many rooms and dividing walls, some marking system must be used to prevent duplication of effort. The identification method should be rapid, practical, and visible or touchable in dense smoke. Some organizations use tags or other objects that they hang on doorknobs or attach to the doors; others have reflective stickers, while still others carry large sticks of chalk and make a distinctive mark on the door. All methods have advantages and disadvantages, but all organizations must adopt some system that is understood by everyone.

Secondary Search

Frequently, the crews assigned to the primary, or initial, search must work under extremely difficult and hazardous conditions. Punishment from heat, smoke, and physical strain, combined with poor visibility, sometimes causes rescuers to miss victims during the early operations. However, when the intensity of the fire has been reduced (even though the fire may not necessarily be fully controlled), conditions will probably be sufficiently improved for a thorough and complete search to take place. It must be remembered that this secondary, or follow-up, search must emphasize thoroughness and not speed as its most important element. A minute and painstaking examination of the premises must be undertaken. If possible, it is best to avoid assigning the personnel who made the primary search to also conduct the secondary search. It is wisest to use fresh crews. The reason for this is that a more thorough search is conducted in this manner and that excessive punishment to certain individuals, because of heat and smoke, is avoided.

The secondary examination of the premises must be painstakingly thorough. It should begin at the roof and then proceed down through the building systematically. Nothing should be taken for granted; every possible hiding place must be searched. This exploration must include not only the immediate fire areas, but also other sections of the structure such as stairways, landings, roof bulkhead areas, cellars, basements, storage rooms, elevator cars, elevator roofs and shafts, and other places where occupants might have sought safety or jumped to escape the flames. The secondary search is not confined to the interior of the structure; outside areas such as rear and side yards, air and light shafts, adjoining roofs, building setbacks, and alleys must also be examined.

At some fires the building may be filled with debris, and a careful examination of tons of material may be necessary. In such cases, sufficient personnel should be assigned to make a thorough check. When examining cellars, basements, and other nonresidential sections of the structure, rescuers sometimes discover indigents, trespassers, and employees who were overcome, even though these areas were not directly exposed to heat and flames. An extra degree of thoroughness is necessary to reduce the number of fatalities in these locations. Rescuers should be impressed with the great importance of making a thorough search, even in buildings that appear to be vacant. Very often children and derelicts wander into vacant structures and are trapped by the flames.

As a rule, the search efforts can be considered successful when both the primary and secondary searches have been completed and all known occupants have been accounted for. However, it has happened that although all known residents had been accounted for, a visiting friend or relative had been overlooked—or had been included in the family count while an actual family member was, in fact, still missing. The best assurance that all victims have been found and evacuated is a professional search.

HIGH-RISE STRUCTURES

The modern trend of constructing buildings higher and larger has created or complicated some problems inherent to fire fighting and rescue. Simultaneously, innovative techniques for erecting tall structures have actually aided fire protection and safety. When a building has been properly designed, constructed, and maintained, any fire should be automatically contained and controlled without its extending beyond the floor on which it originated. The main problem encountered while combating a high-rise fire is a matter of logistics. It is usually a major problem to get sufficient personnel and equipment up to the correct level fast enough to launch a prompt, massive attack.

Rescue and Evacuation

Fire fighters or rescue personnel should be sent to the fire floor and above to reassure and help the occupants. If the hazards are not too great, it is best to request the people to close their doors and remain in their rooms. When there is a danger of the fire's extending to the upper floors, or if the occupants are apprehensive or insistent, they should be evacuated. If the elevators are not available for transportation, rescue teams should caution the people and direct them into the correct stairwell.

When the fire is on the upper floors of a high-rise structure, it may be advantageous to establish a safe refuge area a few floors above or below the fire area, rather than to evacuate the entire building. This will probably be faster, simpler to execute, less demanding on the physical strength of the occupants, and require fewer personnel. A large percentage of the population does not have the stamina to descend twenty, thirty, or forty flights of stairs without collapsing before the bottom is reached. When the top floors of a tall building must be evacuated, a helicopter sent to the roof may offer the fastest method of accomplishing the job.

Stairwells

Large buildings are provided with at least two interior stairways at opposite ends of the structure. The stairwell doors should be identified on both sides with the floor numbers. When only one of the stairwells continues on through to a door in the roof, laws may specify different color codes to indicate that fact. In many of these large structures, entrance may be made from the floor into the stairwell without any difficulty, but a key is necessary to gain access from the stairs into the floor area. This allows occupants to enter the stairwell for evacuation, but it prevents unauthorized persons from gaining access to secured sections of the building. Unless a door must be closed to prevent the passage of heat and smoke, a wedge or rubber lock-stop should be placed in position every time a door is opened, to prevent emergency personnel from becoming trapped in the stairwell.

Until the elevators have been inspected and judged to be safe, the stairways should be used to ascend to the fire floor. If the fire in the structure is well advanced, these stairways will be required for organizing an attack on the blaze. In this case, the stairway that is the more remote from the fire area should be used, to avoid forcing heat and smoke into the clear sections of the floor. When a choice can be made, the stairwell that extends through to a door in the roof should be used for attacking the fire. An open penthouse door over the stairwell will allow heat and smoke to escape; the other stairwell could then be used for occupant evacuation.

Heat and smoke *should not* be deliberately channeled into stairwells unless the fire is on the upper floors of the structure and the volume of heat and smoke does not justify breaking out windows. This method of ventilation is extremely inefficient since the smoke may cool and stratify; blowers would then be required to move it. Diverting heat and toxic gases into the stairwells may create a serious life hazard since the stairwells may be needed to provide rescue and fire personnel with access to the fire floor and the floors above, and to allow evacuation of the building occupants. When a stairwell must be used for ventilation, use the one that extends through the roof.

Under fire conditions or the threat of a possible blaze, the elevators should not be used until they have been placed under manual control; this will prevent them from being automatically called to the fire floor, jeopardizing the occupants. When the fire appears to be extensive, or if the cars and shafts may be involved, the elevators must not be used until they have been checked for safe operating conditions. This will usually require the initial fire attack company to climb the stairs to the fire floor and make a close inspection. Until the elevators and shafts have been inspected and determined to be safe, all cars should be summoned to the ground floor and inactivated with the STOP switch.

The taller structures are provided with two or three banks of elevators; the low-rise elevators service only the lower floors and the high-rise are restricted to the upper floors. The hoisting machinery is generally located on the floor above the highest story serviced by that bank of elevators. When a fire is on the upper floors, and caution in using the elevators is warranted, the low-rise bank may be safely used to its limit. There are usually no openings from the high-rise shaft into the lower intermediate floors. When the fire is in the lower stories, and the shaft is not involved, it may be safe to employ the high-rise bank of elevators to ascend above the fire.

For effective fire fighting and rescue operations during an emergency, all structural components and devices in the building— elevators, stairwells, standpipe hoses and outlets, air conditioning, and alarm systems—must have been inspected during prefire planning sessions. Adequate standard operating procedures for fire fighting and rescue personnel should be adopted during these sessions before the emergency occurs.

QUESTIONS FOR REVIEW

1. What should continually receive the most consideration at every structure fire?
    a. protecting exposed structures
    b. extinguishing of the main body of fire
    c. ensuring the life safety of both building occupants and emergency service personnel
    d. preventing the blaze from spreading into the uninvolved portions of the structure

2. To ensure the safety of the occupants of a burning structure, which of the following should receive first consideration from rescue personnel and fire fighters?
   a. exterior fire escapes
   b. interior stairways
   c. residents leaning out windows and yelling for help
   d. the attack on the main body of the fire

3. A backdraft explosion during fire-fighting operations is usually a result of the ignition of
   a. carbon monoxide
   b. flammable liquids
   c. carbon particles in the smoke
   d. superheated mixture of oxygen and nitrogen

4. The initial search of a burning building by the emergency personnel arriving first should be started on
   a. the top floor and moved downward.
   b. the lowest floor and moved progressively upward.
   c. the fire floor and moved upward.
   d. the floor above the fire and moved both ways simultaneously.

5. When groping through thick smoke where energized electrical wires may be encountered, which of the following methods is least recommended?
   a. feel with the backs of the hands
   b. feel with clenched fists
   c. feel with the backs of the arms
   d. feel with the palms of the hands

CHAPTER 13

Rescue Using Fire Apparatus

Occasionally, there are situations in which victims must be rescued from the roofs or upper floors of buildings, from sides of cliffs, or from canyons or ditches. A long crane can greatly simplify the hoisting and lowering operations required for such rescues, but unfortunately, this type of equipment is rarely available when it is most needed. If there is a construction crane with a skilled operator on the scene, this equipment should be used. While they may not be specifically designed for hoisting procedures, various types of fire department apparatus such as aerial ladders, elevating platforms, and tower ladders can, in most cases, be readily adapted for rescue work (Figure 13-1).

A basket stretcher is very useful for this kind of rescue because the victim is held securely and the stretcher frame is strong. Some organizations carry a specially designed four-point cable bridle that will snap onto the four corners of the basket stretcher and hold the victim level. If the patient is immobilized on a backboard, it may be placed in the basket stretcher without disturbing the casualty. When a stretcher is not available, straps with rings on the ends may be passed through the backboard handholds at each end for secure hoisting. The casualty must be secured to the litter or backboard with restraints so there is no chance of slipping or falling; this will help relieve a conscious victim's fears. Patients are usually more cooperative if their arms and hands are not restricted (Figure 13-2).

AERIAL LADDER

An aerial ladder truck may be used in several ways to remove casualties from buildings, ditches, canyons, and other places where access is difficult. Some manufacturers recommend that their trucks never handle weights in a derrick or crane fashion. Others not only sanction this use, but encourage it. When possible, it is best to avoid placing an undue strain on the tip of the fly extension. Some authorities advocate attaching the hoisting lines to the main section

METROPOLITAN RESCUE

Fig. 13-1. Rescuers using an elevating platform to remove victims from an elevated railroad accident. The basket of an elevating platform, which can be maneuvered with precision, can rapidly and easily evacuate injured or nonambulatory persons from above-ground positions to safety. (Courtesy of the Chicago Fire Department.)

Fig. 13-2. Using an aerial ladder to hoist an injured woman out of a deep canyon. Projecting a horizontally extended ladder over the edge of the canyon and then passing a long rope over the tip of the ladder will allow a Stokes stretcher to be lowered so the injured victim can be lifted out of the abyss by personnel pulling on the rope.

of the ladder for greater structural strength. Manufacturers' recommendations, which depend on the designed sturdiness of the ladder, should be respected when deciding on proper procedures at a fire or rescue incident. A straight ladder can be used to support the aerial ladder at low angles of inclination.

No matter what the rescue conditions are, remember that a ladder is much stronger when the pull is vertical along the center of the rungs, and not in a sidewise direction. First, spot the apparatus so the tip of the ladder can reach the victim. Before starting any rescue operations, the truck must be properly stabilized with outriggers, chock blocks, and jackknifing or positioning. If possible, place the truck so that the ladder extends over the cab or rear end; it is practically impossible to upset the vehicle when it is used in this position. The most unstable position is when the ladder is extended at a right angle to the chassis. The load limitations of ladders must also be remembered.

Removing Victims from Buildings

There are three methods of using an aerial ladder for removing a patient from the upper floors or roof of a structure. The position of the victim and available equipment will dictate which method is the most practical. Crews should be so familiar with all possible techniques that experimenting will not be necessary at the emergency scene.

When the ladder can be placed directly against the building, it may be possible to slide the stretcher down the rungs. Another way to lower a stretcher is to slip a short pike pole through the basket at each end under the top rail. After the victim has been strapped into position, the litter is placed feet downward with the pike poles resting on the ladder beams. A rope is attached to the top of the stretcher so that one person can control the speed of descent, while another person descends below to guide the litter on the way down (Figure 13-3).

A basket stretcher may be attached directly to the tip of the fly ladder and all the raising, lowering, and maneuvering accomplished with the ladder controls. Connect a four-point bridle to the stretcher or thread two large web belts through the litter frame to keep the patient level. If a direct connection to the ladder tip is to be made, the large snap hook on a safety belt provides a very fast and convenient way of doing this.

When a victim must be removed from a building or other high place where an aerial ladder cannot be placed, the apparatus can be used in place of a crane. First, extend the ladder while still nested in the bed until the tip protrudes past the ground ladders and other obstructions. Then stretch a length of 3/4-inch (1.88-centimeter) rope or other lifeline that is at least 150 feet (45 meters) long over the ladder rungs. One end is passed over the top of the ladder and is tied to the basket stretcher harness with a bowline knot. If available, a hose roller should be attached to the top rung to reduce strain and friction on the hoisting rope. The other end of the rope is looped once around the bottom rung of the ladder and the slack is placed on the ground to keep it from tangling. A light 3/8- or 1/2-inch (0.95- or 1.25-centimeter) line is fastened to the foot of the stretcher as a guide

Fig. 13-3. A rope is attached to the stretcher to control the speed of descent when the litter is slid down an aerial ladder.

Fig. 13-4. An aerial ladder is used to remove an injured person from the roof of a building. Passing a long rope over the tip of the ladder and then attaching it to the lifting bridle on a Stokes stretcher will allow the stretcher to be raised and lowered by personnel pulling on the line. An alternate method is to attach the stretcher bridle directly to the ladder tip and then perform all maneuvers with the ladder controls. (Courtesy of the Miami Fire Department.)

rope to keep it from swinging. At least one crew member should be assigned to control the large hoisting rope and one member to handle the light guy line (Figures 13-4 and 13-5).

The ladder is then raised out of the bed and rotated to the proper position. As the ladder is extended, the rescue member controlling the hoisting rope slacks off on the line so the stretcher remains about 10 feet (3 meters) below the tip of the ladder. After the tip of the

Fig. 13-5. Using an aerial ladder to lower a casualty from the upper floors of a building. The rope is passed over the tip of the ladder and attached to the lifting bridle of a Stokes stretcher. Fire fighters at the ladder base will pull the line taut and, after the stretcher has been passed through the window and is hanging free, release the rope gradually to gently lower the victim. To provide safe control, it is best to pass the rope around one of the bottom rungs once or twice; maintaining a slight tension on the rope, as the center fire fighter is doing, will permit a smooth descent.

ladder is positioned directly over the victim, or as close as possible, the litter can be lowered by slacking off on the hauling rope.

After the victim has been secured in the stretcher, the hauling rope is pulled up tight and the victim is raised, rotated, and lowered entirely by maneuvering the aerial ladder. After the victim has been hoisted a few feet (a meter or so), stop and check the fastenings for stability. When there are obstacles that prevent all the patient maneuvering and lowering to be performed by ladder operations, the personnel controlling the hauling rope can slack off on it gradually to assist the ladder operator. The member on the guideline controls the stretcher to prevent it from swinging, spinning, or striking obstacles.

Rescue from Canyon or Cliff

When a climber is trapped on the side of a cliff and cannot climb any higher or descend, it may be possible to spot a truck where the ladder can be maneuvered into a position from which the victim is reachable. Unfortunately, unless there is a road or clearing at the base of the mountain, this simple remedy is seldom one of the options available. In that event, a top hoist is necessary (Figure 13-6).

Fig. 13-6. Using an aerial platform to remove a casualty from an upper story window. (Courtesy of the Baltimore City Fire Department.)

Raising a victim from a canyon, out of an excavation, or up over a cliff edge requires the use of a hose roller and a hoisting line. The vehicle will be in the position of least stress if the truck is placed heading directly toward the cliff or excavation; this will position the wheels at the greatest possible distance from the edge of the bluff. Also, the truck itself will be in its most stable position if the ladder is extended over the windshield with the entire ladder and chassis in a straight line. While the ladder is still bedded, extend it a few feet (about a meter), and secure the hose roller to the top rung. Then pass the hoisting line over the hose roller; one end will be fastened to the stretcher and the other end will pass on top of the rungs to the base of the ladder, where a group of personnel will provide the hoisting power. Some crews pass the end of the rope around a handrail or bumper so one member can tighten up and provide a brake in case of an emergency.

The ladder is then hoisted, rotated, and extended over the cliff edge so that a vertical lift may be made. After hoisting the victim a few feet (a meter or so), stop and check all fastenings for security and make sure that the stretcher is level. Attach one or two guy lines to the litter to steady it and prevent it from spinning. This is a difficult and fatiguing task, so make certain that there are enough personnel on the hoisting line before starting. The efforts of the crew and the ladder operator must be coordinated because every change of ladder hoist and extension will affect the position of the victim and, consequently, the direction in which the rope must be pulled. After the stretcher is clear of the canyon, rotate the turntable until the victim is over a clear spot, and gently release the hoisting line.

ELEVATING PLATFORMS

Elevating platforms may also be used for hoisting and lowering victims; in fact, some of the baskets have booms and winches installed for this purpose. When the apparatus has this equipment, the casualty is placed in the stretcher, a four-point bridle or web straps are attached to the four corners, and the patient is raised or lowered after the platform basket is positioned vertically above the scene.

METROPOLITAN RESCUE

When the apparatus is not equipped with a winch, the platform can be used as an overhead anchor to support a snatch block or similar rigging. The winch and cable from a tow truck or rescue squad vehicle can then perform vertical lifts for hoisting casualties from a caisson, well, excavation, or other deep hole. The load limitation of the boom must be constantly considered (Figure 13-7).

Elevating platforms can be used to rescue trapped or injured victims from the upper floors of buildings, from elevated railroads, and from other aboveground structures in several ways. When the persons are ambulatory and uninjured, several may be placed in the basket and lowered to the ground. In case of a major fire with many trapped victims who must be removed from the heat and flames, use the apparatus to place the people on the roof of an adjacent structure; this rescue procedure is quickly accomplished. Lowering occupants to the ground is a slow method of evacuating many occupants; a ladder is much faster. When the patients must be transported while immobilized in a litter or on a backboard, the platform can be placed on the basket railings.

Fig. 13-7. When it is necessary to lift a victim from a caisson, frozen body of water, well, excavation, or other deep hole, and there is no means of attaching overhead rigging, an elevating platform or aerial ladder can be used as an overhead anchor. Use either a winch or a snatch block to apply the necessary force.

QUESTIONS FOR REVIEW

1. Aerial ladders and elevating platforms may occasionally be used at the scene of a rescue operation to take the place of a
   a. bulldozer.
   b. crane.
   c. power shovel.
   d. back hoe.

2. The best method of hoisting a casualty out of an excavation without causing any further injuries is with a
   a. basket stretcher.
   b. army stretcher.
   c. short backboard.
   d. long backboard.

3. When using an aerial ladder to hoist an injured victim out of a canyon, the most stable position for the truck is
   a. with the ladder extended at right angles to the chassis.
   b. with the ladder fully extended.
   c. with the chassis properly stabilized with outriggers.
   d. with the ladder extended over the truck cab.

4. When using an aerial ladder to remove an injured person from the roof of a tall building, which of the following methods is least recommended?
   a. Allow the patient to descend by himself so operations are expedited.
   b. Slide a basket stretcher down the ladder rungs.
   c. Attach a basket stretcher directly to the ladder tip.
   d. Pass a line over the ladder tip and attach it to a basket stretcher.

5. When an injured person must be hoisted out of a deep excavation that has sides that are likely to collapse from any force, which of the following would be most helpful?
   a. winch-equipped rescue truck
   b. ground ladders from a fire engine
   c. blade of a bulldozer
   d. bucket of an elevating platform

PART 4 ENVIRONMENTAL RESCUE

14 Handling Natural Disasters
15 Rescue from Landslides and Cave-ins
16 Rescue from Water
17 Wilderness Search and Rescue
18 Rescue by Helicopter

CHAPTER 14

Handling Natural Disasters

The American National Red Cross definition of a disaster is "an occurrence such as a hurricane, tornado, storm, flood, high water, wind-driven water, tidal wave, drought, blizzard, pestilence, famine, fire, explosion, building collapse, transportation wreck, or other situation that causes human suffering or creates human needs that the victims cannot alleviate without assistance."

Natural disasters can occur anywhere, at any time. Some catastrophes can be anticipated and prepared for; others are completely unpredictable. In any general disaster, lives can be saved and property damage reduced if people are prepared for the emergency, and know what action to take. Civil authorities, public-service organizations, and emergency rescue groups should cooperate in programs to educate the public on what to do when a disaster occurs.

The National Weather Service (NWS), a component of NOAA, the U.S. Commerce Department's National Oceanic and Atmospheric Administration, has made a detailed study of natural disasters. The Weather Service forecasts and keeps track of their occurrences, issues alerts, watches, and warnings, and prepares comprehensive precautions and safety rules. The timely warnings this agency provides against hazards from the air and ocean—such as hurricanes, tornadoes, severe storms, floods, tsunamis, and pollution-concentrating weather—have prevented great property loss and death. Because of these warnings, the population of this country often has a greater chance to know when disaster is about to strike, and so can take precautions ahead of time.

WEATHER

Weather is never static; it is always dynamic, with changes constantly occurring. Simply defined, weather is the state of the atmosphere surrounding the earth, and is controlled by the variable nature of the atmosphere. Weather conditions are affected by temperature, atmospheric pressure, wind speed, wind direction, humidity, visibility, clouds, and precipitation.

The atmosphere is a gaseous mantle encasing the earth and rotating with it in space. Heat from the sun causes continual changes in this mantle. These variations are interdependent, affecting one another in such a manner that weather is constantly changing in both time and space. The varying moods of the changeable weather found in the lower, denser atmosphere affect all of us. Occasionally, weather is violent and causes death and destruction with hurricanes, tornadoes, and blizzards. Often it becomes balmy with sunny days and mild temperatures. And sometimes it is oppressive with high humidities and hot temperatures.

WINTER STORMS

Most major disasters in the United States result from severe winter storms. In September, the sun leaves the northern hemisphere, its perpendicular rays drifting south of the equator. Until the sun's return in March, polar air rules the northern continental atmosphere, pushing back the tropical warmth of summer. Soon it is autumn, then winter, a season broken by intervals of fine weather, but also by winter storms. These storms trap travelers, take lives, destroy property, and paralyze cities.

Winter storms are generated (as are many of the thunderstorms of summer) from disturbances along the boundary between cold polar and warm tropical air masses, the fronts where air masses of different temperatures and densities wage their perpetual war of instability and equilibrium. The disturbances may become intense low-pressure systems, churning over tens of thousands of square miles in a great counterclockwise sweep.

In some parts of the United States, the Northern Rockies, for example, storms with snow followed by cold are a threat from mid-September to mid-May. During one of the colder months between November and March, it is not unusual for eight separate storms to affect some area across the continent. Intense winter storms are frequently accompanied by cold waves, ice or glaze, heavy snow, blizzards, or a combination of these. Often, in a single storm, the type of precipitation changes several times as the storm passes. The common feature of these storms is their ability to completely immobilize large areas and to isolate and kill persons and livestock in their path. In our northland, the severity of these storms makes them a seasonal threat. Farther south, the occasional penetration of severe winter storms into more moderate climates causes severe hardship and great loss of warm-weather crops.

Probably every winter is a bad year for some portion of the country. Even storms that do not break climatological records can kill. Their danger is persistent, year after year. From 1936 through 1969, snowstorms caused more than 3,000 deaths, directly or indirectly. Of those reported deaths, more than a third were attributed to automobile and other accidents; just less than one-third to overexertion, exhaustion, and consequent fatal heart attack; fewer than 400 to exposure and fatal freezing; and the rest to such causes as home fires, carbon monoxide poisoning in stalled cars, falls on slippery walks, electrocution from downed wires, and building collapse.

As winter begins over North America, Weather Service meteorologists work to detect the disturbances that may become severe winter storms, and develop warnings against the storms' approach. The terms *watch* and *warning* are used for winter storms, as well as for other natural hazards. The *watch* alerts the public that a storm has formed and is approaching the area. People in the alerted area should keep listening for the latest advisories over radio and television, and begin to take precautionary measures. The *warning* means that a storm is imminent and immediate action should be taken to protect life and property.

Freezing Rain

Freezing rain or freezing drizzle is rain or drizzle occurring when surface temperatures are below freezing (0°C or 32°F). The moisture falls in liquid form, but freezes on impact, resulting in a coating of ice glaze on all exposed objects. The occurrence of freezing rain or drizzle is often called an ice storm when a substantial glaze layer accumulates. Ice forming on exposed objects generally ranges from a thin glaze to coatings about an inch (2.5 centimeters) thick, but much thicker deposits have been observed.

A heavy accumulation of ice, especially when accompanied by high winds, devastates trees and transmission lines. Sidewalks, streets, and highways become extremely hazardous to pedestrians and motorists. Over 85 percent of ice-storm deaths are traffic related. Freezing rain and drizzle frequently occur for a short time as a transitory condition between a shower of rain or drizzle and the beginning of snow; this usually happens at temperatures slightly below freezing.

Ice storms are sometimes incorrectly referred to as sleet storms. Sleet can be easily identified as frozen rain drops (ice pellets) that bounce when hitting the ground or other objects. Sleet does not stick to trees and wires; but sleet in sufficient depth causes hazardous driving conditions.

The terms *ice storm, freezing rain,* and *freezing drizzle* warn the public that a coating of ice is to be expected on the ground and on other exposed surfaces. The qualifying term *heavy* is used to indicate an ice coating which, because of the extra weight of the ice, will cause significant damage to trees, overhead wires, and the like. Damage will be greater if the freezing rain or drizzle is accompanied by high winds.

Any road icing condition is extremely hazardous, as most drivers and pedestrians understand. Snow sometimes will provide traction even over an ice layer, and at temperatures just under freezing, traffic churns the snow into slush. But sometimes ice may cause a road condition not readily recognized—traffic melts the thin snow layer, which refreezes as ice and is polished by auto tires into a veritable skating rink. More light snow may obscure the layer of ice, leading unsuspecting drivers into mishaps and causing pedestrians to overestimate how much control drivers have of their vehicles.

Snow

The word *snow* in a National Weather Service (NWS) forecast, without a qualifying word such as *occasional* or *intermittent,* means that the fall of snow is of a steady nature and will probably continue for

Heavy snow warnings are issued by the NWS to the public when a fall of 4 inches (10 centimeters) or more is expected in a 12-hour period, or a fall of 6 inches (15 centimeters) or more is expected in a 24-hour period. Some variations on these rules may be used in different parts of the country. Where 4-inch (10-centimeter) snowfalls are common, for example, the emphasis on heavy snow is generally associated with 6 inches (15 centimeters) or more of snow. In other parts of the country where heavy snow is infrequent or in metropolitan areas with heavy traffic, a snowfall of 2 or 3 inches (5 or 7.5 centimeters) will justify a heavy snow warning.

Snow flurries are defined as snow falling for short durations at intermittent periods; however, snowfall during the flurries may reduce visibility to 1/8 mile (0.2 kilometer) or less. Accumulations of snow from snow flurries are generally small.

Snow squalls are brief, intense falls of snow and are comparable to summer rain showers. They are accompanied by gusty surface winds.

Blowing and drifting snow generally occur together, and result from strong winds that blow falling snow or loose snow on the ground. *Blowing snow* is defined as snow lifted from the surface by the wind and blown about to a degree that horizontal visibility is greatly restricted.

Drifting snow is used in forecasts to indicate that strong winds will blow falling snow or loose snow on the ground into significant drifts. In the northern plains, the combination of blowing and drifting snow, after a substantial snowfall has ended, is often referred to as a ground blizzard.

Blizzards, characterized by strong winds bearing large amounts of snow, are the most dramatic and perilous of all winter storms. Most of the snow accompanying a blizzard is in the form of fine, powdery particles of snow, which are whipped in such great quantities that at times visibility is only a few yards.

Blizzard warnings are issued when winds with speeds of at least 35 miles (56 kilometers) per hour are accompanied by considerable falling or blowing snow, so that visibility is dangerously restricted. These conditions make it very easy to become lost or stranded.

Severe blizzard warnings are issued when blizzards of extreme proportions are expected, involving winds of at least 45 miles (72 kilometers) per hour plus a great density of falling or blowing snow and a temperature of 10°F (-12°C) or lower.

Chill Index

A very strong wind combined with a temperature slightly below freezing can have the same chilling effect as a temperature nearly 50°F (28°C) lower in a calm atmosphere. Arctic explorers and military experts have developed what is called the wind chill factor, which shows the combined effects of wind and temperature as equivalent calm air temperatures. In effect, the wind chill index (see Table 14-1) describes the cooling power of the air on exposed flesh. Table 14-1 shows the cooling power of various combinations of wind and temperature; a table like this helps a rescuer determine how much protection the body really needs.

To determine wind chill, find the outside air temperature on

ENVIRONMENTAL RESCUE

the top line of the table, then read down the column to the measured wind speed. For example, when the outside air temperature is 0°F (-18°C) and the wind speed is 20 miles (32 kilometers) per hour, the rate of heat loss is the same as if the temperature were -35°F (-37°C) under calm conditions. To change Fahrenheit to Centigrade in the table, use the formula °C = (°F - 32)/1.8.

Table 14-1. EQUIVALENT WIND CHILL TEMPERATURES

WIND IN MPH	ACTUAL TEMPERATURES IN DEGREES FAHRENHEIT														
	35	30	25	20	15	10	5	0	-5	-10	-15	-20	-25	-30	-35
	EQUIVALENT WIND CHILL TEMPERATURES IN DEGREES FAHRENHEIT														
Calm	35	30	25	20	15	10	5	0	-5	-10	-15	-20	-25	-30	-35
5	30	25	20	15	10	5	0	-5	-10	-15	-20	-25	-30	-35	-40
10	20	15	10	5	0	-10	-15	-20	-25	-35	-40	-45	-50	-60	-65
15	15	10	0	-5	-10	-20	-25	-30	-40	-45	-50	-60	-65	-70	-80
20	10	5	0	-10	-15	-25	-30	-35	-45	-50	-60	-65	-75	-80	-85
25	10	0	-5	-15	-20	-30	-35	-45	-50	-60	-65	-75	-80	-90	-95
30	5	0	-10	-20	-25	-30	-40	-50	-55	-65	-70	-80	-85	-95	-100
35	5	-5	-10	-20	-30	-35	-40	-50	-60	-65	-75	-80	-90	-100	-105
40	0	-5	-15	-20	-30	-35	-45	-55	-60	-70	-75	-85	-95	-100	-110

Winds above 40 mph have little additional effect. There is little danger of freezing for persons who are properly clothed unless the Equivalent Wind Chill Temperature (EWCT) drops to below -20 degrees Fahrenheit; flesh can freeze within one minute when the EWCT drops to below -20°F.; flesh can freeze within 30 seconds when the EWCT drops to below -70°F.

Reducing Storm Hazards

During winter storms rescue teams may be called not only to the scene of accidents, but also to assist victims of overexertion or exposure, or people lost in blizzards or trapped in stalled automobiles. If the general populace were prepared to cope with heavy snowstorms and knew what to do and what to avoid, the hazards these storms present would be greatly reduced. Emergency service organizations should make every effort to disseminate information to the public along the following lines:

Safety Rules in a Snowstorm

Avoid overexertion and exposure. Exertion from attempting to push a car, shovel heavy drifts, or perform other difficult chores during the strong winds, blinding snow, and bitter cold of a blizzard may cause a heart attack; this can happen even to people in apparently good physical condition.

If your automobile has stalled, remain in the vehicle. Do not attempt to walk out of a blizzard. Disorientation comes quickly in blowing and drifting snow. Being lost in open country during a blizzard is almost certain death; a person is more likely to be found, and is better sheltered, in a car.

Keep fresh air in the automobile. Freezing wet snow and wind-driven snow can completely seal the passenger compartment.

Beware of the gentle killers, carbon monoxide and oxygen starvation. Run the engine and heater sparingly, and only with the downwind window open for ventilation.

Exercise by clapping your hands and moving your arms and legs vigorously from time to time; do not stay in one position for long.

Turn on the dome light at night to make the vehicle visible to rescue crews.

Keep watch. Do not permit all occupants of the car to sleep at the same time.

THUNDERSTORMS

Thunderstorms, generated by temperature imbalances in the atmosphere, are a violent example of convection. Cooling of the cloud tops or warming of the cloud base puts warmer, lighter air layers below colder, denser layers. The resulting instability causes convective overturning of the layers, with the heavier, denser layers sinking to the bottom and the lighter, warmer air rising rapidly.

Mechanical processes are also at work. Warm, buoyant air may be forced upward by the wedgelike undercutting of a cold air mass, or by flowing up a mountain slope. Winds blowing into the center of a low-pressure area may force warm air near that center upward.

In the first stage of thunderstorm development, an updraft drives warm air upwards until the water vapor it contains condenses into visible droplets and a cloud is formed. Continued upward movement produces a cumulus formation: rising mounds, domes, or towers. Air flows in through the clouds' sides, mixing with and feeding the updraft. Strong winds above the developing clouds may produce a chimney effect, drawing air upward to augment the updraft.

As each cloud forms, water vapor changes to liquid or frozen cloud particles, or sometimes to a mixture of both. This process results in a release of heat that becomes the principal source of energy for the developing cloud. Once the cloud has been formed by other forces, this release of heat helps keep it growing.

As the cloud particles grow by colliding and combining with one another, they may become rain, snow, hail, or form a mixture of these. Precipitation begins when the particles become heavy enough to fall, in spite of the updraft.

Having reached its final stage of growth, the cumulonimbus cloud, called a thunderstorm, may have a base several miles wide, and tower to altitudes of 40,000 feet (12,000 meters) or more. Winds shred the cloud top into the familiar anvil form. These cloud towers are sometimes visible as lonely giants, or moving several abreast in a squall line.

This stage is also marked by a change in wind flow within the storm cells. The prevailing updraft that initiated the clouds' growth is joined by a downdraft generated by precipitation. This updraft-downdraft couplet constitutes a single storm cell. Most storms are composed of several cells that form, survive for perhaps twenty minutes, and then die. New cells may replace old ones, so that some storms may last for several hours.

On the ground directly beneath the storm system, this stage is often accompanied by strong gusts of cold wind from the downdraft, or by a heavy precipitation of rain or hail. Lightning always accompanies the thunderstorm, indicating that the thunderstorm is in its most violent stage. Tornadoes may also be associated with the thunderstorm.

Even so, the thunderstorm cell has already begun to die. The violent downdraft, having shared the circulation with the sustaining updraft, now strangles it. Precipitation weakens, and the cold downdraft ceases. The thunderstorm cell, a short-lived creature, spreads and dies.

A thunderstorm's severity can be gauged by observing the intensity of its rain and the strength of the winds, and by watching for the occurrence of certain unique destructive phenomena such as lightning, thunder, and hail.

ENVIRONMENTAL RESCUE

Heavy Rains

Heavy rains mark the mature stage of a thunderstorm, and may produce dangerous flash floods even before precipitation ceases. Stay out of dry creek beds during thunderstorms.

Lightning

People have long marveled at lightning; it was the ultimate weapon of the gods of ancient civilizations. Today lightning is subjected to scientific scrutiny, but it is no less awesome than it was, and it still deserves respect.

Lightning kills more people in the United States than tornadoes, floods, or hurricanes. During the 34-year period that ended in 1974, lightning was responsible for the deaths of about 7,000 Americans—55 percent more than were killed by tornadoes and at least 41 percent more than were killed by hurricanes and floods combined. Because lightning usually kills only one person at a time, the deaths it causes do not attract nationwide attention, as do the deaths caused by such spectacular disasters as hurricanes, tornadoes, earthquakes, and floods, which may kill hundreds and cause millions of dollars in property damage.

Lightning is an effect of electrification within a thunderstorm. As the thunderstorm develops, interactions of charged particles produce an intense electrical field within a cloud. A large positive charge is usually concentrated in the frozen upper layers of the cloud, and a large negative charge along with a smaller positive area is found in the lower portion.

Usually the earth is negatively charged in relation to the atmosphere. But as a thunderstorm passes over the ground, the negative charge in the base of the cloud induces a positive charge on the ground below for several miles around the storm. The positive ground charge follows the storm like an electrical shadow, growing stronger as the negative cloud charge increases. The attraction between positive and negative charges makes the positive current on the ground flow up buildings, trees, and other elevated objects, in an effort to establish a flow of current. But air, which is a poor conductor of electricity, insulates the cloud and ground charges and prevents a flow of current until huge electrical charges are built up.

Lightning occurs when the difference between the positive and negative charges—the electrical potential—becomes great enough to overcome the resistance of the insulating air and to force a conductive path for current to flow between the two charges. Electrical potential in these cases can be as much as 100 million volts. Lightning strokes proceed from cloud to cloud, cloud to ground, or, where high structures are involved, from ground to cloud.

Relatively few people are killed indoors by lightning; the greatest number of indoor deaths are probably due to lightning-caused fires. A smaller proportion of indoor deaths and injuries from lightning happen to people using telephones or standing by or touching fixtures connected to house plumbing or electrical wiring.

City dwellers are somewhat protected from lightning by large steel-framed buildings. These tall structures tend to shield the adjacent areas and safely conduct the lightning into the ground. In rural areas, trees—particularly isolated ones—are likely targets for lightning because of their height. About one-fourth of all fatalities from lightning occur when people have sought shelter under trees. A considerable number of farmers have been killed while operating tractors in the field. In urban areas, the golf course has been a prime target for lightning. Many persons are struck by lightning while swimming, or while engaging in activities near or on beaches, piers, levees, small boats, or water skis.

About one percent of lightning facilities occur among people talking on the telephone. Many people do not consider them lethal instruments, but telephones, especially on rural lines, are capable of acquiring great electrical charges.

Reducing Lightning Hazards

Safety education for the general public is an important function of rescue organizations. Civil authorities and police and fire service organizations should cooperate to publicize safety precautions to follow during natural disasters; this would do much to alleviate the hazards that cost lives and cause injuries. Following are precautions to reduce the hazards of lightning.

Safety Precautions in a Thunderstorm

Develop the habit of keeping a weather eye on the sky and be ready to seek shelter if there is any indication of a thunderstorm approaching. The more severe the thunderstorm, the greater the intensity and frequency of lightning strokes. In almost all cases one can see the towering thunderhead and occasional flashes of lightning at least one-half hour in advance. Usually this is ample time to find shelter or take precautions.

All lightning experts agree that when a thunderstorm threatens, the most important single thing to do is to get inside a home or a large building. An all-metal automobile is an excellent lightning shelter; even if struck, the car allows the current to be discharged harmlessly into the ground. Once inside a structure, avoid using the telephone except for emergencies.

Do not stand underneath a natural lightning rod, such as a large tree in an open area. Avoid locations where you project above the surrounding landscape—do not stand on a hilltop, in an open field, or on the beach, and do not fish from a small boat. Get out of and away from open water; lightning current from a nearby stroke can flow through the water.

Stay away from tractors and other metal farm equipment. Get off and away from motorcycles, scooters, golf carts, and bicycles. Put down golf clubs and other metal objects.

Stay away from wire fences, clotheslines, metal pipes, rails, and other metallic paths which could carry lightning for some distance. Avoid seeking shelter in small isolated sheds or other small structures in open areas. In a forest, find protection in a low area under a thick growth of small trees. In open areas, go to a low place such as a ravine or valley.

If during a thunderstorm you are isolated in a level field or prairie and feel your hair stand on end, lightning is about to strike. Squat or kneel while bending forward to get as low as possible; place your hands on your knees, instead of on the ground. If you are in this position, the chances of lightning using the body as a conductor are minimized. This position is safer than lying flat, because it places a smaller body area in contact with the ground.

Groups of persons, such as hikers or mountain climbers, in exposed situations should spread out, staying several yards apart, so that if lightning strikes nearby the smallest number of people will be affected.

Thunder

The noise of thunder is due to compression waves, which result from the sudden heating and expansion of the air along the path of a lightning discharge. These compression waves are reflected from inversion layers, mountainsides, and the ground surface; this reflection and diffusion of the waves results in a rumbling sound instead of a sharp explosive crack, except when the discharge is very near. Because the speed of light is about one million times faster than the speed of sound, a lightning bolt can be seen before the sound of its accompanying thunder can be heard. This makes it possible to estimate the distance in miles to a lightning stroke by counting the number of seconds between lightning and thunder, and then dividing by five.

Hail

Hailstones are precipitation in the form of lumps of ice that occur during some thunderstorms. They range from pea size to the size of a grapefruit; their size is an indicator of the intensity of the thunderstorm.

Hailstones are usually round, but they may be conical, irregular in shape, or with pointed projections. The layered inner structure of hailstones indicates that hail enlarges by accretion of below-freezing liquid water on a growing ice particle, either by repeated lifting of the particle to levels of the atmosphere where freezing occurs, or by a long descent through strata of supercooled water.

RIVER FLOODS

The flooding of land adjoining the normal course of a stream or river has been a natural occurrence since the earth took its present form. What makes a flood a disaster is human occupancy of the flood plain (Figures 14-1 and 14-2). The economic attractiveness of the level, fertile land along these natural routes of communication has encouraged development of flood-prone areas, despite their potential for disaster. In spite of efforts over the years to provide protection for developed areas, losses from floods continue to mount.

While flooding can result from any accumulation or rise of water on land areas, this section treats only river floods. Flooding resulting from hurricanes, storm surges, and tsunamis is treated in other sections of this chapter. In popular usage, *river flood* describes an overflow of water onto normally dry land. More professionally, hydrologists consider a river to be in flood when its waters have risen to an elevation (referred to as *flood stage*) at which damage can occur in the absence of protective works.

Flash floods are a fact of life—and death—along the rivers, streambeds, and arroyos of the United States. These floods result when rain fills natural and man-made drainage systems to overflowing with water, carrying along with it uprooted trees, smashed structures, boulders, mud, and other debris. Year in and year out flash floods exact their toll of accidental deaths throughout the world.

Fig. 14-1. Fires often burn unchecked during floods, since fire fighters are unable to mount a large-scale attack on the blaze because of the high water. (Official U.S. Coast Guard Photo.)

Fig. 14-2. Flash-flood rescue operations in Montgomery County, Maryland required the combined efforts of the surrounding fire departments. Since water rescue incidents are frequent in this area, the rescue squad is equipped with easily portable boats. (Courtesy of NOAA.)

In 1972 torrential rains on the slopes of South Dakota's Black Hills caused catastrophic flash flooding along a 2-block-wide, 12-mile-(19-kilometer) long stretch of Rapid Creek, which flows through Rapid City. This disaster resulted in the death of more than 200 persons, and caused over $100 million in damage.

Effects of Flooding

A stream bed and the lands adjacent to it are integral parts of every natural watercourse. The ordinary, intermittent overflow of the stream from its normal bed deposits sediment to form the adjacent flood plains, which act as a natural reservoir and temporary channel. Flooding occurs when excess water rises over lands not normally covered by water, and when these lands are used for human habitation or economic enterprise.

The usual sources of excess water within the watersheds of a river are abnormally heavy rainfall or runoff from large accumulations of packed snow. Flooding from a snowpack is caused by rising temperatures (sometimes accompanied by rainfall), which melt the snow unusually fast.

A cause of serious flooding in the lower reaches of larger rivers is the concurrent arrival of flood crests from major tributaries. Flooding can also occur when ice jams are formed that block the river flow. The ice jams are caused by atmospheric and current conditions that induce breakup of river ice but fail to clear the channel. Stoppages will cause flooding, initially upstream and later downstream, when the ice jam breaks.

Primary Effects

The immediate or primary effects of floods result from inundation and the force of currents. Residents and livestock may be drowned, displaced, or injured by the floodwaters and the current-borne debris. Swift currents and debris cause structural damage to homes, buildings, highways, railroads, and bridges. Sanitary, electrical, water, and telephone installations are damaged. Business inventories and personal belongings are lost or damaged. Crops and plants are carried away by currents or destroyed by inundation. Farms can lose stored feed, equipment, and buildings. Farmlands may be deeply eroded by new channels or lose areas of valuable topsoil and vegetation.

The dislocations and disruptions caused by a flood can lead to hunger and disease among the affected population. Many people are left homeless. Obstructed transportation and communications can force evacuation of homes and separation of families. Health hazards are increased or created. Water may be contaminated through damage to water and sanitary systems. Wild animals and snakes may be driven into inhabited areas; rats, when forced from their regular habitat, can cause particularly serious health problems.

Secondary Effects

The force and depth of flood waters can cause disruptions and malfunctionings which in turn cause further damage or hazard. For example, short-term pollution of a river can be caused by chemicals released when a flood engulfs warehouses or other containers. Electrical fires can be caused by short circuits, and fire hazard can result from broken gas mains. These hazards are minimized or avoided by timely preventive measures.

Prediction and Forecasting

Floods are a natural and inevitable part of life along the rivers of our country. Some floods are seasonal, as when winter or spring rains and melting snows drain down narrow tributaries and fill river basins with too much water too quickly. Other floods are sudden, the result of heavy precipitation; these are flash floods, raging torrents that rip through river beds after heavy rains, surge over the banks, and sweep everything before them.

Timely and accurate prediction of the flood event is vital to the effectiveness of emergency measures; the endangered community must also have the training and equipment to prepare for the disaster.

Prediction of river floods involves estimating water runoff into tributary streams and eventually into the rivers. Estimates are based on precipitation in the form of rain or snow that has fallen in the watershed; the degree of soil saturation; the amount of water flowing in tributaries upstream from the forecast point; and atmospheric conditions that influence the continuation and intensity of rainfall or the rate of snowmelt.

Floods are of two types: those that develop and crest over a period of some twelve hours or more, and those that develop suddenly and crest within several hours or even minutes. The term *flood* as used here means those floods with a relatively long period of development. The term *flash flood* is used for those of relatively sudden onset.

The warning time varies by type and location of flooding. In mountainous areas where rapid runoff can lead to flash floods, there may be only an hour's warning at most, and often less. In areas where the terrain will retard the runoff or where flooding results from melting snowpack, the warning time can be days or even weeks.

The amount of warning has a direct relationship to the emergency measures that can be taken. With little warning, it may be possible only to evacuate people to high ground. With longer warning, property can be evacuated or protected, and emergency engineering protective works can be constructed.

Flash flood waves, moving at incredible speeds, can roll boulders, tear out trees, destroy buildings and bridges, and scour out new channels. Killing walls of water can reach 10 to 20 feet (3 to 6 meters) high. There will often be insufficient warning that these deadly, sudden floods are coming, since heavy rainfall, even for short periods of time, may be followed by flash flooding in mountainous or hilly areas.

Ultimately, however, it is the planning by civil authorities and emergency service organizations, training, and education of the local populace that determine the state of readiness and hence the effectiveness of any emergency measures.

Reducing Flood Hazards

People who live in flood areas should be informed about what they can do to protect themselves from the hazards. The authorities and weather services forecast and issue warnings of impending flood conditions, but an individual may still be caught in a dangerous situation, especially in flash flooding. Educational programs should include safety precautions to minimize flood hazards, along the lines of the following example:

Safety Precautions for Flood Areas

When possible flooding is predicted in your area, stay in touch with the news to be on the alert for weather developments and early flood warnings.

When a flash flood warning is issued for the area or there are indications that a flash flood is imminent, act quickly because there may be only seconds to reach a place of safety. Keep alert to signs of wet weather, such as thunder, lightning, or distant heavy rainfall. Watch for indicators of flash flooding, such as an increase in the speed of river flow or a rapid rise in stream level.

Never camp on low ground or in a narrow canyon. Stay out of natural streambeds, arroyos, and other drainage channels during and after rainstorms. Water runs from the higher elevations very rapidly. Remember, a person does not have to be at the bottom of a hill to be a target for flash flood dangers. Do not attempt to cross a flowing stream on foot where water is above the knees.

If driving, know the depth of water in a dip before crossing. The road may not be intact under the water. If the vehicle stalls, abandon it immediately and seek higher ground; rapidly rising water may engulf the vehicle and the occupants and sweep them away. Many deaths have been caused by attempts to move stalled vehicles. Be especially cautious at night when it is harder to recognize flood dangers.

HURRICANES

The king of disastrous weather is a hurricane; no other natural storm can match its size, intensity, duration, and destruction (Figures 14-3 and 14-4). Hurricanes originate in the summer and fall over water not far from the equator where the air is warm and moist.

A hurricane may last several days and ravage hundreds of square miles (square kilometers) of densely populated coastal regions

ENVIRONMENTAL RESCUE

Fig. 14-3. Hurricanes can cause total destruction of structures; this usually results in great numbers of trapped, injured, and dead victims. All debris must be carefully and meticulously checked, survivors should be closely questioned, and every effort made to find and treat all injured persons. The best protection against injury is to carefully heed all warning broadcasts of an impending storm. (Courtesy of NOAA.)

Fig. 14-4. Hurricane Camille ripped structures apart in Biloxi, Mississippi. No other natural storm can match the size, intensity, duration, and destruction that can be caused by a hurricane. Hurricanes create wind forces above 74 miles per hour; gusts may be over 175 mph. Coupled with the destructive winds of a hurricane are other dangers: storm surges that combine with high tides to smash several hundred yards inland; tornadoes that spin off from the walls of thunder clouds; and torrential rains. (Courtesy of NOAA.)

in the Western Hemisphere. Under hurricane-type storms are included all the violent and destructive wind storms belonging to the same family but with different names in various parts of the world. Along the southern and eastern coasts of the United States and in the Atlantic Ocean, Caribbean Sea, and Gulf of Mexico, these storms are known as West Indian hurricanes. Other names used to describe hurricane-type storms include tropical cyclones, typhoons, and baguios.

The disturbances are classified as depressions until wind force reaches 39 miles (62 kilometers) per hour, as tropical storms from 39 to 74 miles (62 to 118 kilometers) per hour, and as hurricanes above 74 miles (118 kilometers) per hour. The hurricane ambles sluggishly forward, usually at a speed under 15 miles (24 kilometers) per hour; sometimes it flies forward at velocities of up to 50 miles (80 kilometers) per hour. The storm increases in size until an area that may exceed 400 miles (640 kilometers) in diameter is blanketed by circular winds whirling with gusts that may be over 175 miles (20 kilometers) per hour.

Coupled with the destructive winds of a hurricane are sister killers: storm surges that combine with astronomically high tides to smash several hundred yards (meters) inland; killer tornadoes that spin off from the walls of thunderclouds that are pushed in front of the hurricane; and torrential rains that flood areas hundreds of miles from the coast.

Effects of Hurricanes

The first indication of a hurricane is usually an extensive mass of unsettled weather. The air begins to move toward and around a central area in which the barometer is falling. As the air moves, it gradually assumes a circular motion around the center of lowest pressure. Then the entire system begins to move, just as a spinning top moves across a floor. The circular or spinning motion becomes more violent as the hurricane develops and often reaches speeds greater than 100 miles (160 kilometers) per hour. A typhoon at Okinawa was recorded with winds up to 155 miles per hour and gusts over 180 miles (288 kilometers) per hour.

A developed hurricane is a vast whirlwind of extraordinary violence. Such storms vary from 25 to 600 miles (40 to 960 kilometers) in diameter. The mass rotates counterclockwise in the Northern Hemisphere and clockwise in the Southern Hemisphere. At the center of the disturbance there is an area called the eye of the hurricane, which may range from 30 to 50 miles (48 to 80 kilometers) or more in diameter. Here the winds are light and the sky is often clear and bright. As the axis of the hurricane passes over a given spot, the wind reaches a maximum velocity in one direction, diminishes to calm, and then increases to maximum force in the opposite direction. The forward motion of the whole storm system is usually 10 to 15 miles (16 to 24 kilometers) per hour while the hurricane is in the tropics, but as the storm moves northward its forward speed gradually increases; sometimes it reaches a speed as high as 30 to 40 miles (48 to 64 kilometers) per hour. As the storm reaches latitudes more remote from the equator, its force diminishes until it blows itself out.

The destructive forces in a hurricane grow rapidly with increased intensity. A 160-mile (256-kilometer) per hour wind is not merely twice as strong as an 80-mile (128-kilometer) per hour gale. If the wind doubles in velocity, the pressure exerted will be in proportion to the square of the difference. When the velocity doubles, the pressure will increase four times; when the velocity triples, the pressure will increase nine times. An 80-mile (128-kilometer) per hour wind can unroof buildings and down large trees. When 160-mile (256-kilometer) per hour winds blow, carrying with them trillions of tons of water weighing almost 1 ton (0.9 metric ton) per cubic yard (0.76 cubic meter), the appalling forces unleashed are almost unbelievable.

Another phenomenon accompanying a hurricane is the flooding of shorelines in its path. This results from the effect of the wind on the water over which it moves. Within the eye is a barometric low pressure, which, with the wind action, causes a water level rise of approximately 1 foot (0.3 meter) for each 1-inch (2.5-centimeter) drop of the barometer. If this action occurs simultaneously with the normal high tide, a disastrously extra-high tide can result.

Prediction and Forecasting

Officials of the National Hurricane Center (NHC) have recently reiterated their concern for the safety of the increasing populations along the Atlantic and Gulf Coasts where many areas lie less than 10 feet (3 meters) above sea level. Storm surges along these flat coasts can spell disaster. Hurricane Camille, for example, had surge heights exceeding 23 feet (7 meters), which were whipped by winds up to 230 miles (368 kilometers) per hour near the hurricane's eye. Extended pounding by massive wind-driven waves had a sledgehammer effect, demolishing any structure not specifically designed to withstand such forces. With the population growth in these vulnerable areas, more and more lives and property are at risk.

NHC meteorologists continuously monitor basic weather data for evidence of a tropical disturbance formation by noting changes in the pressure, wind, cloud, and rainfall characteristics. Once the disturbance is confirmed, attention is focused on determining whether the system will intensify, remain the same, or dissipate. If intensification occurs, a prediction of the path, speed, and direction of the storm is made. When a hurricane poses a threat, hurricane watches are issued at least 36 hours before a storm is expected to affect an area. Hurricane warnings refine the watch information and signify that hurricane conditions are imminent. These warnings are usually issued to allow residents at least 12 hours of daylight to take protective action and plan for evacuation. As the hurricane moves closer to land, advisories are issued every six hours, with intermediate bulletins every two or three hours.

Reducing Hurricane Hazards

For the permanent inhabitants of certain areas of the continental United States, notably the Gulf and East Coasts, there is no practical way to avoid a hurricane.

The greatest loss of life associated with hurricanes is from drowning, by a ratio of about nine to one. A secondary but serious danger is electrocution caused by fallen power lines. Often accompanying hurricanes are tornadoes, adding their characteristically severe winds to those of the storm. After spending its initial force, a hurricane may still bring damaging precipitation and dangerous flash floods.

Certain safety rules can be followed that will help to reduce the perils to human life. Educational programs sponsored by emergency rescue organizations and other public service organizations should make these available to everyone living in areas frequently visited by hurricanes. Following is an example of the kind of information that should be distributed and publicized:

Safety Precautions for Hurricane Areas

At the beginning of each hurricane season, stock up on food and medical supplies.

Before the arrival of a hurricane, board up windows or protect them with storm shutters, secure outdoor objects or place them under shelter, fuel up automobiles to permit rapid evacuation, and closely monitor the progress and position of the storm.

During the storm, remain indoors until notified that the hurricane has passed and that it is safe to go outdoors. If you live in low-lying areas or beach communities, check with local authorities for safe evacuation routes.

Beware of the eye of the hurricane. If the calm storm center passes directly overhead, there will be a lull in the wind lasting for a few minutes to a half hour or more. Stay in a safe place unless emergency repairs are absolutely necessary. But remember, at the other side of the eye, the winds rise very rapidly to hurricane force, and come from the opposite direction.

After the hurricane has passed, observe the following precautions:

Drive carefully along debris-filled streets. Roads can be undermined and may collapse under the weight of an automobile. Landslides along canyon walls and cuts are also a hazard.

Avoid loose or dangling wires; report them immediately to the power company.

Prevent fires. Lowered water pressure may make fire fighting difficult.

Remember that hurricanes moving inland can cause severe flooding. Stay away from river banks and streams. Listen for National Weather Service advisories to keep informed about river flood stages.

TORNADOES AND WINDSTORMS

Tornadoes are the most violent weather phenomena known to us; they are popularly known as *twisters* or *cyclones* (Figure 14-5). Their funnel-shaped clouds, rotating at velocities of up to 500 miles (800 kilometers) per hour, generally affect areas of 1/4 to 3/4 of a mile (0.4 to 1.2 kilometers) wide and seldom more than 16 miles (25.6 kilometers) long. However, they have been known to travel over areas measuring up to 1 mile (1.6 kilometers) wide and 300 miles (480 kilometers) long. Weather conditions that produce tornadoes also manifest themselves as a less violent phenomenon, the severe windstorm. Tornadoes and violent windstorms are treated as essentially the same, since in many ways the prediction and warning systems and preparedness measures apply to both.

If the same atmospheric conditions occur over a large body of water, the vortex phenomenon is referred to as a *waterspout*. Instead of the dust and debris found over land, the funnel cloud usually consists of water spray.

The violently destructive effects of tornadoes and windstorms have made them the number one natural disaster killers in the United States. During the past 50 years, tornadoes have killed almost 9,000 persons, while violent windstorms have slain an additional 9,500. By comparison, the other two prime killers, hurricanes and floods, have killed about 5,000 and 4,000 persons, respectively, during this period.

Tornadoes have been reported in areas outside the United States, but only very infrequently. In the United States they have occurred in all 50 states, with no season being free of them. Normally, the number of tornadoes is at its lowest during December and January and at its peak in May.

Often with little or no warning, a tornado can transform a thriving street in a community into a ruin in a matter of seconds. Death and injury result from such causes as the disintegration or collapse of buildings, debris driven by the high winds, flash flooding caused by the associated heavy downpour, and electrocution following fallen utility lines. The devastation, whether the tornado strikes a rural or urban section, is often freakish: giant trees are twisted and snapped; pieces of wood driven through metal; structures explode; a house destroyed while a neighbor's is untouched; heavy objects moved surprising distances; and dozens of stories of how people miraculously escaped death or injury.

Tornado Effects

Tornadoes are local storms of short duration formed of winds rotating at very high speeds, customarily in a counterclockwise direction. These storms are visible as a vortex, a whirlpool structure

Fig. 14-5. Tornadoes are the most violent weather phenomena known to humanity; they are popularly known as "twisters" or "cyclones." Their funnel-shaped clouds, rotating at velocities of up to 500 miles per hour, generally affect areas one-quarter to three-quarters of a mile wide. No other weather disaster strikes with such suddenness; timely warning is a necessity for saving lives. The abrupt appearance and erratic path of a tornado seldom afford the opportunity for evacuation; under such conditions, people must seek the closest protective shelter. (Courtesy of NOAA.)

of wind rotating around a hollow cavity in which centrifugal forces produce a partial vacuum. As condensation occurs around the vortex, a pale cloud appears; this is the familiar and frightening tornado funnel. Air surrounding the funnel is also part of the tornado vortex. As the storm moves along the ground, this outer ring of rotating winds becomes dark with dust and debris, which may eventually darken the entire funnel.

These small, severe storms form high above the earth's surface, usually during warm, humid, unsettled weather, and generally in conjunction with a severe thunderstorm. Sometimes a series of two or more tornadoes is associated with a parent thunderstorm. As the thunderstorm moves, tornadoes may form at intervals along its path, travel for a few miles, and dissipate. The forward speed of tornadoes has been observed to range from almost no motion to 70 miles (112 kilometers) per hour.

Funnels usually appear as an extension of the dark, heavy cumulonimbus clouds of thunderstorms, and stretch downward toward the ground. Some never reach the surface; others touch and rise again.

Windstorms refer to the damaging effects of winds caused by fast-moving frontal passages, thunderstorms, and squall lines that do not produce tornadoes. Windstorm causes, characteristics, and effects are similar to those of a tornado, except that they lack the extreme violent action and noticeable funnel. Since both of these natural hazards possess these similarities and produce widespread damage and loss of life, they are treated as one for the purpose of this study.

The destructive power of a tornado is due to the combined action of its strong rotary winds and the partial vacuum in the center of its vortex. As a tornado passes over a structure, the winds twist and rip at the outside at the same time that the abrupt pressure in the tornado's eye causes explosive overpressures inside the building. The combined effects of wind and vacuum produce nearly total destruction of structures. Walls collapse or topple outward, windows explode, and debris of this destruction is driven through the air in a dangerous barrage; a piece of straw can become a deadly missile. Heavy objects like machinery and railroad cars have been lifted and carried for considerable distances. Trees may be uprooted. People and animals have been hurled for hundreds of yards (meters).

Vertical and horizontal wind speeds within the vortex of a tornado have never been measured directly by instruments exposed in the tornado funnel. Invariably, the instruments are destroyed by the storm. Engineering studies of tornado damage show that the horizontal wind speed in the center of the tornado may be more than 300 miles (480 kilometers) per hour. The wind speed diminishes rapidly away from the funnel and is relatively light just a short distance away from the area of destruction.

The blast of wind from the rotating funnel is augmented by the tornado's forward speed over the ground, which may be as much as 50 miles (80 kilometers) per hour. This means that a tornado with a 150-mile (240-kilometer) per hour rotational speed, counterclockwise, and a 50 mile (80-kilometer) per hour ground speed will have a wind force on its right-hand side 100 miles (160 kilometers) per hour stronger than on the left.

Most tornadoes come from the southwest. This means that the extreme blast of wind will usually come from the same direction; rooms facing the approaching tornado will be the most dangerous place because they will receive the maximum impact. Conversely, the safest place in a structure with no basement will usually be on the lowest floor, in interior corridors that open only to the east and north, where wind forces usually will be less destructive.

In analyzing the damaging effects of tornadoes on structures, it should be noted that because of the nature of such storms, no one kind of aboveground building is completely safe against damage. Reinforced concrete structures are subject to less damage than the more temporary types. The only type of building that is safe from a tornado is an underground structure, and even these may be subjected to excessive debris loadings, negative and positive pressures, and other forces; these structures require special engineering design considerations.

Prediction and Forecasting

No other weather disaster strikes with such suddenness, making timely warning a necessity for saving lives. In view of the relatively short warning period, immediately available protective shelter becomes an equally critical factor. Residents of flood and hurricane threatened areas generally have ample time to evacuate the danger zone. The suddenness and the erratic path of a tornado seldom afford an opportunity to evacuate. People should be educated and warned to seek nearby protective shelter under these conditions.

At the Federal level, the National Weather Service of the National Oceanic and Atmospheric Administration is responsible for detecting tornado-producing weather conditions and for generating watch and warning messages to be transmitted to the public.

Forecasting that a tornado will or will not occur under given meteorological conditions, or that the funnel will strike at a particular location, is not within the present capabilities. Not enough is known about the basic nature of the phenomenon to permit this type of exact forecast. The NWS can only predict general areas where the probability of occurrence is greatest and notify those sections of the threat. When the area of possible occurrence is established, the tornado is actually detected by radar or human sighting. The tornado develops rapidly and is often short lived. The possibility that it might be obscured by rain or the dark of night is high.

Knowledge of the following general characteristics of tornadoes is useful in tornado preparedness planning:

Time of Day during which tornadoes are most likely to occur is midafternoon, generally between 3 and 7 P.M., but they have occurred at all times of the day.

Direction of Movement is usually from southwest to northeast. Tornadoes associated with hurricanes may move from an easterly direction.

Width of Path averages about 300 to 400 yards (270 to 360 meters), but tornadoes have cut swaths 1 mile (1.6 kilometers) or more in width.

Length of Path averages 4 miles (6 kilometers), but may reach 300 miles (480 kilometers).

Speed of Travel averages from 25 to 40 miles (40 to 64 kilometers) per hour, but speeds ranging from stationary to 68 miles (109 kilometers) per hour have been reported.

The Cloud directly associated with a tornado is a dark, heavy cumulonimbus, the familiar thunderstorm cloud, from which a whirling funnel-shaped pendant extends to the ground.

Precipitation associated with the tornado usually occurs first as rain just preceding the storm, frequently with hail, and as a heavy downpour immediately to the left of the tornado's path.

Sound occurring during a tornado has been described as a roaring, rushing noise, closely approximating that made by a train speeding through a tunnel or over a trestle, or the roar of many airplanes.

Reducing Tornado Hazards

The extremely high winds, overpressure, heavy rains, and hail associated with tornadoes and severe thunderstorms are the major causes of damage to human life and property. At present, there is no known method of totally eliminating these hazards. However, certain recommended safety rules can be followed that should reduce or lessen the perils to individuals. In the interests of public safety, civil authorities and emergency rescue organizations should make these rules known throughout the community. Because a tornado may strike at any time, this should be a continuing educational effort. Following is an example of the kind of tornado survival information that should be publicized:

Tornado Safety Rules

Seek shelter in a building of reinforced concrete construction, in a storm cellar, or in a small room or closet in the center of a house without a basement.

Keep windows open, but stay away from them.

Secure outdoor objects, such as garbage cans, garden tools, signs, porch furniture, and other loose items if time permits; this will prevent them from becoming missiles in the high winds.

Evacuate mobile homes. Structures can be protected only by being designed to withstand the high winds and pressure differentials associated with tornadoes. Mobile homes present a special problem because of their vulnerability to overturning during high winds. They should be firmly anchored with cables.

Evacuate auditoriums, gymnasiums, and other structures with wide, free-span roofs.

If driving a vehicle, abandon it and seek shelter or take cover outside in the nearest depression, ditch, or ravine.

In the open country, lie flat in the nearest depression, ditch, or ravine.

TSUNAMIS

The seismic sea wave, or tsunami, is not a tidal wave; it is not caused by the gravitational pull of sun and moon as the tide is. The phenomenon we call tsunami is a series of traveling ocean waves of extremely long length and period; they are generated by disturbances associated with earthquakes occurring below or near the ocean floor. As the tsunami crosses the deep ocean, its length from crest to crest may be 100 miles (160 kilometers) or more, its height from trough to crest only an insignificant distance. It cannot be felt aboard ships in deep water, and cannot be seen from the air. But in deep water, tsunami waves may reach forward speeds exceeding 620 miles (1,000 kilometers) per hour.

As the tsunami enters the shoaling water of coastlines in its path, the velocity of the waves diminishes and wave height increases. It is in these shallow waters that tsunamis become a threat to life and property, for they are transformed into terrifying monsters that can crest to heights of more than 115 feet (35 meters) and strike with devastating force (Figures 14-6 and 14-7).

Tsunami waves are a relatively rare phenomenon, neither completely understood by scientists nor fully appreciated by the inhabitants of vulnerable coastlines. Tsunamis are generally confined to the Pacific Basin and occur in almost all cases as a result of earthquakes along the Pacific seashore. Historically, a tsunami of major proportions has struck the United States or its Pacific possessions on the average of once every eight years.

Tsunami waves are extremely destructive to life and property. They may impact as a result of a distant earthquake on the opposite shore of the Pacific Ocean, or one may strike in the immediate earthquake epicenter region, as in Alaska in 1964. In 1960, tsunami damage from an earthquake off the Chilean coast caused $25.5 million damages in Hawaii and the west coast of the United States; 61 lives were lost at Hilo, Hawaii; major damage was caused in Los Angeles harbor.

Fig. 14-6. Devastation caused in Hilo, Hawaii, by a tsunami. Note the huge boulders carried inland by the power of the waves. A tsunami is a series of traveling ocean waves of extremely long length and period; they are generated by disturbances associated with earthquakes occurring below or near the ocean floor. In shallow waters tsunamis become a threat to life and property, for they can crest to heights of more than 100 feet (35 meters) and strike with devastating force. This particular tsunami was caused by a South American earthquake. (Courtesy of the International Tsunami Information Center, Honolulu, Hawaii.)

Fig. 14-7. Tsunami damage at Kodiak, Alaska, following the 1964 Alaskan earthquake. Waterfront structures were demolished and boats were driven up on shore. The tide in a harbor can vary tremendously as the result of a tsunami, even though the actual seismic wave in the ocean is relatively low in height. Unlike an ocean breaker, a tsunami does not batter or pound; it wreaks destruction by rapidly changing the level of the water within a constricted channel or harbor. (Courtesy of NOAA.)

Effects of Tsunamis

As a result of a tsunami, the tide in a harbor can vary tremendously, even though the actual seismic sea wave at sea is relatively low in height; however, the height is not necessarily a measure of the extent or power of a seismic sea wave. In fact, it should be emphasized that this is not a wave in the conventional sense, but rather a phenomenon that primarily affects the tide. Unlike an ocean breaker, it does not batter or pound; it wreaks destruction by rapidly changing the level of the water. The current can assume unbelievable speed in restricted channels, as much as 15 or 20 knots in the space of a few minutes. The current also reverses itself in the same short period of time; less than two or three minutes at the peak of intensity.

The effects of these waves on the coastal areas of the Pacific Ocean are characterized by maximum destructive force at the water's edge. Damage further inland is potentially high, even though the force of the wave has diminished, because of the floating debris that batters the inland installations. Ships moored in harbors often are swamped or left battered and stranded high on the shore. Breakwaters and piers collapse, sometimes because of scouring actions that sweep away their foundation material and often because of the sheer impact of the waves. Railroad yards and oil tank farms situated near the waterfront are particularly vulnerable. Oil fires frequently result and are spread by the waves.

Prediction and Forecasting

The prediction of tsunami generation is even more difficult than the prediction of earthquakes. Not only must the earthquake potential for a given area be predicted, but also the consequent potential for tsunami generation. However, due to the difference in velocity between earthquake waves (20,000 miles or 32,000 kilometers per hour) and tsunamis (600 miles or 960 kilometers per hour), it is possible to locate an earthquake once it has occurred, to analyze its potential for generating a tsunami, and to issue warnings in time for protective measures to be taken.

Within a few minutes after an earthquake occurs, seismologists analyze seismograms and report the information to the Tsunami Warning Center (TWC) in the Hawaiian Islands where the location and magnitude of the earthquake are determined. If a possible seismic sea wave is indicated, a tsunami watch is established and tide stations outward from the epicenter are queried for confirming water wave recordings. When positive wave action is reported by a tide station, the Honolulu center issues a tsunami warning, which includes any reported wave heights and the expected times of arrival for threatened areas. At the shoreline a brief warning is offered because an abnormally low tide immediately precedes a seismic sea wave.

Reducing Tsunami Hazards

Tsunami hazard reduction is of lesser scope than for other types of natural disasters because the affected section is usually limited to about 1 mile (1.6 kilometers) inland from the coast. If warning is received early enough, area authorities can take hasty preventive action; people can be evacuated, ships can clear harbors or seek a safer anchorage, airplanes and vehicles can be moved, buildings can be closed, shuttered, and sandbagged. Paradoxically, the very warning to alert the residents of a potential tsunami may cause people to endanger themselves so that they may watch such a spectacular event.

VOLCANOES

Worldwide, volcanoes are a very rare but extremely potent menace. Active volcanoes are few in number and the areas near most of them are sparsely populated. The greatest number of the largest eruptions have occurred in uninhabited sections; if such an eruption happened in a densely populated area of the country it would produce a catastrophe of enormous proportions.

Volcanic activity results in a number of different phenomena. Ground water is heated by molten material (magma). Geysers and hot springs are caused by this heating process. In addition, solid material such as cinders, bits of lava, and ashes are ejected into the atmosphere. The flow of lava down the volcano slopes often accompanies spectacular and violent eruptions. The rise of lava to the surface from depths of tens to hundreds of miles (kilometers) is the ultimate cause of volcanoes. Much of this material is deposited close to the vent or vents from which it was ejected and, in time, forms a hill or mountain—a volcano.

The chief dangers posed by volcanic eruptions are lava flows, airborne clouds of volcanic debris, and pyroclastic rocks and flows. Lava flows are the most familiar product of volcanic eruptions. They are essentially the overflow of magmatic material from the lava lake in the center of the volcano or an associated vent. Lava flows can be nonexplosive (effusive) or explosive (eruptive). Suboceanic volcanoes are usually the effusive type. Those on islands and continental land masses are generally eruptive.

EARTHQUAKES

Most Americans think earthquakes occur only in California. However, much of the United States has an earthquake potential (Figure 14-8). While it is true that most of the earthquakes do occur on the West Coast, other states that have experienced major earthquakes include Missouri, Montana, Nevada, New York, Oklahoma, South Carolina, Tennessee, and Washington.

An estimated one-half billion dollars of damage was caused by both of the most recent major earthquakes: the San Fernando Valley, California, in 1971 included 64 deaths; the area in and around Anchorage, Alaska, in 1964 took 131 lives. The San Francisco earthquake and fire of 1906 took the most lives, an estimated 500. The worst East Coast earthquake hit Charleston, South Carolina, in 1886, killing 60 persons.

Earthquakes, of all the natural disasters in this country, can inflict the greatest loss of life and property. They are the most difficult disaster phenomenon to prepare for because they may occur without warning at any time of the day, during any day of the year. In addition to the dangers of ground shaking and surface faulting, earthquakes often trigger the disastrous secondary effects of fire, floods caused by dam failure, landslides, and tsunami.

Most natural hazards can be detected before their threat matures. But a seism, from the Greek word *seismos* (earthquake), has no known precursors, and so it comes without warning. For this reason, earthquakes continue to kill in some areas at a level usually reserved for wars and epidemics. Nor is the horror of a lethal earthquake completed with a heavy death toll. The homeless living are left to cope with fire, looting, pestilence, fear, and the burden of rebuilding what the planet so easily shrugs away. Aftershocks of lesser intensity usually continue for many days.

Causes and Characteristics

The earth is not truly solid, but is in a state of constant flux. It is acted upon by periodic forces of the solar system that produce stresses and movement of the earth's surface. The relatively small portion of the crust at which the stresses are relieved by movement is the focus of an earthquake. From this point, mechanical energy is propagated in the form of waves that radiate from the focus in all directions through the body of the earth. When this energy arrives at the surface of the earth, sometimes from as deep as 434 miles (700 kilometers), it forms secondary surface waves of longer periods. The frequency and amplitude of the vibrations thus produced at points on the earth's surface, and hence the severity of the earthquake, depend on the amount of mechanical energy released at the focus, the distance and depth of the focus, and the structural properties of the rock or soil on or near the surface of the earth at the point of observation.

Intensity

Intensity is an indication of an earthquake's apparent severity at a specified location, as determined by experienced observers. Through

Handling Natural Disasters

Fig. 14-8. Earthquake damage in Anchorage, Alaska, following the 1964 disaster. Of all the natural disasters that occur in the United States, earthquakes can inflict the greatest loss of life and property. Since they may occur without warning, they are also the most difficult disaster to prepare for. In addition to ground shaking and surface faulting, earthquakes often trigger the disastrous secondary effects of fire, floods caused by dam failure, landslides, and tsunami. (Courtesy of NOAA.)

interviews with persons in the stricken area, damage surveys, and studies of earth movement, an earthquake's regional effects can be systematically described. For seismologists and emergency workers, intensity becomes an efficient shorthand for describing what an earthquake has done to a given section.

The Modified Mercalli Intensity Scale generally used in the United States grades observed effects into 12 classes ranging from I, felt under especially favorable circumstances, to XII, damage total. Rating earthquakes by intensity has the disadvantage of being always relative.

Magnitude

Magnitude expresses the amount of energy released by an earthquake as determined by measuring the amplitudes produced on standardized recording instruments. The persistent misconception that the Richter Scale rates the size of earthquakes on a scale of ten is extremely misleading, and has tended to mask the clear distinction between magnitude and intensity.

Earthquake magnitudes are similar to stellar magnitudes in that they describe the subject in absolute, not relative, terms; they refer to a logarithmic, not an arithmetic, scale. An earthquake of magnitude 8, for example, represents seismograph amplitudes 10 times larger than those of a magnitude 7 earthquake, 100 times larger than those of a magnitude 6 earthquake, and so on. There is no highest or lowest value.

Risks and Consequences

Thousands of small earthquakes occur in the United States every year. Moderate or severe earthquakes are relatively infrequent, but they pose a significant threat for which special hazard reduction and preparedness measures are needed.

In recent years, the United States has experienced one severe and one moderate earthquake. The severe earthquake in Alaska in 1964, which was between 8.3 and 8.7 on the Richter Scale, released energy equivalent to 100 underground 100-megaton nuclear explosions on one line. The Alaska earthquake caused $500 million in property damage, 131 deaths, and hundreds of injuries. The moderate earthquake in 1971 at San Fernando, California, was rated at 6.6 on the Richter Scale; it released only about 1/100 as much energy as the Alaska earthquake, yet it caused about the same amount of property damage and half as many deaths.

Most deaths caused directly by an earthquake are the result of structural collapse. The actual movement of the ground in an earthquake is seldom the direct cause of death or injury. The largest number of casualties result from falling objects and debris because the shocks can shake, damage, or demolish structures and other buildings. Earthquakes may also trigger landslides and generate huge ocean waves (seismic sea waves known as tsunamis), each of which can cause great damage.

Earthquake Effects

A large earthquake is one of nature's most devastating phenomena. The energy released by a magnitude 8.5 earthquake on the Richter Scale is equivalent to 12,000 times the energy released by the Hiroshima nuclear bomb. While these cataclysms have their foci well below the earth's surface, in the past, cities have been destroyed and thousands of lives lost in a few seconds as the result of great earthquakes.

Primary Effects

The onset of a large earthquake is initially signalled by a deep rumbling or by disturbed air making a rushing sound, followed shortly by a series of violent motions in the ground. The surroundings seem to disintegrate. Often the ground fissures and there can be large permanent displacements. Buildings, bridges, dams, tunnels, and other rigid structures are sheared in two or collapse when subjected to this movement. Vibrations are sometimes so severe that large trees are snapped off or uprooted. People standing have been knocked down and their legs broken by the sudden lateral accelerations.

As the vibrations continue, structures with different frequency response characteristics are set in motion. Sometimes resonant motion results. This is particularly destructive, since the amplitude of the vibrations increases (theoretically without limits) and usually structural failure occurs. Adjacent buildings of different frequency response can vibrate out of phase and pound each other to pieces. In any event, if the elastic strength of the structure is exceeded, cracking, spalling, and complete collapse may result. Chimneys, high-rise structures, water tanks, and bridges are especially vulnerable to vibrational motion.

The walls of high-rise buildings without adequate lateral bracing frequently fall outward, allowing the floors to cascade down one on top of the other, crushing the occupants between them. In the poorer countries, where mud brick and adobe are used extensively in construction, collapse is often total even to the point of returning the bricks to dust.

Water in tanks, ponds, and rivers is frequently thrown out of its confines. In lakes, an oscillation known as "seiching" occurs, in which the water surges from one end to the other, reaching great heights and overflowing the banks.

Secondary Effects

Often as destructive as the earthquake itself are the resulting secondary effects such as landslides, tsunamis, and floods. Landslides are especially damaging, and often account for the majority of lives lost. The fire damage frequently increases because of the loss of fire fighting equipment destroyed by the quake and the breaking of the water mains essential to fire extinguishment. Blocked highways can hinder the arrival of outside help.

Other secondary effects include the disruption of electric power and gas service, which further contributes to fire damage. Also, highways and rail systems are frequently put out of service, presenting special difficulties to rescue and relief workers.

Earthquake Preparedness

Although there is no way to eliminate all earthquake dangers, injury and damage can be reduced substantially, if preparedness measures are taken. The proper authorities to institute these measures are civil officials in cooperation with rescue and emergency service organizations and the police and fire services. In every community with earthquake potential people should be reminded of what to do in an earthquake. Because they come without warning, this should be a continuing educational effort. Following is an example of earthquake safety precautions that the public should know:

Earthquake Safety Procedures

The most important thing to do during an earthquake is to stay calm. Persons who remain cool and composed are less likely to be killed or injured; also, others in the vicinity have a greater tendency to stay calm, too. No moves or actions should be taken without thinking about the possible consequences. Motion during an earthquake is not constant; commonly, there are a few seconds between tremors.

During an Earthquake

If you are inside a structure, stand in a strong doorway or get under a desk, table, or bed. Watch for falling plaster, bricks, light fixtures, and other objects. Avoid tall furniture, such as china cabinets, bookcases, and shelves. Stay away from windows, mirrors, and chimneys.

Do not rush outside. Stairways and exits may be broken or may become jammed with people. Power for elevators and escalators may have failed and the cars could be stalled. Elevator cars may have been jolted off their tracks, so it is best not to trust them; it is safest to descend by the stairways. If in a crowded place such as a theater, athletic stadium, or store, do not rush for an exit because many others will do the same thing; it would be ironic if a person were to survive a quake and then be trampled by a stampeding mob. If it is necessary to leave a building, choose the exit with care and, when going out, take care to avoid falling debris and collapsing walls or chimneys.

If you are outside when an earthquake strikes try to stay away from high structures, walls, power poles, lamp posts, or any other construction that may fall.

Avoid falling or fallen electrical power lines.

If possible, go to an open area away from all hazards; do not run through the streets.

If you are in an automobile, stop in the safest possible place, which, of course, would be in an open section; remain in the car.

After an Earthquake

After a quake, the most important thing to do is check for injuries. Seriously injured persons should not be moved until medical help is available unless they are in immediate danger of further injury.

Watch for fires and fire hazards. If damage is severe, water lines to hydrants, telephone lines, and fire alarm systems may have been broken or disrupted.

Investigate utility lines for damage. If there are gas leaks, shut off the main valve, which is usually at the meter. Do not use matches, lighters, or open flame appliances until it is certain that there are no gas leaks. Do not use electrical switches or appliances around gas leaks or flammable liquid spills; they could emit sparks that might ignite the gas or vapors. Turn off the electrical power if there is damage to the wiring or equipment; the main switch is usually in or next to the main fuse or circuit breaker box. Spilled flammable liquids, medicines, drugs, and other harmful substances should be cleaned up as soon as possible.

Check fireplace chimneys carefully for cracks and other damages; is there any instability? When inspecting a chimney for damage, always approach it cautiously because weakened structures may collapse with the slightest of aftershocks. Particular investigations should be made of the roof line and in the attic because unnoticed damage can lead to a fire.

Stay away from beaches and other waterfront sections where seismic sea waves (tsunamis), sometimes incorrectly called tidal waves, could strike.

Avoid steep landslide-prone areas if possible, because aftershocks may trigger a landslide or avalanche, especially if there has been a lot of rain and the ground is saturated.

QUESTIONS FOR REVIEW

1. The official definition of a disaster is:
   a. Any calamity that causes the death of innocent victims.
   b. An accident that interferes with the orderly progress of civilization.
   c. A natural occurrence that creates havoc beyond the ordinary ability of authorities to cope with.
   d. A situation that causes human suffering or creates human needs that the victims cannot alleviate without assistance.

2. In this country, the greatest number of major disasters results from:
   a. Winter storms.
   b. Lightning and thunderstorms.
   c. Tornadoes.
   d. Hurricanes.

3. Which of the following has the greatest effect on the Chill Index?
   a. Humidity.
   b. Wind force.
   c. Wind direction.
   d. Dew point.

4. When a thunderstorm threatens, the best way to avoid being struck by lightning is to:
   a. Stand under a tall tree to ground the charge.
   b. Stand in the center of a large flat meadow.
   c. Get inside a building.
   d. Swim in the closest body of water.

5. The wildest and most disastrous weather is created by a:
   a. Hurricane.
   b. Tornado.
   c. Tsunami.
   d. Volcano.

CHAPTER 15

Rescue from Landslides and Cave-Ins

Rescue personnel are commonly summoned to accident scenes where casualties are trapped by landslides, cave-ins, and other incidents occurring in mines, excavations, tunnels, and other below-ground enclosures. Death or injury may be caused by a moving mass of dirt or rock, fires, explosions, or asphyxiation.

CAVE-INS AND LANDSLIDES

Locating and removing victims trapped under a landslide or cave-in present a number of difficult problems for the rescue team. Excavation and trench cave-ins account for a growing number of fatalities and serious injuries in construction work every year. Often the casualty is dead, but it is always possible that even after hours of being trapped, the victim may still be alive. The collapse of the soil or other material might have left a small void that permits the chest to expand and the entrapping substance might be porous enough for a little air to filter down to the victim. Therefore, quick action must always be taken (Figure 15-1).

Causes of Cave-ins

Improper shoring or no shoring at all is the leading cause of trench cave-ins. A ditch collapse is likely to occur when the material taken from the excavation is deposited too close to the trench edge. Improper sloping or undercutting of ditch walls can increase the risk of cave-in if equipment operators do not know what they are doing. Governmental regulations require that all trenches or excavations deeper than 5 feet (1.5 meters) be shored. The requirement also applies to some ditches less than 5 feet (1.5 meters) deep where soil conditions are unstable. Regulations also state that the spoil, which is the material taken from the trench, should be placed at least 2 feet (0.6 meter) from the edge of the excavation, and that it be retained by effective barriers. Contractors are inviting an accident if they

Rescue from Landslides and Cave-ins

disregard underground utilities such as water, gas, or oil pipelines, or if they do not check the shoring of the excavation each morning. Shoring may loosen overnight due to changes in weather conditions, temperature, or soil movement. Excessive erosion caused by rain or rivers may destroy building foundations. Structural collapse results in many casualties (Figure 15-2).

Fig. 15-1. A rescuer sets a hydraulic jack into position to prevent further collapse of the excavation. Oxygen is administered as soon as the victim's head is uncovered. (Courtesy of the New York City Fire Department.)

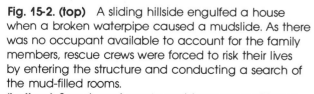

Fig. 15-2. (top) A sliding hillside engulfed a house when a broken waterpipe caused a mudslide. As there was no occupant available to account for the family members, rescue crews were forced to risk their lives by entering the structure and conducting a search of the mud-filled rooms.
(bottom) Search and rescue problems are multiplied when a landslide engulfed a residence. Rescue workers used timbers to stabilize the wreckage so they could enter, although it was still being demolished by a sliding mass of earth, rocks, and other debris. Unless there is a strong indication that there are live victims to be saved, the lives of rescue workers should not be jeopardized. (Courtesy of the Los Angeles County Sheriff's Department.)

ENVIRONMENTAL RESCUE

Size-Up

At the scene there is usually confusion and excitement. The first-responding personnel should contact the person in charge of the job or a responsible eyewitness and obtain as much pertinent information as possible. There may have been more than one person involved in the accident. To make sure, the person in charge must account for all of the workers.

The officer in charge of rescue should find out how long the victim has been buried before committing the crew to an immediate and frenzied rescue operation. If the victim has been buried for a long time, the chances for a successful rescue are remote, but cannot be disregarded. The working environment must be evaluated and the chances of another earth movement occurring should be estimated. After these facts and probabilities are obtained, then the officer can decide if there are sufficient personnel and equipment on the scene; if not, additional units should be requested.

Preliminary Operations

It is essential to avoid the danger of further cave-ins; the weight of heavy equipment and workers increases this hazard. Rescuers must wear helmets and protective clothing. Although it may appear awkward to work in a confined area while wearing these bulky items, they may save someone from injury or suffocation if another cave-in should occur. Be careful when working to avoid injuring other persons or the victim. Hand tools are usually the most effective means of rescue, unless the exact location of the casualty is known and it can be determined that there is no danger of inflicting further injury. Any mechanical digging equipment used at the scene must have a member assigned to offer close observation and guidance for the operator to prevent further injuries. Where there is danger of further collapse or slides, it may be possible to dig a hole a short distance away from the cave-in and then tunnel to the victim. Rescuers must be aware of any other hazards that might exist at the scene, such as underground electrical wiring, waterlines, explosives, or toxic or flammable gases.

Civilian construction workers at the scene may be able to help with mechanical and hydraulic shoring equipment, but their help should be enlisted only voluntarily. If construction personnel are attempting to find and remove the victim, do not insist that they allow rescue personnel to take their place. They are probably well trained in using hand tools and have a good idea of the casualty's location. Instead, ask them if they would like to be relieved and try to offer direction if it seems required. If untrained neighbors or bystanders are attempting the rescue, they should be replaced by the rescue crew.

Again, only the minimum number of personnel should be used in the danger area, but rotate personnel frequently. When members become fatigued, accidents become likely. Paramedics and other medical personnel will remain out of the immediate cave-in area, but be ready to move in as soon as the victim is located and exposed. Vehicles, bystanders, and heavy equipment must be kept well away from the danger area because their weight could cause additional landslides or cave-ins. Special care should be taken that rescuers do not stand directly over the location of the victim, as this weight could make the difference between life and death.

Digging and Shoring

While the digging and shoring operations are being carried out, there is usually great risk to the rescue crew. It is not practical to suspend operations until there is no danger, but care must be taken to observe as many safety precautions as possible.

In any cave-in rescue operation the safety of the rescue personnel must be assured by shoring up the excavation walls to prevent further collapse. As the digging activities progress, the shoring material is set up as quickly as possible. Only a minimum number of rescuers should be exposed to the hazards of another cave-in until the shoring is in place. Shoring should be placed on both sides of the cave-in area, and above the victim if possible. Timbers, planks, plywood, or other material may be positioned to help prevent slides. Tarpaulins can be used to keep sand and other small material from leaking through when shoring timbers are not placed very close together. The shoring should be braced with horizontal trench jacks or timbers at the top. Braces that hold the shoring in place must be as perpendicular to the shoring as possible. Bracing should start at the top and work downward. Rescuers must use caution that these braces are not accidentally displaced by the careless swinging of a tool (Figure 15-3).

Where there is a great amount of loose sliding material, such as dry sand, the material may be stabilized by spraying it with water. Moist sand will not slip as easily as dry sand. Use caution not to dampen it too much because the sand will turn into mud, which is more difficult to control. An oil drum with both ends cut out may be used to protect a casualty by gradually forcing it downwards, while loose material is removed from the inside.

Ladders should be lowered into the excavation on both sides of the cave-in area to provide entrance and a quick exit. They should extend above the trench. Stake and tie the ladders to prevent slipping; they might provide some support in case of further slides. Remove dirt by the use of buckets tied to ropes, which are pulled to the surface and dumped. Care must be taken when operating at the edge of an excavation in order to prevent a further collapse (Figure 15-4).

A horizontally extended aerial ladder or crane may be used to remove debris or the victim without placing pressure on the banks of the excavation. Two ladders or planks may be lashed together at the top and placed over the hole to serve as an A-frame; a block may be placed at the top with a rope run through it in order to make it serve as a crane.

Rescue members should be rotated often. In operations where there is danger of oxygen depletion, or toxic gases, it is necessary to circulate the air, employ breathing apparatus, or provide oxygen for both the workers and the victim. The oxygen hose from a welding torch, or a flow from an air compressor, has been successfully utilized to provide an adequate supply of oxygen.

Recovery of Victims

As soon as any portion of the casualty's body is uncovered, take immediate steps to locate and uncover the person's face and chest, and start administering oxygen. It is important that the rescuers expose the victim's full chest and head in order to allow for adequate chest expansion and a good respiratory exchange when oxygen is provided. When working on the victim, the rescuer should keep as much weight as possible away from the trapped person to

ENVIRONMENTAL RESCUE

relieve pressure on the chest cage; a plank to stand on may be advantageous. One member may be placed behind the victim to administer oxygen and support the person's head and body as they are exposed.

Rescuers must never pull on a part of the body; dig it out and lift it free. It is impossible to see through the dirt and other debris to ascertain if the victim is trapped by timbers or rocks; the slightest pull might further injure the person or collapse the excavation. If the casualty is alive and conscious, reassurance and as much comfort as is possible should be provided. A physician or EMT-P may administer a sedative if the rescue will take some time.

The victim should be removed in a basket stretcher or backboard in a horizontal position. Always assume that there are back and neck injuries in this type of accident.

MINE ACCIDENTS

A tunnel collapse is one of the worst kinds of mine accidents. A piece or mass of rock, coal, or ore falling from the roof of a tunnel

Fig. 15-3. The victim should be removed in a basket stretcher or backboard, in a horizontal position. Always assume that there are back and neck injuries. (Courtesy of the Los Angeles City Fire Department.)

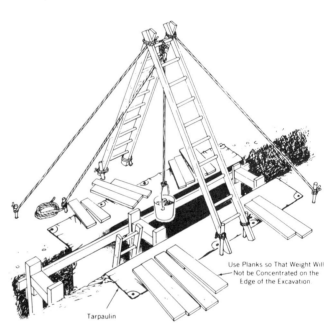

Fig. 15-4. Two ladders may be lashed together at the top and placed over a hole or ditch to serve as a sheerleg (or A-frame) for hoisting relatively heavy loads vertically. Passing a rope through a block that is fastened to the top lashings will provide an improvised crane. Tarpaulins may be used to keep sand and other small material from sliding into the ditch.

may kill a person instantly or pin the victim under a heavy mass of rock or debris. Fires and explosions also cause casualties to workers in coal and ore diggings.

Causes

The basic cause of mine fires and explosions, except those due to spontaneous combustion and friction, is an igniting agent in contact with combustible material, explosive gas, or flammable dust. A gas explosion is possible only if an explosive mixture of gas is permitted to accumulate and be ignited. Coal dust cannot detonate unless enough powder is allowed to collect so that it can be ignited and propagate or extend flame and violence.

Hazards

Fires and explosions are an ever-present hazard; in fact, they are the greatest dangers that confront officials and workers in the operation of mines. They not only jeopardize the lives of every person in the workings, but the survival of rescuers who participate in lifesaving operations. The most common peril in connection with fires is the asphyxiation of workers by smoke and fumes. Often this occurs in distant parts of the mines, as well as in proximity to the fire. Injury or death of rescuers while fighting the fire or removing casualties from a falling roof, tunnel collapse, or ignition of an explosive mixture of gas or dust may occur at any time. To those who must work on them, mine fires are far more hazardous than are explosions, because the explosion has usually already done its harm, whereas the fire may at any time cause a blast or other disastrous occurrence.

Undoubtedly, a fire in a dry, dusty bituminous coal mine that is liberating explosive gas in the vicinity of the fire area is the most hazardous to combat and extinguish. There is not only the danger of a gas explosion, but the greater hazard that even a small detonation may stir up and ignite coal dust. Falling roof material may throw a cloud of coal dust into the air that could come in contact with the flames of the fire and produce a widespread and destructive explosion.

Changes in Mine Atmosphere

In an explosion the gases are formed very quickly by rapid burning of gas or dust; the gases evolved by a fire usually result from slow burning of carbonaceous material. The gases from a fire are much like those from an explosion. The oxygen percentages will decrease, carbon dioxide and nitrogen concentrations will increase, and probably some carbon monoxide will be formed. Under some conditions, in both fires and explosions the carbon monoxide content may be very high and the oxygen concentration decidedly low.

Survival in a Mine Accident

When a fire or explosion occurs in a mine the first impulse of the survivors is to rush toward the nearest exit. To avoid such impulsive action, emergency rescue organizations should cooperate with mine safety officials in a continuing program of education. Mine workers should be instructed in underground safety precautions and proper procedures in case of accident or fire. If the workers encounter smoke and gases, especially if it is advancing upon them, they must retreat and attempt to find another possible means of escape; if no way out is found, they finally come to the point where they must do something to protect themselves. Usually there is at least one cool-headed person to take charge of excited and more or less terror-stricken people, and for the common good that person should be obeyed. When escape has been cut off because of smoke and gas, and the workers have retreated to an isolated section of the mine where there is comparatively good air, they should give prompt, careful thought to making an attempt to close themselves away from the oncoming toxic atmosphere by building a bulkhead in such a manner as to prevent poisonous gases from entering the region of refuge.

Rescue Operations

Notwithstanding the strain they are under, rescue personnel should try to remain cool and not rush aimlessly around. They should carefully check the maps of the mine workings and then decide the best routes of ingress and the most likely places where surviving workers may have taken refuge. It is best if only personnel with experience in that type of mining activity be used in the actual rescue operations. An adequate number of self-contained breathing apparatus, preferably of long duration, must be available for all personnel (Figure 15-5).

RESCUE FROM TANKS, PITS, AND HOLES

Rescue crews are often called to locations where workers have been overcome by toxic vapors and gases in tanks, pits, cesspools, tunnels, tank cars, or other below-grade or confined spaces. Entering such a structure without first considering the potential dangers can result in death. There are three possibilities under which the enclosed area could contain a hazardous environment: the atmosphere can be flammable or explosive, toxic, or oxygen deficient. The problems can be either singular or in combination. Therefore, the officer in charge must consider the probability that one or more of these three hazards is present. In addition to these perils, there is the possibility that an electrical danger or a flooding condition could also be present.

Preliminary Operations

Workers and other persons on the scene who may be familiar with the situation should be questioned to obtain information about the incident. Required information would include the number of casualties, the type of gases or vapors that are in the atmosphere, the length of time the victims have been incapacitated, and any other hazards that might be in the area.

Every rescuer entering the irrespirable atmosphere must wear a self-contained breathing apparatus. If the device has an optional pressure demand position on the regulator and mask, this feature should be used to prevent any possibility of inhaling toxic gases because of mask leakage. Where the manhole or other opening is too small to allow a person equipped with a respiratory protection device to pass through, the rescuer can don the mask and the rest of the unit can be lowered separately. Extreme caution must be exercised to avoid pulling the facepiece from the member's face.

It is a good precaution to attach a lifeline to each member working in the toxic area; a bowline knot should be used, provided it is snugly applied around the chest and under the armpits.

Ventilation

When convenient, placing a smoke ejector over the entrance to blow fresh air into the area will help ventilate. Compressed air hoses

Rescue from Landslides and Cave-ins

Fig. 15-5. A smoke-shrouded team of rescuers equipped with breathing apparatus, lights, and other equipment carry a victim of a coal-mine disaster. Respiratory protection devices and rescue techniques for mine accidents and fires are highly specialized. Since the emergency scene is usually a long distance from the mine entrance, breathing apparatus must be capable of providing complete protection for several hours. (Courtesy of the Bureau of Mines, U.S. Department of the Interior.)

lowered to the bottom of the compartment will also allow fresh air to flow into and purge the rescue area of suspected hazardous air contaminants. Air hoses may require tying to a stationary object or having a heavy weight attached to prevent whipping when pressure is released.

Lighting

Adequate lighting is necessary. If there is a chance that the atmosphere may be explosive or flammable, only explosion proof lights may be taken into the area.

Entering the Area

Most tanks and tank cars will have a permanently installed steel ladder on the inside wall; pits and cesspools probably will have no means of descending. If available, utilize the installed facility; otherwise, a ladder may be positioned through the opening. If neither of the above options is feasible, then it may be necessary to lower the rescuers with the lifelines. When this is done, it is best to place a hose roller or other padding on the edge of the entrance to reduce friction and wear on the rope and to decrease the physical strain on the members handling the hoisting line.

Victim Care

Rescuers should avoid any attempt to administer oxygen to the victim while still in the irrespirable atmosphere unless the person is trapped and cannot be promptly extricated. If there will be a delay, it is better to lower a spare breathing apparatus for the use of the victim rather than for the rescuer to share one unit by buddy breathing.

Personnel working close to the opening should be wearing breathing apparatus as there could conceivably be a hazardous concentration of gases and vapors for some distance around the hatch.

ENVIRONMENTAL RESCUE

Removing the Victim

Most methods of hoisting unconscious persons are complicated and time consuming, but when a victim is overcome by fumes the casualty must be removed as quickly as possible from the irrespirable atmosphere and resuscitation started. The best method of removal is to tie a clove hitch around the ankles. This knot ties easily and quickly, even under difficult conditions. Hoisting feet first, instead of head first, is preferable for the following reasons: There is less chance of further injury to the casualty; the victim's body, especially the head, will be less likely to catch on projections and will pass more easily through small openings; and mucus and other liquids may begin to drain from the respiratory passages while being hoisted. The bowline knot under the armpits can also be used to securely attach the victim to the hoisting rope, but the head may flop over and catch while an unconscious person is passing erect through a small opening.

When the casualty is in a tank, pit, or other relatively shallow enclosure, the person can probably be lifted out by several persons pulling on a lifeline, especially if a hose roller or other padding is used on the rim to prevent chafing of the rope. However, when victims must be hoisted out of deep shafts and tunnels, it may be more practical to use a construction crane or fire department aerial ladder as a portable elevator. Position the vehicle so the tip of the boom or ladder is directly over the opening. An improvised A-frame over the opening will allow a winch and cable to be used.

RESCUE FROM A PRESSURIZED TUNNEL

Many times tunneling operations require internal atmospheric pressurization to prevent the inflow of underground water or the collapse of unstable soil. To accomplish this, pressures as high as 50 pounds per square inch (psi) above the normal sea level atmospheric pressure of 14.7 psi may be needed. Rescuers responding to incidents in this type of environment will encounter several problems.

Pressures

To avoid possible confusion it is desirable to define what is meant by increased atmospheric pressure. In pressurized tunnels the air pressure usually is expressed in terms of pounds per square inch gage (psig). Gage pressure readings show the pressure above the normal atmospheric pressure of 14.7 pounds per square inch (psi) at sea level. Absolute pressure is equal to the gage pressure plus atmospheric pressure. For example, a gage reading of 14.7 psi is equivalent to 14.7 plus 14.7, or 29.4 pounds per square inch, absolute (psia). There are times when atmospheres are used. One atmosphere is the equivalent of 14.7 psi, which is the pressure at sea level. Gage or absolute pressures expressed in psi can be converted to atmospheres of pressure by dividing by 14.7.

Compression and Decompression

The exposure to increased pressure is divided into three phases. The first is undergoing compression, the second is working under the increased pressure, and third, is undergoing decompression.

Compression occurs when an individual enters a caisson or tunnel, which is at a pressure greater than atmospheric, through a lock or steel container, one end of which is connected to the caisson or tunnel under increased air pressure. The lock has two doors, one at the pressure end that opens outward into the caisson and the other at the atmospheric end. To enter the higher pressure atmosphere a person passes into the lock when it is at atmospheric pressure. The door at the pressure end is held tightly closed by the high pressure in the caisson. After the entrance door is closed, air from the caisson is admitted; the lock atmosphere equalizes with the pressure in the caisson or tunnel. This force holds the entrance door closed tightly. The interval of time required to produce the pressure equalization varies, depending on volume and rate of inlet of the higher pressure, but usually one to three minutes is required for equalization. When the pressures have equalized, the door leading into the tunnel can be opened readily and the person can enter the tunnel.

The only problem usually encountered in compression is equalizing the pressure in the middle ear with the external pressure. Should a person be unable to equalize the pressures because of an upper respiratory tract infection or other cause, intense pain will result and the eardrums may be ruptured. A spray to relieve nasal congestion may facilitate clearing the ears. Swallowing or yawning normally helps equalization and holding the nose and exhaling forcefully with the mouth closed may also help. A fire fighter or rescue member who has a cold or sinus problem should not attempt compression because an inability to equalize inner ear pressures may delay the rescue or fire fighting operation. There is a small percentage of the population that cannot adjust to pressurization and may suffer some harm; these persons should be designated as not to be assigned to pressurized tunnel emergencies.

After one has entered the pressurized tunnel, activity is about the same as at normal pressure. The nitrogen level of the body increases in both the tissues and fluids. The nitrogen increase depends on the pressure, length of exposure, and physical activity. To prevent absorption of excessive amounts of nitrogen, which can produce bends on decompression, the length of time a worker can be in a pressurized atmosphere is controlled by various state and local regulations. Basically, the permissible working period is decreased as the pressure is increased.

The situation in decompression is essentially the reverse of compression since the individual must readjust from an abnormal environment to a normal one. This is done by entering the lock, closing the tunnel door, and bleeding the excess pressure to the outside atmosphere until there is a pressure balance. Decompression must be done slowly to avoid nitrogen bubble formation, which causes the malady known as "the bends." The interval of time required varies, depending on the pressure to which the person has been exposed and the duration of exposure.

It is imperative that the rescue personnel or fire fighters have guidance and assistance from those knowledgeable persons who are in charge of the construction project. This discussion has only covered the basics; the experts on the scene must be relied upon for specific guidance.

Breathing Apparatus

The regulators on self-contained breathing apparatus are designed to admit respirable air to the lungs at the pressure of the surrounding atmosphere. Therefore, if the pressure is twice as high as normal, air at a pressure of 29.4 psi will be admitted to the mask and the duration of the respiratory protection device will be correspondingly reduced. A mask with a rated duration of 30 minutes will probably have a maximum life of 15 minutes. Tripling the atmospheric pressure will reduce the duration to one-third of normal. Simply stated, the service life of open circuit demand-type breathing apparatus is inversely proportional to the pressure of the atmosphere in which it is used. Because the duration of the respiratory protection device is reduced so significantly, the use of this type of equipment in a pressurized atmosphere is not safe unless a large compressed air cylinder or other source of breathing air can be utilized.

Closed-circuit breathing apparatus may be used because the increased atmospheric pressures do not affect the mask's duration. However, there is some controversy over the medical effects of breathing pure oxygen at elevated pressures.

Transporting Victims

After a worker has labored for a lengthy time in a highly pressurized atmosphere, it may require hours of controlled decompression to physically arrive safely back to normal atmospheric conditions. Should a serious accident or illness during working hours require the person's immediate hospitalization, the problem of rescue personnel promptly transporting the victim out of the pressurized tunnel and to the medical facility may create problems, because too rapid decompression may cause the patient to suffer from the bends or other malady. To accomplish this, some agencies maintain small portable decompression chambers for transporting casualties. The patient is positioned in the hyperbaric chamber, placed under controlled pressure, and transported.

Fires

Because of the need to maintain the internal air pressure if tunnel wall collapse and water intrusion is to be prevented, ventilation of an underground fire is impossible. Another factor that complicates tunnel fire fighting is that the fires may be a long distance from the access shafts which means that protection time for demand-type breathing apparatus under pressure is reduced drastically. Therefore, where fire fighters may be required to respond to tunneling operations, it is a good idea to develop auxiliary respiratory protection of long duration.

Drilling

Because there are problems in compression and decompression with or without breathing apparatus, and some other difficulties appear when masks are worn, organizations located in areas where tunneling operations under pressure are being conducted, with the consequent possibility of combatting a fire or performing a rescue operation at increased atmospheric pressures. Personnel should meet with the contractor doing the tunneling to work out procedures to be followed in case of accidents. This will eliminate, or at least significantly reduce, the potential problems. Untrained persons without previous experience should not try to be heroes. If advance planning has not been possible, then be sure to heed the advice of the contractor's personnel.

QUESTIONS FOR REVIEW

1. The leading cause of persons being trapped by a cave-in of a construction site ditch or trench is:
   a. Improper shoring.
   b. Loose, soft earth.
   c. Moist, muddy soil
   d. Improper sidewall slope.

2. When responding to a reported trapped victim in a ditch call, the first-arriving rescue personnel should initially:
   a. Jump into the ditch and attempt to uncover the victim's face.
   b. Properly shore the ditch so no further slides occur.
   c. Contact the job supervisor or responsible eyewitness.
   d. Remove the loose soil with the hands so the victim will receive no further injuries.

3. When rescue personnel are digging to uncover a buried victim, which of the following statements is least correct?
   a. Administer oxygen as soon as the patient's face is uncovered.
   b. Rescuers should stand on planks to relieve pressure on the victim's body.
   c. It is a vital necessity to expose the victim's full chest.
   d. As soon as the victim is uncovered down to his waist, he can usually be dragged free by lifting on the arms and body.

4. The greatest danger that confronts miners is one of the following:
   a. Cave-ins.
   b. Fires and explosions.
   c. Accumulations of toxic gases.
   d. Preignition of blasting materials.

5. Rescue personnel entering mines, tanks, pits, or other confined areas must always be equipped with:
   a. Lifelines.
   b. Flashlights.
   c. Self-contained breathing apparatus.
   d. Forcible entry tools.

CHAPTER 16

Rescue from Water

The number of lives lost through drowning is increasing every year because a greater number of people engage in water sports and recreation. In addition to deaths, a large number of victims become permanently disabled as a result of swimming, diving, and other water-related accidents. The survival techniques described in this chapter could form the basis of education and training programs conducted by emergency rescue personnel in the interests of promoting public safety. Educational materials could also be introduced into schools, athletic clubs, the YMCA, YWCA, and other similar groups. Missionary work of this nature would go a long way toward saving lives and avoiding injuries in water-related accidents.

If recovery of the victim is prompt and proper resuscitative measures taken, many lives may be saved. In any water accident, the alertness and preparedness of the rescuer is essential. Knowledge of the correct techniques for the various types of mishaps will aid in a quick recovery of the victim and prevent further harm to both the victim and rescuer in an already distressed situation.

Normally, the first person to see someone in trouble in the water is the one who should perform the rescue, if that person has the skill. When the initial observer is unable to offer aid, the exact location of the victim should be noted and help should be sought. The safety of the rescuer must always be considered. Many double drownings occur when untrained volunteers attempt swimming rescues.

A person does not have to be a swimmer to assist anyone in trouble on the surface of the water. The person on shore can often hold out a pole or stick to the drowning victim, or throw such things as an empty gasoline can, picnic jug, plastic bottle, or other flotation. If a rope is tossed, it is necessary to take time to properly coil the line and hold onto one end so it will not snarl. If the distance is too far to throw, or nothing is available, often a boat, log, paddle board, inflated inner tube, or other large floating object can be shoved out to the distressed victim. It is tragic enough to lose one person without also suffering the loss of a well-intentioned, but incapable, rescuer.

A near-drowning is similar to other types of emergency situations requiring resuscitation. The body must have oxygen and normal circulation for life to continue. In all cases, the drowning victim must be brought to the surface within a very few minutes to have a chance to survive. After rescue, artificial ventilation or cardiopulmonary resuscitation measures should be started immediately and continued until the person is either revived or pronounced dead by a physician (Figure 16-1).

DROWNING

Drowning occurs when the air passages are submerged in liquids. Drowning victims die because of asphyxia (lack of oxygen) or complications related to taking water into the lungs. Lack of oxygen leads very quickly to loss of consciousness, cardiac arrest, and brain damage.

The first reaction of a person when water enters the mouth and nose is to cough and swallow while trying to catch a breath. In doing this, large amounts of water may be taken into the stomach. During the struggle for breath, some water will enter the air passages, thus causing a spasm of the larynx; this will effectively seal the air passageway. The spasm will therefore result in a lack of oxygen to the lungs and this will rapidly cause unconsciousness. A very little amount of water, perhaps just a teaspoonful, will have reached the lungs at this stage. If rescue were to take place at this instant, immediate clearing of the airway and artificial ventilation would in most instances revive the victim.

Fig. 16-1. A tank ship explosion in a crowded harbor will undoubtedly result in many victims who require rescue. The entire ship, both above water and below, must be thoroughly searched to find all casualties.

ENVIRONMENTAL RESCUE

Heart arrhythmias are caused directly by hypoxia or by changes in the chemical composition of the blood. Fresh water absorbed into the body in large quantities through the lungs tends to dilute blood chemicals and cause the red cells to swell slightly; in some cases they burst. Salt water is more concentrated than blood; it tends to pull fluid out of the blood, thus causing pulmonary edema. Dirty water, chemicals, and salt residue all cause irritative lung problems that must be treated at a hospital. Rescuers should keep careful watch over the victims saved from drowning to be certain that breathing continues once it has been restored.

The fate of a victim threatened with drowning is determined by five factors:
1. The speed of rescue.
2. Prompt clearing of the airway.
3. Immediate artificial ventilation or cardiopulmonary resuscitation.
4. The kind of water in which the victim was submerged.
5. The victim's own ability to recover.

Emergency Care

The basic life support procedures that follow drowning are the same as those for other accidents, and CPR should be performed as quickly as possible. When attempting to rescue a drowning victim, the rescuer should get to the scene rapidly with some conveyance, such as a boat or surfboard; if a conveyance is not available, a flotation device should be carried. The rescuer always must exercise caution when attempting to aid a drowning person; under no circumstances should attempts be made that would endanger the rescuer's own life.

External cardiac compression should never be attempted in the water because it is impossible to perform it there effectively. Mouth-to-mouth or mouth-to-nose ventilation, however, can be performed in the water, although in deep water it is difficult and often impossible unless the rescuer has access to some type of flotation device on which to support the victim's head. Artificial ventilation should always be initiated as soon as possible, even before the victim is moved out of the water, into a boat, or onto a surfboard. As soon as the rescuer can stand in shallow water, work should begin.

In cases of suspected neck injury, the victim should, if at all possible, be floated onto a back support before being removed from the water. If artificial ventilation is required, the routine head tilt or jaw thrust maneuvers should not be used. Artificial ventilation should be accomplished with the head maintained in a neutral position, using the modified jaw thrust maneuver.

As soon as the victim is removed from the water, standard artificial ventilation or CPR should be performed. Drowning victims swallow large volumes of water and their stomachs usually become distended, which may impair ventilation; it may be necessary to remove excess water by turning the victim onto one side and by compressing the upper abdomen. The person may be turned over quickly into the prone position and lifted with the rescuer's hands under the stomach to force the water out; this is referred to as "breaking" the victim.

Fresh water absorbed into the body through the lungs will dilute the blood chemicals, thus causing the red cells to swell or burst. This change in the chemical composition of the blood may cause the heart to go into ventricular fibrillation. Salt water is more concentrated than blood; it tends to draw fluid out of the blood vessels into the lungs, which causes pulmonary edema. This alteration of the blood chemicals may also cause heart irregularities. There should be no delay in moving the victim to a life support unit where advanced life support facilities are available. Every submersion victim, even one who requires only minimal resuscitation, should be transferred to a medical facility for follow-up care.

Cold Water Drowning

There have been a large number of authenticated cases of drowning victims who have been submerged in cold water for up to 30 minutes and then were resuscitated with no brain damage or other serious aftereffects. Normally, 4 or 5 minutes is considered the upper limit of time the brain can go without oxygen before being irreversibly damaged. There are two natural defenses of the body that will extend the survival period in cold water drownings.

One key defense is that cold water below 68°F (20°C) lowers the body temperature; this slows the blood flow and the brain will consequently require less oxygen. Such cooling of the body is frequently employed in open heart surgery to prevent brain damage.

Another defense is the "mammalian diving reflex" that permits air-breathing seals and porpoises to stay under water for 30 minutes or so. The reflex slows the heartbeat. Less blood is pumped to tissues such as muscle, skin, and internal organs, which can survive with little oxygen; more goes to the brain and tissues needing more oxygen. The colder the water, the more profound the diving reflex, with, perhaps, more protection for the brain. The reflex is probably stronger or more active in children and young people; there is much less reflex effect in warmer waters. The precise role the diving reflex plays in humans is unclear; the reduction in the brain's oxygen requirements induced by the cold water is probably the more important contributor to the protective effect of cold water immersion.

In cold water immersion, the heart rate slows. There may be no detectable breathing and no palpable pulse. The skin gets little blood, so it becomes blue (cyanotic) and cold. The eyes assume a fixed and dilated state. To the lay person, and even to some experts, the victim appears clinically dead. These victims may be saved with quick and aggressive resuscitative efforts, even after apparently prolonged submersion. Resuscitation must begin immediately upon bringing the victim to the surface; it should include cardiopulmonary resuscitation with 100 percent oxygen if it is available.

SUBMERGED AUTOMOBILES

Highway accidents occasionally cause automobiles to become submerged in lakes, streams, or reservoirs. Unless the vehicle strikes another car or object before it plunges into the water, the occupants could remain conscious and uninjured. Normally an automobile will float momentarily, sink rather slowly, and take several seconds to reach a bottom 20 to 30 feet (6 to 9 meters) deep. The sinking time depends on several factors: weight of the car in relation to its size, the load it is carrying, and how airtight the passenger compartment is. The vehicle will descend heavy end first; the engine is usually the heaviest unless there is a heavy load in the trunk or rear seat.

ENVIRONMENTAL RESCUE

When the car hits the water on all four wheels, it will probably remain upright on the bottom. If the windows are open and the passengers are knowledgeable about the situation and capable of helping themselves, they can simply open the doors and float to the surface.

If the windows are closed and the passengers are able to help themselves, they must wait until the pressure equalizes inside and outside of the vehicle, take a deep breath, open the doors, and rise to the surface.

When the occupants are still in the vehicle when the rescuers arrive, rescuers must immediately dive and remove the victims. An air bubble will be found in all submerged vehicles for many minutes after the accident, so hope of life must be maintained. The air bubble will be at the highest point in the passenger compartment. If forcible entry must be made, it should be done at the lowest point so the trapped air will not escape. If the rescuers are unable to open the door because of the pressure or damage, then the glass must be smashed. Tempered glass in the side and rear windows can be shattered with a screwdriver or other sharp pointed tool with little effort. Divers should remain oriented so directions will not become confused; this will enable them to return to shore in the darkness.

Survival in a submerged automobile is very common. In fact, authorities in Holland regularly conduct courses in how to survive if an automobile runs off a dike.

DIVING ACCIDENT RESCUES

Incorrect diving techniques or an unfamiliar body of water can cause serious injuries to even the most experienced swimmers. People who dive into shallow water, strike their heads on underwater obstacles or the diving board, engage in horseplay, or enter the water at the wrong angle may suffer spinal injuries or other trauma. Vertebrae may be displaced or fractured, but with no injury to the spinal cord; if the neck can be kept from falling forward, backward, or rotating from side to side, injury to the spinal cord and the consequent paralysis can often be prevented.

The basic rescue rules are:
1. Do not remove the injured from the water.
2. Keep the victim floating on the back.
3. If the victim is not breathing, immediately start artificial ventilation.
4. Always support the head and neck level with the back.
5. Wait for sufficient help so the victim can be handled without further injury.
6. Immobilize the possible fracture by splinting the victim while floating.

Normal emergency medical care techniques require that accident victims with possible fractures be splinted where they lie. In water-related accidents, many rescuers forget this important principle because drowning is uppermost in their minds. When the victim comes to the surface, usually floating face down, the first thought is to get the victim out of the water quickly. No thought is given to the possibility of a broken neck or other spinal injuries. It is not uncommon to see excited and hysterical parents or bystanders, being untrained in the proper methods of rescue, turn the victim over quickly and roughly. This would cause the head to twist and drop back. The injured person is then lifted out of the water with no support to the head or back and carried to the side of the pool.

Floating the Victim

The correct handling of a diving accident victim can be divided into two parts: (1) the proper handling of the victim in the water and (2) the proper removal of the victim from the water. Prevention of further injury involves the correct way to turn the injured person over and the best flotation technique to be utilized until help arrives with a rigid body support to immobilize the victim. The water is a good support and the victim can be maintained afloat with minimum support and the neck will be in line with the body.

A person coming to the surface of the water with a neck injury may be floating face up or face down in the water. Usually the victim is unable to move the arms and legs, depending upon the severity of the blow to the head and the spinal cord injury. If floating face down, it is necessary to turn the person over quickly but with care so as to prevent drowning. The rescuer should approach the victim at the head. The rescuer then places the palm of one hand on the victim's back between the shoulder blades so that the rescuer's forearm rests gently against the crown of the victim's head. The palm of the other hand is then placed under the injured person's shoulder. The rescuer then pushes up on the shoulder and gently turns the victim; as the victim is rotated, the hand and arm on the victim's back and head is kept in place to act as a pivot and a support. With this motion, the body, including the head, rotates as a unit. There is no twisting of the head.

Once the victim is turned over, the rescuer then takes a position at the side of the body and slides one hand from the victim's shoulder down to the small of the back. This should be done in waist-deep water, if possible, as here the rescuer can stand and float the victim until help arrives. Only light pressure is required against the small of the back to float the injured person. As soon as possible, a spineboard or other rigid support should be placed under the victim's body.

It is not necessary to touch the victim's head while afloat because the buoyancy of the water acts as an adequate support and floats the head in its natural position. However, there are two exceptions to this rule: (1) lacerations to the head or face with severe bleeding and (2) cessation of breathing.

When a victim sustains injuries with severe hemorrhage, the rescuer should gently support the back of the head with one hand while direct pressure is applied to the open wound with the other hand. Simultaneously, attention should be devoted to maintaining the head, neck, and spinal column in a straight line. If there is more than one rescuer, the victim should continue to be floated while the second rescuer cares for the wound.

In diving accidents, a blow to the head may be hard enough to cause cessation of breathing or the victim may have floated face down too long. Once the victim is turned face up, check the breathing. If breathing has ceased, artificial ventilation must be started immediately. The rescuer then has two problems with which to be concerned: breathing and spinal injuries. With extreme care, both can be accomplished by a trained rescuer. In cases of suspected neck

ENVIRONMENTAL RESCUE

injury where artificial ventilation is required, the routine head tilt or jaw thrust maneuvers should not be used. Mouth-to-mouth breathing should be accomplished with the head maintained in a neutral position, using the modified jaw thrust maneuver.

Should the victim surface in deep water, the rescuer should approach at the head. The rescuer, treading water, should turn the victim face up and assist the person to float. By placing a hand firmly under each shoulder, the rescuer would use a frog or scissors kick while swimming on the side or back to slowly tow the injured person into shallow water, being careful not to strike the victim's body while kicking. The rescuer's hands must not cradle the head.

In rough water or beyond the surf breakers, it is preferable to float, support, and move the victim in a direction parallel to the waves in the water beyond the breaker line. Instead of attempting to take the victim in through the surf, it is better to wait for a lifeguard boat to take a spineboard out to the injured swimmer.

Removing the Victim from Water

The second part of the proper handling of diving injuries is concerned with the correct removal of the victim from the water. There is no haste in properly handling and removing an injured swimmer from the water because suspected neck injuries should be carefully treated, However, extremely low water and air temperatures, very rough water, or lacerations of the face and head may accelerate the hazards or traumatic shock; thus, rescue attempts require reasonable haste and care (Figure 16-2).

Removing the victim from the water can best be accomplished by means of a rigid support, which, when placed under the injured swimmer in the water, will maintain the body and head in a straight line. A rigid support need not be elaborate—a backboard, plank, door, or a water ski will do the job. Styrofoam surfboards are not recommended as they are often brittle and will tend to break when a person is lifted on them. Their extreme buoyancy makes them difficult to position under the victim without excess motions and effort.

When the injured swimmer has been correctly turned over and floated, two other rescuers should position the rigid support parallel to the victim's body and hold it on edge, vertical to the water surface; one rescuer should be at each end of the board. The primary rescuer then places one hand under the small of the back and the other hand under the legs so that the victim is kept in a straight line; this enables the board to be placed under the body without striking the feet. The board is then forced down in the water a short distance, leveled off, and brought up slowly until it makes contact with the body. Once the support is in position, the rescuer may slide hands out from under the victim. The injured person's head should be immobilized to keep it from moving by packing rolled up towels or clothing against each side; this prevents the head from rolling from side to side. To further immobilize the victim's body and assure staying on the board, some kind of tie can be used around the chest area and the upper thigh.

The rescuers should then gently transport the victim on the board to the side of the pool or body of water; a side step in unison will prevent excessive movement of the support. When the side of the pool is reached, the board and victim are lifted in a horizontal plane.

Fig. 16-2. A flash flood in Montgomery County, Maryland created a raging torrent of water that resulted in many injured victims. A rope stretched across the creek provided support for the victims and rescuers until safety could be reached. These victims of violent water accidents must be handled carefully, since the tumbling they received, and the rocks and other objects they may have collided with, could have resulted in fractures, spinal injuries, and other trauma in addition to the respiratory distress caused by being under water. All rescuers at water rescues should be equipped with some type of flotation device. (Courtesy of NOAA.)

SCUBA DIVER RESCUE

A qualified diver should know the difficulties that may arise in diving and to be alert and prepared for any mishap. However, during the summer months and vacations, persons who have been insufficiently trained, or are diving under unfamiliar conditions, encounter problems with which they are unable to cope. Overexertion is the most frequent cause of distress among inexperienced divers; this quickly results in panic and the diver quickly forgets everything he has been taught about survival. By the time a person puts on a wet suit, a weight belt, and a 35 to 40 pound (16 to 18 kilogram) tank, and then walks out on the rocks or to the beach, the person is usually tired even before beginning the dive. Divers entering the water from a boat frequently forget to calculate the current drift; they may ascend to the surface when their air tanks are depleted and discover that they are a considerable distance from their boat. Accidents occur when a diver, being physically out of shape, overexerts. There are other diving diseases and accidents that can also incapacitate a person.

Panic is the prime killer; too many divers use most of their air during the dive and when it is time to surface, they attempt to snorkel their way in; water is easily breathed in this way. When breathing in water, their first response is to raise their heads above the surface; the buoyancy of the body is insufficient to remain in this position. When divers become frightened their immediate response is to position themselves vertically in what is called the "ladder

ENVIRONMENTAL RESCUE

climbing syndrome" and try to keep the chest above the water. This requires a maximum work effort at which point most persons can only last about 30 seconds to one minute before they submerge and drown. The best thing to do when in distress is to float on one's back to safety. The empty air bottle will provide some positive buoyancy and offer support. This action reduces panic and lets the diver breathe with a minimum amount of effort.

Some actions that would signal a SCUBA rescue situation are a diver surfacing without the buddy's showing up, a steady stream of bubbles coming to the surface, a diver staying on the surface with no movement, or a diver struggling in the surf line.

Rescuing Diver on the Surface

A diver waving an arm above the head is calling for help. When rescuing a diver on the surface, the rescuer should:

1. Release the victim's weight belt. If so equipped, both the victim's and the rescuer's flotation vests should be inflated to aid buoyancy.
2. Raise the victim's face out of the water and remove obstructions such as face mask, mouthpiece, or snorkel.
3. If the victim is not breathing, turn the person over onto the back and immediately start artificial ventilation.
4. Alert other nearby swimmers and divers to aid in getting the victim to safety.
5. Keep moving toward shore or the boat at all times. If the current or wind is especially strong, utilize it to save energy.
6. When the surf is rough, remain clear of the victim's air tanks.

Rescuing a Submerged Diver

When rescuing a diver on the bottom, the rescuer should:

1. Cut the victim free of kelp or other entanglements.
2. Release the victim's weight belt and pull the person to the surface by using the air tank or harness as a hand hold. Occasionally, it may be advantageous for the rescuer to drop weights also.
3. If so equipped, inflate both the victim's and the rescuer's flotation vests.
4. Alert other nearby swimmers and divers to aid in the rescue.
5. Turn the victim onto the back and administer artificial ventilation if the person is not breathing.

SCUBA DIVING HAZARDS

Skin diving with the use of self-contained underwater breathing apparatus (SCUBA) is becoming more popular every year. This pastime has been put to practical use by many rescue squads for underwater search and recovery. A discussion of the psychological and physiological aspects of SCUBA diving, together with first aid care, is important in any rescue manual since victims of diving accidents may be encountered in any area (Figure 16-3).

Air and Gas Laws

When a diver goes to even a moderate depth, most of the troubles encountered are caused by the air breathed. This incongruous statement is explained by the physical laws governing the behavior of

Fig. 16-3. A scuba-equipped rescuer administering artificial respiration. Notice that the rescuer does not take the victim to shore for treatment. To be effective, artificial respiration must be administered quickly.
(Courtesy of the Los Angeles County Sheriff's Department.)

air and water. Humans are not amphibious animals. Therefore, any time they go beneath water, they must take an air supply with them. This air can be supplied through a hose or from tanks carried on the diver's back. As the diver descends into the water, pressure on the body increases nearly 1/2 pound per square inch (0.23 kilograms) for each foot of descent. At a depth of 33 feet (10 meters), pressure on the diver amounts to two atmospheres or 29.4 psi (pounds per square inch); at 66 feet (20 meters), pressure increases to three atmospheres or 44.1 psi. Therefore, every 33 feet (10 meters) of additional depth will subject the diver's body to an increased pressure of 14.7 psi. The air breathed must be supplied at the same pressure as the water pressure at the depth the diver has reached. The most significant single cause of a diver's difficulties arises from the fact that air is compressible and water is incompressible.

All gases are compressible and follow Boyle's Law, which states that at a constant temperature, the volume of a gas is inversely proportional to the pressure. This means that if pressure on the gas is doubled, volume is decreased to one-half; also, the converse is equally true and a decrease in pressure will result in a corresponding increase in volume. This law is of fundamental importance to a diver because it relates the quantity of air in his cylinders to the depth and duration of the dive.

When a mixture of gases is under pressure, each gas exerts a "partial pressure" in proportion to its percentage of the mixture. Thus the partial pressure of nitrogen in the air at atmospheric pressure is 14.7 X 78/100, which equals 11.8 psi. An increase in the overall pressure of a mixture of gases increases the individual partial pressure of each gas in the mixture. Thus, if the air pressure is doubled to two atmospheres, the partial pressures of all the constituent gases will be doubled, partial pressure for nitrogen in this case being 23.6 psi. When calculating partial pressures the absolute, not the gage, pressure of the gas should be used. In effect, this law means that if air at atmospheric pressure contains one percent carbon monoxide, when the air is compressed to a pressure of two atmospheres the toxic effect on the body will be the same as atmospheric air containing two percent of the gas. The same rule applies to all mixtures of gases.

If the volume of a gas is kept constant and the pressure in it is increased, such as when charging cylinders, the temperature of the gas will rise; similarly, if the pressure in it is reduced the temperature will fall.

If a gas is brought into contact with a fluid, such as air in the lungs coming in contact with the blood, some of the gas will be absorbed by the fluid. The rate of absorption depends on many factors, but the quantity absorbed will vary in proportion to the pressure of the gas, increasing when this pressure is increased until an equilibrium is obtained, and being liberated from the fluid when the pressure is decreased.

These gas laws account for many of the physical and physiological problems associated with diving; a knowledge of them will help understand the effects of breathing compressed air.

Changes of Pressure and Volume with Depth

If a flexible bag containing air at atmospheric pressure is taken underwater, it will be squeezed progressively smaller as it descends. This is due to the increasing pressure of the water which acts on the air inside the bag, compressing it to the same pressure as the water surrounding it. When the bag is brought to the surface, the air will expand until it is again at a pressure of one atmosphere and its original volume. Conversely, if the flexible bag is filled with compressed air at a depth of 99 feet (30 meters) in the water, and expanded to the size it previously occupied at the surface, then allowed to ascend, it will progressively expand until it arrives at the surface four times larger.

It is important to realize that the relative changes in pressure and volume are greater near the surface than at a depth. For example, when taking the same flexible bag from the surface to 33 feet (10 meters) the water pressure on it is doubled and its volume halved, but in descending from 99 to 132 feet (30 to 40 meters) the pressure is increased and volume reduced by only one-fifth, although the depth increase is the same in each case. This has an important practical application. Equalizing the pressure on the eardrum (clearing the ears) must be done more frequently when descending nearer the surface. There is also a greater risk of the occurrence of air embolism during ascent when nearing the surface.

In order to permit the diver to breathe under water the air supplied by breathing apparatus must be the same pressure as that of the surrounding water. If this condition of balanced pressurization is fulfilled the diver will be able to inhale and exhale normally since the pressures inside his lungs and on the outside of his body are balanced. The pressure of air in the SCUBA cylinder must, of course, always be greater than the water pressure at the diver's depth at any time. In SCUBA diving, if the regulator is on the same level as the lungs, the regulator diaphragm will supply air at the same pressure as the surrounding water; this internal air pressure will keep the diver's lungs properly inflated. If this air pressure is less than that of the surrounding water, external pressure will squeeze the body and reduce the volume of the lungs.

Endurance of Compressed Air Cylinders

In order to provide the volume of air required for breathing under water, the actual mass of air needed will be greater than at the surface. At 33 feet (10 meters), where a diver's air must be supplied at 2 atmospheres pressure, 1 cubic foot of air would be equivalent to 2 cubic feet at a surface pressure of 1 atmosphere. Correspondingly greater requirements apply at lower depths. This means that the endurance of a compressed air cylinder is reduced as the depth increases in the same ratio as the volume of air in the flexible bag was reduced upon its descent. A compressed air cylinder which would supply a diver with air for 30 minutes on the surface would last only 15 minutes at a depth of 33 feet (10 meters); at 66 feet (20 meters), it would be used up in 10 minutes. Care must be taken to allow for this shortened duration as the diver descends.

Psychological Aspects of Diving

A difficult problem to evaluate and handle is the psychological aspect of diving. Any psychological disorder, a history of epileptic episodes, losses of consciousness from any cause, or any known neurological disorder makes diving highly inadvisable. Individuals who tend to panic in emergencies may find occasion to do so in diving. Any recklessness or emotional instability in a diver is a serious liability for companions as well as for the diver.

Diving in really low visibilities can be thoroughly unnerving and must not be taken on lightly. The possibilities of developing claustrophobia (fear of being confined in limited spaces) are high, and the risk of panic is always present. Underwater movements should be slow, deliberate, and exploratory. The person who is psychologically unsuited for such work must overcome the desire to try it for its glamorous aspects.

Physiological Aspects of Diving

An understanding of the elementary anatomy of the ear is a help in preventing damage from the pressure changes involved in underwater swimming. Anatomists divide the ear into three parts: the outer, middle, and inner ears. The outer ear consists of the visible part and a short tube running down to the eardrum. The inner ear contains the organs of hearing and balance; the middle ear lies between the two and is a roughly cubical space. The eardrum is a membrane that is stretched across the entrance of the middle ear and seals it off from the outer ear. It is sensitive to pressures and vibrations entering the outer ear passage; to function properly and to avoid being damaged, it must be balanced with equal pressures on both its outer and inner surfaces.

Vibrations of the eardrum are transmitted by a series of small bones to the inner ear where the organs of hearing are situated. There is also an opening called the eustachian tube that burrows through the bone in the base of the skull and emerges near the back of the nose, close to the adenoids. This opening is usually closed, but opens with the action of swallowing; the act of swallowing ensures that the pressure in the eustachian tube is the same as the pressure in the nose. The result is that the middle ear contains air at a pressure equal to the pressure in the outer ear, so that no strain is exerted on the eardrum. If the eardrum is perforated, from inflammation or injury, there is nothing to stop water in the outer ear from entering the middle ear and causing infection. For this reason people who have perforated eardrums are advised to keep their heads above water.

Deafness and earache often accompany the common cold. Nasal catarrh blocks the opening of the eustachian tube and the air trapped within is slowly absorbed. Pressure in the middle ear falls below that in the outer ear and the eardrum inclines inward. Satisfactory equalization of pressures is impossible under these circumstances and persons with a cold should not dive. Should mucus be forced by pressure from the eustachian tube into the middle ear it may cause infection, temporary deafness, and acute discomfort which may not abate for several weeks.

Under normal circumstances, pressure differences in the atmosphere are so small that the work of the eustachian tube goes unnoticed. The functioning of the eustachian tubes is often noticeable in mountain driving while ascending or descending; ear popping may occur because of a difference in pressure between the middle ear and the outer ear. Difficulties are greatly magnified in water because of the larger increase or decrease in pressure for each foot of depth. In the case of divers who are subject to large and rapid changes of pressure, the eustachian tubes have to be opened very often and equalization obtained by a conscious effort. Water entering the ear of a diver during descent presses against the eardrum, deflecting it inward. By opening the eustachian tubes, air being breathed at a pressure equal to the water pressure flows from the throat into the middle ear cavity; this restores the equilibrium of the membrane. This process has to be repeated at intervals throughout the descent. Pain in one or both ears indicates that this pressure balance is not being achieved and that the descent must be checked; otherwise, ear damage is inevitable. It must be emphasized that ear clearing exercises should commence upon entering the water and should be practiced continually throughout the descent. It is a great mistake to wait until pressure buildup on the eardrum is felt before attempting to open the eustachian tubes. If pressure pains are experienced upon descending, always ascend a few feet to obtain relief before attempting to clear the ears.

Before diving, blowing the nose is strongly advised. Check the ears by closing the mouth, holding the nose, and attempting to breathe out. In this manner air is forced into the eustachian tubes and the eardrums can be felt to click or pop as they bulge outward. Attempt to breathe in and the process is reversed. Failure to achieve this ear popping at the surface usually indicates that trouble will be experienced below water. Use of a nasal inhalant or nose drops may clear up the nasal congestion.

Equalization of pressure on the ear is just as important when coming up as when going down. Occasionally pressure pains are felt during an ascent because the air, at a higher pressure than that of the surrounding water, is trapped in the middle ear cavity; this forces the eardrums outward. The diver should descend slightly and equalize pressure by swallowing.

When a diver breathes compressed air from a SCUBA tank at any depth, the air in his lungs is at the same pressure as the water pressure at that depth. If the diver were to stop breathing for any reason and make a quick ascent to the surface from 33 feet (10 meters) the pressure on the lungs would be reduced to one-half and, consequently, the volume of air inside the lungs would double. Alveoli in the lungs will become ruptured when subjected to a few pounds of excessive air pressure; holding the breath when ascending or descending can cause serious lung, brain, or heart damage and death can result.

Diving Diseases and Accidents

Aerotitus, Media (middle ear squeeze). This is a hemorrhage within the tympanic membrane and middle ear caused by failure or inability to clear the ears through neglect or by blockage in the eustachian tubes. It may be prevented by equalizing ear pressures properly during descent. Divers must not dive with head colds or infection which blocks the eustachian tubes.

The symptoms include extreme pain in the ear upon descent, redness and swelling of the eardrum, bleeding into the drum or middle ear space, and spitting up blood. Rupture of the eardrum will often result in bleeding outside the drum and a dull ache that usually continues for several hours. When the drum is ruptured, water can enter the middle ear and inner ear through this hole. The water, particularly if cold, may upset the organs of balance; this causes vertigo and complete loss of sense of balance. The diver is unable to control movements up or down. In these cases the diver should attempt to hold on to some stationary object until water in the inner ear warms up and balance is restored.

No attempt at first aid should be made. Competent medical aid should be sought immediately. Avoid pressure until the damage heals; a small rupture of the eardrum will heal in about ten days provided there is no ear infection. There are usually no ill effects following a slight tear of the membrane, but repeated ruptures may cause permanent deafness.

Aerodontalgia. Aerodontalgia is an acute pain in a tooth caused by an increase in pressure on a tooth. An air pocket under a faulty filling, inlay, or dental cap becomes subject to increased pressure and pain results. Dental assistance should be sought. Do not continue diving until the condition has been cleared up.

Air Embolism (traumatic air embolism, overexpansion of the lungs). Air embolism is one of the most serious and most easily developed physiological complications occurring in diving and is due to a relative excess of air pressure in the lungs. When using compressed air breathing apparatus, a diver breathes air at the same pressure as that of the surrounding water. Should for any reason descent be made while holding the breath, air in the lungs will expand due to the decreasing external water pressure. If air is not allowed to escape through the mouth or nose, it will force its way through the walls of the alveoli into the capillary blood vessels, rupturing the alveoli in the process. Air, in the form of small bubbles or emboli, will pass through the heart and block small blood vessels throughout the body, including those in the heart, brain, and spinal cord. In more severe cases large bubbles may gather in the heart where they will prevent it from beating properly; circulation fails and the diver dies. Always breathe normally when using SCUBA; never hold the breath during ascent. If making a free ascent without SCUBA, exhale continuously.

The danger of air embolism is greatest near the surface where the largest pressure variations occur. It is possible to be afflicted with air embolism in an ascent of only 6 to 8 feet if the diver commences the ascent with full lungs and does not exhale.

The symptoms usually occur within seconds but may occur up to five minutes after surfacing; they may occur before the diver reaches the surface. Prior to unconsciousness the diver may experience weakness, a tightness in the chest, dizziness, numbness or paralysis of the arms or legs, visual disturbances, convulsions, and unconsciousness. A bloody froth is often visible at the mouth. In less severe cases a choking feeling in the throat, with a hoarse voice, rattling in the chest, accompanied by a sensation of air under the skin of the chest, may be experienced.

Should air embolism occur, no emergency first aid treatment is entirely effective. Recompression as soon as possible in a recompression chamber is necessary to diminish the size of the bubbles in the blood. Even so, death may follow in severe cases or if some time has elapsed before recompression. Artificial respiration or oxygen therapy should be administered if necessary. The victim should be treated for shock and transported to medical attention immediately.

Anoxia (oxygen deficiency, hypoxia). Tissues fail to receive enough oxygen to maintain life when a diver does not receive enough oxygen. This may be caused by an equipment malfunction, improper use of SCUBA, or any other inadequacy or loss of air supply. Breath holding or skip breathing to extend the use of compressed air tanks is dangerous.

The diver will not receive adequate warning of this difficulty. The pulse rate and blood pressure increase. There is a slowing of responses, confusion, clumsiness, and foolish behavior resembling the effects of alcohol. Cyanosis of lips, nailbeds, and skin is apparent. Anoxia may result in unconsciousness and stoppage of breathing.

The diver should get to the surface and fresh air immediately and breathe pure oxygen if it is available. Artificial respiration should be administered when indicated.

Bends (nitrogen absorption, caisson disease, decompression sickness). During normal breathing, blood flowing through the capillaries of the lungs absorbs air through the alveoli. The main gases of the alveolar air are dissolved in the blood plasma in proportion to their partial pressures, which depend upon the concentration of the gases in the alveolar air and the pressure of the air. In considering the bends, the only concern is the simple solution of nitrogen in the blood plasma.

If a diver makes a descent and the pressure of the air he breathes is increased to two atmospheres, the partial pressure of nitrogen breathed will be twice as great as it was at the surface with an air pressure of only one atmosphere; twice as much nitrogen can therefore be dissolved in the blood. Some of the nitrogen dissolved into the blood passes into the fatty tissues, particularly those of the brain and spinal cord which contain a high proportion of special forms of fat. When the partial pressure of nitrogen in the alveolar air is increased as the diver descends, the nitrogen dissolved in fat and other tissues will also be increased. The amount of nitrogen in the tissues depends on the depth, duration of the dive, and exertion of the diver. If a diver breathes air in which the partial pressure of the nitrogen is doubled for a sufficient time, the total nitrogen dissolved in the tissues will also be doubled.

As pressure is gradually decreased during ascent, the body tissues will no longer be able to retain the extra nitrogen and it will subsequently be liberated into the blood and then breathed out through the lungs. If, however, a higher partial pressure of nitrogen has been maintained for some time, and is then suddenly reduced, the nitrogen will come out of solution quicker than it can be carried away by the blood in the lungs and will form bubbles in the tissues and in the blood.

A clearer understanding of the process may be gained from an analogy. In the manufacture of carbonated drinks, the bottle is pressurized with carbon dioxide gas which is absorbed by the drink. If the cap were removed slowly the gas would be heard issuing under pressure, with little disturbance of the liquid. Should the cap be removed quickly the carbonated drink would immediately bubble and froth in its effort to give up the dissolved gas. As long as the diver stays under pressure the blood will resemble the soda pop with the cap secure. Should descent be made too quickly, allowing the nitrogen to be released too quickly, the blood will bubble and froth in a manner similar to the suddenly uncapped soda pop.

Bubbles so formed may interfere with the normal workings of the body. For instance, bubbles in the fatty tissue and synovial membranes around the joints lead to stretching of the nerve endings in these tissues. The ensuing pain gives rise to the condition known as the bends; it gets its name from the fact that one of the symptoms of persons suffering from this complaint is severe pain in the limbs which causes the victim to double up in anguish and be unable to straighten the affected joints.

Bubbles liberated in the spinal cord are even more serious and may cause paralysis. They may also be liberated in the brain, giving rise to cerebral bends. In the more severe cases, large bubbles are formed in the heart chambers and the person dies.

The rate at which nitrogen is absorbed and given off through respiration varies with individuals and with the same person on different occasions. About 12 hours is usually required to get rid of all excess nitrogen absorbed in the human body. Il order to allow the dissolved nitrogen to be liberated from the blood, gradual decompression is necessary. A slow and controlled ascent to the surface in order to allow sufficient time for the excess nitrogen to be removed from the system is required. The U.S. Navy Standard Decompression Table and the Repetitive Diving Tables are the authorities to be referred to for guidance. Never rise faster than 60 feet (18 meters) per minute; this slower speed will usually prevent bends when SCUBA diving. This speed may be attained and gauged by limiting the speed of ascent to the same speed at which the small bubbles of air released by the regulator rise.

The main symptoms of the bends may occur immediately upon ascending or within any time up to 6 hours after the dive. In the majority of cases there is localized pain in the joints of the lower limbs (knee and ankle), and sometimes in the arms. Dizziness, paralysis, shortness of breath, itching, swelling and skin rash, and convulsions may also occur.

Should a diver contract the bends, the treatment is to force the bubbles in the blood back into solution by means of recompression, followed by a long decompression according to special tables drawn up for this purpose. This must be carried out under expert medical attention. Any inept attempts to give emergency treatment in the water by taking the stricken diver down deep for recompression and then gradual decompression may seriously compound any injuries and cause permanent disability or death. Treatment should be carried out as soon as possible after symptoms have been observed since permanent damage may follow. Oxygen or artificial respiration should be administered during transportation if respiratory difficulty is encountered. If the victim is to be transported by aircraft the pilot should be instructed to fly at low altitudes; high altitudes will aggravate the victim's condition. Information about the maximum diving depth attained, amount and frequency of dives, and any other information pertinent to the victim's activity should be supplied.

Emphysema, Subcutaneous. Air may be forced through the lung coverings and appear under the skin of the chest and neck where it may be felt as bubbles. The cause is the same as for air embolism, overexpansion of the lungs causing air to leak through the alveoli and bronchial tubes into the surrounding tissues.

The symptoms are not usually extreme; the victim may have only a feeling and a look of fullness in the neck and a change in the voice. Difficulty in breathing and swallowing may occur.

Medical advice should be sought immediately. If there is breathing difficulty, administer oxygen or artificial respiration.

Emphysema, Mediastinal. This is the presence of air in the tissues around the heart, lungs, and large blood vessels in the middle of the chest. The cause is the same as for air embolism and subcutaneous emphysema; overexpansion of the lungs causes air to leak into the surrounding tissues.

Symptoms include pain under the breastbone, shortness of breath, shock, cyanosis of the skin, lips, and fingernails.

Immediate medical advice should be sought. If air embolism has occurred, it should be treated accordingly. Oxygen should be administered if breathing difficulty occurs.

Nitrogen Narcosis (rapture of the depths). Although nitrogen forms nearly eighty percent of the atmospheric air, it normally has no effects on the human body, either good or bad. If, however, the partial pressure of nitrogen is increased, as when breathing compressed air under water, its concentration in the blood will rise; this produces a weak anaesthetic effect very similar to that caused by nitrous oxide (laughing gas) and has effects similar to mild alcoholic intoxication. As a diver descends and the pressure increases, symptoms first appear when air is breathed at about 4 atmospheres pressure and becomes severe at 8 to 10 atmospheres.

The onset and intensity of these effects will vary with individuals and circumstances, just as the effects of alcoholic drink differ from person to person. The effects are likely to be greater in persons who are nervous, tired, or already under the influence of alcohol. Nitrogen narcosis is a very real hazard to divers on deeper dives, which should never be undertaken without regard to its possible occurrence. It has been responsible for the death of many divers who have attempted to establish depth records.

If, in special circumstances, the diver breathes a mixture of oxygen and helium, instead of the normal atmospheric air composed of oxygen and nitrogen, narcotic effects do not occur. Decompression times are extended, however, and helium bends require special treatment.

The first symptoms, which may commence at about 100 feet (30 meters), are loss of judgment, poor coordination, and increased hearing perception. The symptoms will gradually increase with depth and the air may take on a coppery taste. There is a feeling of well being, lack of concern for safety, increasing dizziness, difficulty in accomplishing even simple tasks, and foolish behavior. Highly susceptible divers may be near unconsciousness. At about 250 feet (75 meters) even the simplest actions become difficult, confused, and muddled.

Intensity of the effects is directly related to the partial pressure of nitrogen, so that the effects of nitrogen narcosis diminish and finally disappear upon ascent. There are no subsequent complications or aftereffects. Thus, a diver who realizes that nitrogen narcosis is setting in should ascend at once to regions of lower pressure.

Pneumothorax. Pneumothorax is caused by the presence of air between the lungs and the chest wall in the pleural cavity. The cause is the same as for air embolism. Symptoms include cyanosis of skin, lips, and fingernails, and pain in the side of the chest. There may be great difficulty in breathing.

The victim should be treated as for air embolism. In severe cases, trapped air in the pleural cavity should be removed by a doctor since it interferes with action of the lungs.

Oxygen Poisoning. Although oxygen is essential to life, it can, on occasion, act as a poison. At normal atmospheric pressure pure oxygen can be breathed without any ill effects up to a few hours; 60 percent oxygen concentration can be breathed indefinitely without harm.

If pure oxygen is breathed for too long, it has two harmful effects: it causes a constriction of the small blood vessels in the brain, thus interrupting the flow of blood to the brain, and it poisons certain chemical substances called enzymes that regulate cell metabolism. These enzymes also regulate the use of oxygen by the cells; if they are not functioning, the cells cannot use oxygen. Too high a level of oxygen has the strange effect of causing the same symptoms as anoxia (oxygen lack).

The harmful effect of breathing pure oxygen is greatly accentuated if it is breathed under pressure. If pure oxygen is breathed at a pressure of 4 atmospheres, serious symptoms occur in less than 1 hour; and if breathed at higher pressures, symptoms appear earlier. In practice, even the onset of mild symptoms may be fatal underwater; the use of underwater breathing apparatus employing pure oxygen should be limited to 33 feet (10 meters) of depth (equivalent to 2 atmospheres of pressure). Even at shallower depths than this, symptoms sometimes occur. Varying with the individuals, the deeper the dive, the sooner the symptoms will appear.

Even when using compressed air containing only 21 percent oxygen, as with SCUBA, oxygen poisoning can occur. At a depth of 297 feet (89 meters), which is equivalent to 10 atmospheres, the partial pressure of oxygen is equal to 2 atmospheres; therefore, at this depth, or even less, the diver may be subject to the effects of high pressure

ENVIRONMENTAL RESCUE

oxygen. Air cylinders should never be filled with pure oxygen.

Symptoms include facial twitching, particularly of the lips, trembling of the fingers and arms, coughing and hiccups, nausea and vomiting, sudden and violent convulsions, and unconsciousness.

A victim of oxygen poisoning should be brought to the surface and allowed to breathe atmospheric air. If conscious, have the victim hold the breath or confine breathing to a paper bag; this will wash out the oxygen concentration from the blood and allow the carbon dioxide level to rise.

Sinusitis. The sinuses are bony cavities in the skull situated in the forehead, cheekbones, and above the nasal passages. They are lined with a membrane which is continuous with the nasal tract and connected by small ducts; blockage of these ducts will prevent equalization of air pressures in the cavities.

Symptoms include severe pain during descent, relieved upon ascent. This should not be ignored or rupture of the sinus membrane, followed by bleeding from the nose and mouth, will occur. Severe sinus trouble may cause swelling and discoloration under the eyes following a dive, and the sinus area will be tender to the touch.

A victim of sinusitis should discontinue diving until medical advice has been obtained. Nose drops, spray, or inhalant should be used before diving to permit discharge from sinus ducts. In general, one should not dive when suffering from a head cold or sinus blockage.

ICE RESCUE

In most states, at any time during the winter a glittering expanse of ice may entice the adventuresome, the sportsminded, the unwitting, and especially the young into severe difficulty or disaster with little or no warning. The safety of anyone on the ice depends upon knowing and being able to recognize safe ice conditions. The first stages of ice formation and the last ones of disintegration are the two really dangerous periods of the winter season.

Precautions

As with other rescue problems, thorough training and speedy application of proven procedures bring the most beneficial results. Unnecessarily endangering the life of a rescuer must be avoided. Frequently, a relatively simple rescue operation is made unduly complex because an uninformed person attempted the rescue and suddenly was caught in the same predicament as the first victim. This is particularly true of the individual who dives into any ice hole to seek a submerged person without making adequate preparation. The minimum precaution that must be taken is to tie a stout line securely about the waist of the rescuer and to have the other end in the hands of persons staying a safe distance from the hole. Failure to do this will imperil the life of the rescuer.

Contrary to popular belief, there is very seldom air space between the water and the underside of a layer of ice. Furthermore, once the rescuer goes underwater and swims away from the hole in the ice, there will be difficulty relocating the point of entry. With breath giving out and the cold water sapping strength, the rescuer thus committed ends up a victim. Loss of body heat in the water is 25 times greater than loss of body heat in the atmosphere. The body will shiver when the skin temperature drops to 86 degrees. Once the human body is chilled and shivering commences, muscle strength is reduced by 50 percent. Thus the ability of an unprotected person to survive in really cold water is limited. Some victims will die in about 1 hour in water at 40°F (5°C).

The immediate goal of a rescuer must be to keep the victim from sinking. It must be realized that the victim may have become incapable of self-help. Every effort should be made to calm and reassure the victim so that strength will not be expended in thrashing about; this will also weaken the ice by cracking it further.

Self-Rescue

Self-rescue is entirely possible in a majority of the cases, but is often defeated by unreasoning panic on the part of the victim. When a person is stimulated by the exercise of skating or other exertion, there is some shock caused by sudden submersion in icy water. Quick, vigorous, but thoroughly preplanned action will avoid needless exhaustion and unnecessary breaking of the ice.

Upon submersion, the correct procedure is to extend both arms on the surface of the ice surrounding the hole to support the weight of the body while calling for help. The endangered person's legs naturally tend to come up forward under the ice. This should be prevented by executing a crawl swimming kick. The weight of the lower part of the body will be supported by the water instead of the ice, thus enabling the victim to get support from ice otherwise too thin to hold body weight. This position will facilitate being pulled to safety on the ice.

For self-rescue, the victim should assume the planing position and try to crawl forward flat on the stomach until the hips are at the edge of the ice, then quickly swerve sideways, and with arms extended above the head, roll quickly away from the possibly breaking edge of the ice. The initial pullout on the stomach can be facilitated by the use of some sharp object, such as large nail, spike, knife, or the heel of a skate blade used as an ice pick; this will enable the person to obtain a pulling purchase on the smooth surface of the ice.

The endangered person should not attempt to climb straight up out of the hole since body weight is likely to break the edge of the ice. After crawling out or pulling out on one's stomach, the person should not attempt to kneel or stand until well away from the hole, even if the ice is apparently holding well. Thin or bad ice will often support weight when distributed over a large area, as in a spread-eagle position, but will crack and break when the weight is concentrated on a small point, such as the knees or feet.

Equipment

Life Preservers. Life vests should be standard equipment for rescue squads and all personnel should be equipped with them at all times when engaged in any type of water or ice rescue.

Ladder. Probably the best rescue device is a light ladder and line. The ladder is laid flat upon the ice and shoved to the person in the hole (Figure 16-4). A line may be used as an extension, if necessary. The ladder rungs are excellent hand grips for the victim. Even if the ice breaks beneath the weight of both the victim and the rescuer, the ladder will always angle upwards out of the broken ice area and support their weight.

ENVIRONMENTAL RESCUE

Stretcher. A Stokes or folding-type stretcher will make an excellent rescue device. By attaching a line and lashing lifejackets to it, a rescuer will have a device that distributes weight over a broader area and permits sliding up to the ice hole, effecting the rescue, and being pulled back to safety with the victim.

Ring Buoy. For ice rescues in which the rescuer must remain at some distance from the endangered person because of the weakened condition of the ice, a ring buoy with line attached may be scaled along the ice to a victim with considerable accuracy. A line weighted at one end with a hockey stick, piece of wood, or any other type of weight can be skidded out across the ice. A loop should be set up on the weighted end so that the numbed victim can place it under the armpits.

Plank or Board. A one-person rescue operation involves the use of two planks or wide boards of as great a length as it is possible to obtain. The rescuer should lie at full length on one plank and then shove the other one ahead. The rescuer then worms out onto the second plank, draws the first one alongside, and shoves it forward to the victim.

Rope. The rescuer must be equipped with skates or creepers, or must be supported in some other way to enable pulling out the victim;

Fig. 16-4. A light ladder is probably the best ice rescue device, because it will distribute the weight of the rescuers and victim over a wide expanse of ice and because it has excellent hand grips. The ladder is laid flat upon the ice and shoved to the immersed victim. Even if the ice breaks beneath the weight of the rescuer and victim, the ladder will always angle upwards out of the broken ice area and support their weight. (Courtesy of Spring 3100, the New York City Police Department.)

Fig. 16-5. A small light boat can be used quickly and safely to rescue a person who has broken through ice. A flat-bottomed punt is particularly adaptable to ice rescue, since it can be slid across the ice rapidly and will support the rescuers on bad ice or on open water. A long wide plank of lumber can also be used to spread a rescuer's weight over a wide expanse of ice, thus lessening the chances of endangering the rescuer. (Courtesy of Spring 3100, the New York City Police Department.)

otherwise the rescuer will slide toward the hole. If the end of the rope is bunched and knotted to about the size of a person's fist, it will carry better and farther, especially in a wind. A knot or loop tied in the line will help prevent the rope from slipping through the victim's hands; a piece of wood or short tree branch tied to the line will serve the same purpose.

If the victim cannot grasp the line at all, a noose 30 inches (75 centimeters) in diameter at the end will enable slipping the head and one arm through. A completely helpless victim requires that a rescuer go out for the person with a rope tied around the rescuer's body and held by assistants on shore. The helpers must be supported in some way to facilitate the rescue.

The precaution of fastening a rope around the rescuer should be taken even when the rescuer attempts to reach the edge of the hole by means of planks or a ladder. Another variation involves the tying of a large loop in the center of a line which is long enough to reach across the expanse of ice. Securing one end, or having it held by another rescuer, and then walking around the edge of the ice to the other side will allow the loop to be brought close to the victim.

Pole. A light long pole with a short length of line attached, ending in a large loop, is excellent for making ice rescues when a rescuer can get close enough with safety to the hole into which the victim has plunged.

Boat. Occasionally during the spring breakup of the ice on rivers, bays, and large lakes a small rowboat can be utilized quickly and with safety (Figure 16-5). A flat-bottomed punt is adaptable to this usage since it may be slid across the ice rapidly and will support the rescuers on bad ice or on open water.

Human Chain. Four or five strong individuals may form a chain; this is recommended when no equipment is available. The first person should have both hands free, but each succeeding person in the file seizes an ankle of the person ahead with one hand, thus forming a chain. When the members are spread out on the ice, they should slide themselves forwards with their free hands until the first person can seize the victim by the wrists. Slowly and cautiously the whole line then wriggles back to safety. It is important that all persons in the line keep their weight spread out.

Under-Ice Body Recovery

Recovering the body of a person who has fallen through the ice and drowned is one of the most difficult and hazardous jobs facing rescue personnel. This is a task that should only be attempted by the most skilled and experienced SCUBA divers; lives should never be jeopardized in rescue operations when it is obvious that the victim has perished (Figure 16-6).

It will probably be very obvious where the victim penetrated the ice; keep everyone away from that area until the safety hazards have been ascertained. If the ice is too thin to safely hold the weight of the rescue team, then a boat may be the best answer. When there is a doubt, utilize planks or ladders to spread out the weight over a larger surface. Keep all spectators and everyone else who is not actively participating in the operations back on the shore where they will not create problems.

Take time to analyze the facts before committing the squad to any one rescue method. How thick is the ice? How fast is the current? How deep is the water? How large was the victim and how was he

ENVIRONMENTAL RESCUE

dressed? Normally, a drowned person of average build who is dressed in a bathing suit will not drift further than one and one-half times the depth of water from the point of entry. However, dense cold water and the thick clothing of winter will cause a body to be more buoyant until the clothes become soaked. Also, small children and obese persons float better than thin people. These facts will dictate the most likely location of the body.

When the ice is thick and the victim fell through a small hole such as used for fishing, the best method would probably be to make a tripod over the hole and suspend a rope and anchor from it; this will form a vertical guideline. Divers would then snap a guideline to this rope so they could again find the opening from underneath; it is practically impossible to find a hole in the ice when diving. Besides the guideline, divers should have tenders on the surface attending the individual safety lines attached to their harnesses. If the body is not found within the area that can be safely searched through the opening, then additional holes should be cut downstream and the procedure repeated.

If the ice is too thin to safely support the weight of the rescuers, it may be best to break a channel and utilize a boat. In this case, the

Fig. 16-6. Ice may complicate an underwater search for a body; under some conditions the only practical way to recover the body of a person who has broken through ice and drowned is to wait until after the spring thaws. Contrary to popular belief, there is very seldom any air space between the water and the underside of a layer of ice; the length of time a victim can survive after breaking through ice is therefore very short. When the ice is thin and the water is shallow, divers in protective suits and rescuers in boats, dragging hooks through the water and probing with long poles, may discover the body. Under no conditions should a bystander immediately dive through the hole in the ice and attempt to locate the victim; an ice hole is almost impossible to relocate from below. (Courtesy of the Denver Fire Department. Photo by Terry Brennan.)

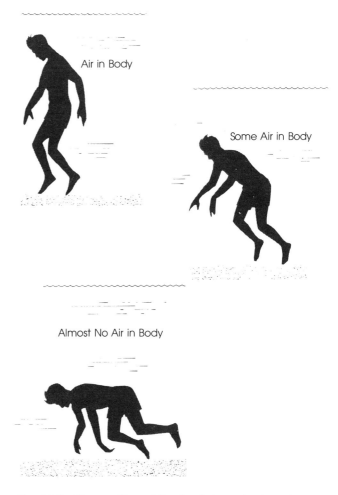

Fig. 16-7. The position of the body in water largely depends upon the amount of water in the lungs and stomach.

boat would proceed slowly while divers attached to the boat by safety lines would fan out and search under the ice.

When the ice is thin enough that it may be readily broken up, break up the area well downstream and conduct a thorough search. Then slowly proceed upstream, clearing the ice and letting it float down into the searched area. Be sure to check the underside of all ice slabs for the body.

If the water is not too deep and there are sufficient personnel, a line formed across the stream can slowly proceed upstream, removing ice and searching as they move.

UNDERWATER SEARCH AND RECOVERY

As long as there is the slightest possibility that the victim of an underwater accident is still alive, no effort should be spared to accelerate rescue. Unfortunately, many rescue operations are merely body recovery efforts due to delay in finding the victim. Artificial respiration should be started immediately upon recovery of the body, unless it is definitely known that the victim has been under water continuously for a great length of time. Resuscitation must be continued until the victim is either revived or pronounced dead by a physician.

The Human Body in Water

The average body has about the specific gravity of water. This means that the body will displace its own volume of water and the volume of water displaced will weigh about the same amount as the body. Therefore, the average body will almost float; sometimes bodies of victims who are fat and bodies of small children do not sink, but remain floating on the surface. As long as the body is totally submerged in the water, it will weigh approximately 1 pound (0.45 kilogram); for this reason, heavy tackle is not needed to make a recovery. The slightest hook in the clothes or body will bring it to the surface provided gentle pressure is used and the hook is not torn out.

When a person falls into the water, the momentum of the fall will make the body sink. The victim will be holding some air in the lungs, and there will be some trapped in their clothes; instinctive swimming movements will bring the victim back to the surface. Some may gasp, take in air and water, and sink again; this cycle may be repeated until they finally sink to the bottom.

The position of the body in the water largely depends upon the amount of water in the lungs and stomach (Figure 16-7). In approximately 10 percent of the drowning cases there is little water or no water in the lungs; in such cases, the body will be in an almost upright position. With less air in the lungs and stomach, the body will be in a crouched position, and with almost no air in the body it will be in a crawling position.

A body will rise slowly to the surface when sufficient gas is formed in the intestinal tract to make the body buoyant; this gas is the result of bacterial decomposition. The time necessary to generate the necessary gas will depend upon the temperature of the water and contents of the victim's stomach when drowned. In summer, the average time is 18 to 24 hours. In winter, or when the water is very deep and cold, the time will be much longer. A body will not rise suddenly from the bottom, but rises gradually as more gas is formed and the body becomes buoyant.

ENVIRONMENTAL RESCUE

Locating a Body in Water

A body will usually remain in the general area where it submerged and will likely be found within 25 to 50 feet (7.5 to 15 meters) of that location. Even with a strong current, it will probably be found within 100 to 150 yards (90 to 135 meters) of where it went down. It has been established that the average body under average conditions will be within one and one-half times the depth of the water; for example, if the water is 20 feet (6 meters) deep the body will probably be found within 30 feet (9 meters) of where it went down. Opportunity for quick body recovery may depend upon accurately determining the exact location where the victim disappeared (Figure 16-8).

Where a current exists, a body may drift to the first eddy or deep hole, depending upon the force of the current and obstructions on the bottom. If the body is floating, it may hang up on some obstruction down current or downstream. If a recovery operation has sufficient manpower available at the scene, it is good policy to dispatch personnel down current on the chance that the body was floating.

When a rescue squad arrives at the scene immediately after submersion and the water is very calm, the victim may be located by a thin stream of air bubbles coming from the body. These bubbles are caused by water pressure on the chest and abdomen forcing out the air remaining in the body. Even under these circumstances, the body may be 10 to 12 feet (3 to 3.6 meters) from the spot where the bubbles are breaking the surface of the water. Other things can give off similar streams of air bubbles, but any such leads should be checked out.

Water Rescue Equipment

The minimum equipment for a rescue squad is a boat, life jackets, oars, drag hooks, rope, and a vehicle to transport the squad and

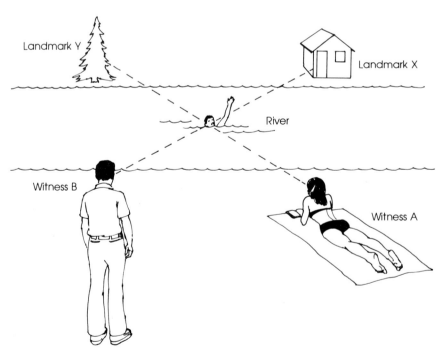

Fig. 16-8. How to locate a body: A witness at point A saw victim in line with tree Y. A witness at point B saw victim in line with landmark X. The body will be near the place where line AY crosses line BX.

equipment to the scene. For efficient operation under varied circumstances, a squad should have more than the bare minimum. The amount and type of equipment will depend upon local needs and available funds.

Boat. The boat must be of such size that two rescuers can operate it safely and yet be able to reach shore in case of foul weather. It must handle easily with oars since most dragging operations will be carried out by rowing; a motor will not slow down sufficiently to permit the drag hooks to sink to the bottom, nor does it allow the operator to stop the boat immediately when the drag operator feels a strike. However, there should be a motor on the boat that has sufficient power to get the crew to the scene of operations and to operate the boat in stormy weather. All personnel in the boat should wear life jackets at all times.

Drags. There are many types and styles of drag hooks now in use by rescue squads and there is considerable difference of opinion as to which type is best to use. Because of the large variety of conditions that may be encountered in different localities, all commercial types should be considered and the models selected that best suit the anticipated situations. In the selection of a type or style of drag, it is well to remember that the body is usually on the bottom in a position which may be anywhere between crouching and upright. The body will weigh about 1 pound (0.45 kilogram) as long as it is fully submerged; therefore, heavy tackle is not needed to make a recovery. A drag that will sink, sweep along the bottom, and lift about 1 pound (0.45 kilogram) is sufficient. A 5 or 6 foot (1.5 to 1.8 meters) long towbar with treble hooks spaced along it makes an efficient drag.

A large four-way hook attached to a pole is excellent for areas which are filled with rock, stumps, or other obstructions. It can be used in water up to 25 feet (7.5 meters) deep and is relatively easy to use when the handle is joined in 4-foot (1.2-meter) sections. It is used in an up and down tamping motion over the search area. The hook can be used to secure a body when it is near the surface or after it has been hooked by light drags.

Buoys and Markers. A sufficient number of buoys and markers, or other floats, complete with anchors and anchor lines should be carried to mark the area of operation adequately.

Dragging Techniques

On arrival at the scene of a drowning, the rescue squad must first determine as accurately as possible the exact point where the victim submerged. A marker buoy should then be set at this location and other floats of another color set approximately 50 to 100 feet (15 to 30 meters) in all four directions from the marker buoy. When the area of operation is so marked, then the method or technique necessary to make a recovery can be determined.

There are many methods of body recovery; the method used depends on the area and conditions of the water. A study of every body of water in the area should be made beforehand and pertinent information entered in a master book, with a copy carried in the water rescue vehicle. This book should contain the size, shape, and type of each body of water in the area, the depth, currents, bottom conditions, and the most practical route to reach them. Such areas should be rechecked often and pertinent changes entered in both books. As new lakes are established, a map of the bottom should be prepared before the lake fills. Such maps will prove of much value in dragging operations.

ENVIRONMENTAL RESCUE

It would be foolish to launch a boat and attempt to drag a body of water that is only 3 or 4 feet (1 to 1.2 meters) deep. A recovery could be quickly made by rescuers lining up shoulder to shoulder and shuffling across to the other side and then returning; areas of search should overlap the previous search path. It is well to set up a guide rope so that the squad will be positive of the search area and no part of the bottom will be missed. An exception may be made in cold weather when it would be a risk to the health of the personnel to get into the icy water without protective clothing.

Shore-to-Shore Grappling (Figure 16-9). In flowing water a body may be carried with the current for a short distance; the dragging operation should start downstream from where the victim was last seen. The rope length should be over twice the width of the area to be covered and the drag secured to the center of the rope. The rescue crew will take positions on opposite banks of the stream and slowly pull the hooks across the area. If nothing is hooked, the crew should move upstream a distance equal to one-half the width of the drag and pull the drag back across. This procedure should be continued until the entire area is covered, and then reversed so the area is covered twice. If the body is not then recovered, the search area should be extended in both directions and the procedure repeated.

Boat-to-Shore Grappling. This method is similar to shore-to-shore grappling, except that one crew is on shore and the other crew is in an anchored boat. The crew on shore should move along the shore after each drag for some distance, then the boat should be moved, reanchored, and the operations repeated.

Boat-to-Boat Grappling. This procedure is similar to shore-to-shore grappling; two boats should be anchored both bow and stern, and facing in the same direction.

Grappling from a Moving Boat. The least advantageous method of dragging for a drowned person is from a moving boat since movements cannot be controlled perfectly; the entire bottom may not be thoroughly covered. However, dragging from a boat is the principal means of recovering bodies from large, deep lakes and rivers. Untrained volunteers often place the most powerful motor that they can obtain on the largest boat and race back and forth in the area; they usually work for hours and have nothing to show for their efforts.

It must be kept in mind that a body submerged in water weighs approximately 1 pound (0.45 kilogram) and is resting on the bottom; therefore, the grappling hooks must sink to the bottom if they are to make a strike. Fast movements will not permit the hooks to stay down where they can make a solid contact if they hit the victim's body. Examination of many drowned persons shows that the body was hit many times with the drag hooks, but the hooks were moving so fast and with such force that they only slashed the body and did not remain fastened.

A motor may be used in rough water. It is best to utilize the motor to power the boat upstream with the hooks out of the water. Then allow the boat to float slowly downstream with the hooks on the bottom. If there is no current, rowing into the wind will give the oarsman the best chance to keep the boat on a straight course.

Buoys and floats should be used to establish the body location as nearly as possible. If there is no recovery within one hour, expand the search area and continue systematic dragging operations. Always be sure that each drag sweep overlaps the one just completed.

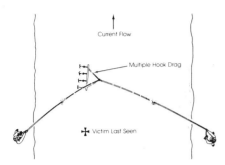

Fig. 16-9. Grappling from shore to shore. In flowing water a body may be carried with the current for a short distance; the dragging operation should start downstream from where the victim was last seen. Rescue personnel will take positions on opposite sides of the stream and slowly pull the hooks across the area. If nothing is hooked, following drags will slowly progress upstream.

Additional markers dropped at the end of each sweep will give a positive indication of the area which has been covered.

When a number of boats are in use, the boats should form a V-formation and make sweeps in the same manner as is recommended for one boat. Special care must be taken to see that each drag path overlaps the one next to and ahead of it.

Where there is a current, drop a bag of old clothes or other such material into the water, placing a weight in the bag so that it will sink, and attaching it to a float so that it will be suspended a foot or two off the bottom. Place this bundle in the water where the victim was last seen. Follow the float and drop markers along the path that is established; this route should be followed with the grappling hooks. It is most probable that the body will be found in the first eddy or deep hole where the marker stops.

Pike Pole Hook. If the bottom is rocky, or full of tree stumps and brush, it is impossible to drag and the use of a pike pole hook is recommended. This type of grappling hook is usually constructed so that it may be extended by adding four-foot sections of handle until a length of 20 to 25 feet (6 to 7.5 meters) is obtained. The pole should be held in a vertical position and the bottom tamped with quick up and down strokes. This probing action enables the operator to determine what is being struck by the bottom end of the pole, and lessens the possibility of the hook becoming fouled with debris on the bottom. The boat should be rowed slowly so the area can be thoroughly covered. When the body is located, the boat must be stopped and held stationary until the body has been securely hooked and raised to the surface.

Search and Recovery Diving

In many areas skin divers can be used to make rapid recoveries of drowning victims. Only experienced divers should be used. All rescue divers should be certified by competent authorities as to their ability before they are assigned and permitted to dive on an emergency assignment. Divers must be coordinated in their efforts.

Recovery diving in murky water is very hazardous. Water in a harbor that contains many boats has a layer of black oily silt which may vary in depth from a few inches to 4 or 5 feet (1.2 to 1.5 meters). The silt is easily disturbed, causing a loss of visibility. Behind dams where there was originally a river bottom, or in a rock quarry, the bottom is generally filled with trees, grass, and moss that grew among the obstructions when the water was at a lower level. This vegetation will hang up on a diver's air tank and face plate. Many times these bodies of water are murky even though they are quiet. In such a situation, it is best to use the maximum number of divers and make as few sweeps as possible, keeping hand-to-hand contact if possible.

Equipment. Besides the basic SCUBA gear needed by each diver, special equipment may be needed for the search. An inflatable safety vest should be worn by each diver to provide flotation in case of emergency. An efficient knife should always be carried in an accessible location in case the diver becomes entangled in underwater debris. The following equipment is also recommended:

1. **Surface Floats.** These markers should be painted a distinctive color for maximum visibility.
2. **Anchors and Bottom Guidelines.** Anchors should be heavy enough or so designed that the lines will be secured to the bottom without drifting. Guidelines should be thin enough so

ENVIRONMENTAL RESCUE

that they will not have much resistance when pulled through the water, and yet strong enough to be stretched taut.

3. **Safety Lines.** A light strong line should be attached to each diver, unless diving takes place in waters containing entangling debris.
4. **Buddy Lines.** A short buddy line fastened to the SCUBA harness may be gripped by adjacent divers when diving in murky water; this will allow equal spacing.
5. **Underwater Lights.** In most night diving operations, underwater lights or flares are essential.

Preparations for Diving. If several divers are to participate in a search, each should understand exactly what the other members are supposed to do. With this information, a minimum of communication is needed under the water. All divers must operate in pairs (the buddy system) if maximum safety is to be obtained; this rule should not be deviated from unless an extraordinary circumstance prevents it. Two persons will work as a team and will be responsible for each other's safety. Visual or touch contact should be maintained at all times while in the water.

Divers should attach sufficient weights to their belts so that they will sink easily with a full air tank. Each member should check the partner's equipment before entering the water. Before entering the water, all divers should clear their ears by holding the nostrils, closing the mouth tightly, and blowing slowly and moderately (not suddenly and forcibly) until the ears are clear; this is indicated by a popping of the ears. If both ears cannot be cleared by this procedure, the diver should not try to force them. Neosynephrine solution, or other nose drops, may be placed in each nostril with the head hanging back; the diver should remain in this position for 20 seconds, taking long slow inhalations. If the ears still do not clear easily, the process may be repeated. If both ears cannot be cleared, one should not dive except as a last resort to perform a rescue.

Diving Tenders. At least one member must be constantly stationed at the edge of the shoreline, wharf, pier, or boat gunwale to handle safety lines and serve as a tender for each team working in the water. Tenders are responsible for relaying communications between divers and supporting units. They should be constantly alert for hazardous conditions and must not leave the divers unattended. Tenders should maintain traffic control within the immediate area so the divers are not endangered by boats or strong propeller currents. It is important to maintain an open space around the diving area.

Diving Operations. When possible, the buddy system should be adhered to with two divers working together as a team. When the search area is small, or one diver is experiencing ear trouble, one diver may descend alone. When there is a single diver in the water, there should be another diver on standby prepared to dive if needed. In murky water, each diver should maintain contact with the partner by either harness or a buddy line, or they may hold hands.

Care must be taken to avoid fouling fins, harness, or air tanks on underwater obstructions. A diver who becomes entangled should remain calm and attempt disengagement in a slow and deliberate manner. Divers should always back out of narrow or obstructed areas on the assumption that the way in was clear.

Everything possible should be done to prevent disturbing the bottom and stirring up the silt. Divers should not tread water in a vertical position when near the bottom or touch the bottom with the hands, fins, or body any more than absolutely necessary.

It is beneficial to keep a record of the areas covered in the search and the teams which covered them. In the event of two or three day operations, a log of this type helps to prevent needless dives in areas already searched. When a search operation has been completed, a brief critique should be held to make notes of the operation for further reference. Information about currents and bottom conditions may prove beneficial if future dives are necessary in that area.

General Search Methods

A search is generally conducted by divers swimming in a definite pattern. Different patterns may be better suited to one type of search than to others; this will depend upon the type of the bottom, depth of water, and the base of operations. Several search techniques are listed; an imaginative combination of two or three of these patterns will prove effective in almost any instance.

Visual Search. A minimum of three feet visibility is necessary for an effective visual search. Stop above the bottom, turn in a complete 360-degree circle slowly, and visually scan the area. Then swim a search pattern by keeping well above the bottom, but within visual range. The search interval should be less than twice the visibility limits. Swim, continuously scanning from side to side, at a speed calculated to give complete visual coverage. A compass may be utilized when visibility is good; a search line should be used when visibility is poor.

Contact Search. Contact searching is used when conditions of little or no visibility are encountered. The diver should descend to the bottom and search with the hands and arms. The search interval should be somewhat less than an arm span (usually 5 feet or 1.5 meters). The diver should propel forward at a speed calculated to give sufficient time to check the area by simultaneously sweeping half circles ahead and to the side with both hands.

Search Patterns

Circular Sweep Pattern (Figure 16-10). The circular search pattern can be participated in by any number of divers, yet is effective for a single diver. The number of circles made from one reference point is determined by the length of the guideline.
 1. Place the anchor, which is attached to the guideline, in the center of the area to be searched.
 2. Space the divers along the line, starting close to the anchor, at a distance permitted by the visibility of the water.
 3. Divers should keep abreast and slowly swim in a full circle; they should keep the guideline taut.
 4. When a 360-degree sweep has been completed, the diver farthest from the anchor remains in position and the other divers take their places beyond that person on the line and swim another circle. It is important that one diver remain in position; otherwise, the area covered by the first circle will not be known and the object of the search might be missed.

Semicircular Sweep Pattern (Figure 16-11). The semicircular sweep pattern is most effectively conducted from a wharf or a shoreline. The number of sweeps made from one reference point is determined by the length of the guideline.

ENVIRONMENTAL RESCUE

1. Tie the guideline near the bottom of a wharf piling or anchor it on the shore.
2. Space the divers along the line, starting close to the anchorage, at a distance permitted by the visibility of the water.
3. The divers should keep abreast and slowly swim in a semicircle, keeping the guideline taut.
4. When a 180-degree sweep has been completed, the diver farthest from the anchorage remains in position; the other divers take their places beyond that member on the line and swim a second semicircle. It is important that one diver remain in position; otherwise the area covered by the first semicircle will not be known and the object of the search might be missed.

Straight Sweep Search Pattern (Figure 16-12). The straight sweep from a single guideline is generally used when there are many divers participating in the search and a good visibility exists.
1. Place the anchor considerably beyond where the object of the search is thought to be.
2. Pull the baseline taut and fasten it to the shore.
3. Attach a short line to the baseline with a loop. The length of this guideline will depend upon the number of divers utilized.
4. The divers should grasp the guideline and swim abreast until they reach the anchor.
5. The divers will then swim a 180-degree arc so that they end up headed toward shore.
6. When the divers are again abreast of each other, they will swim toward the shore.
7. If the object of the search has not been found, the baseline may be moved to a new location.

Straight Sweep Along a Shoreline (Figure 16-13). An effective method of searching a quiet body of water, such as a lake or a pond, is to have the divers spaced along a guideline. This type of search is effective when searching for small children who have drowned, since they will generally be close to shore.
1. Space the divers along the guideline at the distance permitted by the visibility.
2. The guideline is held by one man on the shore who walks along the edge of the water with the divers as they search; the divers should keep abreast of each other.
3. If the object is not found, the last diver should hold position and the other divers will take a position farther out on the guideline for another sweep.

Straight Sweep in Current (Figure 16-14). The straight sweep may be used in locations where there is a current in one direction. It is difficult to use more than one diver at a time in a strong surf or current, but other divers may be used to ensure a thorough search. The additional divers should swim the same sweep as the first diver.
1. Fasten one end of the guideline to an off-shore anchor or to an object on the opposite shoreline.
2. Attach the other end of the line to an object on shore.
3. The guideline should be played out from shore a given distance each time the diver swims out to the anchor, or to the opposite shore, and back.
4. If the object of the search is not found, move the two anchors to a new location and repeat the search pattern.

Rescue from Water

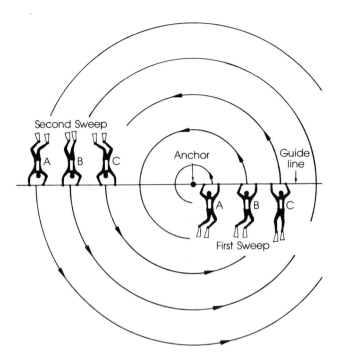

Fig. 16-10. A circular sweep pattern can be participated in by any number of divers, yet it is effective for a single rescuer. This method can efficiently and thoroughly check a large area in a relatively short time. An anchor is placed in the center of the search area, divers are spaced along a line fastened to the anchor, and a 360-degree circular sweep is made; the line coordinates the swim pattern and ensures that no space is missed. After a complete sweep, the divers move out further from the anchor and repeat the circular sweep. Instead of using a long rope to make large circular sweeps, it is more effective to relocate the anchor.

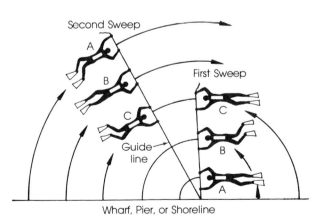

Fig. 16-11. The semicircular sweep pattern is most effectively conducted from a wharf or shoreline. The number of sweeps made from one reference point is determined by the length of the guideline. The guideline is tied to a wharf piling or anchored on the shore, divers are spaced along the line, and 180-degree sweeps are made. After a sweep has been completed, divers move further out on the line and repeat the operation. If the search is not successful, the guideline is moved and the operation repeated.

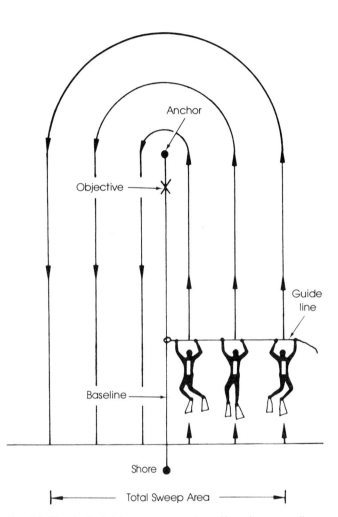

Fig. 16-12. A straight-sweep search pattern is generally used when there are many divers participating in the search and good visibility exists. A line is stretched taut between the shore and an offshore anchor. A guideline with a sliding loop coordinates the divers, so that a thorough search of the area is made.

ENVIRONMENTAL RESCUE

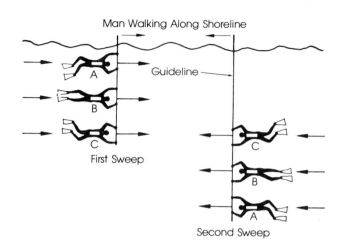

Fig. 16-13. A straight sweep along a shoreline is an effective way to search a quiet body of water, such as a lake or pond. This method is effective when searching for children, since they will generally be close to shore. Divers are spaced along the guideline and they keep abreast of a coordinator who slowly walks along the edge of the water. In some cases, a rowboat on the outer end of the guideline will aid in coordinating the divers, so that no spot is missed.

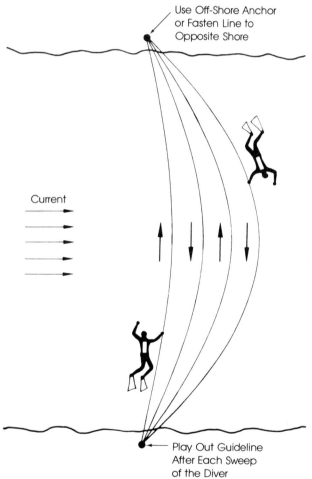

Fig. 16-14. A straight-sweep search in a current is generally inefficient, since it is difficult to thoroughly check an area of fast-moving water. It is best to fasten one end of the guideline on shore and the other end to an offshore anchor or the other bank of the stream. Each time the diver completes a swim along the guideline, the rope is played out so the following swim will be over a different route. The anchorings at the ends of the guideline are frequently moved to new locations and the routine is repeated.

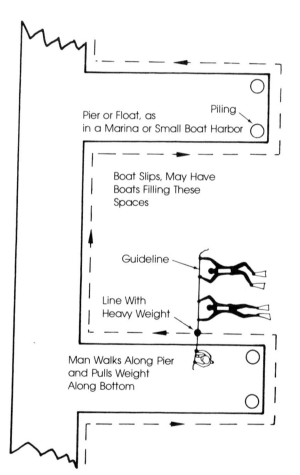

Fig. 16-15. A search with a weighted line is effective when it is necessary to check around a pier or wharf that is oddly shaped, or one that has pilings to provide boat slips. A line with a heavy weight is slowly pulled or carried along the edge of the wharf to coordinate the search pattern. Divers are spaced along a guideline that is fastened to the weight. In most cases it is best to keep the weight off the bottom, so mud and sand will not be disturbed and reduce underwater visibility.

Rescue from Water

Search With a Weighted Line (Figure 16-15). When a search is required around a pier or wharf which is oddly shaped, or has pilings to provide boat slips, the divers may swim from a guideline which is held by a person on the pier.

1. Fasten a weight near the end of a length of rope, Leave approximately 5 feet (1.5 meters) of line extending past the weight for each diver.
2. Lower the guideline until the weight barely touches the bottom. If the bottom is muddy, raise the weight so that it will not stir up the mud and reduce the visibility.
3. The divers should position themselves along the guideline and swim abreast, keeping the guideline at right angles to the pier. Sufficient divers should be used to completely sweep the area with one pass since the weight and the divers' fins will stir up the bottom and reduce the visibility.
4. The person on the pier should pull the weight along the bottom slowly and guide the divers through the search area.
5. The divers should swim slowly so that they do not get ahead of the guideline and start to move in a circle.

DROWNPROOFING

A normal adult human body is just slightly buoyant; a full deep breath into the lungs will increase buoyancy, exhalation will decrease the ability to float. When a swimmer is attempting to get a breath of air, his initial response is to want to get his head out of the water. The average human head weighs approximately 17 pounds (8 kilograms) which is far too heavy for the body buoyancy to support above the water. A swimmer who has sufficient buoyancy should float on the back to safety if in distress; this reduces panic and lets the swimmer breathe with a minimum of effort.

Treading water, floating on the back, or swimming any of the various strokes can lead to various degrees of exhaustion and consequent panic. Few people can hold their heads above water for any length of time. A swimmer who stops treading water will sink. In rough water, back floating can be impossible.

Drownproofing is a technique by which a swimmer can survive for hours by conserving energy and allowing the body's natural buoyancy to furnish flotation. This method requires the swimmer to float face down in a dangling position, head beneath the water; upward ascent for air periodically is done by performing an easy, modified version of the breast stroke.

With head face down in the water, the drownproofed swimmer can rest while the back of the neck stays at or near the surface at all times. The body becomes a natural life jacket in a crouched position. The arms and legs hang down loosely. After 6 to 10 seconds, it becomes necessary to obtain an exchange of air. The swimmer brings hands upward, folded in front of the face, and lifts the legs in preparation for a scissors kick. Exhaling is done by tilting the head upward, out of the water, until the chin is at the surface. To keep from sinking while inhaling, the swimmer should thrust the arms sideways and downward, also doing the scissors kick. Then the head drops forward into the water as before.

Unless the water is sufficiently cold that hypothermia becomes a problem, this method of water survival can be used for a long period of time because little energy is required.

COLD WATER SURVIVAL

Four-fifths of the earth's surface is covered by open water. Although accounts of survival incidents are often pessimistic, successful rescue operations are very possible in the majority of cases. The sinking of the Titanic in 1912 provided a dramatic example of the effects of cold water immersion. Partially due to a lack of preparedness with protective clothing, adequate flotation equipment, and a knowledge of survival procedures, none of the 1,489 persons immersed in the 32°F (0°C) water was alive when rescue vessels arrived 1 hour and 50 minutes after the sinking. Countless lives could have been saved had the victims known more of how to cope with cold water. In other instances, there have been countless victims of cold water exposure resulting from overturned pleasure craft, water sports mishaps, and the sinking of oceangoing vessels.

It is important to realize that a person is not helpless to effect survival in cold water. Body heat loss is a gradual process and recent research shows that in 40°F (5°C) water a normally dressed person has a 50 percent chance of survival after 2 hours exposure. Simple techniques can extend this time to 3 hours or more if the person is wearing a lifevest.

Formerly, it was recommended that people could survive longer in cold water by swimming or otherwise exercising in order to increase the body's heat production. It is now evident that while exercise does increase heat production, it increases heat loss even more, the overall effect being to shorten survival time. As the body areas that show the greatest heat loss are the lower neck, chest, sides, and groin, the best survival method in cold water is to wear a flotation device and hold the arms firmly against the sides of the chest, keeping the thighs together, and raising the knees to protect the groin area. Use of this posture at a water temperature of 50°F (10°C) can result in a predicted survival time of about 3-3/4 hours, an increase of 50 percent over that of the passive position (Table 16-1).

Table 16-1. ESTIMATED COLD WATER SURVIVAL TIME

Survival Method	Estimated Survival Time, in hours, When the Water Temperature is:		
	40°F (5°C)	50°F (10°C)	60°F (16°C)
No Flotation Device:			
Drownproofing	1.08	1.44	2.26
Treading Water	1.46	1.96	3.07
Flotation Device:			
Posture to Lessen			
Heat Escape	2.87	3.80	5.96
Passively Floating	1.96	2.62	4.11

Ship or Airplane Abandonment

If forced to abandon a ship or aircraft at sea, preplanned survival procedures will increase the chances for a successful rescue.

Hypothermia, the loss of body heat to water, is a major cause of deaths in water accidents. Often the cause of death is listed as drowning; but, most often the primary cause is hypothermia and the secondary cause is drowning. After an individual has succumbed to hypothermia, the victim will lost consciousness and then drown. Table 16-2 gives the stages of survival in cold water.

The following pointers for survival in case of forced ship or airplane abandonment should be part of the general water safety education program in every community:

1. Locate and wear a personal flotation device as quickly as possible. This is probably the single most important item of survival equipment. Survival in cold water is tough enough without having to contend with staying afloat too. Know how the flotation device is worn and used before the emergency.
2. Try to enter the water in a lifeboat or raft. By doing so, a person will avoid wetting his insulation and losing valuable body heat to the water. Avoiding entering the water, thus keeping dry, will greatly increase the chance for living.
3. Wear several layers of clothing. If a person is fortunate enough to stay dry and enter the water in a lifeboat or raft, the trapped air within the layers of clothing will provide excellent insulation. However, if the victim becomes wet in abandoning the ship, the layers of clothing, although wet, will slow down the rate of body heat loss. Clothing having a thickness, such as waffle-weave underwear, woolens, and pile-lined garments are excellent for building up layers. An ideal outer layer of clothing would be a waterproof or coated fabric windbreaker or foul weather garment. If afloat in the water, a waterproof outer garment having close fitting wristlets, anklets, and collar would reduce the passage of water through these openings and help to cut down body heat loss.
4. Protect the head, neck, groin, and the sides of the chest for these are areas of rapid heat loss to cold water. In addition, waterproof insulated gloves or mittens with a snug fit at the wrists reduce the water exchange at the hands, preserve body heat, and prolong the usefulness of the hands for performing helpful maneuvers. Insulated boots and layers of socks impart the same protection to the feet. A great amount of body heat can also be conserved by covering the head with an effective hood. The neck can be protected with a towel, scarf, or other cover.
5. If conditions prevent abandoning the ship or aircraft in a lifeboat or raft, and it is necessary to enter the water directly, try to minimize the shock of sudden cold water immersion. Rather than jumping into the cold water, it is better to enter gradually. A sudden plunge into cold water can cause rapid death or an uncontrollable rise in the breathing rate; this could result in an intake of water into the lungs. If jumping is necessary, hold the breath by pinching the nostrils and avoid swallowing water during the plunge.
6. Once in the water, whether accidentally or by ship abandonment, a victim should orient himself and attempt to locate the plane, ship, lifeboats, rafts, or other floating objects. If there was insufficient time to properly prepare the clothing, take time now to button up as much as possible. Often motion pictures depict the victims disrobing themselves so they may swim and float better; this is directly opposite to the best survival techniques. In cold water it is natural to experience violent shivering and great pain in the limbs; these are natural body reflexes that are not dangerous. It is necessary, however, to take action as quickly as possible to button up, turn on signal lights, and other survival necessities before dexterity is lost in the hands.
7. Stay upwind of the ship or aircraft, but in the vicinity, until it sinks. Avoid fuel covered waters.

ENVIRONMENTAL RESCUE

8. Make a thorough search for missing passengers and crew if in a lifeboat or raft. Carefully patrol the entire area near the crash especially in the direction toward which waves are moving. Look very carefully; some of the missing persons may be unconscious and floating very low in the water.
9. While afloat in the water, no attempt should be made to swim unless it is to reach a nearby craft, a fellow survivor, or a floating object which will offer support. Unnecessary swimming motion will pump out the warmed water between the body and the clothing layers; this will increase the rate of body heat loss. In addition, unnecessary movements of the arms and legs send warm blood from the body's inner core to its cold outer layer; this results in a very rapid heat loss. Hence, it is important to remain as still as possible in the water, however painful it may be. *Remember*, pain will not kill, but heat loss will.
10. Do not use *drownproofing* in cold water. Drownproofing is a survival technique whereby a person relaxes in the water and allows his head to submerge between breaths. It is an energy saving procedure to use in warm water when not wearing a lifevest. However, the head and neck are high heat loss areas and must be kept above the water; this is why it is extremely important to wear a lifevest in cold water. If not wearing a flotation device, tread the water only as much as necessary to keep the head out of the water.
11. The body position that is assumed in the water is very important in conserving body heat. Tests show that the best body position is one where the knees are held up against the chest in a doubled up fashion with the arms tight against the sides of the chest. This position minimizes the exposure to the cold water of the groin and chest sides, which are both areas of high heat loss. Try to keep the head and neck out of the water. Another heat conserving position is to huddle closely to one or two others afloat, making as much body contact as possible. It is necessary to be wearing a flotation device to be able to hold these positions in the water; also group closely together in a lifeboat or raft to conserve body heat.
12. Attempt to board a lifeboat, raft, or other floating platform or object as soon as possible in order to shorten the immersion time. Remember, body heat is lost about 25 times faster in water than in air. Since the effectiveness of clothing insulation has been seriously reduced by water soaking, a person must now try to shield himself from the breeze to avoid a wind-chill effect (convection cooling).
13. Keep a positive attitude about survival and rescue; this will improve the chances of extending a victim's survival time until rescue comes. The will-to-live does make a difference.

Table 16-2. COLD WATER SURVIVAL STAGES

Water Temperature		Exhaustion or Unconsciousness	Expected Time of Survival
°F	°C		
32.5	0.3	Under 15 min	Under 15–45 min
32.5–40	0.3–4	15–30 min	30–90 min
40–50	4–10	30–60 min	1–3 hr
50–60	10–16	1–2 hr	1–6 hr
60–70	16–21	2–7 hr	2–40 hr
70–80	21–27	3–12 hr	3 hr–indefinite
Over 80	Over 27	Indefinite	Indefinite

Person Overboard

When a person falls overboard, it will probably catch everyone who can help by surprise. Whether a passenger on an ocean liner or a crew member on a sailboat, some useful advice to follow for preventing this situation is more valuable than attempting to locate a victim in a large body of water under the most horrible conditions that usually exist.

1. Always wear a lifevest when working in a hazardous position or when going on deck for any reason in poor weather.
2. Use extreme care when walking along decks, up and down ladders, and across gangplanks. Use the hands to grip rails and other support devices. Do not lean on rails and lifelines. Keep the hands out of pockets so they are free to grab for support in the event of a slip or fall.
3. Be particularly careful at night or in fog. Poor visibility increases both the chances of an accident and the difficulty in locating a victim in the water.
4. Common sense dictates that dress should be appropriate to the weather, but some care in the selection of clothes can make a difference in survival. If possible, wear many layers of clothing, including a waterproof outer layer. Make certain that the neck, wrist, and ankle portions of the clothing are snug fitting. This will reduce the exchange of water within the clothing if a person should go into or under water. Also, woolen clothing are better insulators than cotton, especially when wet.

DEEP WATER RESCUE

Regardless of how a person becomes a water casualty victim, once the person is in the water the immediate need is to be rapidly located and rescued. Since water entry may occur at any time of the day or night and in various degrees of fog and poor visibility, it is vitally important to have some type of distress signal to attract attention. Types of signals that are effective are whistles, lights, mirrors, and even a white handkerchief. When using a flashlight, a good point to remember is that cold temperatures reduce the effectiveness of even a fresh battery. Therefore, give the light and battery a little body warmth to improve the battery performance.

The waving of outstretched arms or splashing of the water is an effective and simple signaling technique, but it should only be used when rescue is nearby.

A survival victim in the water or afloat in a lifeboat or raft may be rescued by any one of several different ways. The person may be sighted by a surface vessel and helped aboard by personnel of that vessel.

Rescue may also come in the form of a helicopter. The most popular retrieval device in use by the Coast Guard for hoisting a water survivor is a mesh screened rescue basket. The use of the basket is rather obvious; it has an open top and all that is required of the survivor is to climb in and stay put as he is raised to safety. Although the other services have a variety of pickup devices, ones commonly utilized for nonmilitary situations are the rescue sling (horse collar) and the rescue net. The rescue sling is a padded belt that is wrapped around the back and under the arms of the survivor. The two ends of the sling are attached by their looped rings to a hook at the end of the

ENVIRONMENTAL RESCUE

helicopter cable. The Navy rescue net is a rope-like cage with an open side for the survivor to enter and be seated while being raised to safety. No matter which approach is used, a victim may be assured that rescuers will enter the water to assist a person in getting into the retrieval device if he is not able to help himself (Figure 16-16).

If victims are located by a landplane that is unable to land on the water, flotation devices and rafts may be dropped so the victims can survive until other rescue operations can be carried out.

Fig. 16-16. Helicopter rescue of a victim from deep water. Note the rescuer in the water who assisted the casualty into the retrieval device. (Courtesy of U. S. Coast Guard.)

QUESTIONS FOR REVIEW

1. Normally, the rescue of a drowning person should be initiated by the:
   a. First observer of the victim's plight.
   b. Best swimmer in the vicinity.
   c. Lifeguard at the body of water.
   d. Most concerned bystanders.

2. When a swimmer is having difficulty in staying afloat, which of the following is the least important attribute of a rescuer?
   a. Initiative.
   b. Adaptability.
   c. Eagerness to get involved.
   d. Swimming ability.

3. The greatest majority of drowning victims are least affected by:
   a. Water in the stomach.
   b. Water in the lungs.
   c. Spasm of the larynx.
   d. Coughing and gagging.

4. The fate of a victim who is threatened with drowning is least determined in the majority of cases by the:
   a. Speed of rescue.
   b. Prompt clearing of the airway.
   c. Swimming ability of the rescuer.
   d. Kind of water in which the victim was submerged.

5. When a diver is knocked unconscious in the water, which of the following is the worst technique?
   a. Keep the victim floating on the back.
   b. If the victim is not breathing, immediately start artificial respiration.
   c. Immediately drag the victim out of the water so body heat is not lost.
   d. Immobilize the victim on a backboard before attempting removal from the water.

CHAPTER 17

Wilderness Search and Rescue

Every year a greater number of people enter the wilderness areas of this country in search of recreation, riches, exercise, and adventure. As these hordes of travelers increase, a correspondingly greater number of accidents and other misfortunes occur. There is a very good possibility that anyone who participates in outdoor sports, or who lives in or adjacent to a rural or wilderness area, may be called upon to participate in a rescue operation or may become a victim (Figure 17-1).

Fig. 17-1. In a relatively small area (near Mt. Whitney, California) there may be many types of rescue problems. Note fairly large lake, extensive snow (in July), heavy forest, and jagged rock. The snow line and tree line are well defined. (The darkest area in photo is trees.) (Photo by Ron Schneiders.)

Search and rescue (SAR) operations are based on the humanitarian principle which compels people to render aid to those in trouble. These distress situations may cover the spectrum from looking for a small child who has wandered away from his parents in a campground to conducting an extensive search covering several thousand square miles (kilometers) for a missing aircraft. Rescuing a trapped or fallen hiker or mountain climber, offering emergency medical care, evacuating an injured or critically ill victim, and a myriad of other problems are important duties of SAR personnel. Some element of jeopardy to the individual's life must be present either by reason of the victim's physical condition, the situation from which rescue is required, or both. In general, an SAR incident is considered imminent or actual when it is apparent that persons are, or may be, in distress or when there is a transmitted request for assistance.

Rescues in difficult terrain usually require specialized experience and knowledge that can only be obtained by years of study and practice. It would be impossible to condense all of the information needed to perform mountain, desert, and other wilderness search and rescue operations into one chapter of a book. For this reason, specific instructions on skiing, snowshoeing, mountain climbing, desert travel, spelunking (caving), and other specialized techniques have been omitted.

Each SAR mission will have its own particular circumstances; the responsibility and response required will vary widely. It is not possible to develop detailed, comprehensive procedures that can be applied unvaryingly at all times. A general sequence of actions and events is provided. Some of the procedures are performed simultaneously, in different order, or are skipped entirely to suit particular circumstances. Because of the many variables encountered in SAR operations, procedures and techniques should be tempered with judgment, having due regard for conditions existing at the time which may require caution or resourcefulness.

SAR ORGANIZATIONS

Most of the SAR incidents may be grouped into three categories: search, rescue, or medical evacuation. Each situation is unique and requires investigation to determine the last known location of the victims, the urgency of the incident, and the number of people involved. There are many organizations which may be called upon to participate in rescue operations that require different specialized skills and techniques from those in which the local personnel have been trained. Mature consideration by local authorities will be necessary to determine if their personnel can safely and effectively perform the necessary duties. If there is a reasonable doubt, it is best to delay the operations until adequately trained and equipped personnel can be obtained; in these days of rapid transit, it may not take long. To do otherwise might drastically increase the number of victims requiring rescue.

Although there are many agencies involved in search and rescue in the United States, the federal government assumes overall responsibility. Traditionally, the military forces and Coast Guard have enjoyed a respected reputation for assisting the populace when an accident or disaster occurs. The federal government encourages the lower levels of government to assume SAR responsibility within their geographic boundaries and capabilities. The response to local civilian SAR incidents, such as lost individuals or injured vacationers, is, first, the responsibility of the community, is, second, the state, and is, third, the federal government. All civil forces and resources should be utilized to the fullest extent consistent with the demands of the situation prior to requesting military assistance.

The SAR forces must be logically selected to render aid. The SAR unit selection involves two major evaluations: (1) the operational capability of the organization and (2) the training and experience of the crew manning the unit. The crew selected should be able to reach the scene quickly.

The initial location of overdue, lost, or downed aircraft is the primary responsibility of the Civil Air Patrol (CAP), which is a corporation created by federal statute and established by law as a voluntary civilian auxiliary of the United States Air Force.

Private volunteer organizations such as units of the Mountain Rescue Association, National Jeep SAR Association, National Association of Search and Rescue Coordinators, Explorer SAR, National Ski Patrol, and other groups of highly skilled and thoroughly trained paraprofessional wilderness experts provide an extremely valuable service to the American public. These dedicated groups, whose primary concern is to provide the necessary personnel, expertise, and equipment to aid the victims of wilderness tragedies, are invaluable at any incident. Without their unique capabilities, the successful conclusion of many wilderness accidents would be impossible.

Safety Education

The most significant step a rescue organization can take is education. Making the people who use wilderness areas aware of safety precautions that have been developed for various activities will cut tremendously the number of people needing rescue. The accidents that do not happen and the travelers who do not become lost or injured will not require the services of a rescue group.

Although it is practically impossible to list all the precautions for each of the many kinds of wilderness activities, there are a few general rules that apply to almost any wilderness situation. The first is: never travel alone. Anyone who hikes, climbs, or explores by himself is asking for trouble, and even the smallest accident can be fatal. On longer and more rigorous trips, three or even four persons is a minimum. If one person is injured, someone can stay with the casualty, while others go for help.

Persons who are in a weakened condition, who have neglected their physical conditioning, or who have physical disabilities should stay well within their limitations. Young children and elderly people should be closely watched to prevent their becoming lost or injured in hazardous terrain. People should know that while traversing wildland areas, it is most important to travel methodically, to take time to plan the safest route. Travelers must learn to emphasize safety, not speed, and to conserve energy. There is an old mountaineering adage which states, "The first man up the mountain is usually the last one down."

It is safest to confine traveling to patrolled areas or paths. If a more extensive trip is planned, information is provided by forest rangers and other responsible persons; someone should be informed about the anticipated length of the journey and the route to be traveled. It is important to carry sufficient equipment in case an overnight camp must be made; this should include food, water, extra clothing, a map, a flashlight, a compass, and other survival gear.

Even if one is traveling on well-marked trails, and especially if a cross-country trip is planned, a map is the best guarantee against getting lost. The best maps for wilderness travel are the topographic maps provided by the U.S. Geological Survey, which show the position and shape of land forms. Even the most inexperienced reader, with some practice, will have little trouble identifying all major features of the terrain, and the roads and trails in a region. Other organizations, such as the U.S. and state forestry and park services, automobile clubs, travel clubs, and recreational groups, also print excellent maps. They are usually quite inexpensive and a traveler's surest protection against becoming lost in the wilderness.

Rescue organizations should take pains to advise travelers to watch the weather conditions and head for civilization at the first sign of an impending storm. It is hazardous to be caught on exposed terrain during a wind, rain, snow, lightning, or any other type of storm. An improvised shelter may be safer than attempting to travel during inclement weather.

Communications

Proper communications is the key to a successfully coordinated SAR incident. If participating SAR units do not follow correct communications procedures, the result will be a series of individual maneuvers, rather than a team effort. The ideal communications system is one that allows a mission coordinator to communicate directly with each element in the unit. A compromise must be reached wherein the maximum of reliable communications is afforded while still remaining within the limits of technical, economic, and atmospheric conditions. Functions required to accomplish the mission should be homogeneously grouped or assigned responsibilities according to their capabilities and ability to communicate. Only through planned organization and proper utilization can the SAR force fully realize the most benefits and advantages of a well-planned and modern communications system. Therefore, the successful operation of any communications system depends upon how well activities at one location are coordinated and integrated with similar activities at other locations and a provision for management of the organization.

USE OF AIRCRAFT IN SEARCHES

Aircraft are normally the most satisfactory units for conducting a preliminary search because distant areas may be reached quickly and a larger section of the countryside may be covered within a given time (Table 17-1). However, along with the benefits, there are definite limitations on their usefulness.

Rescuers should not request nor expect pilots to fly under some limitations of altitude, weather, and terrain. All aircraft are limited in their working altitude and may not be capable of maintaining a reasonable margin of safety while flying in high mountains. Turbulence may be a serious problem while flying in rough country. Mornings are usually the best time for flights because there is less of the turbulence associated with the afternoon heating by the sun. Cumulus clouds indicate turbulence and strong updrafts and downdrafts. Upcurrents often exist over ridges. Terrain features are obviously of great concern to a pilot.

A fixed-wing aircraft should not be flown low into a blind canyon as this will leave no escape if anything goes wrong and the airplane cannot gain any additional altitude. It is usually safer to fly near one side of a canyon as this will allow making a U-turn if it becomes necessary.

A light aircraft may be able to land safely at a high altitude, but may not be capable of taking off again with a heavy load.

The speed and higher altitudes of aircraft, which increase the rate of area coverage, are not so advantageous when searching visually for small targets such as hikers in brush or forested sections. Looking for a person in a forest from an airplane is similar to searching for an ant in tall grass. A victim who is cooperative may seek out a clear area to become visible, but an uncooperative person will be hard to detect.

When using aircraft for visual search the following points should be kept in mind: Slower aircraft are better for the smaller targets; search effectiveness is more efficient at the lower altitudes; a maximum number of observers in the plane is desirable; helicopters are better for small search areas; flight durations must be relatively short due to high fatigue factors affecting the crew; and aircraft use may be curtailed by icing, hail, turbulence, or low visibility when bad weather conditions exist.

Table 17-1. THIS TABLE SHOWS THE ELEVATION NECESSARY TO OVERCOME THE EFFECT OF CURVATURE OF THE EARTH WHEN SEARCHING THE HORIZON. THE HIGHER THE SEARCHER, THE LONGER THE VISUAL RANGE.

HEIGHT OF EYE VERSUS HORIZON RANGE							
Height		Horizontal Distance		Height		Horizontal Distance	
Feet	Meters	Miles	Kilometers	Feet	Meters	Miles	Kilometers
1	0.3	1.3	2.08	200	60	18.6	29.76
3	0.9	2.3	3.68	250	75	20.8	32.28
6	1.8	3.2	5.12	300	90	22.8	36.48
9	2.7	4.0	6.4	350	105	24.6	39.36
12	3.6	4.6	7.36	400	120	26.3	42.08
15	4.5	5.1	8.16	450	135	27.9	44.64
18	5.4	5.6	8.96	500	150	29.4	47.04
20	6	5.9	9.44	600	180	32.3	51.68
30	9	7.2	11.52	700	210	34.8	55.68
40	12	8.3	13.28	800	240	37.3	59.68
50	15	9.3	14.88	900	270	39.5	63.2
60	18	10.2	16.32	1,000	300	41.6	66.56
70	21	11.0	17.6	2,000	600	58.9	94.24
80	24	11.8	18.88	3,000	900	72.1	115.36
90	27	12.5	20	4,000	1,200	83.3	133.28
100	30	13.2	21.12	5,000	1,500	93.1	148.96
150	45	16.1	25.76	6,000	1,800	102.0	163.2

IMMEDIATE ACTION

Persons at the scene of an accident must give immediate consideration to the advisability of attempting a rescue operation. What is the nature of the problem? Can persons at the scene perform the necessary operations? If a victim is injured or endangered, there may be no alternative but to attempt rescue even against overpowering odds. At other times, conditions may be so hazardous, and the chances of success so slim, that rescue should be delayed until additional equipment and personnel have been obtained. Unduly jeopardizing rescuers to retrieve or transport a dead body is never justified. Self-sacrifice is spiritually commendable, but under some conditions it can have the effect of sacrificing everyone.

When a person has been injured or trapped, the first response of companions often invites further accidents. The victim must, of course, be reached quickly; this should not be done at the expense of further casualties. When there is a hazard of an avalanche or a rock slide, those persons who are not needed immediately should take cover but remain alert. On difficult terrain, only one or two rescuers should be dispatched to reach the victim (Figure 17-2). If the victim is in a crevasse, over a cliff, or trapped in a subterranean tunnel out of sight or hearing, then all signals and actions must be predetermined before a rescue is attempted. Remaining calm is important, since excitement only contributes to further accidents.

If for any reason an injured person must be left unguarded in a hazardous position, tie the person securely to prevent falling. Victims who are left alone occasionally become confused and some have fallen or wandered to their deaths.

Summoning Aid

The same remoteness that makes the wilderness so attractive to so many people creates problems when aid must be summoned. When a telephone is readily available, the local operator will be able to contact the appropriate emergency service. Usually, finding assistance will not be so simple.

In most areas where specialized rescue operations are likely to be needed, organizations with the necessary skills, training, manpower, and equipment may already have been formed. They can frequently effect searches and casualty evacuation more effectively than an untrained group, even considering the time required to contact them.

When time and conditions permit, the rescuer should carefully evaluate the types of necessary skills and equipment. Groups trained in one area of rescue operations may be ineffectual in other rescue situations. For example, a ski patrol may operate effectively on snowy slopes, but be little better than novices if the rescue is likely to involve work on bare rock faces. Knowing which rescue organization to contact may save a lot of time and even lives.

Government Agency Assistance

Some situations, however, don't allow for this kind of careful analysis. When a person has become lost, or travel to summon aid is impractical, a small fire may alert forest rangers, aircraft, or other help. Rescuers must be careful to build the fire where it cannot spread. A small fire made with green brush will produce a great volume of smoke and be more effective than a large flame. All wilderness areas are under the supervision of some governmental agency, however remote their representative might be. The U.S. National Park Service, the U.S. Forestry Service, state park or forestry services, local law enforcement or fire departments, or some other organization can usually be contacted by use of a signal fire. When time permits, a more direct contact with the responsible organization is preferable, but the physical and psychological pressures of an emergency situation may make the difficulty of such contact impractical. A good practice is for travelers to know before entering

Fig. 17-2. A member of a mountain rescue team rappelling down a cliff to reach a trapped victim. Whether referred to as sliding or rappelling, the procedure involves being safely suspended in a belt or harness and then controlling your descent down a line by changing the friction on the rope with some device. This is a quick, reasonably safe way to get down from high places with a minimum of equipment. (Courtesy of the Los Angeles County Sheriff's Department.)

wilderness areas who is responsible for that area and to notify that agency that they will be in a certain area for a number of days. The National Park and Forest Service both require overnight hikers to procure a fire permit, which serves as a method of notifying officials of their route. Knowing where a hiker will be and for how long is a fine precaution against being stranded for long periods of time.

EVALUATION OF THE RESCUE SITUATION

The ability to analyze and evaluate a rescue situation is a vital requirement for all persons who become involved in a rescue incident. This process may be divided into three major steps: (1) the definition of the task, (2) the circumstances under which the rescue team will be required to operate, and (3) planning a course of action.

Definition of the Task

Too often rescue attempts end in either embarrassment or disaster because the task was not properly defined at the start. Nothing can undermine the morale of a rescue team more than to be rousted out of bed in the middle of the night, and spend two back-breaking days searching for a person who isn't even lost. Even worse is to have rescuers injured or killed trying to rescue a corpse. This, then, becomes the first question: Is a rescue attempt really required or justified? Will there really be any practical chance of success under the existing conditions? Should the rescue effort be restricted to gathering information until sufficient knowledge has been accumulated, or should a rescue team be dispatched immediately?

Before a systematic rescue effort can be made, the following information is necessary:

1. Who, what, where, when, why, and how the accident occurred.
2. The number of casualties, the nature and the seriousness of the injuries.
3. The distance of the accident or victim from a road or a trail, and the type of terrain; these factors will determine the probable difficulty of evacuation.
4. How many people are still at the scene, and their strength. Will they be able to help a small nucleus of trained rescue personnel, or must the rescue party figure on supplying the necessary manpower?
5. Whether the people at the scene will wait at that spot, or move to a safer location. A definite rendezvous point must be determined.
6. What method of evacuation will be necessary: carrying by stretcher, sliding out on a toboggan, lowering down a steep cliff, lifting out of a deep mine, etc. Sufficient equipment must often be carried in, so the method of evacuation must be predetermined (Figure 17-3). Similarly, it is necessary to consider the need for support facilities such as relief manpower and the establishment of a base camp. If an air ambulance service is available, it must be determined if a light aircraft or a helicopter can land in that region. In the summertime horses and mules can save much physical labor; snowmobiles and other winter vehicles will traverse snowy terrain with speed and ease. It is most important to evaluate what is required and what is available, and then make the decisions.

These questions are extremely difficult to answer objectively because emotional responses get tangled up with logical thought processes. For example: the pilot of a light airplane radios that he is about to crash and gives his location; his wife and child are with him; the crash site is a mountain peak which is 12,000 feet (3,600 meters) in altitude; there is a blizzard blowing and the temperature is -20°F (-29°C). The rescue team determines that the minimum length of time that would be required to reach the crash site would be six hours, under the most optimistically estimated conditions. Even if the family survived the crash, they would not be able to survive six hours of numbing cold since they are above the tree line, and no fuel or shelter will be available; the deep snow would make travel impossible. These arguments will make little impression on the family's relatives; they will probably want to organize and lead a rescue party themselves. A case like this is very difficult because there is no way to prove that the loved ones are dead. The same situation in the summer, under good weather conditions, at lower altitudes, or other circumstances would undoubtedly alter the conditions under which survival would be possible. But the determining factors are that the rescue attempt would be extremely hazardous for the rescuers and the chances of success would be pitifully small.

At the other extreme is the case of the person who does not require rescuing because he is not in trouble. Often this is brought

Fig. 17-3. A rescue team transporting a victim under hazardous conditions. A large force of rescuers is required to evacuate an injured patient from a rural environment. A wintery accident on steep snow-covered mountain slopes involves the constant threat of snow avalanches, rock slides, slippery footing, and a multitude of other complications. In this photograph the victim has been treated and his condition stabilized before sliding him out to civilization on a toboggan. (Courtesy of the Los Angeles County Sheriff's Department.)

about by a lack of consideration for other people. A good example is the occasion when two persons hike in to a high mountain lake to fish and arbitrarily decide to extend their stay for a few extra days. This poses the most severe problem for anyone considering a rescue effort. These hikers could be in very serious trouble, or they could simply be very inconsiderate. Often the decision as to whether to attempt rescue or not must rest on someone's evaluation of the hikers' character and experience, an insubstantial basis for evaluation at best.

Circumstances of the Operation

When the question of whether there is to be a rescue attempt or not has been resolved, the question then arises as to the specific job of the rescue team, and the circumstances under which it will have to operate. Reduced to its essentials, a person or a group of persons may be lost, injured, stranded, trapped, or a combination of these. Each of these conditions carries its own implications. A person who is lost, especially a small child, will require a very large search organization, even if the area is relatively small. If a person is hurt, but his location is known, a very small team is usually sufficient. One of the rescue party should be a doctor, unless the rescue personnel are sufficiently trained in emergency medical care to offer adequate treatment. If the person is stranded or trapped, a small team will probably be sufficient, but specialized skill and equipment may be required.

Some situations are so well-defined and occur so frequently that organizations have formed to deal almost exclusively with them. One such example is that of the skier who is injured and is in a known location; almost every ski area has a ski patrol to handle this type of rescue. Another example is that of the downed aircraft; the Civil Air Patrol has adopted this particular rescue problem and has become very adept at handling it.

Another important circumstance to consider is the time element. Obviously, it is always important to achieve a rescue as quickly as possible. There is usually a compromise involved here; the more time that is available, the more thoroughly the rescue attempt can be planned and the more certain will be its success. If eight or ten hours are available for a search, it might be possible to mobilize a large group of trained personnel, bring in rescue vehicles, secure aircraft, and set up an elaborate communications network. If the rescue must be accomplished in two hours or less, none of these will probably be utilized. The determining factors are the condition of the victim and the environment. In mild weather at low altitudes, a person who is lost, even if slightly injured, can survive for a long time. If the victim has food and water available, and is reasonably cool-headed and competent, a survival period of weeks is not impossible. On the other hand, if the person is seriously injured, or the weather is bad, survival time might be measured in minutes. If it is raining, snowing, or the wind is strong, survival conditions are much worse than the mere temperature might indicate. Unless it is known that the victim is an experienced outdoorsman, it should not be assumed that person will do anything significant to ensure survival. Unfortunately, the rescues that are the most difficult to perform are usually the ones that must be accomplished in the shortest possible time.

Often it is best to get several experienced persons together and attempt to make a realistic estimate of the amount of time available for the rescue operation. One of these should be a medical doctor who can make an estimate of the victim's probable survival period; this should be based upon the weather and the person's physical condition, such as age, disabilities, and medical history. After the time required for rescue efforts of various degrees of complexity and thoroughness has been analyzed, then the best overall schedule can be worked out.

Planning a Course of Action

After it has been determined that a rescue operation is necessary, and the surrounding circumstances have been taken into account, the actual implementation of the operation depends on careful planning concerning several factors: the time available, the terrain and weather situation, the condition and needs of the person to be rescued.

The time available will be determined primarily by the ability of the person needing rescue to endure. If sick, or either very young or very elderly, or if exposure is a serious problem, then the time available is much less than it would be for an injured hiker in relatively good condition, with ample provisions, and under good weather conditions. Allowances will have to be made when speed is critical, and frequently a somewhat riskier operation will have to be undertaken rather than taking the time to organize a more complete operation. Rescuers must also balance the time factor against considerations of available rescue crews and equipment.

The terrain will also affect the nature of the operation. At altitudes above the timberline, visibility is very good, so that air search may be possible. However, flying conditions in mountainous areas are frequently less than perfect, so air search is not always one of the options. Also, at high elevations, altitude sickness becomes an important factor. The symptoms may vary from mild headache and upset stomach to vomiting, diarrhea, severe dehydration, and hemorrhaging. The difficulties of both travel and rescue at high altitudes are lessened somewhat by the extreme inaccessibility of these regions; usually, only experienced hikers are able to penetrate this far into the wilderness, so that the person needing rescue is likely to be able to offer some assistance to his rescuers.

It is at lower altitudes that most rescue situations will arise. Large numbers of inexperienced and ill-equipped vacationers, hunters, fishers, hikers, and the like enter such areas each year, and a sizable number become lost or injured. Although the rescue operation is much less hazardous than at higher altitudes, the people involved are likely to be less able to render any assistance in their own behalf, are more likely to panic, and may not know even the simplest precautions or signals to aid rescuers. Trails and roads are usually available for access to the general area, and four-wheel drive vehicles, horses, and pack animals may be available for less accessible areas.

Winter conditions can complicate what would otherwise be a relatively simple operation. At higher altitudes, where winter will mean extreme cold, high winds, and deep snow, rescues are most frequently a forlorn hope. Unless the person needing rescue was fairly well equipped, the chances for surviving more than a few hours are practically nil. Unless air rescue units can penetrate the area fairly quickly, risking a rescue unit under such dangerous conditions is likely to be a vain and foolish endeavor.

At lower altitudes, however, conditions will probably be less severe. Most of the rescue operations here will involve skiers, who are likely to be fairly close to rescue units. An injured skier should not attempt to walk through the snow, since this is extremely tiring. Rather, the skier should plant skis in the form of a cross in the snow at his head, in order to warn other skiers and to signal the ski patrol.

The best conditions for air search and rescue operations are in low altitudes where there is little forestation. The combination of low altitude and fairly stable wind conditions maximizes the handling ease of either fixed-wing aircraft or helicopters, and the lack of cover provides wide visibility. The use of both fixed-wing aircraft and helicopters has revolutionized rescue procedures. Helicopters can pluck stranded individuals from cliffs, glaciers, and other inaccessible locations and rush them to hospitals in a short time. Both types of aircraft have limitations, however, in that they need sufficient air density in which to operate, and are thus ineffective at high altitudes. Furthermore, fixed-wing aircraft can only be used in areas where they can land in order to effect a rescue, and thus require at least a lake or large meadow near the rescue site. Helicopters require a large horizontal sweep for their rotors, and thus cannot be operated where there is heavy cover. In short, air travel is a vital means of achieving wilderness rescue, but it is not useful in all situations, and should never be depended upon as a substitute for careful planning.

SAR SUSPENSION OR TERMINATION

All reasonable actions should be taken to locate distressed victims, determine their status, and effect their rescue as rapidly as is humanly possible. SAR units are responsible for taking whatever action they can to reduce disability and save lives at any time and place their facilities are available and can be effectively used. Nevertheless, there is a definite limit beyond which SAR services are not expected and cannot be justified. Known and inherent risks must be carefully weighed against the mission's chances for success and the gains to be realized. Personnel and equipment should not be jeopardized, nor the mission attempted, unless lives are known to be at stake and the chances for success are within the capability of the personnel and equipment available.

The probability of finding survivors and their chances of survival diminish with each minute after an incident occurs. All SAR activities should therefore take prompt and positive action so that no life is lost or jeopardized through wasted or misdirected effort. Records have indicated that the life expectancy of injured survivors decreases as much as 80 percent during the first 24 hours following an accident, while the chances of survival of uninjured survivors diminishes after the first three days. These figures are averaged from overall experience. Individual incidents will vary with local conditions such as terrain, climatic conditions, ability and endurance of the survivors, emergency survival equipment available to the victims, and the SAR forces available.

The decision on whether to conduct extended search operations must be based upon the probability of survivors. All reasonable actions should be taken to locate the victims and effect their rescue and recovery. During any search operation where the objective has not been located within a reasonable time, the decision to continue or suspend search efforts must be made. After all probability of locating survivors has been exhausted, further extended opera-

tions become uneconomical and unwarranted. The authority to suspend a search mission rests with the controlling agency which has jurisdictional authority and/or has assumed SAR responsibility. The following considerations will influence the decision to continue or suspend further search efforts:
1. Extensive search coverage has produced negative results.
2. All leads and sources of information have been thoroughly investigated and have produced negative results.
3. Prevailing weather or forecast storm causes deteriorating safety conditions.
4. Personnel and/or resources are exhausted.
5. Lack of any further information for guidance leaves SAR forces in a quandary.
6. The possibility of detecting the victim is minimal.
7. The probability of finding survivors is not feasible due to climate or time lapse.

EXPOSURE AND EXHAUSTION

The environment in which the rescuers and victims are exposed is a definite determining factor that affects the practicability of immediately launching an SAR effort and it can severely limit the time necessary to complete a successful rescue. In some cases, the environment will be the most critical factor of all. If the current weather will not allow an SAR operation to proceed without endangering additional lives, the SAR effort should normally be delayed. If the weather is currently good but is forecast to deteriorate in the near future, more rapid action is necessary and detailed planning and organization may suffer due to the reduced time available. If the weather is good and is forecast to remain fair, more extensive planning may be accomplished.

Exposure and exhaustion are caused not only by physical conditions, but by mental ones as well. Both of these conditions are increased by the psychological effects of height, difficulties, and exertion on a person, as well as the physiological effects of cold, heat, dampness, and fatigue on the human body. The endurance of a person depends mostly upon the innate characteristics of the individual, physical strength, and mental outlook.

A warm person is usually an efficient one, because neither the physical nor mental energies are being sapped by the effects of cold. Cold blunts the senses and interferes with movement, making it difficult to think clearly or to control the limbs correctly. A large proportion of the energy resources of the body become committed to the task of keeping a person warm, and therefore are not available for movement; under these conditions a person will tire quickly. A cold and tired person will rapidly sink into a state in which self-help is impossible; an exhausted person will rapidly succumb to exposure.

Mountains make their own weather; on long routes weather changes can be anticipated. Wet or icy rock can make an otherwise easy route impassable; cold may reduce climbing efficiency; a cloudy night may be too dark to travel. Atmospheric temperatures drop 3 to 5°F for every 1,000 feet (300 meters) of rise in elevation; clear, warm air in a valley can gradually become chilled, its moisture condensed into a cloud, and then precipitated as a violent blizzard on a high crest. Steep slopes can accelerate balmy zephyrs into chilling gales.

ENVIRONMENTAL RESCUE

Weather Forecasting

Travelers in the wilderness areas should keep alert for early warnings of deteriorating weather conditions. Birds and other wild animals are very sensitive to temperatures and atmospheric pressure changes; travelers should watch their actions as they are reasonably accurate weather predictors. Seagulls and other ocean birds will invariably flock inland hours before a marine storm materializes. Wild animals will bed down or become restless and birds will stop singing and instinctively seek dense cover for protection.

Different sections of the country have various significant indications that the weather conditions may change. Some signs which are important in one area may have no significance in another. Generally, it is best to keep an eye on the skies for a gathering cloudbank, distant lightning and thunder, a towering thunderhead, or swift moving low clouds as all of these signs predict storms or deteriorating weather.

There are several sayings which our forefathers utilized to predict weather; these may have some value in wilderness areas and out at sea where modern civilization does not appreciably affect the environment with air pollution. Anyway, remembering a few of the rhymes will make one more aware of the changing weather conditions:

Red sky in the morning,
  Is the sailor's sure warning;
Red sky at night,
  Is the sailor's delight.

Evening red and morning gray,
  Send the traveler on his way;
Evening gray and morning red,
  Bring down showers on his head.

When the grass is dry at morning light,
  Look for rain before the night.

Whenever the clouds do weave,
  It will storm before they leave.

When clouds appear like hills and towers,
  The earth is refreshed by frequent showers.

Actually, these old sayings are predicated on easily observed signs of weather change as when a bright red sun rises and then is quickly hidden in a cloudbank, rain can be expected. It the sun is obscured by clouds when rising, but then breaks clear, good weather is indicated. A heavy dew in the morning forecasts a fair day. If the campfire smoke rises almost straight up in a long, thin vertical line, good weather is fairly certain for the next 12 to 24 hours as the barometric pressure is high. If the smoke is heavy and tends to settle and drift away, look for rain as the barometric pressure is low.

Therefore, travelers in the wilderness areas should remain constantly aware of any environmental changes as they could foretell either a bettering or a worsening weather condition. There are several definite indications of a weather change:

1. An abrupt, definite change of wind direction or force.
2. An obvious barometric pressure change. A fast dropping barometer is a sign of a coming storm while a rising barometer is an indication of fair weather.

3. Cloud color, shape, and direction of travel are important weather indicators. White, wispy clouds at a high elevation bring good news; heavy gray or black clouds are definitely ominous.
4. An abrupt temperature change may bring good news or bad news.
5. Any noticeable humidity change may be important.

Lightning

Lightning has killed and injured many mountaineers; any indication of a thunderstorm should signal an immediate evacuation of the mountain peaks and ridges. Lightning is more prevalent in some areas than in others; it is likely to be seasonable. A group of persons should not congregate together if there is a potential lightning hazard. If they are widely dispersed, it is unlikely that several persons would be affected simultaneously.

Lightning creates two hazards: the direct strike and ground currents. An electrical discharge always follows the path of least resistance. A direct hit is more likely to strike peaks than valleys, crests of a ridge than the sides, lone trees or upright persons than a surrounding meadow, tall snags than low vegetation. Ground currents radiate out from the bolt, flowing sheetlike along the easiest surface path.

An advantage can be taken of the presence of a prominent peak, or other likely location for a direct hit if it is not too near. When a person gets close under a pinnacle to obtain shelter, his body may become an alternate path for the ground currents. If far out from under the peak, the lightning is just as likely to strike the victim as it is the pinnacle. The most obvious way of avoiding a lightning strike during an electrical storm is to keep away from peaks, ridges, and unprotected flat expanses.

The current in the bolt of lightning does not immediately dissipate; it tends to radiate out from the bolt. The current flows sheetlike along the easiest paths of electrical conduction on the surface of the ground. These ground currents are strongest near the point of the strike; they rapidly diminish in intensity with distance. Even at a remote point which is well away from the strike, the ground currents can be lethal. To avoid ground currents, one should stay away from the best conductive paths of electricity, such as anything damp or wet. This includes wet ropes, earth, puddles, cracks and crevices filled with water, and rocks covered with moss or lichens.

Caves may not be entirely safe as a small cave or overhang may offer a false sense of security. At the mouth of a cave, a charge may jump across the opening and pass through a person's body.

Damage to the human body is exactly the same as the injury from any other electrical source. The extent of the damage depends upon the amplitude and duration of the exposure and on the path of the current through the body. A relatively small quantity of electricity flowing from one hand to the other would probably be far more serious than a greater amount flowing from one foot to the other because a larger number of vital organs, such as the heart, lungs, and spinal cord, will lie in the upper path.

Electrical currents are forced through the body by potential differences in strength; the points where the currents could enter the body should be kept as close together as possible. If a person is caught out in the open, the best stance is a deep crouch. Since ground currents burn the body on entering and leaving, a sitting person

ENVIRONMENTAL RESCUE

could be burned on the buttocks and feet. Therefore, crouching with the feet close together and the hands off the ground reduces the possibility of currents passing through the vital organs of the body.

Contrary to popular belief, metal objects such as backpack frames, ice axes, and fishing poles will not contribute to the hazards unless they project above the person's head. Actually, sitting on a backpack, bedding roll, or a coil of rope could offer significant protection against ground current injury.

SEARCH FOR A MISSING PERSON

An organized search for a person lost in any kind of terrain is ordinarily not started until it is reasonably certain that he has not been delayed for some normal reason. If the person is healthy, well equipped, experienced, and has been traveling where injury is unlikely, then the search may even be deferred overnight. Under some conditions there should be no hesitation. Immediate search is mandatory if the person is either very young or extremely old, has a physical or medical disability, is very fatigued, is inexperienced, or is poorly equipped.

Obtaining Information

Complete and factual information about the missing person is necessary. It is important to ascertain where the person was last seen, clothing worn and equipment carried, the direction in which headed, the purpose of the trip, physical condition, mental attitude, experience, and the features of the surrounding terrain. Collecting these pertinent facts are important in planning the search operation and predicting its ultimate outcome.

The persons who are the most knowledgeable should be interviewed, but no one should be ignored who could possibly have any information. Often the complete story must be pieced together from many sources. Persons who are close to the victim will probably be excited and concerned; possibly they may be on the verge of panic. The rescuer's approach to these persons is most important. They should not authoritatively be ordered to calm down nor told not to worry as most people cannot completely control their emotions. Asking general questions to determine their involvement will help take their minds off the problem, but SAR personnel should not appear unconcerned. The rescuer's approach should be one of concern and understanding. Interviewers should be constantly aware that relatives or friends may be deliberately omitting important information by not volunteering that an older person is senile, a child is mentally deficient, an adult has been drinking heavily, there has been a family argument and the disappearance may be deliberate, or any other situation that could affect the search tactics. Although the complete questioning of the witnesses may take some time, it frequently turns out to be time well spent as it actually reduces search time. When there are sufficient personnel, some searchers can start to look after receiving a brief description while the more intensive witness interrogation proceeds.

On the initial contact with a group, and periodically while interviewing, the crowd should be visually surveyed for persons who show an undue interest or appear to have information. Someone may be obviously disagreeing with the information which is being offered. Later they may be taken aside and questioned.

Questions should be asked that require a definite answer, instead of a simple yes or no. Sufficient time to answer completely should be allowed. Often it is a good practice to pause briefly after they finish as the silence may prompt them to tell more. The pause will also allow the questioner an opportunity to phrase a better question to pursue the subject. Questioning chronologically will help all concerned as it will aid the witnesses' memory and make any omissions obvious. SAR personnel should be sure that each point is explored completely before moving on to the next subject.

After the questioning, the witnesses should be thanked for their help, a deep concern for their problems should be expressed, and an assurance made that they have been of great help.

Search Procedures

The type of terrain to be searched will obviously affect the ease with which the victim will be detected. The more level the terrain, the more effective will be the search. The more trees, vegetation, rock outcroppings, and other surface irregularities that exist, the greater the difficulties.

There are three search procedures, and the method selected will depend upon the conditions: the nature of the terrain, the number of personnel readily available, and the size of the territory which might require combing. A small party may be able to make no more than a *Rapid Search* along a limited track, trail, or path. With more people available there can probably be a *Contour Search* throughout a section of the countryside. If these searches prove ineffectual, a large group can then proceed with an *Area Search* in order to thoroughly comb a considerable amount of territory. If the missing person may be anywhere within a large area, then an area search should be initiated immediately. Aircraft may be utilized to scan the entire section, but personnel on the ground are needed to thoroughly comb the terrain.

Regardless of the adopted method, before beginning the search, each rescuer must know the entire plan. If the party is to be divided, then each division should know the location of the others. Communication should be maintained with portable radios, audible signals, or lights. All members must know the rendezvous point, and at what time they are to meet. They should rendezvous at the agreed-upon time regardless of whether or not their search has been successful.

Rapid Search. The initial search is usually made by the first persons on the scene or a team of well-trained, self-sufficient, and highly mobile searchers whose primary responsibility is to quickly scour the last seen section in hopes of locating the lost person before being able to wander too far away from that point. SAR personnel should check the areas most likely to produce a victim first, such as trails, roads, campsites, lakes, clearings, and other points of attraction. The initial crew's efficiency and usefulness is primarily based upon speed of response and accuracy of the first-hand information available at the scene.

The most logical and reasonable locations should be checked. A heavy drinker may be found in a nearby bar. After a family argument, the lost person could be at a bus terminal or motel. A lost child may have wandered to a nearby playground or other attraction. A small child or senile person could be confused and just wandering. Stationing volunteers who are not capable of arduous searching at the most likely crossroads or attractive locations would be wise.

ENVIRONMENTAL RESCUE

Rescuers participating in the rapid search should be acutely clue conscious so they will do more good than harm; every person who enters a search area is a potential clue destroyer. Rescuers should be somewhat schooled in clue analysis. Evidence has shown that a diligent, rapid search has a great chance of locating the lost person as most of the survivors are found during the initial search effort.

This search group should be kept small to allow maximum travel efficiency and to prevent these searchers from getting in each other's way. Also, the smaller and more highly trained the crew, the fewer clues will be destroyed before they are observed. The purpose of this rapid search phase is a quick, decisive search to reduce the time lag between the person's becoming lost and the searchers' finding him.

For safety, the searchers should travel in pairs, or at least remain within earshot of each other. This means that a team of two or three persons can cover only a narrow strip of ground and their best chance for success is to outguess the lost person. They should put themselves in the person's place and visualize probable errors, such as blundering into the wrong one of two ravines which are similar in appearance, or onto the wrong ridge. The most likely places, such as the exits of wrong gullies and ridges should be checked first. If the victim is not found in these places, the rescue group should retrace its original trail while looking for tracks showing where the missing person wandered off. Once such a point is found, the searchers should proceed swiftly along the most likely path, watching for footprints. When checking terrain where tracks may be intermittently visible, the party should fan out in broad intervals, calling to each other frequently to maintain contact, and pausing often to listen for calls from the lost person. If the missing person is either very young or old, the rescuers must carefully check every clump of brush, cave, and other hidden place since he may be unconscious or terror-stricken. These victims will often hide motionless when strangers are approaching, so they will not cooperate in their own rescue. They will not necessarily answer when their names are called. If after several hours no clues have turned up, then it is probably time to seek additional help.

Contour Search. Contour patterns are used to search along and around peaks, razorbacks, steep slopes, and other mountainous terrain features. This method is also effective along irregular shorelines of lakes, rivers, oceans, and other bodies of water.

SAR personnel should keep in mind that lost persons often fight topography and are likely to be found in the most rugged portion of the surrounding country. Persons who follow natural routes are seldom lost for long periods of time. Children under five years of age frequently travel uphill all the time.

The basic idea behind the contour search method is to intercept the missing person's trail by traveling at right angles to his expected direction. Several teams should travel on different contours of the terrain; if any team crosses a track which appears to be the one sought, the other searchers should be notified. Once a promising track is found it should be carefully preserved; it should be marked clearly with sticks, poles, rock cairns, or anything else which is suitable. The track may then be followed by walking a few feet to one side, not directly on it. If the track fades out it can often be rediscovered merely by choosing the most logical route or by dividing up the group to follow several possible trails.

Area Search. When no tracks are found, or in heavily traveled country where there are many distractions, an area search usually becomes necessary. An area search could be conducted simultaneously with the rapid search if sufficient personnel are available, or immediately following the rapid search if crews are limited. This method requires a large number of searchers and a very exact organization. The leader first establishes the search area by determining a perimeter within which the lost person is most likely to be found. Once the perimeter has been established, a continuous search of this area is made. In order to accomplish this objective, it is best to divide the area into sections. Teams of two or three members are allotted a section of reasonable size, defined by natural boundaries or by grid lines drawn on a map. Each team then conducts a contour search of its assigned area. If this fails, then the entire area should be carefully and systematically examined in minute detail so that no signs left by the lost person are missed. Sometimes the area of initial search must be broadened and airplanes or helicopters employed.

To establish the perimeter boundaries, the potential rate of travel for the missing person should be estimated. A three-year-old girl will not cover as much ground as a healthy ten-year-old boy. An eighty-year-old may be the slowest of all. If it is estimated that the victim could have traveled a maximum of 5 miles (8 kilometers) since last seen, then a 5-mile (8-kilometer) radius should be checked.

MOUNTAIN RESCUE

Mountain rescue is an extremely specialized field because probably the widest range of variables will affect the necessary operations. Mountain wilderness areas are difficult to penetrate, and even harder to get out of if anything goes wrong. Even under the most ideal conditions, it takes a long time to send word out that help is needed; it requires a long time to organize a rescue party and travel to the accident scene; and it takes a long time to transport the victim to safety.

The terrain will vary in steepness and accessibility. The slopes may be covered with snow and ice, they could be wet and slippery, or they may be dry. The weather may be miserable, or it could be beautiful. Conditions at the bottom may be quite different from the rescue area, so the skill and equipment requirements may change as an ascent is made.

The rescuers themselves may be in danger of an accident or injury. It is often physically difficult to approach the victim safely on steep mountain slopes, and caring for injuries under these conditions may be virtually impossible. A fall can occur anywhere; the length of time required for the rescue operation, and the large number of people needed, simply multiplies the chances of an accident occurring to one of the rescuers.

Equipment

Four factors of outdoor safety and comfort are: weather, terrain, altitude, and body limitations. Each will affect the equipment necessary for the trip, the activities enjoyed, the safety to be enjoyed, and the hazards encountered.

Equipment needs may vary, but any group which is planning to operate in mountainous terrain should be aware of the variety of

ENVIRONMENTAL RESCUE

situations in which it might find itself, and should be prepared for as many as possible. An adequate amount of rope is an absolute requirement for climbing, towing rescue toboggans, securing an injured person in a safe position, and other uses. Parties operating at night should carry adequate reliable lighting. Footwear is especially important. Good multipurpose boots will make rescue operations easier under many conditions, and crampons and other climbing equipment may be necessary. Where snow is a factor, skis which will permit the user to ascend slopes are necessary. In general, a rescue operation should be sure that it knows the conditions it is likely to encounter and that it has the appropriate equipment; failing to take this elementary precaution could endanger the whole operation and subject the rescue operation to a need for rescue itself.

Avalanches

Whenever a person is on snowy or icy mountainous slopes, it should be borne in mind that there is always some degree of hazard of a mass of snow or ice sliding down the mountainside. Every precaution must be observed at all times.

Avalanche Causes. An avalanche occurs whenever the pull of the gravity on a mass of snow or ice is sufficient to overcome the anchorage or the internal cohesion. To determine the probability of a slide on any suspected slope, the following factors should be considered:

1. Steepness of the slope. The probability of avalanche increases with the slope angle up to a certain steepness, and then decreases as the slopes get steeper and approach the vertical. The reason for this decrease is the failure of the snow to adhere in large quantities to extremely steep slopes unless it is packed by the wind; it simply runs off in small, harmless slides before any great depth can be accumulated. Some authorities have set an arbitrary critical slope angle for avalanche formation at 22 degrees, with slides apt to occur when slopes exceed this figure; avalanches have been observed on slopes as moderate as 15 degrees, although this is an unusual condition.
2. Shape of the slope. Overhanging cornices are extremely hazardous. It should be remembered that snow is tautly stretched over protuberances and is firmly compressed in hollows; convex slopes are thus more prone to avalanche than concave hollows.
3. Anchorage to the ground surface and the bond between the snow layers. Slides rarely start in a dense forest or closely spaced rock projections on slopes, but such obstructions offer little protection when the avalanche starts from above.
4. Possible existence of a buried snow slab. Wind-packed snow, called wind slab, snow slab, or slab, breaks loose rapidly. When it is sliding, it is the worst killer of all.
5. The internal cohesion of visible and buried snow layers. Be suspicious of any slope where there is some indication that the snow layers are not adhering to each other.
6. Recent weather as affecting the last three factors. Temperature has more ramifications in avalanche formation than any other single factor; both air temperature and the temperature within the snow cover must be considered. As a broad rule,

avalanche hazard varies inversely with the temperature during the winter when snow and air temperatures are generally below freezing. As temperatures decrease, the snow tends to remain in an unstable condition for a longer period after a snowfall, and the snow pack becomes increasingly brittle and liable to fracture. The most commonly encountered slides occur in new snow; this may happen during or soon after a snowfall. Spring weather and slopes facing the sun favor avalanches produced by thawing. Water forms at the snow surface, percolates downward, and rapidly warms the snow to the melting point. Excess free water provides a lubricating action that causes wet slides.

7. Evidence of prior avalanches. Slides usually follow existing channels. Ravines and gullies are many times more dangerous than adjacent slopes, since they tend to become natural chutes.
8. Depth of the snow. The hazard of an avalanche increases as the depth of the snow increases. The dangers are particularly acute just after an exceptionally heavy snowfall.

However, there is no rule-of-thumb to determine whether a snow slope is in danger of avalanching; all of the foregoing considerations should be taken into account.

Avalanche Precautions. Although avalanches are usually thought of as a menace which come sweeping down on the unwary mountaineer from above, relatively few victims are claimed by such slides. Most avalanches that involve a climber or a skier are triggered by the person.

Avalanches are not to be trifled with. Any safe alternate route, even if it is longer and more difficult, is preferable to one which is in doubt. Occasionally, though, there may not be any reasonable alternative route; the dangerous slope must be traversed. In this case, every precaution must be taken. Only one person should be on the slope at one time, with the remainder of the party keeping vigilant in a safe place; in case the traveler is caught in a slide, they should mark the probable location carefully and start immediate rescue procedures. It is considered a good safety measure for each person traveling in slide areas to fasten a long length of brightly colored cord or rope from the waist, which is referred to as an "avalanche cord." This marker will remain on or near the surface and will facilitate finding a buried traveler.

There is the least amount of danger on the ridges, more on the valley floor, and the most on inclined slopes. In general, a dangerous slope should be crossed as near the top as possible. An exception to this rule, however, must be made when the slope is considerably concave so that the upper portion is the steepest. The advantage of crossing high is that the person will then be above most of the sliding snow mass if an avalanche does occur, and therefore least likely to be dragged down by the slide and buried. When several persons are traversing a slope, they should not follow in each other's tracks. On hazardous slopes, each bit of force will help shear the snow mass from the surrounding snow strata; repetitive forces on the same trail may start an avalanche. It is usually much safer to travel along ridges than in gullies, a flat-topped ridge being ideal; but it is imperative to be sure that what looks like a flat-top ridge is not an overhanging cornice or the skiers' weight may break the edge off. During the periods of high danger the only practical time to travel may be

ENVIRONMENTAL RESCUE

early in the morning, late in the afternoon, or at night when the freezing cold has stabilized the slopes.

Avalanche Survival. Avalanche victims are usually overwhelmed in one of two ways: they may be struck by a developed avalanche descending from above or they could have been instrumental in triggering the release of the snow mass. In either case, the victims may be able to take some action for self-preservation.

If an avalanche is seen descending down a slope from above, threatened mountaineers can seek shelter behind large trees, massive rocks, or other bulwarks which have a chance of offering protection. Everyone should remember that the force of an avalanche could be similar to the force of an explosion. It may be possible to avoid the slide by darting to the side, but the chances of directly outrunning an avalanche are extremely poor.

When actually caught in an avalanche, it is crucial to avoid becoming buried. This can best be done by lying on the back or on the stomach, with feet downhill, and then maintaining a powerful swimming motion with both arms. The importance of removing skis, pack, and other equipment before becoming caught in the full sweep of the slide cannot be overemphasized. In a wet snow avalanche, it is particularly important to exert every ounce of strength to get one's head above the snow; if this is not possible, a space should be made around one's face and chest as soon as the mass comes to rest since a wet snow avalanche frequently freezes solid as soon as it stops moving.

Avalanche Rescue. Experience has shown that the average survival limit of a person caught in an avalanche is two hours, though there are records of survivors who have lived as long as 72 hours while buried. Ordinarily, victims are either killed instantly, or they die within a short period from suffocation, bodily injury, or shock. Snow is porous and usually contains enough air to support life. An ice mask, formed by the condensation of the victim's breath, may form in about two hours; this causes suffocation by forming an airtight seal around the face. Rescue efforts are therefore designed to remove the victim within the two hour limit. Due to special circumstances, such as an air pocket, which may prolong the life of the trapped person, rescue efforts should not be abandoned for at least 24 hours.

Eyewitnesses of an avalanche accident play a crucial role in search and rescue. Anyone seeing an accident should try to observe the locations where the victims were caught and the points where they were last seen, thereafter marking the points or noting the locations in reference to fixed prominent locations.

When one or more members of a party are caught in an avalanche, other members in safe positions should carefully observe the path of those being swept away in order to ascertain where they may be found when the avalanche comes to rest. Then, unless there is an extreme danger that the undermining effect of the first avalanche will cause another slide, the members who best observed the probable position of the victims should remain in place to direct the others; markers should be placed so as to aid the survivors in determining the best place to look.

The same conditions that triggered the previous slide may still be present. If there is the slightest possibility of another avalanche, guards should be posted to warn the searchers, and the rescuers must know where to go in case of an alarm. If the danger becomes critical, rescuers must not hesitate to suspend the operations.

A metal detector could be an invaluable aid to avalanche rescuers if it has controls which would enable the operator to tune out ground mineralization without affecting the sensitivity of the unit. This would allow only metallic and mineral objects with a concentration different from the earth to cause a reaction. There are metal detectors that are sensitive enough to detect a zipper or other small metallic object, but many of them are adversely affected by the background interference from the large mineralization present in most mountain soil. Users of these instruments must remember that batteries are affected by the cold temperatures.

Mountaineering rescue groups often utilize trained dogs to track and locate lost or trapped persons; time and again they have located buried avalanche victims rapidly and efficiently. Experience has shown that buried victims have sometimes been located by untrained dogs which happened to be on the scene and instinctively joined in the search. Therefore, whenever possible, a dog should be taken on rescue operations, whether trained or not.

Rescuers should make a rapid search for the slide victim or any visible equipment. The debris should be scanned for clues. If none are apparent, then the slide should be searched with a scuff line, which is composed of a single line of rescuers marching shoulder to shoulder. This line moves back and forth across the slope scuffing the snow with their feet and scanning it for clues that may be at or just below the surface. A human body is bulky and all other things being equal, it is likely to be thrown toward the surface or the sides of the slide. Statistics indicate that the majority of victims were in a relatively prone position and were buried 2 feet (60 centimeters) or less below the surface.

If any indication of the victim's position is found, probing should be started. Long poles or rods are best for this purpose, but it is frequently necessary to use ski poles with the baskets removed, or reversed with the wrist straps removed; skis may also be utilized. Probing should not be done in a haphazard manner, but should be well-organized under a leader. A rapid search should be made through the areas of the greatest probability, striving for a probing interval of 2 to 2½ feet (60 to 75 centimeters) between insertions. Often the body is caught by protuberances such as trees, boulders, and ridges; check these carefully. Pay special attention to the tip and the edges of the slide.

If the initial rapid search does not find the victim, then the area to be probed should be carefully, but quickly, laid out and the probing done systematically to cover each square foot of snow surface. The searchers are usually arranged in a line, spaced shoulder-to-shoulder, and are kept in position by a guide cord held by a person at each end of the probe line. The line moves slowly forward, keeping in alignment, with each member sounding with a probe from left to right, and making soundings 1 foot (30 centimeters) apart in each direction. The presence of buried objects is indicated by an abrupt change of penetration depth. An indication of the type of object encountered is given by the degree of yield in the obstruction. When a suspicious object is struck, the probe should be left in place and a check made by detailed probing. If the victim is not located in the anticipated position, the same systematic search should be extended, either by moving the initial point outwards in all directions or by starting from the top or the bottom of the avalanche tip, whichever seems the most likely to succeed quickly. It is important to mark the probed areas upon completion so that efforts are not duplicated. Shovel crews should accompany the probers to dig in any likely spots. To dig out an object, shovels are used until it is believed that discovery of the victim is imminent; the final exposure is made by hand digging.

If the avalanche is quite deep and the victim is not found by probing, it may be necessary to dig trenches parallel to the contour down to the ground level or to the level of undisturbed snow at intervals of six to ten feet, depending upon the length of the probes. The digging should start at the top of the slide and proceed uphill, throwing the snow from one trench into the one just dug. The snow between the two trenches should be probed horizontally.

When there are a large number of survivors, it may be desirable to send one or two rescuers for additional help. If the number of survivors is small and help is some distance away, a search should be conducted for at least one hour before going for help. When the conditions still appear to be hazardous, it may be unwise to start immediate rescue operations before the slope stabilizes. The site of an accident should never be left without carefully identifying the place with markers that will not be obliterated by a storm.

Rock Slides

Just as snow avalanches are the most common mountain danger for a skier, falling rocks are the most common hazard for a rock climber. Weather is a frequent cause for a rock fall; rocks are brought down by changes in the temperature and the resulting splitting action of intermittent freezing and thawing, as well as by heavy rain.

Rock falls may occur on any steep slope, particularly in gulleys, chutes, and where there are overhanging cliffs. Rock slides are most likely to occur on slopes where there is an accumulation of small slippery rock debris, called scree; talus slopes are similar in danger to scree slopes, except that the pieces of rock are larger. These areas should be avoided since the chances of a climber starting a rock avalanche are great.

When traversing hazardous terrain, climbers should constantly watch both above and below them. If a rock is accidentally kicked loose, the shout, "Rock! Rock! Rock!" warns others who may be below. It may, however, prove dangerous to look around upon hearing this warning. The climber's safest course of action is to flatten up against the nearest rock wall, thus allowing the falling rock to pass over. If a falling rock is seen approaching, watch the object and present the narrowest profile; the jeopardized person should not move until it is clear which way the rock is headed, then move to avoid peril. Otherwise, any move may be into the path of a falling rock by blindly or prematurely trying to avoid it.

Unless traversing a scree or a talus slope where the danger of starting a rock avalanche is high, climbers should walk closely together so that a dislodged rock falling between two persons will not gather momentum.

CAVE RESCUE

Spelunking, which is the study and exploration of caves and other underground caverns, is becoming popular among a large segment of the population. Any time a person enters unfamiliar or difficult terrain, whether it is above or below ground, he may encounter unforeseen difficulties and require finding, rescue, and evacuation. Because of the precipitous and rugged terrain that may be encountered underground, standard mountaineering equipment and techniques are usually required.

Death and injury in caving accidents are most commonly caused by drowning, by being crushed by falling boulders, by becoming inexorably stuck, and by exposure. Spelunking is a dangerous sport, so rescue parties should be experienced and well equipped. Claustrophobia, the fear of enclosed places, is a frequent cause of panic, and a constant danger.

Route Finding

Most of the large underground caverns have been thoroughly explored and surveyed; therefore, charts or maps should be available. However, caves containing a labyrinth of passages, or newly discovered caverns, may offer a multitude of opportunities for even experienced spelunkers to become lost or trapped. People with photographic memories have an advantage when it comes to route finding because they have the facility of fixing things in their minds in terms of pictures. It is not usually considered wise to depend entirely upon the memory; notes should be taken and markers placed so that the return route or previously explored passages may be identified.

Route finding involves more than just following maps or surveys; it is a complete technique aimed at finding the way into and out of any cave, charted or not. The basis of this technique is careful observation. While moving into a passageway or cavern, one should remember to look back at frequent intervals so that the passages are seen as they will appear on the return journey. This is particularly important after negotiating climbs because what may seem obvious enough on the way in may be quite easily missed on the way out.

It is probably best to assume that any junction on the route out will not be obvious. At abrupt turns, particularly where junctions with other passages are involved, errors may be easily made. Every person in a group should be concerned with the problem of route finding; it must not be left solely to the leader. In unfamiliar caves, no one man can remember all of the details at every point.

When searching underground caverns for lost or missing persons, the most accessible portions should be explored first. Rescuers should call out at frequent intervals and then quietly listen for responses. They should look about for recent footprints, overturned stones, or other indications of prior travel. Many caves are visited frequently by large numbers of persons, and in consequence, the routes are marked from the heavy travel. When this happens, follow the paths until fresh tracks, or the absence of signs, indicate the advisability of continuing. Passages which have been checked out should be marked prominently so that later teams will not duplicate efforts.

Route finding during the initial exploration of caverns consists of choosing the path of least resistance at all times. The idea is to select, where alternative routes exist, the easiest way which is consistent with safety. The initial safeguard is to observe before acting; it is extremely fatiguing to be continuously entering passages that turn out to be wrong, and then retracing one's steps.

There are no hard-and-fast rules that can be given as general advice on route finding, other than the need for constant watchfulness and observation. It is not true, as is sometimes supposed, that the way into caves is downward and the way out is upward. In many cases the entrance is at the bottom of the system and, therefore, a spelunker will move upwards throughout the trip. It is equally common to find caves in which the route runs first down, then up, and then down again. Nor is it true to say that the best route will be the most obvious passage at any junction. Far too many mistakes in route finding are made because people attempt to make all caves fit into their preconceived notions of what a cave should be like; the best approach is to expect the unexpected.

Labyrinth Travel

Cave passages rarely run level for very long; there are frequent breaks in the contours of the caverns. A knowledge of mountain climbing techniques is therefore an essential part of a person's general background before entering the more intricate portions of a cavern. Equipment carried should include a generous number of ropes and rope ladders. Of course, lighting equipment is indispensable.

Narrow passages, referred to by spelunkers as "squeezes," may vary in both length and in tightness. Naturally the size of the person makes a difference, and in severe caves it is not uncommon to find squeezes so tight that only small people can get through. When tackling a narrow passage, it is important to inspect it first and note its general shape. Then the squeeze should be entered in such a way that the broadest parts of the body fit into the widest parts of the passage. Generally it is easier to move on one's front than on one's back, and to go head first rather than feet first. The human body bends forwards and sideways easily, but hardly at all backwards; the aim should be to go around corners so that the curve is a forward bend for the spelunker. Where the squeeze slopes downwards it is best to go feet first, because in so doing the proper control of one's movements may be had by bracing against the sides and roof of the passages.

Where the person to be rescued has become stuck in a narrow passage, it is necessary to keep the person as calm as possible, because panic stiffens the body and makes rescue more difficult. If calming does not permit the trapped person to get out, bulky clothing should be removed, by cutting away if necessary. Sometimes rescuers can help by giving instructions to aid in gaining leverage for pushing. Exhaling is useful, since it collapses the rib cage and creates additional space. Care is essential, however, since a person can get wedged in even worse than at the beginning and be totally unable to breathe.

Water in Caves

Where there is water in the caverns, it should be avoided as much as possible; a wet man is a cold man, and a cold man is a liability. Where water cannot be avoided, it may be wise for all members to be equipped with wet suits, such as are worn by SCUBA divers. At those times when the stream has to be followed, the party should avoid wading as much as possible and traverse at a higher level if the ledges allow. Added to the physiological hazards of a wetting are the practical disadvantages of wading. One's movements are hampered by the water and the footholds are obscured (Figure 17-4).

One of the major hazards of caving, especially where active streams flow in the cave, is that of being trapped by rising water. Heavy rainfall, such as from a sudden thunderstorm, is the most common cause and its consequences are variable. If the full fury of the storm should break over the catchment area that feeds the cave, a rapid and very great rise in the level of the stream may result. Even a prolonged, heavy rain can cause streams to rise dangerously, though, in this case, there should be adequate time for a retreat from the cave.

In a cave the onset of dangerous weather conditions on the surface cannot, of course, be known immediately. There is always a time lag, which means that conditions have time to become worse before effective action can be taken. When a party is following a stream, any rise in the level will be noted fairly soon after it occurs; the noise of the water will also increase. A change from a clean stream to one that is dirty or cloudy, or an increase in the amount of floating debris, should be considered a warning of impending danger. Air drafts are a reliable warning; an increased air flow should be noticed.

Owing to the general increase in the quantity of water flowing underground from the surface, normally dry routes may suddenly be invaded by water. A stream flowing in what is usually a dry passage is a sign that all is not well.

The best action to take under those circumstances is to get out of the cave if it is still possible to do so. However, this advice cannot always be followed because conditions may have already deteriorated too far. A retreat should be made immediately, keeping a constant watch on the water level. Persons trapped by rising water in a cave should move with all possible speed to the highest point available, and wait there until the flood subsides.

Another type of problem connected with water in caves is the sump, which is a place where the passageway is totally submerged. Sumps that are regularly traveled usually have a guide wire or rope permanently attached which the caver can follow through the water by holding the breath. Another method of passing through these sumps is to lower the water level by bailing the water out. Unexplored sumps should not be free dived; SCUBA diving equipment should be used, since the length and difficulty of sumps cannot be predetermined.

MINE RESCUE

Rescue efforts in established and commercially working mines will not be covered in this chapter as there will usually be established and experienced organizations already formed and well trained. To cover the myriad situations which may be encountered, and the vast amount of knowledge that must be mastered, would take an entire book for this subject alone. However, there are many small mining operations being conducted at remote locations, and adventurous explorers are constantly entering old abandoned shafts and passageways. Therefore, a few peculiarities of mine rescue should be covered (Figure 17-5).

Mines differ from caves in that they are man-made, instead of being naturally formed. In most cases they must depend upon timbers to shore them up to prevent collapse; in old mines the lumber may be decayed and weak, and even in new mines the construction may be amateurish and unsafe. Water may have flooded the lower passages, or the entire mine. Breathing apparatus may be necessary as toxic gases are quite common. Fires may provide a lethal volume of heat and gases. Ladders in shafts may be old, damaged or otherwise unsafe.

Rescuers entering mines must be constantly aware of their own safety. They must check the structural soundness of ladders and shoring. If there is any doubt that operations can continue safely, then all activities should halt until structural reinforcement can be made.

DESERT SURVIVAL AND RESCUE

The desert and semiarid sections of the world are frequented by travelers throughout the year. These regions are characterized by brilliant sunshine, a wide temperature range, sparse vegetation, a scarcity of water, a high rate of evaporation, and low annual rainfall. Some regions are flat and sandy, some mountainous and rocky, while others may be salt marsh or sand dunes. A summery day may generate intolerably high temperatures, and a wintery night could produce freezing cold. A temperature variation of 80°F is not uncommon between noon and midnight. Altitudes vary widely also; Death Valley is considerably below sea level, while other desert areas are 5,000 feet (1,500 meters) or higher in altitude. Travel in the desert can be an interesting and enjoyable experience, or it could be a deadly or nearly fatal nightmare.

The scarcity of water produces sparse vegetation, or a complete absence of it, with the accompanying lack of shade. Dangers from wildlife are probably greater in these areas than in most others because of the poisonous characteristics of many; these include snakes, scorpions, tarantulas, gila monsters, and others. It is easy to become lost because the terrain is usually monotonous and repetitious. Dangers from being caught by a flash flood in a canyon or ravine, in case of a sudden rainstorm, are serious at certain times of the year.

Fig. 17-4. Water in a cave means that a change in weather conditions could flood the cavern.

Fig. 17-5. Abandoned mines are extremely hazardous; they cause many casualties every year. These mines were fairly safe while they were being worked, since the miners used timbers to shore up tunnels in places where there was danger of collapse. This old gold mine in the deep recesses of Death Valley, California, was abandoned many years ago. Time and the elements rotted and weakened shoring, ladders, and other construction materials. A rock fall can be viewed through the mine entrance. Prospectors, explorers, and other unwary travelers are constantly being trapped, injured, and killed in the many abandoned mines that exist in the rural sections of this country.

Desert Survival

A person who becomes lost should not panic. It is best to sit down for awhile, survey the area, and take stock of the situation. The lost person should try to remember how long it has been since having lost the way. Then a decision can be made on a course of action. It may be best to remain in that location and let companions or rescuers do the searching and locating. This is especially true if there are water and fuel nearby, or if there is shelter from the sun and other elements.

Travelers are often stranded when their automobile breaks down, or when an airplane must make an emergency landing. When this happens, they should remain by their vehicle since it will be easier to spot from a searching or passing airplane or automobile. The stranded people should arrange markers to attract attention and prepare to light a smudge fire in case an aircraft is seen, they should climb to the tops of nearby hills to look for signs of civilization, and they should carefully ration out the available supplies to last the longest possible period of time. If the stranded persons decide that the chances of an accidental discovery are remote, it may be best for the two strongest and most experienced hikers of the group to go for help. If they wait until it becomes evident that no help is coming, there may be no one who is strong enough to seek assistance.

It is far easier to become lost in the desert regions than in any other type of terrain. Often the only distinguishing characteristics in a hundred square miles of territory are the roads and telephone poles that have been put there; otherwise it would be one rolling, sand covered hill after another. Even in those areas where there are prominent landmarks, such as well-defined hills, boulders, and scrubby trees, the inexperienced person will have a great amount of trouble in orientation. Not only do the hills change shape due to erosion, but they look different from various angles and light conditions. A mountain seen at sunrise will not seem to bear the slightest resemblance to the same mountain viewed at noon. A person who is used to traveling in forested or in high mountain regions will receive a real sense of security from a compass and a map. The same person traveling in the desert regions may find that they offer a false sense of security since the compass is less reliable due to large mineral deposits and the reference points shown on a map may not be readily distinguishable.

Desert Travel

The only safe method by which an inexperienced person may travel is to stay on the roads or trails. Persons who find themselves in the position of having to walk out of the desert, and cannot find a road or a familiar landmark, should be very careful to attempt to walk in a straight line in what seems to be the most desirable direction. A straight line can be achieved to some extent by picking out a faraway object and walking toward it. Before the object is reached, another object farther ahead and in line should be picked; this now becomes the aiming point. Thus the person proceeds in a kind of leap-frog fashion. This sounds better than it actually is; in practice impassable canyons and chasms will be encountered, which will necessitate a detour. Unless extreme caution is exercised, the bearing will be lost. Whatever happens, the lost traveler should travel in as straight a line as possible since sooner or later a road or some other sign of civilization will be met. Persons who change their minds

ENVIRONMENTAL RESCUE

several times, or start walking in circles, will succumb to exhaustion long before help is found. It is important to remember that distances are deceiving in the desert due to the clear dry air. A mountain which looks a mere 2 miles off might in fact be 20 miles away and, of course, in any situation where a hiker is anxious to get somewhere, walking 2 miles (3 kilometers) will seem much longer.

There are special rules and techniques for walking in the desert. By walking slowly and resting about 10 minutes in each hour, a person in good physical condition can cover about 12 to 18 miles (19 to 29 kilometers) per day at the start; less will be accomplished after the person becomes fatigued or lacks sufficient water or food. Night traveling, however, has many drawbacks; stumbling and falling is inevitable; water and food sources may be overlooked. On the hot desert it is best to travel in the early morning or late evening, spending the midday in whatever shade is available. The position of the sun early and late in the day will give a better sense of direction. When walking, the easiest and safest path should be selected. Hikers should go around obstacles, not over them. Instead of going up and down steep slopes, a zigzag course will prevent undue exertion. It is best to go around gullies, ravines, and canyons instead of through them. A steady, easy step will conserve energy.

Just as important as finding assistance is being able to lead the rescuers back to the other survivors. Prominent landmarks and the direction traveled should be remembered (Figure 17-6).

Desert Rescue

Because of the extreme temperatures and scarcity of water which greatly limits the survival time, rescue efforts should be started rapidly. Aircraft can rapidly and efficiently scour a large expanse

Fig. 17-6. Rescuers should memorize prominent landmarks and the direction traveled so a direct route can be traveled on the way back. There are special rules and techniques for walking in the desert. On the hot desert it is best to travel in the early morning or late evening, spending the midday in whatever shade is available. The position of the sun early and late in the day is most useful for gaining a sense of direction.

of desert areas because of the sparse vegetation. Luckily, except for sandstorms, the climatic conditions of sun and dry air will usually furnish excellent flying conditions. Simultaneous ground search operations should be started on the most likely roads and trails.

SWAMP, MARCH, AND SLOUGH RESCUE

Rescue operations in shallow water or marshy areas offer problems unique to these regions. Many of the rescue procedures previously discussed can be utilized or adapted, while other situations may require improvising.

Since this type of terrain may be found throughout the entire world, the hazards will vary widely. Wildlife could vary from poisonous reptiles, to blood-sucking leeches, to carnivorous crocodiles, to man-eating fish, and to innumerable other animal forms. Insects may be a simple nuisance, a hazard to life because of blood sucking or chewing habits, or carriers of diseases.

The major difficulties encountered in rescue operations involving rescues in swampy areas will involve transportation. Often a combination of foot and boat travel is necessary. Amphibious vehicles have been devised, and are popular in these sections. Airboats, which are light boats driven by an airplane engine and propeller, will rapidly traverse terrain which is too shallow for even a canoe. There have been aircraft crashes and other accidents in a swampy section where rescue efforts and evacuation of the injured survivors had to be delayed until helicopter transportation could be secured. Conservation groups, governmental agencies, and sportsmen clubs will usually have vehicles which have been adapted to the local conditions.

QUICKSAND AND QUAGMIRES

Quicksand, which is a deep mass of loose sand mixed with water, may trap the unwary traveler. It may have the appearance of smooth dry sand, but the water underneath lubricates the grains and allows them to flow easily. Quicksand is one of the most widely publicized outdoor dangers, and one of the least understood.

There is nothing mysterious about quicksand; it acts as any thick liquid would. Pits of quicksand are natural death pits for animals because they cannot think. In their struggles, they only dig themselves deeper into the mire. A person who reacts sensibly can escape it. Humans are comparatively light in weight and can float on water. Therefore, they can float on quicksand. It has no power to suck down bodies, but frantic struggling to free the feet and legs will create a forceful downward movement that causes the sand first to move away, and then quickly return to pack around the legs. The result is a firmer and deeper hold on the body. Further struggling repeats the process until the body is engulfed completely.

The most important rule is not to panic. It must be remembered that basically the victim is in water and can swim or crawl free. No attempt should be made to pull one leg free as this will force the other leg downwards with all of the body weight. Sudden motions should be avoided. If caught in quicksand, one should immediately throw oneself flat on one's back, arms and legs spread as far apart as possible to aid buoyancy. If the legs are caught, an attempt to gently extricate them one at a time should be made. When free of entrapment, the

ENVIRONMENTAL RESCUE

victim should roll slowly to firm ground or turn onto the stomach and do a slow breast stroke. If all motions are made slowly and carefully, it will be possible to swim to safety. The main thought is to prevent overexertion.

Quicksand and quagmires are found all over the world. Quicksand may be most often found around areas where water rises to the surface of the earth. These hazards may be found in stream and river beds, washes, wells, and around springs. One must be especially careful when walking on sandy soil around these conditions. Quagmires offer the same hazards and are equally difficult to detect. A quagmire is composed of decayed material or mud suspended by underground water and are often found near old water holes, in muskeg country, and around swamps, marshes, and tidal flats. When traveling in quicksand or quagmire country, it is a good idea to probe the path ahead with a long pole.

QUESTIONS FOR REVIEW

1. There is an old mountain climbing adage that states, "The first man up a mountain is usually the last one down." This statement emphasizes that:
   a. Safety should be heeded.
   b. Energy must be expended to make speed.
   c. The most enthusiastic climbers linger to enjoy the beauty of nature.
   d. Fast climbers usually travel further than slow ones.

2. When searching a vast ocean or desert area for lost travelers, the greatest hindrance to seeing long distances is:
   a. Natural terrain.
   b. Waves and trees.
   c. Curvature of the earth.
   d. Fog or air pollution.

3. Before a systematic rescue effort can be started in a rural wilderness area, which of the following is least important?
   a. The number of casualties and the seriousness of the injuries.
   b. The relief rescue personnel that are available.
   c. The type of terrain.
   d. How many uninjured persons are at the scene to help.

4. According to statistics, the life expectancy of injured survivors in an accident in a wilderness environment decreases as much as 80 percent during the first _____ following an accident. The missing time is:
   a. 24 hours.
   b. 72 hours.
   c. 5 days.
   d. 10 days.

5. When a small child is reported to be lost in a wild mountainous section of the wilderness, the first arriving rescue personnel should conduct:
   a. A contour search.
   b. A generalized search.
   c. An area search.
   d. A rapid search.

CHAPTER 18

Rescue by Helicopter

Experience has shown that helicopters and other aircraft can offer invaluable help to the emergency services. A helicopter can rapidly deliver an emergency medical technician (EMT) to the scene of an injury or illness where an evaluation of the degree of danger to the victim can be made and emergency medical care may be provided. An airborne ambulance can transport a patient rapidly from point to point, accompanied by an EMT to provide emergency medical care while en route. A helicopter can provide rapid transportation of personnel and blood, drugs, or other material requirements to the place of emergency need from the location of supply. Aircraft are valuable assets when searching for lost or missing persons in rural and wilderness sections of the country. A helicopter can provide the above services to some geographical areas not accessible to ground vehicles.

However, there are some times and places where the helicopter has no advantage over the ground ambulance for response to the scene of an accident or illness. At other times and places aircraft offer no advantage over the ground ambulance for conducting interhospital patient transfers. There are certain categories of injury or illness in which the patient's condition will receive no additional benefit, or may actually be worsened, by helicopter use. Experience has shown that the use of a helicopter as an emergency ambulance can provide a medical advantage in only a small percentage of the total number of sick and injured in any given operating area. If a surface ambulance can reach the scene in a reasonable length of time, it can usually accomplish the task better than a helicopter.

FIXED-WING AIRCRAFT

The most significant role played by fixed-wing aircraft in rescue operations is that of providing immediate assistance to survivors and serving as the eyes of approaching rescue units. Orbiting the survivors, dropping survival equipment, and confirming the position all serve to improve the morale of the survivors, provide for their immediate emergency needs, fix the location to prevent additional long searching, and save valuable time in getting a rescue unit on the scene.

The role of the landplane in actually performing a rescue is usually limited to instances where there is a suitable landing runway at or near the distress scene or where the aircraft is designed to operate from rough or improvised strips. When climatic conditions permit, fixed-wing aircraft may effect a rescue by using frozen lakes and rivers as runways. If equipped with skis, landplanes are able to land and take off on snow and ice. These operations are hazardous and the urgency of the situation should be carefully considered before they are attempted.

SAFETY RULES

All rescue personnel must remain constantly aware that many persons have been killed and injured while boarding or working around helicopters. Rescue crews assigned to ground duty and passengers should know the rules for flight safety, procedures for loading and stowing equipment, and ground safety rules. These practices, coupled with mature judgment and alertness, will result in safe and effective air operations (Figures 18-1 and 18-2).

Always approach or leave a helicopter from the front or downhill side within the view of the pilot. Walk in a slightly stooped over, head down position. Remember that the slower the rotor is moving, the lower it will dip. Never approach from the rear because there is danger from the tail rotor. During takeoffs and landings all personnel should remain well away from the aircraft. The main rotors may dip to one side as the craft moves and the tail rotor may swing around.

Do not approach the helicopter until a signal is received from the pilot or crew member. Passengers should not enter or leave the helicopter without the pilot's permission. Personnel must not approach the craft until the engine is off and the rotors stopped, or the pilot signals permission. Even after touchdown the pilot may want to shift the helicopter's position.

Personnel in the crew should wear hard hats with chin straps fastened, goggles, and bright jackets or vests. Soft hats and caps must be held by hand.

Long-handled tools, skis, litters, and similar items should be carried low and parallel to the ground when approaching or leaving the aircraft so that the rotor blades will not be struck.

HELICOPTER LANDING SITES

The helicopter has limitations. Rescue personnel must always remember that the pilot best knows the ability and limitations of the aircraft. The pilot is in charge of, and responsible for, the safety of the aircraft and all the pilot's decisions are final and must be obeyed.

Rescue personnel who may be involved in helicopter operations must be aware of what conditions combine to make a good landing site. If the entire crew is knowledgeable, the pilot's job will be made easier and the entire rescue operation can be completed effectively, efficiently, and safely. When a helicopter arrives on the scene, invariably the first item of business is to reconnoiter the area and evaluate the landing site. There are several considerations in selecting

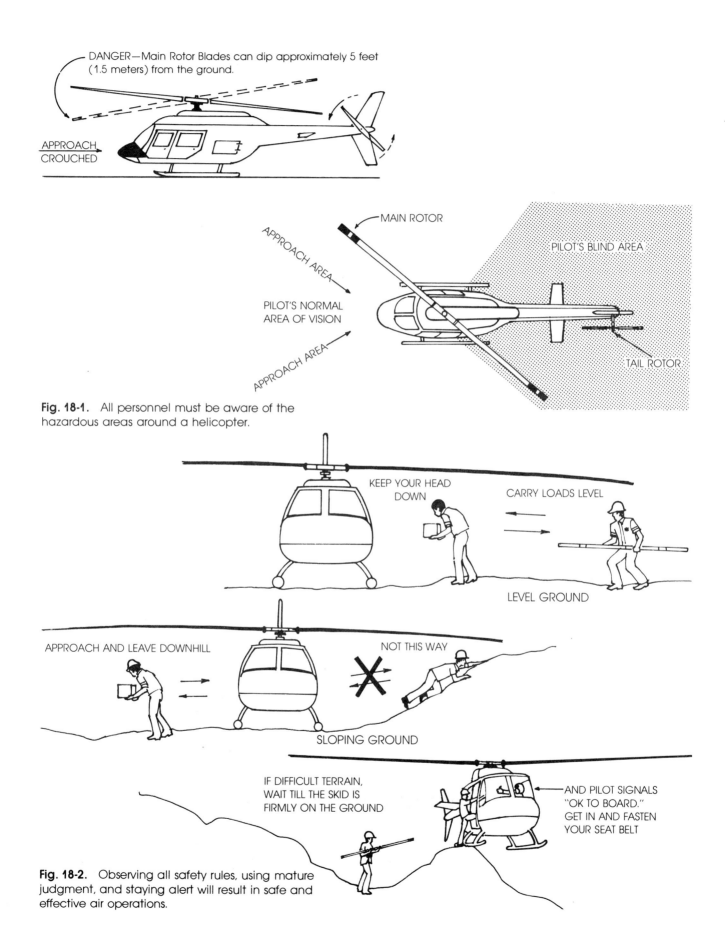

Fig. 18-1. All personnel must be aware of the hazardous areas around a helicopter.

Fig. 18-2. Observing all safety rules, using mature judgment, and staying alert will result in safe and effective air operations.

ENVIRONMENTAL RESCUE

a suitable landing site. These include, but are not limited to, wind, obstacle clearance, surface lighting if at night, security, and communications (Figure 18-3).

Pilot Considerations

Probably the first thing that a pilot does is to determine the direction from which the wind is blowing. This is done by observing windsocks, smoke, trees, wind streaks on the water, etc. The direction of the wind and obstacle clearance are the two most important factors when making an approach into and landing in an area other than an established airport or heliport. Ideally, the pilot likes to land directly into the wind. A tail wind or crosswind affects both the performance and the maneuverability of the aircraft.

Having evaluated the wind conditions, the attention of the pilot is directed to the obstacles that surround the proposed landing site. Pilots are reluctant to make approaches into areas where tall trees, power lines, or buildings lie in the flight path. The inherent danger is obvious. Not only is there an increased risk of striking an object, but should a power loss occur the pilot has no choice but to attempt a landing into the barriers. The approach then, if possible, should be over a clear area and into the wind. Very few cases arise, though, where the approach is directly into the wind and over a perfectly clear area. However, the pilot selects an approach that provides the best combination of these two variables. Generally, the more open the proposed landing site is, the more suitable it becomes.

Another item of significant importance is the general condition of the landing surface. Clear, level pavement or other hard surface is preferable. An accumulation of debris in the immediate vicinity presents a hazard in that rotor wash can propel loose objects into the engine intake. Engine failure is the obvious result. While heli-

Fig. 18-3. Mountain rescue group evacuates an accident victim. Because of steep terrain, the helicopter could touch down on only one skid. (Courtesy of the Los Angeles County Sheriff's Department.)

copters certainly can land on sand, it is undesirable. Sand blown by rotor wash can at times reduce the pilot's visibility, and it accelerates engine wear. Wetting down the landing site will reduce the problems caused by blown sand and dirt.

If the evacuation occurs during hours of darkness, lighting can be important. Well-lit areas improve the pilot's depth of perception, but personnel should guard against improper or careless use of lights. Make sure no floodlights are pointed upwards into the approach path. Also, take care not to shine lights directly into the cockpit as this can blind a pilot. A useful suggestion might be to form a large square with cars and turn the headlights on. This lights the landing area quite well.

Rural Helispots

A rural helistop consists of a landing area cleared to safely accommodate a helicopter on the ground. The best helistops are located on exposed hills or knobs because these offer a 360-degree choice of landing and takeoff directions. Preferably choose a spot where a drop-off is possible for takeoffs. The higher the elevation, the more important the drop-off becomes. If a helicopter has to make a vertical takeoff, it does so on power alone. With a drop-off the craft may use less power, carry a larger payload, and have a greater safety margin. Remove all obstacles such as brush and trees within at least a 50-foot (15-meter) diameter around the touchdown pad for small helicopters and at least 100 feet (30 meters) for larger models.

On level terrain a truly vertical takeoff should not be considered safe at any elevation. The safest takeoff path in a level situation should be at least 300 feet (90 meters) long and slightly downhill.

Around rivers and lakes the water furnishes a poor ground effect for hovering. River currents move the ground cushion and can cause pilot disorientation. Ice should be used only if no other location is available because a great many variables are associated with an ice-covered surface.

Canyon bottoms are treacherous because of dead air spaces and downdrafts. If the canyon is deep, the helicopter will need a long forward run to pull out or a wide enough area in which to circle in order to gain altitude.

Depth perception on snow and glacial ice is often poor, so it is important to clearly mark the helispot with objects of a contrasting color. Smoke grenades will also aid in orienting the pilot. On icy surfaces it is wise for the ground personnel to stay well clear of the helicopter during landings and takeoffs because the torque may cause the tail rotor to swing around.

AIR AMBULANCE OPERATIONS

When a helicopter is being used as an air ambulance, a number of patient care and evacuation procedures should be considered that may differ from the usual methods. Helicopters in emergency medical service are intended to:
1. Provide quick identification and response to accidents.
2. Sustain and prolong life through proper emergency medical care measures, both at the scene and in transit.
3. Provide the coordination, transportation, and communications necessary to bring the injured and definitive medical care together in the shortest practicable time, without simultaneously creating additional hazards.

Travel Vibration

Traveling by helicopter imposes a certain degree of roughness and vibration on the passengers. When comparing this to the roughness and vibrations encountered when riding in ground ambulances, it is necessary to consider other factors such as smoothness and contours of the roadway. Relative to the ground ambulance, experience has shown that the helicopter is less detrimental on rural or remote missions and more harmful on metropolitan trips.

Any rough movement, vibrations, or sudden jerks of a patient cannot help but bring further discomfort and could add to the severity of the victim's condition. If at all possible, intubation of the airway must be performed and intravenous (IV) administration of fluids should be started before taking off, because, once airborne, these operations are complicated by the vibration and turbulence of flight.

Medical Problems

Some types of illness and injuries are adversely affected by the decreased atmospheric pressure encountered as an aircraft gains altitude. Because helicopters normally cruise at lower altitudes than fixed-wing aircraft, the medical problems pertaining to altitude are normally not as serious. At altitudes under 1,000 feet (300 meters), the considerations pertaining to oxygen administration are the same, and difficulties encountered in case of vomiting, sucking wounds of the chest, and injuries to the sinuses, ears, and brain are dealt with in the same manner as during ground ambulance transportation.

A victim with a penetrating chest wound and a tension pneumothorax (air trapped between the lungs and the chest wall) should be transported by a ground vehicle if at all possible. The victim may be barely able to breathe sufficient air at ground level for survival because the lung capacity has been considerably reduced by the pocket of air within the chest wall. The entrapped air may cause serious complications to the casualty by expanding as the altitude increases and the atmospheric pressure is correspondingly decreased. Similar problems may occur when patients have gas trapped in their abdominal cavities by an intestinal obstruction, hernia, or similar trauma. Air trapped in other body cavities, such as the sinuses and the inner ear, will likewise be adversely affected by decreased atmospheric pressures.

Persons suffering from severe blood loss, heart disease, anemia, or any other injury or disease that affects the respiratory or circulatory systems must be closely watched while being transported in an unpressurized aircraft. With increasing altitude, the air is thinner and oxygen becomes more scarce. At 5,000 feet (1,500 meters) the lungs are supplied with approximately 80 percent of the volume that is available at sea level; this concentration decreases to about 60 percent at 10,000 feet (3,000 meters). Therefore, persons who have respiratory or circulatory problems on the ground will experience additional complications as the altitude increases. When the victims must be transported by aircraft, the pilot should be instructed to fly as close to the ground as possible. If the patient shows signs of increasing pain, additional difficulty in breathing, or intensified confusion and mental deterioration, a lower altitude is necessary. Administering oxygen may offer some relief.

The noise and vibration of air travel interferes with communications and the evaluation of vital signs. Excessive dust in the air from prop wash or rotor wash requires that extra care be devoted to covering open wounds and securing the bandages.

Cabin temperatures decrease as the flight altitude is increased. Normally, cabin heaters are adequate to compensate for this, but periodic assessment of the patient's thermal comfort should be made and blankets or additional covering provided as required.

Patients who have had fractured limbs immobilized with pneumatic pressure splints must be watched closely because pressure changes associated with the altitude will affect the splint's rigidity. These devices have a tendency to expand and tighten at the higher altitudes and loosen at the lower levels. Additional precautions must be taken during cold weather to ensure against frostbite because of the reduced circulation caused by the air splint.

The potential for airsickness and subsequent vomiting is enhanced by the presence of injury, illness, or apprehension. Observe the patient closely for the development of this complication, especially if the person is unconscious. Be prepared at all times to restore an adequate airway by the removal of vomit or other fluids.

If the patient is to be transported on an externally mounted litter or if a hoist is to be used, before the arrival of the helicopter the victim should be informed of the method of the pending evacuation. The patient must be made aware of the noise, wind, and turbulence factors that may be expected.

A patient in an outside litter must be well secured and the person's hands completely restrained to prevent any of the limbs from reaching outside the confines of the litter. Face and eye protection will prevent further injury from blowing debris caused by rotor winds. The victim is loaded with the head forward so observation of the person by paramedics is possible at all times.

SEARCH AND RESCUE OPERATIONS

The helicopter is one of the most efficient SAR vehicles in use. Its slow speed and ability to hover make it suitable for search as well as rescue operations, particularly where small targets are sought or close scrutiny of the terrain or ocean surface is required. Its ability to land in a confined area and to operate from ships enables it to remove persons from inaccessible areas and rough seas, or to rescue and administer emergency medical care long before a land vehicle or marine craft could do so. Many SAR helicopters are equipped with a hoisting winch and cable for accomplishing a rescue from a hovering position. Non-SAR craft usually are not equipped with this type of hoisting capability and must perform a rescue either by landing or by hovering just off the surface of the terrain and permitting the victims or survivors to be brought aboard.

Scanning Procedures

Scanning the terrain from an aircraft while searching for a lost person is generally conducted at a height of 300 feet (90 meters) or less and at a speed no greater than 60 miles (96 kilometers) per hour. Ground cover will have an important effect on the most efficient speed and altitude. Observing from an aircraft can be tiring and boring, so it is important to use highly motivated persons for this duty. If possible, it is best to employ observers with previous flight experience as it is easy to become airsick while concentrating for long periods of time at the terrain rushing by, especially if the air is turbulent.

Although all personnel will have a description of the subject, they must be aware that the person may have discarded an outer garment, or otherwise altered appearance. A child may have taken refuge under brush or be hiding because of fear or confusion. In fact, it is often impossible to detect lost victims in rough terrain unless they cooperate and are seeking rescue. Therefore, observers should concentrate on looking for persons or anything unusual, instead of a particular individual, and not focus their search on one certain item of clothing or article. Binoculars are used only to thoroughly check out a possible sighting.

Victim Evacuation

Helicopters accomplish rescues by either landing or hoisting. Landings are usually required at high altitudes, as in mountains, because of limitations of helicopter power for maintaining a hover. Landings are made for all rescues by helicopters when a suitable landing site is available. Hovering the helicopter and hoisting the survivor aboard not only requires more engine power, but it also presents a greater hazard to both the aircraft and survivor. Also, the victim may suffer from the increased noise and excitement of a hovering rescue.

There is danger if helicopters are operated close to collapsed parachutes. Parachute inflation by rotor downwash can cause injury to an attached survivor or the parachute can be sucked into the rotor of the helicopter.

There is a considerable potential for the use of helicopter ambulances on throughways and freeways that pass through metropolitan areas. Surface congestion can be avoided because helicopters may have direct access to an accident site that is blocked to ground ambulances. On divided highways, the roadway past an accident scene will probably be clear of vehicles and can provide an excellent landing pad. When necessary, police officers can temporarily block traffic long enough for the casualty to be evacuated.

The factors to be taken into account when selecting a site for a hovering evacuation are generally the same as those for choosing a helispot except that a smaller ground area, rougher terrain, and steeper slopes are permissible. On the other hand, it is extremely important that there is plenty of room for both the main and tail rotors because that pilot may have to turn the aircraft if there are wind changes. Wind conditions and air density may affect safe hovering, but only the pilot can be the judge.

A helicopter in flight builds a static charge that must be removed before contact between any portion of the aircraft and an individual on the ground. During hoist or aerial delivery operations, the most effective method of grounding this electrical charge is to ensure that the metal hoist cable or basket makes contact with the water or ground on the approach to the hover position.

Rescue Basket. A rescue basket (Figure 18-4), of which there are several models, is the preferable method of hoisting any victim with a winch and cable, especially if the person is unconscious or can offer little or no help.

Stokes Stretcher. The Stokes stretcher (Figure 18-5) is generally used only when the condition of the victim is such that the person must be in a prone position. When either a rescue basket or a Stokes stretcher is employed, a light steadying line should be utilized to guide the device and prevent it from spinning or swinging.

Rescue Sling. The rescue sling (Figure 18-6), or horse collar, may be used at any time when, in the opinion of the pilot, the use of the basket is inadvisable and the employment of the litter is unnecessary.

Rescue by Helicopter

The use of the sling may be difficult if the victim is not assisted by a trained ground crew. The excitement of a sling hoist may increase shock or heart problems. In cases of chest, arm, or shoulder injuries, further damage may be done. The victim may slip out of the sling if all precautions are not observed.

Penetrator. The penetrator hoist is a bullet-shaped device that penetrates the tree canopy when lowered. When unfolded, it resembles an anchor with three prongs, each prong providing a seat for one person. When sitting on the seats facing the center or hub, a safety strap is used to secure each passenger to the penetrator. The penetrator hoist is an excellent method for landing rescue personnel or the removal of ambulatory victims from inaccessible locations.

FIRE FIGHTING OPERATIONS

There are many uses for helicopters at fires. The practicality of using helicopters for transporting personnel and equipment to blazes in

Fig. 18-4. Helicopter rescuing survivors of a stranded fishing vessel by lowering a rescue basket with a winch cable. After being placed in the basket, the patient can be transported quickly and safely to the nearest hospital. A helicopter in flight builds a static charge of electricity; this charge must be removed before any contact between the aircraft and an individual on the ground, so the individual will not receive a painful, and possibly dangerous, shock. The most effective method of grounding this electrical charge is to be sure that the metal hoist cable or basket makes contact with the water or ground on the approach to the hover position. (Courtesy of the United States Coast Guard. Photo by Steve Ginesi.)

Fig. 18-5. Hoisting an injured victim from an inaccessible location in a Stokes stretcher. (Courtesy of the United States Coast Guard.)

ENVIRONMENTAL RESCUE

Fig. 18-6. An open sea rescue with a horse collar sling. (Courtesy of the United States Coast Guard.)

high-rise structures or buildings covering an extensive area, and the evacuation of occupants and injured victims, should receive advance consideration during preplanning inspections and fire fighting drills so that problems can be anticipated and standard operating procedures may be adopted.

Roof Landings

If there is no established heliport on the roof of the building, or if the craft is heavier than the roof can support, the helicopter may have to hover slightly above the deck. These aircraft are very heavy and can do enormous damage to a roof that is not structurally designed for them. Capable pilots can hover a very short distance above a roof and accomplish all objectives without the risks of landing.

Ground Landings

It is best to establish the ground landing spot at some distance from the fire so that the noise and confusion attendant on the air operations will not have a detrimental effect on fire fighting procedures. Police assistance is needed for crowd control and the streets between the heliport and the fireground should be kept clear of bystanders and spectators in case of an aircraft accident.

Size-up of the Fire

Helicopters have demonstrated their importance as command centers to allow officers an opportunity to observe patterns of fire, heat, smoke, and other conditions in the burning building. When fire fighters are delayed in arriving on the fire floor, aerial observers can help size up the fire. They can determine the fire floor or floors, obtain an estimate of the involved area, observe the exterior upward spread of fire, spot fires on adjacent roofs, and warn personnel on the ground of falling shards of glass and other debris.

Aircraft are also extremely helpful at forest and brush fires because the overall extent of the blaze is readily apparent and fire fighting crews and equipment can be easily deployed and placed in the most effective positions.

Transporting Personnel and Equipment

Aircraft can transport personnel and equipment to the roofs of buildings in which the elevators are unsafe or not operational, or access to the fire floors may be excessively delayed. When crews are conveyed to a roof for fire fighting purposes, all members should be fully equipped with breathing apparatus, forcible entry tools, and hose packs. Some departments anticipate that the standpipes may be inoperative and not be capable of supplying effective fire streams to the hose lines. In this event, the aircraft can airlift portable tanks of water and pumps to the roof for a rapid attack on the fire. Aircraft aid is also utilized for rural forest and brush fire fighting.

Hose lines may be airlifted directly out of apparatus hose beds and flown to building roofs or up mountain slopes. This method of laying hose lines is far faster than stretching the hose by hand. Caution must be exercised, however, so that an excessive amount of stress is not placed on the couplings so that the hose is damaged.

Ventilation

Under some conditions, the rotor downdraft can either aid or hinder fire fighting and ventilation operations. A strong downdraft driven through roof openings can create such a strong pressure in the stairwells that it may be difficult to open doors. Blowing fresh air may aid ventilation, or the smoke may be forced back into the building. The course of the fire spread may be changed.

Evacuating Occupants

Where occupants must be evacuated from the upper floors of a tall building, there may be advantages to airlifting them off the roof. Time may be saved by transporting the people to an adjacent structures, instead of to the ground. When aerial evacuation is necessary, sufficient personnel should be assigned to the roof to control the people and to maintain communications with the pilot.

Personnel aiding in rooftop operations must be aware that hot gases venting from roof openings could pose a problem to the pilot because they may deprive the helicopter of needed lift.

QUESTIONS FOR REVIEW

1. Experience has shown that helicopters can be utilized to replace ground ambulances for the rapid transportation of victims of accidents and sudden illness successfully:
   a. Without exception.
   b. Usually, if the hospital has a helipad.
   c. In the majority of cases, unless the victim has respiratory problems.
   d. Very seldom.

ENVIRONMENTAL RESCUE

2. When approaching a helicopter when the rotor is turning, which of the following is least recommended?
   a. Remain within view of the pilot.
   b. Approach and leave from the front.
   c. Travel on the downhill side.
   d. Walk with head upright and stand straight so all hazards may be detected.

3. When landing a helicopter, the pilot's first consideration is usually devoted to:
   a. Wind direction.
   b. Cleared-off area.
   c. Landing hazards such as trees.
   d. Adjacent structures.

4. Which of the following is the most likely to adversely affect a victim who is being transported in an aircraft?
   a. Rough movements and vibration.
   b. Air turbulence.
   c. Decreased atmospheric pressure.
   d. Motion sickness.

5. Using aircraft in fire fighting operations, such as forest fires and in high-rise structures, is least complicated by:
   a. Lower atmospheric pressure at high altitudes.
   b. The necessity of hovering above the terrain because of the lack of a cleared area.
   c. Heated gases above a fire interfering with engine operation.
   d. Hot gases venting from roof openings may deprive the pilot of needed lift.

PART 5 RESCUE FROM HAZARDOUS MATERIALS

19 Respiratory Protection
20 Managing Hazardous Gases, Liquids, and Chemicals
21 Handling Nuclear Radiation Incidents
22 Handling Bomb Threats and Explosives

CHAPTER 19

Respiratory Protection

To enable fire fighters and other emergency service personnel to survive in atmospheres contaminated by the smoke and gases generated by fires, and also work at incidents where toxic gases and vapors have been released, respiratory protection is essential. Breathing apparatus are designed to protect personnel from the inhalation of heated smoke and toxic gases, and from an oxygen deficient atmosphere. The apparatus are also intended to protect the wearer's eyes from the smoke irritants and particles that not only cause soreness and pain, but may interfere with a person's faculties at critical times when all of the senses must be operating at top efficiency. This equipment protects the user's life and allows the person to stay longer and work better in contaminated atmospheres.

Fire fighters are acutely aware of the hazards from smoke, toxic vapors, and oxygen-deficient atmospheres faced in rescue and fire suppression operations. Every year the number of materials that give off toxic products of combustion increases. Current breathing apparatus offer a high level of protection from such hazards, but because the units are heavy, bulky, and hard to don and wear, they are often not fully utilized by the working rescue personnel and fire fighters. Very often emergency personnel will feel that the discomfort and loss of mobility caused by using modern breathing apparatus is more significant than the protection it provides in marginally dangerous situations.

The increased use of respiratory protection devices has allowed rescue and fire fighting personnel to get into hazardous predicaments that they could not have reached without the equipment. Although the accelerated employment of self-contained breathing apparatus has reduced the number of deaths and injuries from smoke inhalation, the incidence of fatal accidents from burns and structural collapse is growing each year.

In the early days of combating fires, a fire fighter was judged by an ability to work in thick smoke with no respiratory protection; leather lungs were a prerequisite for the job. Personnel were ridiculed and scorned if they hesitated to enter a hot smoky atmosphere to combat a blaze or search for a victim. Originally, fire fighters grew luxuriant beards to help solve the problem. They were actually instructed to soak their whiskers with water, and place the wet mass of hair into their mouths, before entering a burning building. This procedure was faster, and just about as effective, as covering the nose and mouth with a wet cloth.

Breathing apparatus is undoubtedly the most important safety equipment supplied to emergency service personnel since oxygen-deficient and toxic atmospheres are commonly encountered in rescue and fire fighting activities. These conditions are very dangerous. It is important to understand at the beginning that no breathing apparatus yet designed provides foolproof, guaranteed protection under all fire and gaseous conditions that may be encountered. Each type has its limitations and there have been many occasions in which rescuers, even trained and experienced personnel, have lost their lives while attempting to remove endangered or asphyxiated victims from gaseous or oxygen-deficient areas. Many of those killed died because they did not thoroughly understand the operation or the limitations of the equipment they were wearing. Others who were asphyxiated knew, but gambled that they could exceed the limitations and survive.

With respiratory protection, a rescuer will be capable of searching for victims more thoroughly because a compelling need for a breath of fresh air would not be a vital factor. Personnel will be able to concentrate on the search instead of wondering how much longer they can remain in the hot choking smoke. Rescuers equipped with masks will not avoid entering the deep recesses of large structures where a breathable atmosphere could not be obtained.

Fire fighters equipped with breathing apparatus can make an immediate attack on the blaze, instead of waiting for effective ventilation to provide a respirable atmosphere. When the crew taking the initial hose lines into the building are provided with masks, a faster attack on the actual fire is attained. The blaze is therefore knocked down quicker and the damage caused by heat, smoke, and water is held to a minimum.

Respiratory protection will allow more efficient interior ventilation. No longer will a fire fighter have to grope through the smoke, force open or break out a window, lean out and take a deep gasp of air, and stumble on to the next window for the next breath of air. Breathing apparatus has revolutionized working in contaminated atmospheres to the same extent that SCUBA equipment has aided underwater divers (Figure 19-1).

RESPIRATORY HAZARDS

The cardinal rule for anyone who plans to enter a confined area is *never trust your senses*. What may look like a harmless situation may indeed be a potential threat. An atmosphere that may smell strange at first can impair the olfactory sensitivity and make a person careless. Indeed, some of the deadliest gases and vapors have no odor at all. One must always anticipate that a hazardous atmospheric condition might exist in fire structures, tunnels, utility manholes, vaults, sewers, subcellars, excavations, tanks, sump pits, silos, cold storage facilities, ships' holds, stacks and chimneys, ductwork, mines, abandoned wells, sewage treatment plants, and chemical plants and

RESCUE FROM HAZARDOUS MATERIALS

Fig. 19-1. Fire fighters with self-contained breathing apparatus and a chain saw preparing to ventilate an apartment-house fire to release the confined smoke and heat. (Courtesy of Surviv-Air Protective Breathing Equipment, a division of U.S. Divers Company. Photo by Chester M. Houswerth, Jr.)

warehouses. In addition, hazards sometimes appear unexpectedly in normal situations due to inappropriate industrial disposal or leakage of toxic substances into earth strata or conduits.

Physiological and chemical hazards are often compounded by physical factors such as the tendency of gases lighter and heavier than air to pocket in low places, against the ceiling, or other irregular surfaces. Certain gases and chemical compounds have the potential to build up static electricity. Humidity and temperature can also transform a normal environment into a hazardous one. It should be remembered that caution must always be exercised, and that the unexpected can occur, when functioning in confined area environments. Three of the most common atmospheric conditions that constitute hazards are:

1. *Oxygen deficiency.* Normal fresh air contains 20.9 percent oxygen. Life ceases quickly without adequate quantities of this vital gas. Often the oxygen content of the air can become fatally low in a brief period of time. The particular danger of asphyxiation is the inability to detect and diagnose the problem. As the

oxygen content of the atmosphere drops, euphoria (a false sense of well-being) develops and a person is lulled into inactivity.
2. *Combustible gases and vapors.* Some atmospheres may explode or ignite if a source of ignition appears. This category of environmental hazards includes naturally occurring gases and the vapors of a large group of liquids used as fuels and solvents.
3. *Toxic gases and vapors.* These atmospheres contain contaminants that even in a low concentration can cause serious injury or death. Toxic substances are commonly found in industry as well as being generated by natural processes.

HUMAN RESPIRATION

Of all the gases contained in the atmosphere, oxygen is the only one used by humans in the process of respiration. The nitrogen is not absorbed or changed in the lungs and may be breathed over and over. The carbon dioxide is expired in the exhaled air. Oxygen is necessary for the support of life and of combustion. It is colorless, odorless, tasteless, nonpoisonous at ordinary concentrations and pressures, and the most important to life of all the gases in the atmosphere (Tables 19-1 and 19-2).

When inhaled into the lungs, oxygen is absorbed by the blood and carried to the tissues of the body where it is used in chemical reactions in which foodstuffs are metabolized to form heat and energy. The union of oxygen and carbon results in the generation of carbon dioxide and the emission of heat; this process sustains the body temperature and life itself. Carbon dioxide is then transported by the blood to the lungs where it is excreted.

When breathing in a normal atmosphere, a person usually consumes between 10 and 35 percent of the oxygen inhaled into the lungs. The exhaled air contains 2.6 to 6.6 percent carbon dioxide and some moisture. Everyone breathes easier and works best when the atmosphere contains about 21 percent oxygen, the concentration usually present in clean fresh air. One can live and work, though not so well, when there is less oxygen. When the atmospheric oxygen is reduced to 17 percent, workers will breathe a little faster and

Table 19-1. COMPOSITION OF CLEAN DRY AIR AT SEA LEVEL.

GAS [a]	CHEMICAL SYMBOL	VOLUME %
Nitrogen	N_2	78.08
Oxygen	O_2	20.95
Argon	Ar	0.93
Carbon Dioxide	CO_2	0.03
Others	—	0.01

[a] Others include neon, helium, krypton, xenon, hydrogen, methane, nitrous oxide, and ozone.

Table 19-2. APPROXIMATE RATE AND VOLUME OF RESPIRATION BY HEALTHY HUMANS OF AVERAGE SIZE. [a]

DEGREE OF ACTIVITY	RESPIRATIONS PER MINUTE	AIR INHALED, LITERS PER MINUTE	AIR INHALED, CUBIC FEET PER MINUTE
Resting	16	up to 10.2	up to 0.36
Light Work	30	up to 30.9	up to 1.09
Heavy Work	40	up to 45.3	up to 1.60

[a] Extremely heavy work or vigorous exercise may cause the above figures to be exceeded.

deeper; this would be similar to ascending from sea level to an altitude of 5,000 feet (1,500 meters).

CONSERVATION OF AIR

For any given person the rate of breathing and the volume of air required for respiration is a variable factor governed primarily by exertion; it is also affected by nervous state, tension, cold, lung capacity, illness, age, and other indeterminants. It is, therefore, impossible to give exact figures for the average consumption of air. However, the quantity of air required under normal conditions may be established with some degree of accuracy. For this purpose, it is assumed that an experienced and relaxed person in good physical condition and wearing a breathing apparatus, with little exertion, will consume approximately 1 cubic foot (30 liters) of air each minute. The same person, under identical emotional conditions, performing arduous labor will probably require 1 1/2 to 2 cubic feet (45 liters) or more of air every minute to satisfy respiration needs.

Users of breathing apparatus should not expect to obtain exactly the same service life from the equipment on every use, nor should they anticipate routinely receiving the factory-rated duration of the unit. The work being performed may be more or less strenuous; emotional factors will probably be different. Where the labor is more difficult, the unit's duration may be shorter, possibly as little as one-half the rated duration. A person swinging an axe will require more oxygen than a fire fighter directing a hose stream. A large person will use more air than a smaller person. Officers should anticipate this and be prepared to relieve the harder laboring individuals more frequently. When possible, personnel should not be left in a hazardous concentration of toxic smoke and gases until the duration of their breathing apparatus has been exhausted.

Extending the Duration

The service life of the unit will depend on such factors as the degree of physical activity and physical condition of the user; the degree to which the wearer's breathing is increased by excitement, fear, claustrophobia, or other emotional states; and the amount of training or experience that the user has had with this or similar equipment. The condition of the apparatus and whether or not the air or oxygen cylinder was fully charged at the start of the work period are, of course, very important. The presence in compressed air of carbon dioxide concentrations greater than the 0.03 percent normally found in the atmospheric air will increase the rate and depth of respiration. In training sessions it is important to call attention to these limiting factors and to develop operational and maintenance procedures that will allow the wearers to obtain the maximum use from this protective equipment.

Breathing apparatus should not be donned immediately after running, working hard, or while breathing deeply. It is possible that the equipment will offer such sufficient restriction that the wearer cannot obtain sufficient air; this might cause a belief that the mask is defective. Also, the extraordinary demands on the apparatus would greatly decrease the anticipated duration. Persons should rest and compose themselves before donning any type of respiratory protection equipment.

An easy method of extending the duration of any demand-type apparatus is to delay coupling the mask hose to the regulator until the wearer is near the contaminated atmosphere. Conversely, the hose should be uncoupled immediately after leaving a toxic environment. Besides conserving the mask's duration, fresh clean air is cooler and less restrictive.

Breathing Control

In order to keep the consumption of air close to the average volume so that the rated duration of the breathing apparatus can be attained, or improved upon, proper breathing control is necessary. A slow, yet free, breathing rhythm should be adopted and practiced. Normal inhalation without attempting to fill the lungs, and then slow complete exhalation, will allow the lungs sufficient time to absorb the maximum amount of oxygen from the air. It is imperative to relax, remain calm, and make a conscious effort to maintain a normal breathing rhythm. Lack of breathing control can easily reduce the duration of a breathing apparatus by one-half through excessive air consumption.

Wearers should not attempt to prolong the use of the equipment by holding their breath, skip breathing, or by taking extremely shallow breaths as this can cause anoxia. SCUBA divers have drowned while attempting to extend the duration of their apparatus by using these methods; fire fighters and other rescue personnel have likewise gotten into trouble with these unwise practices. Remember—there is a definite minimum quantity of oxygen that is necessary to stay conscious and maintain life. However, it would be wise to practice these survival techniques in a clean fresh atmosphere to develop the ability to greatly extend breathing apparatus duration in case a person became trapped and could not escape. Under these circumstances, inactivity would cause the body to require less oxygen if excitement and panic were controlled.

Breathing apparatus imposes a slight restriction to breathing. When a person is overexerting or in a panic, the wearer may commence taking rapid, deep breaths with the consequent possibility of breathing beyond the mechanical capabilities of the equipment. An inexperienced person would believe that the equipment was faulty when it was unable to fully supply the demands. Fear and confusion would further increase the respiratory rate, and the wearer would probably become convinced of being in trouble and head out for fresh air. There is a definite maximum quantity of air that any breathing apparatus can supply; no attempt to exceed this amount should be made.

There are two other important factors of breathing control. A worker must limit physical exertion so the body's oxygen requirements are minimized and an attempt to control mental emotions such as fear, apprehension, and claustrophobia must be made as they will automatically increase the respiratory rate. A slow and steady breathing rhythm at all times will help to avoid most of the respiratory difficulties that a breathing apparatus user may encounter.

Unfortunately, there is one more factor that occasionally shortens the effective duration of a mask. Some personnel succumb to the overpowering desire to escape from a toxic and hazardous atmosphere by deliberately overbreathing because they know that when they are out of air, they can retreat to the fresh cool outside

RESCUE FROM HAZARDOUS MATERIALS

atmosphere. A proper training program by the concerned officers should ensure that all members of the crew would obtain approximately the same duration from their equipment.

Buddy Breathing

When the air supply of a breathing apparatus becomes depleted while the wearer is still working in a poisonous atmosphere, it is possible to escape by buddy breathing with another person. With demand-type equipment, the best method is for both persons to keep their facemasks in place to keep the interiors from becoming contaminated. Both should uncouple the inhalation tubes from the regulators and then take turns attaching the hose to the regulator that still contains air. They may alternately take a few breaths until an escape has been made. A hand is held over the uncoupled hose or the breathing tube is squeezed to close it off so toxic gases are not inhaled.

With other types of self-contained breathing apparatus, it may be necessary to remove the facemasks and alternately apply the mask to the face so a few breaths may be obtained. If this procedure is used, care must be taken to drive as much of the toxic atmosphere as possible out of the facepiece by forcibly exhaling before a breath is taken.

Buddy breathing should be frequently practiced at drills as it could mean the difference between life and death at some unanticipated moment where there is no time for indecision or hesitation.

RESPIRATORY PROTECTION IN A HIGH-PRESSURE ATMOSPHERE

Occasionally, rescue personnel and fire fighters must use breathing apparatus in an environment where the atmospheric pressures are greater than the normal sea level pressure of 14.7 psi. If fires have to be extinguished or persons rescued in underground installations that have to be kept under a high pressure to prevent flooding from the infiltration or an inrush of water, respiratory protection will cause many problems. Anyone entering or working in such areas under high pressure must enter and leave through an air lock system.

Compressed Air Breathing Apparatus

It must be remembered that using demand-type breathing apparatus under conditions of high atmospheric pressure will reduce the duration of use. In order to enable the user to breathe, air is supplied by a demand valve at the pressure of the surrounding atmosphere. Thus, to maintain 1 cubic foot (28 cubic decimeters) of air at the surrounding atmospheric pressure, a larger mass of air is required when the atmospheric pressure is greater than normal. The mass is directly proportional to the pressure when it is measured in atmospheres. Thus, twice the amount of air is used when the pressure is two atmospheres (29.4 psi) as when the pressure is one atmosphere (14.7 psi), and the available air supply will be halved at this pressure. When the atmospheric pressure is three times normal (44.1 psi), the air will be used three times as rapidly. From these facts, the air consumption at any pressure may be calculated. For example, a 30-minute demand-type breathing apparatus may offer a trained and

Respiratory Protection

experienced wearer only 10 minutes of respiratory protection when working in an atmospheric pressure three times greater than normal.

Closed-Circuit Breathing Apparatus

It is best to use closed-circuit breathing apparatus under high-pressure conditions because the atmospheric pressure has little effect on the duration. Human beings can breathe pure oxygen safely for only a limited length of time at the higher pressures. When the pressure reaches 1.8 atmospheres or over, the wearer may start feeling the detrimental effects of oxygen intoxication.

Filter Breathing Apparatus

Cannister masks are out of the question. Since a caisson is an enclosed area, the dangers of oxygen deficiency cannot be excluded. Moreover, in the case of rescue work, the inhalation resistance would become uncomfortably high because of the greater air density.

TRAINING PROGRAMS

Persons who expect to wear breathing apparatus in irrespirable atmospheres and perform work under extremely hazardous conditions must be physically fit and be expertly instructed in the construction, use, care, and maintenance of such equipment. Moreover, after receiving the initial instruction, periodic refresher training will allow them to remain proficient. Situations that require the use of masks are usually dangerous and a wearer's life may depend on a thorough familiarity with the equipment.

Authorities recommend a continual training program for all personnel and they specify that supervisors as well as the lower ranking members be taught by competent instructors. Training and inspection procedures for the use of self-contained breathing equipment should become closely integrated. Ideally, a training program will so thoroughly familiarize a fire fighter or rescuer with the equipment that procedures for donning, checking, and use become automatic. Training programs should include:

1. Instruction in the nature of hazards that may be encountered and what could happen if the device is not used properly.
2. A discussion of why self-contained breathing apparatus is the proper device for a particular purpose.
3. An explanation of the device's capabilities and limitations. Instruction and training in the actual use of the device with close supervision to ensure that it continues to be used properly.
4. Continuing classroom and field training to recognize and cope with emergency situations.
5. Training in the care and cleaning of equipment as recommended by the manufacturer.
6. A continuing instructional program to provide rescue and fire personnel an opportunity to handle respiratory protection devices, have them fitted properly, test tightness at all connections as well as the mask seal, wear it in normal air for a familiarity period, and, finally, depend upon it in a simulated toxic environmental situation.

If breathing apparatus is to be used to its full potential, drill sessions will be more thorough than just practice donning the units in a minimum amount of time. All personnel must drill while wearing the masks under conditions that closely simulate actual emergency conditions. Full protective clothing, including gloves, are worn. Activities that include climbing stairs and ladders, handling loaded hose lines, simulating physical rescue, and searching buildings for victims are valuable practice. Smoky conditions can be imitated by taping covers over the facemasks to obscure vision.

Thorough practice of fireground and rescue operations under realistic conditions while wearing full protective clothing and breathing apparatus will allow personnel an opportunity to become thoroughly familiar with the equipment. After they are acquainted with the benefits and the safety factors engineered into the masks, the crew will have a greater respect for and can obtain the full potential advantages that the devices offer. Many persons have encountered difficulties, with serious and even fatal results, while wearing masks. In almost every instance the difficulties could be attributed to the poor physical condition of the wearer, poorly assembled or otherwise defective equipment, lack of proper training, or improper procedures. Therefore, the need for adequate specialized instruction and training in the use and care of all types of respiratory protection cannot be emphasized too strongly.

Thorough training will allow the personnel an opportunity to gain full confidence in the breathing apparatus; they will also appreciate their personal limitations in the exertion of fire fighting and rescue. The duration of supply will vary according to the metabolic requirements of the individual, the amount of exertion, and the mental state of the wearer; these variable factors could easily cut the average protective duration of a mask in half. It takes concentrated effort and self-discipline to use such equipment to its greatest capacity. The wearer must be in excellent physical condition and must be able to dominate emotions in order to control exertion and breathing rhythm.

Officers responsible for setting up training programs must consider the need for developing skill in the use of breathing apparatus and yet be careful not to create situations that may cause loss of life. In the past, several unfortunate training exercises have led to deaths of fire fighters. The best drills include realistic training in a building where fire and smoke can be generated; they must be properly supervised and controlled. Burning down old structures that need to be demolished offer valuable practice. The basic purpose of such training is to teach the crew to use masks as safely and efficiently as they would employ any other type of equipment. Since all breathing apparatus requires a little more exertion for the wearer to obtain sufficient air or oxygen, each person has to develop an individual breathing cycle in order to make the maximum use of this protective equipment. In training, one should learn to space breathing and exertion over a specific time period while performing hard work with visibility obscured and wearing full protective clothing as one would at a fire or other incident. It accomplishes little when a member trots around a well-lighted structure and calls it a mask drill. Personnel should be taught the extent of protection available while doing different types of work under various conditions.

Individual reactions to wearing breathing apparatus vary considerably, but most persons, when they first try on this equipment, experience some discomfort, perhaps even claustrophobia, when they put on the facepiece and realize that they are breathing an atmosphere provided by mechanical equipment. Such feelings are to be expected and can only be overcome by prolonged, frequent training as well as actual experience. It takes considerable effort to wear the mask, climb a ladder, and go through fire fighting or rescue routines until the air supply in the cylinder is exhausted. This training is necessary if fire fighters or rescuers are to understand their own capabilities and the peculiarities of the apparatus. One must develop an individual sense of timing when using the equipment under conditions of severe physical exertion and must be able to identify the low-pressure warning signal from one's own breathing apparatus and the sounds from the equipment used by other fire fighters nearby.

Practical Drills

Search and rescue drills are the most popular training exercises; they are valuable practice, are easily observed and critiqued by the officers, and they may be made interesting by varying the conditions. Rearranging the furniture so that searchers will receive no familiarity advantage and placing one or more simulated victims in the room or structure will increase interest. Besides receiving instruction and experience in using masks under realistic conditions, the wearers can be instructed in the correct methods of rescue and victim recovery.

Fire streams should be maneuvered, hose lines handled and advanced, ladders raised and climbed, and other strenuous activities conducted so that personnel can learn the value of conserving energy and controlling their breathing to avoid exceeding the air volume limitation of the equipment. They will also have an opportunity to feel the difference in balance that will be experienced because of the weight and bulk of the apparatus.

There are many places in buildings and ships that a person cannot pass through with a breathing apparatus strapped on one's back; these include attic scuttle holes, under structure crawl spaces, ventilation ducts, and other restrictions. Entrance openings into tanks, underground vaults, ships' holds, and other confined spaces are rarely large enough to allow a person with full protective allotment of clothing and mask to pass through. The actual scene of an emergency when lives of victims are involved is the wrong place to attempt to figure out an answer to such a dilemma when it suddenly arises. Drills will allow procedures to be formulated and practiced to assure that a realistic, effective, and safe method will be adopted. At times it may be best to take off the apparatus and pass it through first, leaving the facemask on so breathing can be continued. Sometimes just moving the ladder away from the crawl hole will provide sufficient room to enter.

Facepiece Fitting

Every respiratory protective device wearer must receive fitting instructions. This will include demonstrations and practice in how the device should be worn, how to adjust it, how to determine if it fits properly, and how to remove it correctly. The sealing effect of a mask is greatly dependent on the shape of the face to which it is to fit tightly. The principal influencing factors are the dimensions of the face, the bone structure, the type of skin, and the growth of hair and beard. These items can modify to a greater or lesser extent the shape of the sealing line or the tight fit of a mask.

Facemasks should not be worn when physical conditions

prevent a good face seal. Such conditions may be a growth of whiskers, sideburns, unusual facial contours, a skull cap that projects under the facepiece, temple pieces on glasses, or clothing that prevents proper adjustment. The absence of one or both dentures can seriously affect the fit of a facemask. The helmet or winter liner chin straps should not be fastened in such a way that loss of the helmet would pull the facemask away from the face. Diligence in observing these factors are to be evaluated by the officers while conducting periodic drills. To assure proper protection, the facepiece fit will be checked by the wearer every time the mask is put on.

Providing respiratory protection for individuals wearing corrective glasses is a serious problem. A proper seal cannot be established if the temple bars of eyeglasses extend through the sealing edge of the full facepiece. As a temporary measure, glasses with short temple bars or without temple bars may be taped to the wearer's head. Wearing of contact lenses in contaminated atmospheres with a mask should not be allowed. Systems have been developed for mounting corrective lenses inside full facepieces. If corrective spectacles or goggles are required, they must be worn so as not to affect the facemask fit.

Every wearer must know how to quickly adjust a facemask to ensure a proper seal around the face. The best way to don a facepiece is to make certain that all straps are loose; attempting to fit the facemask when the head harness straps are too tight will cause unnecessary strain and result in not only worn, but occasionally broken straps. The head harness should always be slipped in place with all straps fully extended and flipped over the lens so they will not become entangled. Placing the mask under the chin first, then pulling the head harness over the back of the head will expedite operations.

Tightening the straps in the correct order will ensure a correct seal most of the time. The two straps closest to the temples, or slightly behind the eyes, should be grasped first; grip one in each hand and pull back along the head with the fingers holding the straps against the head until a firm resistance is felt. This practice will seal the mask evenly and firmly to the face.

Then both hands should be dropped to the two bottom straps located near the neck or behind the jawbone. Pull back and slightly upward with the fingers touching the neck until the straps are snug.

Always tighten the top strap(s) last. Allow the pulling fingers to remain in contact with the top of the head until firm resistance is felt. Some manufacturers recommend tightening the bottom straps first, then the middle straps, and the top straps last. With some models of facepieces, this sequence might be more effective.

The wearer then smooths out any twists and turns in the straps and takes up the slack in the chin and temple straps to ensure a good fit. The headstrap on top is used to adjust the position of the facepiece lens. The straps must not be overtightened.

The facepiece seal is then checked by crimping the breathing tube to close it off or by sealing the hose inlet against the palm of the hand and inhaling gently. If the mask is properly sealed, it will collapse against the face and remain there as long as the breath is held. If the facepiece does not collapse, adjustments for a better seal can be made by pulling the headstraps tighter. Some authorities recommend against holding the palm of the hand over the tube opening because an irregularity of the palm can prevent an effective

Respiratory Protection

seal unless the person making the test is quite careful. The airtightness of the facepiece may be helped by grasping the bottom of the mask and wiggling it once or twice for final adjustment.

SAFETY PRECAUTIONS

Most fire departments and rescue organizations insist that all personnel work in pairs at any emergency that requires the use of breathing apparatus. This arrangement allows the partners an opportunity to watch over each other's welfare and are instantly available in case one or the other becomes injured or endangered.

Officers must have some method of determining what personnel are working in a structure, where their approximate location is, and when they entered the building. Someone outside the contaminated atmosphere must know when the breathing apparatus is approaching its limit of duration. Personnel should be equipped and ready at all times to launch a rescue operation if there is the slightest indication that anyone is in trouble. Some departments utilize a display board and markers with names on them. The marker for each person is posted on the board upon entering the building, together with the time and anticipated location; it is removed when the wearer again emerges. Other organizations use a clipboard or other system.

The welfare of mask wearers must be considered when hose streams are directed into a structure or the building is ventilated to assure that personnel will not become unnecessarily jeopardized or subjected to injury or punishment. Water and steam can greatly hinder visibility by condensing on the facemasks.

Practically all modern facepieces have speaking diaphragms to facilitate communication between wearers of breathing apparatus; this is a valuable accessory because close cooperation is necessary between crew members. It is not necessary to remove the facepiece in a noxious atmosphere to talk when no speaking diaphragm is supplied because, with a little practice, wearers can talk well enough to be understood with masks on.

If mask wearers become nauseated and have to vomit because of heat exhaustion, breathing foul air, or some other trouble, they should start for the fresh air, pull the mask to one side, vomit, and immediately replace the facepiece. The person must refrain, if possible, from inhaling while the mask is off.

Some gases and chemicals can penetrate a human's system through the skin and, therefore, it must be constantly remembered that self-contained breathing apparatus will not completely protect anyone against this type of hazard. When these types of gases and vapors are encountered, the additional protection of gloves, boots, and special impenetrable clothing is required in these atmospheres to be reasonably safe.

In atmospheres containing irritant gases, such as ammonia and sulfur dioxide, rubber bands placed around the sleeves and pants legs will help prevent burns at the armpits, crotch, and other moist parts of the body. It is a good practice under these conditions to smear grease or an ointment on the tender and vulnerable portions of the body.

A particularly dangerous time occurs when the fire is knocked down and some of the crew are in different parts of the structure or area. Because the need for fire fighting or rescue action may be over, there is a natural tendency to remove the facepieces to obtain a breath of fresh air. This is a simple common occurrence, but far too many casualties have resulted because fire fighters were standing in an area that lacked oxygen, had a concentration of carbon monoxide, or contained some other toxic vapor. Company officers must maintain continual communications with their personnel to assure that they move into safe areas before they remove their facemasks. When possible, this is done outside the involved building or contaminated area.

Lifeline

Whenever possible, a lifeline of light rope should be secured to members using breathing apparatus, but the very nature of fire fighting and rescue often makes this difficult. A lifeline offers added safety to the members because it provides constant communication with personnel on the outside; serves as a safety line in areas where pits, elevator shafts, open hatchways, and other hazardous openings may be encountered; can be used by the wearer to find a route out; and may be employed by rescuing members to find their way to personnel in distress.

Some organizations place a lifeline on each member entering a toxic atmosphere, while others only tie one to the first person. Both methods have their good and bad points. When the rope is tied to only the lead member, all of the others grasp the line with one hand so they can follow but not be pulled down if the leader should fall. By not being tied to the lifeline, one or two members can step up quickly to assist the leader, go back for help, or leave the line for a moment to open a window or do some other job without upsetting the movement of the other personnel on the line. The line should be tied around the waist with a bowline knot; this knot will not slip and cause unwanted constriction if a heavy pull is made. Also, a bowline will not jam and it can be easily untied if one has to free oneself in an emergency.

When a lifeline is used, pulling on the rope can provide an elementary communication system. The signals should be simple and easily memorized so everyone will remember them. One system uses the word *oath* as the memory key to the rope pulls (Table 19-3).

Table 19-3. LIFELINE SIGNALS

CODE	TUGS	MEANING	EXPLANATION
O	1	OK	Everything is all right. This can be either a question or a reply.
A	2	Advance	Proceed, if made from the outside. Moving forward, give more line, if made from the inside.
T	3	Take Up	Turn back, if made from the outside. Take in line, backing out, retreating, or moving position, if made by the inside member. Also, keep slack out of the line.
H	4	Help	Emergency signal that is not to be ignored. Exit at once, if made from the outside. Assistance needed, if made by the inside personnel.

To be certain that the emergency signal may be fully understood, and the urgency emphasized, the line pulls should be continued instead of only giving four; all four may not have been received. If the *help* signal originates from outside, the mask wearers are to exit at once. If the *help* signal originates from inside, personnel equipped with breathing apparatus must be immediately sent in to assist.

The lifeline signals must be pronounced and definite; a full swing of the arm helps to prevent confusion. Slack is not allowed in the line. The signals are answered by repeating the signal in the reverse direction. The signals may originate from either the mask wearer or the line tender, and the OK signal should be given frequently. Failure to receive an answer to the OK signal is sufficient cause to start immediate rescue action.

The line tender must remain alert at all times to watch for the possibility of the member inside being cut off or trapped by a mine cave-in, a spreading fire, a building collapse, or some other sudden change. The tender is not to leave the line unattended. One person should not attempt to tend more than one line. It is not a good practice to have all the available breathing apparatus in use at the same time. It is best to have at least one or two members ready to go to the aid of any mask wearers who get into trouble.

Warning Devices

Extreme caution must be continuously exercised by all wearers of breathing apparatus to be certain that they do not use up the entire protective supply of air or oxygen while still in a contaminated atmosphere. All types of masks have some method of alerting the wearers when the duration of the equipment is almost exhausted.

The point at which the wearer should retreat to fresh air is determined by considering the time or pressure remaining, the rate at which the duration is being depleted, and the distance to safety. It is wise to allow ample time for escape because delays in the retreat could occur from obstructions or difficulty in finding the way.

If difficulty in breathing is encountered, or if one becomes disoriented, confused, or lost, one should resist the temptation to pull off the facepiece because one breath of a contaminated atmosphere could result in immediate death. The wearer must remain calm, stop and rest, organize mental faculties, and then start out again.

Compressed gas units are equipped with a regulator gage that indicates the pressure remaining in the cylinder. The pressure gage should be observed frequently, and a wearer who cannot see it in heavy smoke should withdraw to a location where it can be read. It is important to constantly be aware of the pressure remaining in the cylinder, even if the unit is equipped with a low-pressure warning device, as the mechanism may fail to work.

When laboring, workers have a tendency to forget to check their pressure gages. Alerting devices warn the wearers when it is time to leave the contaminated atmosphere and replace their air cylinders. Personnel should never escape the working area without letting the rest of the crew know they are leaving; their absence could cause unwarranted apprehension.

Automatic warning of an impending air supply depletion may be provided by either an alarm bell or a restriction that alerts the wearer by causing more difficult breathing when there are four to six minutes of breathing time left under average conditions. In some models the air used to operate the bell is not wasted, but is returned to the breathing circuit. For this reason, these bells ring during inhalation, but stop during exhalation. Other models ring continuously. Some organizations have a requirement that when any alarm bell starts ringing, all personnel should leave the contaminated atmosphere. This procedure is based on the thought that if all personnel entered the structure at the same time, the rest of the masks are approaching exhaustion. Also, it is difficult to determine which bell is actually ringing. This requirement could really cause confusion in a location where several persons with similar masks are working together, and there is a continual rotation of personnel.

This system does not take into consideration the fact that, depending on the person's work rate, lung size, and other factors, one mask wearer may consume air at a faster or slower rate than the average. Requiring a complete exodus of breathing apparatus users every time any one bell sounds will not allow any organization to obtain the maximum duration from their masks. Of course, teams paired together will stay together. Everyone should be sufficiently familiar with the equipment and the pressure remaining in the air cylinder that the wearer will be instantly alerted when the bell sounds.

BREATHING APPARATUS TESTING AND APPROVAL

The approval or certification of personal protective devices, performed at this time by the National Institute for Occupational Safety and Health (NIOSH), is based on regulations originally developed by the U.S. Department of the Interior, Bureau of Mines. From 1919 to 1973, the Bureau of Mines performed tests and issued approvals on respirators. On March 15, 1973, the Bureau officially transferred testing of these devices to the NIOSH Testing and Certification Laboratory at Morgantown, West Virginia.

Respirator approvals issued by the Bureau under older regulations are scheduled to expire on the dates listed below. After the expiration dates, users should purchase only devices approved under the newer regulations and manufactured under strict quality control requirements, although they may continue to use Bureau-approved respirators until the terminal approval dates that were established to permit orderly replacement with new devices.

Expiration of Bureau of Mines Respirator Approvals. Respirators purchased with Bureau of Mines approvals, before NIOSH took over, will be approved for use until the following dates (unless, in the case of certain self-contained breathing apparatus only, they are upgraded to meet the new requirements and are relabeled):

Self-contained breathing apparatus	March 31, 1979
Supplied air respirators	March 31, 1980

TYPES OF BREATHING APPARATUS

Respiratory protective equipment includes all types of apparatus that enable a wearer to live and work in an atmosphere containing irrespirable or poisonous gases, smoke, dusts, fumes, vapors, or mists, or an atmosphere partly or wholly depleted of oxygen. Such devices differ rather widely and may be divided into two main groups: (1) those that afford protection by removing contaminants and (2) those by which a respirable atmosphere from an uncontaminated source is supplied to the wearer.

Breathing apparatus is generally classified as follows: the filter-type canister mask; the hose mask, which is a unit supplied with air or oxygen from an outside source through a hose; and self-contained breathing apparatus. The self-contained apparatus may be further divided into three basic types: demand-type, self-generating type, and oxygen cylinder rebreathing apparatus. The demand apparatus operates on an open-circuit principle since the exhaled air is discharged through an exhalation valve to the outside atmosphere. The self-generating type and the oxygen cylinder rebreathing apparatus both operate on the closed-circuit principle since a captive atmosphere is circulated within the unit and is constantly enriched with fresh oxygen. The carbon dioxide is removed chemically.

Self-contained breathing apparatus are the only devices that will protect the wearer against irrespirable or poisonous gases, regardless of the concentration, and against atmospheres containing little or no oxygen.

Demand-Type Breathing Apparatus

The majority of fire fighting and rescue organizations are equipped with self-contained, demand-type breathing apparatus. Through years of experience, this equipment has proved to be safe and dependable (Figure 19-2).

Description of the Apparatus. Each breathing apparatus has four main parts: the facepiece assembly, the regulator assembly, the compressed air cylinder and valve assembly, and the cylinder-holding frame and harness assembly.

The molded rubber facepieces on modern units have built-in ducts that cause the air entering the mask from the inhalation tube to flow across the inner surfaces of the lenses to prevent fogging. The facepiece is equipped with an exhalation valve through which the exhaled breath is discharged to the outside atmosphere; these valves also serve as secondary safety valves since they prevent excessive pressure from building up in the mask.

The regulator assembly consists of a regulator shutoff valve, a valve-locking device, a bypass valve, a pressure gage, and a high-

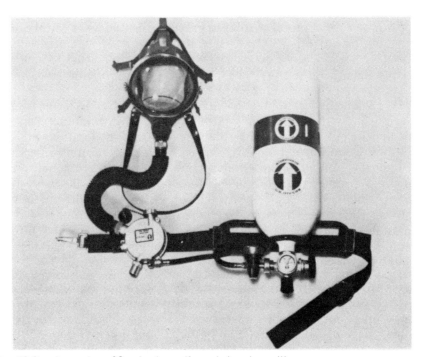

Fig. 19-2. A modern 30-minute, self-contained, positive pressure, demand-type breathing apparatus. (Courtesy of Surviv-Air Protective Breathing Equipment, a division of U.S. Divers Company.)

pressure air hose. The primary function of the regulator is to reduce the pressure of the compressed gas and to supply it at a normal ambient pressure to the wearer upon demand. A safety valve is provided to relieve excessive pressure if the reducing valve fails to operate. The mainline, or regulator, shutoff valve provides a means of closing off the flow of air if the reducing valve or admission valve sticks open. The regulator shutoff valve locking device prevents an accidental closing of the valve.

The bypass valve—the red handwheel mounted on the side of the regulator—supplies a flow of air from the high-pressure cylinder into the facepiece and is to be used in the event of the failure of any part of the regulator. By opening the bypass valve and closing the regulator valve, the wearer may adjust the mask pressure to the minimum comfortable breathing pressure; air may be conserved when operating the bypass valve by opening it when inhaling and closing it when exhaling.

The pressure gage, which is attached to the regulator where it can be seen at all times by the wearer, indicates the pressure of the air supply in pounds per square inch (psi). When the gage needle reaches the colored section, the wearer should immediately start for fresh air because approximately 400 psi is still available; this is roughly sufficient for three to five minutes. Some makes of breathing apparatus warn the wearer of this danger point by ringing a bell; others increase the inhalation effort, but moving a reserve lever will restore normal air flow for about five minutes.

The cylinder-holding frame and harness assembly holds the cylinders and regulator in position for the greatest amount of comfort. The harness straps are equipped with adjustable buckles so that the harness can be adjusted to fit the wearer.

Air cylinders are made of high-carbon steel and are tested in accordance with Department of Transportation (DOT) specifications. The 30-minute bottles have a capacity of from 40 to 45 cubic feet (1.2 to 1.3 cubic meters). A full capacity is indicated when the pressure gage reads 2,216 psi (2,015 psi + 10 percent). All compressed gas bottles should be hydrostatically tested every five years to meet the requirements of the DOT. The month and year of manufacture or the test is stamped on the neck of each cylinder.

With the use of technology developed by NASA, new lightweight cylinders are being offered by many manufacturers. A composite cylinder is made without seams from an aluminum alloy. It is then wound over its entire surface with high-strength fiberglass filaments inpregnated with epoxy resin. This unique construction process produces an air bottle that is not only lighter, as it is approximately one-half the weight of a steel cylinder, but is also stronger and more resistant to abuse and corrosion. The new bottles may be operated more safely at higher pressures than conventional cylinders. Consequently, the same volume of air can be carried in a smaller size bottle.

The NASA Breathing Apparatus. The National Aeronautics and Space Administration (NASA) recently cooperated on designing a new mask incorporating features that offer advantages over the older designs. It is a self-contained, open-circuit, demand-type breathing apparatus. The backplate is made of solid plastic and is designed to accommodate either a 40- or 60-cubic foot (1.2- or 1.7- cubic meter) cylinder. The bottles are constructed of an aluminum liner and are fully overwrapped with a resin-impregnated fiberglass, providing a lighter and stronger cylinder than those made of steel. The bottles are charged to over 4,000 psi.

Using the 40-cubic foot (1.2-cubic meter) cylinder, the mask weighs 20 pounds (9 kilograms) and provides at least 20 minutes of operating time. With the 60-cubic foot (1.7-cubic meter) bottle, the unit weighs 26 pounds (12 kilograms) and provides a minimum of 30 minutes operating time. It should be understood by every person who uses a mask that the actual duration period of a fully charged cylinder is predicated upon the wearer's breathing rate, which is the result of exertion, excitement, training, physical condition, and other factors. The cylinder reducing assembly, which decreases the high cylinder pressure, is mounted on the backplate and supplies a lightweight demand regulator attached to the facepiece.

Principles of Operation. Self-contained, demand-type breathing apparatus all use the same open-circuit principle of operation, with some minor differences. This type of respiratory protection consists of a high-pressure cylinder of air mounted in a sling or strapped to the wearer's back. The pressure of the compressed air is controlled by a reducing valve and is supplied to the wearer through a regulator assembly, an admission valve, and a rubber facepiece. The exhaled air passes through an exhalation valve into the outside air. The majority of these units is rated to offer full respiratory protection for 30 minutes; comparable masks with smaller bottles are rated for 5 to 15 minutes and are recommended for emergency escape use only.

In some highly toxic atmospheres, a small amount of facemask leakage can be dangerous. In these cases, the pressure demand model is used. A special feature built into the regulator and facemask permits air to fill the system to a pressure greater than atmospheric, but retains it so none is wasted. A vacuum can never occur inside the mask, even under extremely high demand conditions. Leakage into the mask is impossible because, through any leak path, air is blowing outward.

Straight demand-type breathing apparatus and the positive pressure demand-type both operate identically; only the inhalation and exhalation pressures will vary between the two types. Demand breathing apparatus maintains a pressure inside the facepiece in relation to the immediate environment that is positive during exhalation and negative during inhalation. Pressure demand, or positive pressure, breathing apparatus are designed so that the pressure inside the facemask in relation to the atmosphere is positive both during inhalation and exhalation.

Demand and pressure demand breathing apparatus look almost exactly alike externally. Pressure demand masks have either a spring-loaded or a pressure-balanced exhalation valve and a control on the regulator which will permit air to flow into the facepiece at all times. A slight positive pressure is kept in the mask by the pressure demand devices.

The operation of pressure demand apparatus requires special operational procedures to ensure maximum duration of the unit. If the apparatus has fixed pressure demand devices, the main control valve must be shut off when the facepiece is removed from the face or air will continue to flow from the mask. Breathing apparatus equipped with control selectors allow the equipment to be used either as a demand or pressure demand unit. Apparatus so equipped should be placed in the demand position when the facemask is taken off the face or the main control valve must be shut off to preclude the loss of air through the mask in the pressure demand position. Pressure demand apparatus requires care in the adjustment of the facepiece as an improperly fitted mask can cause outward leakage, thus reducing the available air supply.

Emergency Operation. If a malfunction occurs in a breathing apparatus regulator, emergency procedures dictate that the wearer return to clean air as quickly as possible. Therefore, all personnel must be thoroughly familiar with the location and use of the unit's bypass system so no delays will be experienced if troubles occur.

During normal operation the mainline regulator valve is kept open and the bypass control valve closed. If the automatic demand regulator becomes inoperative so that no air passes through it, opening the emergency bypass valve (red knob) by turning the regulator counterclockwise will provide a continuous flow of air to the mask through bypassing the demand regulator mechanism. If failure or damage to the demand valve causes a continuous flow of air through the regulator, the bypass valve should first be opened, and then the mainline valve closed, to prevent an excessive leakage of air. The bypass flow is then adjusted to the wearer's requirements. It is usually necessary to open the valve only slightly to avoid excessive loss of air through blow-by.

In most models of breathing apparatus, the mainline valve handles are yellow and the bypass valve handles are red. In some models, the valve handles are touch coded so that the correct knob can be selected by its shape when vision is impaired by smoke or darkness. It is important to practice selection of valves solely by touch to assure that no delay will be experienced if an emergency arises. When the bypass valve is used with the mainline valve closed, the regulator gage will indicate a zero pressure.

This procedure is for emergency exit use only. During normal demand operations there is no reason to use the bypass valve. The duration of the air supply is greatly reduced when the unit is operating through the bypass system.

Checking the Mask Before Each Use. Fire fighters have perished when they ran out of air or a malfunction occurred while wearing breathing apparatus. Everyone is convinced that the equipment must be carefully restored to a fully serviceable condition after every use, and it is usually accomplished, but that does not relieve personnel of the obligation to perform a quick, but thorough, check before the mask is donned on the fireground. The need to enter a structure at the earliest possible moment to perform urgently required rescue or fire fighting operations often tempts dedicated personnel to skip some of the safety checks that are necessary before entering a toxic atmosphere. This occasionally results in a fireground statistic.

While inspecting a breathing unit would seem to be a simple procedure, steps should be taken in a specific sequence with the same automatic checkpoints each time the equipment is used. A good training program will go far to ensure that the continuity is consistently followed in the interest of speed and safety.

Before donning the apparatus, the high-pressure hose should be checked to be certain that it is securely attached to the regulator and the cylinder valve. The cylinder valve is then opened, the cylinder valve safety lock engaged, and the lock checked by attempting to

Respiratory Protection

close the valve. A full bottle will indicate approximately 2,216 psi (2,015 psi + 10 percent at 70°F or 21°C), unless the wearer's organization has standardized on a pressure that is either higher or lower than normal. If, locally, a different pressure is considered a full cylinder, that figure should be used. The bottle must be replaced if the pressure gage indicates insufficient air.

When the equipment has been properly serviced, the cylinder gage will read *Full*. After the cylinder valve has been opened, the wearer should compare the pressures of the cylinder gage and the regulator gage. Since this type of pressure gage is normally accurate to only within a plus or minus 5 percent of the full gage scale (plus or minus 150 psi), there may be 300-psi pressure difference between the two gages without exceeding the normal inaccuracy. If there is a significantly greater difference between the two gage readings, the apparatus must not be used until it has been checked by a qualified technician.

The airtightness of the facepiece is checked by crimping the breathing tube to close it off or by sealing the hose inlet against the palm of the hand, and then inhaling gently. If the mask is properly sealed, it will collapse against the face and remain there as long as the breath is held. If the facepiece does not collapse, adjustments for a better fit can be made by pulling the headstraps tighter. Holding the palm over the tube opening may not provide as effective a seal as grasping the breathing tube because of skin surface irregularities.

It is important that the exhalation valve be checked. This test is made by inhaling, then squeezing the breathing tube and exhaling. If the exhalation valve is not operating properly, a heavy blow-by of air will be felt in the temple area. A small blow-by of air should be disregarded since it might be caused by exhaling too sharply.

If these tests prove to be satisfactory, the breathing apparatus is ready for use and the inhalation tube coupling can be connected to the regulator. The wearer is now ready to work in a contaminated atmosphere for a specified period of time, which depends on the rated duration of the unit.

Cleaning. For sanitary reasons, the facepiece and inhalation tube are cleaned and disinfected after each use, even if they do not appear dirty. This also offers an opportunity for inspection. It is recommended that a cleaning solution be made by mixing water that has been heated to between 140 and 160°F (60 and 71°C) with any detergent that contains effective disinfectants.

Next, the inhalation tube is removed from the facepiece. Care must be observed not to lose the gasket located in the coupling nut. The facemask and inhalation tube are washed in the cleaning solution, using a sponge or soft brush for hard-to-get-at places. When cleaning the lens, it must be remembered that it is susceptible to scratching; avoid rubbing grit or any abrasive substance against the surface. Dirty water is then flushed away and the two pieces rinsed inside and out with clean running water. Stretching the tube will remove any water lying in the convolutions and will allow a careful inspection for any signs of small cracks, perforations, or wear along the tube corrugations.

The facepiece and inhalation tube are then submerged in a germicide solution for the time recommended by the manufacturer, usually about two minutes. They should be sponged off with the germicide solution and thoroughly rinsed in clean water. The last step is to allow the facemask and the inhalation tube to air dry. Carefully stretch the breathing tube to release excess water. Any attempt to force dry the parts by placing them near a heater, or in direct sunlight, will cause the rubber to deteriorate. The use of straight alcohol as a germicide also will cause deterioration.

Routine Inspection. The self-contained breathing apparatus is a vital part of the fire fighter's personal protective equipment. It is a safeguard against hazardous atmospheres. But this apparatus is just like any other piece of professional equipment. It requires proper inspection and maintenance.

Authorities recommend that the cylinder pressure of all masks be checked daily; it is best if the person who is assigned to wear the unit has the responsibility for the inspection. When the bottle gage reading is less than the standard maximum pressure, the bottle should be replaced with one that is fully charged. Pressure in the high-pressure system is released by breathing through the facemask until the air is exhausted. This relieves pressure on the demand regulator and the high-pressure hose. If so equipped, the alarm bell will ring. Most manufacturers specify that the bypass valve must never be used to exhaust air pressure.

A complete inspection of all breathing apparatus at least once a week is necessary to ensure that the units are constantly ready to use.

Facepiece. The facemask skirt and headstrap should be inspected for pliability and signs of deterioration. Stretching and manipulating the rubber with a massaging action will keep it pliable and flexible, and prevent it from taking a set during storage. All parts, especially the lens, should be clean and free of dirt and dust. Examine the buckles to see that they function properly and are free of excessive rust. Check the facepiece for leaks. Special attention should be given to the exhalation valve and the joint between the lens and the skirt. Exhalation should be smooth, with no sticking of the exhalation valve.

Low-Pressure Hose. Examine the end fittings of the low-pressure hose for damaged threads and freedom of movement. Check the O-rings for damage. Check the hose for holes, cuts, and tears.

Regulator. To check the regulator, prepare the unit for use, don the facepiece, and breathe. The regulator should deliver ample quantities of air without fluttering or free-flowing. Also, check for leaks and tightness of all fittings. Examine the inlet filter, if so equipped, for cleanliness.

The regulator should be tested for leaks with a fully charged bottle. The cylinder valve is turned on and the regulator allowed to build up to the full cylinder pressure, after which the cylinder valve may be shut off. With the bypass valve closed, the regulator gage is then watched approximately one minute for a drop in pressure. If there is a noticeable drop in pressure, the regulator should be repaired. If no leaks are found, the regulator valve is then shut off and the cylinder valve turned on. Air trapped in the demand valve is breathed out. If the regulator valve is operating properly, no air may be breathed through the regulator until the regulator valve or the bypass is opened.

Another test for leaks utilizes a soap and water solution; if bubbles appear, then a leak is indicated. This soapy water solution is placed around the threaded connections of the cylinder, the regulator, and the gage.

High-Pressure Hose. Inspect the high-pressure hose for leaks, cuts, cracks, and abrasions. Check the end fittings for tightness.

Alarm Bell. To check the alarm bell, connect the regulator and alarm to a full air cylinder. Close the bypass valve. Open the cylinder valve. The bell should ring briefly as pressure builds up in the lines. Close the cylinder valve. While watching the regulator pressure gage, slowly open the bypass valve and bleed the pressure down. The bell should ring briefly as pressure vents from the lines.

Cylinder Valve. Check the cylinder valve for leaks and excessive operating torque.

Pressure Gages. Check the cylinder and regulator gages for leaks, cracks, or excessive scratches in lens, and case damage. Also check for a loose or bent pointer and obviously inaccurate pressure reading.

Backpack. Check the backpack for broken, twisted, or excessively frayed straps, defective or excessively rusty buckles, defective stitching, and other deterioration.

Recharging Air Cylinders. It is always advisable to purchase extra cylinders with every breathing apparatus so that empty bottles can be replaced immediately. Cylinders can be recharged with either a compressor or a cascade system, or a combination of both.

In the majority of masks, 2,216 psi is considered to be a full charge. Some of the newer bottles, however, are constructed so they will withstand a maximum pressure of 4,500 psi. When using the 2,216 psi cylinders, approximately 1 1/2 minutes of use time are lost with each 100 psi reduction in pressure. Bottles with higher pressures will experience a comparable diminished duration. This fact emphasizes the importance of charging air cylinders to their maximum working pressures. When recharging a cylinder, the air should be controlled so that pressure builds up slowly; this will reduce heat buildup from compression to a minimum. The bottle may be overfilled 10 percent above the pressure rating stamped on the cylinder neck. A slight top-off may be necessary after the cylinder cools.

Compressed air cylinders should always be filled while submerged in a water bath, but do not immerse the pressure gage. Filling the cylinders behind a heavy steel enclosure will protect the worker in case of a ruptured bottle.

If a cascade system, which is a manifold consisting of 300-cubic foot (9-cubic meters) cylinders of compressed air, is used, the number of storage cylinders is generally based on the number of self-contained breathing apparatus in service and the frequency with which they are used. In most applications, five storage bottles are sufficient. The breathing apparatus bottle is connected to the manifold and the storage cylinder with the lowest pressure opened. When the pressure in the manifold and the bottle equalizes, the valve on the storage cylinder is turned off and the valve on the cylinder with the next highest pressure opened. This procedure is repeated until the breathing apparatus bottle is fully charged. Pressures continue to be equalized cylinder by cylinder until the desired working pressure in the apparatus bottle is obtained.

When demand-type breathing apparatus will be used for extended periods on the fireground, it is necessary to provide replacement air cylinders. Fire departments in large cities have various ways of accomplishing this. Some have special vehicles with cascade systems of air cylinders so that breathing apparatus can be replenished on the fireground. A few have portable compressors mounted on a truck. Many organizations merely send an extra supply of bottles to an emergency where self-contained breathing apparatus will be in prolonged use. It is essential that a fire department have more than an estimated adequate supply of cylinders for the duration of an emergency.

Recharging High-Pressure Bottles. The 4,500 psi operating pressure of some models of the newer breathing apparatus requires a higher pressure supply of air than is currently available to most fire departments. There are several possible alternatives for providing a source of pure, high-pressure air for cylinder refilling.

Outside Vendors. Many firms, particularly those engaged in supplying the SCUBA market, may have the equipment necessary to fill cylinders at the required purity and pressure. If such a firm can be located in the local area, using this source would avoid incurring capital outlays for charging facilities.

Cascade Systems. Large pressure vessels of 6,000 psi breathing air are commercially available from air supply companies. A cascade and regulator system for filling breathing apparatus bottles to 4,500 psi from these vessels can be readily fabricated. This method of filling requires low capital outlays and may be suitable where the usage rate of respiratory protection equipment is relatively low.

Pressure Intensifiers. It is possible to use a low-pressure charging system (2,400 psi) to drive a pressure intensifier. This is a device that uses two pistons of different sizes to boost pressure. These systems, when used with the current compressors, are relatively inexpensive, but give reduced charging rates.

High-Pressure Compressor and Air Purification System. A department that is committed to using a high-pressure breathing apparatus, and has a high use rate for the equipment, may find it cheaper, and certainly more convenient, to install a high-pressure compressor, an efficient air purification system, and an air storage arrangement to provide large volumes of pure breathing air.

Respiratory Air Stations. Respiratory air stations must be capable of producing, with consistent reliability, air that at least meets, and preferably surpasses, a specified minimum standard. The quality of the safe, pure, breathable air that is yielded is of prime importance. Equally as important is the operating safety of the equipment itself. Components of sturdy quality construction must comply with all applicable safety codes. It is also highly desirable that the unit incorporate safeguards against mechanical failure and automatic features to compensate for human forgetfulness or error.

The use of inadequate equipment can result in unsatisfactory performance and high operating costs. Poor compressor performance may impair the quality of the output air. This will have an adverse effect on the people who have to breathe the air. They may face temporary discomfort, reduced efficiency, permanent health impairment, or even death.

There are two sources of contamination to be concerned with when producing respiratory air: those present in the atmosphere and those generated by the air compressor. High-pressure respiratory air stations consist of four basic components: the compressor, purification system, storage system, and filling station.

Compression. In the process of producing pure compressed air, the compressor, usually driven by an electric motor, pumps air at a high pressure into the rest of the system. This makes it possible to store the air in a reduced space for use when needed. Compression also prepares the air for the next step, purification, which is accomplished by squeezing out the contaminants. The higher the pressure, the more impurities are forced out; this action is similar to wringing out a wet cloth or sponge. It has been discovered through practical experience that the most efficient pressure for processing air to the purity level required for breathing purposes is around 3,000 psi.

The compressor should be capable of producing a higher pressure than the minimum necessary because the higher pressures aid the purification process and also make it possible to store a greater volume of air in a smaller space. A compressor rated at 3,500 psi is ideal for most fire department and SCUBA shops, but some situations and equipment may require a pressure rating of 5,000 psi or more.

The compressor's capacity, or volume output, will depend on anticipated usage and air storage facilities. This is usually expressed in standard cubic feet per minute (SCFM). There are three major determinants to consider when ascertaining the required output. Time is probably the most important factor. For example, a fire department with a small compressor could spend much of the night refilling bottles after a training session or a fire. The second item is the number of cylinders that will be refilled on a regular basis, and the third determinant is the amount of air that is capable of being stored.

The selection of a good environment for the compressor intake is important because the air drawn in must be as clean as possible. It is generally recommended that the intake be above floor level, preferably 8 feet (2.4 meters) or higher, and well removed from all possible sources of contamination such as vehicle engine exhausts, furnaces, smoke, sewers, and other operations and conditions that might emit pollutants.

Purification. Purification is the removal of all harmful matter and consists of filtering out solid particles, removing all liquid contaminants, lowering vaporized and gaseous impurities to established levels, and eliminating all unpleasant odors.

The compressor intake filter usually removes particles 10 microns and over in size. A micron is a unit of length measurement used to express the dimensions of very small particles. One micron is equivalent to one millionth of a meter, or approximately 0.0004 of an inch (0.01 millimeter); for practical reference, 25 microns equal 0.001 inch (0.025 millimeter).

The purity of high-pressure air used in respiratory protection equipment must exceed the normal environmental level. The atmosphere in an average metropolitan area contains over four million dust particles in every cubic foot (28 cubic decimeters) of air. Other harmful liquid and gaseous contaminants present include hydrocarbons, carbon monoxide, carbon dioxide, ozone, and oxides of nitrogen and sulfur. Add to these the impurities generated by the compressor itself, such as oil, water, carbon, and wear particles and the total purification process can be appreciated.

Storage. Once the air is compressed and purified, it is generally stored in a large stationary receiver or individual cylinders. The purpose of the storage system is to save time in refilling empty bottles; it will allow a quantity of air to be compressed before needed and then when it is necessary to fill a large number of bottles, high-pressure air is instantly available.

Stationary air receivers are classified as *unfired pressure vessels* and must conform to ASME (American Society of Mechanical Engineers) Codes which specify a 4 to 1 safety factor. ICC (Interstate Commerce Commission) or DOT (Department of Transportation) cylinders do not meet the requirements of the Code. The air receivers should be painted yellow in accordance with the standard color identification code for breathing air.

Filling Station. The filling station should be located away from the compressor and storage system, if possible. This allows for safer and quieter working conditions. A filling station consists of the following items: fill panel equipped with control valves and gages, cooling tank filled with water that will allow a faster filling rate by absorbing the heat generated by compression, and a fragmentation deflector that provides protection for the operator in the event of a tank rupture.

Holders. Brackets and holders have been designed to securely store breathing apparatus on fire apparatus to protect the masks from damage and yet keep them instantly available. Some departments prefer that the breathing units be held by brackets fastened to the backs of seats so that the crew can put them on before arrival at the scene. Other installations hold the units on the sides of the apparatus or in compartments to allow the wearers to back into them while standing on the ground. Regardless of the type of bracket or where the holder is fastened, these devices facilitate the rapid donning of the equipment.

Closed-Circuit Breathing Apparatus

Closed-circuit equipment, or rebreathers, recycle the user's breath, removing carbon dioxide and adding oxygen. Because of their efficiency, rebreathers are light in weight, small in size, and long in duration. Rebreathers are supplied by various manufacturers with rated durations that range from 45 minutes to 4 hours. Oxygen breathing units render the wearer's respiration completely independent of the ambient atmosphere (Figures 19-3 through 19-5).

These units are effective wherever complete respiratory protection is required, such as in mines, industrial plants, fire fighting, and rescue operations. They are capable of supplying their full rated duration in environments pressurized up to two atmospheres, absolute, as may be encountered in tunnels, caissons, and deep submergence vessels. Although designed for use in atmospheres immediately dangerous to life, they must be used with adequate skin protection when worn in an environment containing gases or vapors that poison by skin absorption.

In the closed-circuit systems, oxygen stored in a cylinder passes through a pressure reducer and into the breathing bag from which the desired amount of oxygen is inhaled. The exhaled gases pass through a carbon dioxide absorber. Here the carbon dioxide from the wearer's breath is removed and the unused oxygen flows into the breathing bag. Fresh oxygen is added, the renewed breathing gas passes on to the user, and the cycle is continued.

All rebreathers contain a breathing bag that acts as a reservoir for the user's breath. Because the closed-circuit breathing apparatus

Respiratory Protection

Fig. 19-3. Closed-circuit breathing apparatus with a rated duration of 45 minutes. Recirculating the exhaled breath permits the unit to be much smaller and lighter than conventional open-circuit equipment. (Courtesy of the BioMarine Industries, Inc.)

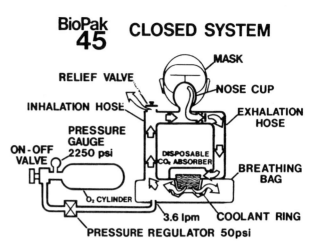

Fig. 19-4. Schematic drawing showing a closed-system breathing apparatus with a duration of 45 minutes. (Courtesy of the BioMarine Industries, Inc.)

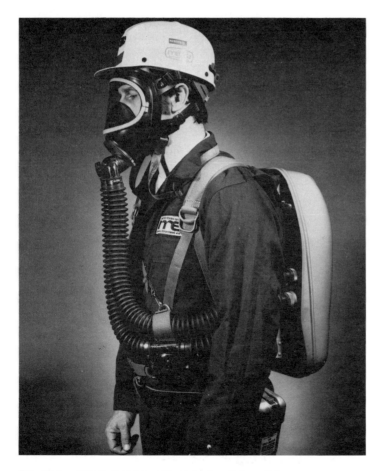

Fig. 19-5. Modern 4-hour closed-circuit breathing apparatus being tested by personnel of the Mining Enforcement and Safety Administration, United States Department of the Interior.

removes the carbon dioxide from the gas that the wearer exhales and replaces the oxygen that is consumed, the gas that is breathed is almost 100 percent oxygen.

Self-Generating Oxygen Breathing Apparatus

The Chemox oxygen breathing apparatus is a closed-circuit self-contained breathing unit operating independently of outside air. It employs a replaceable canister containing a chemical that removes exhaled carbon dioxide and moisture, and automatically evolves oxygen to satisfy breathing requirements (Figure 19-6).

The oxygen supply in the Chemox mask is generated in a canister containing potassium superoxide (KO). Moisture in the exhaled breath of the wearer causes the potassium superoxide to liberate oxygen and form an alkaline salt. This oxygen passes through the rebreather bag to the wearer. The alkali produced by this chemical reaction removes carbon dioxide from the next inhaled breath and more oxygen is added. The rate of oxygen liberation is governed by the wearer's rate of breathing.

Because this reaction cannot be accurately controlled, the equipment is designed to produce more oxygen than is needed for metabolism. The excess oxygen is vented to the ambient air through a dump valve.

This mask provides complete respiratory protection in oxygen-deficient areas of highly toxic atmospheres for 1 hour. The 1-hour service life is based on test procedures; in use, a longer or lesser protection period may result, depending on the individual user and the level of exertion.

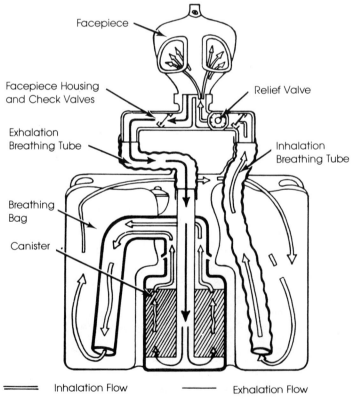

Fig. 19-6. The flow diagram of a self-generating oxygen breathing apparatus.

Masks are equipped with a speaking diaphragm that permits easy conversation. An automatic timer rings a bell at the end of a preset time to warn the wearer that the interval of protection is becoming short and the person should return to fresh air. The canisters now manufactured permit immediate, vigorous activity by the wearer and will efficiently supply adequate amounts of oxygen at temperatures as low as $-20°F$ ($-29°C$).

This simple equipment design has the primary advantages of low initial cost, light weight, compact design, and a rated duration of 1 hour. The main drawback is the unit cost of the canisters is high. Once the chemical reaction is started, it cannot be interrupted. Regardless of need, the entire chemical charge must be used or discarded.

Airline or Hose Masks

Air respirators supplied by an outside source are probably the most comfortable type of breathing apparatus available. The facepiece of an airline mask is attached to an air supply by a hose. A cool, clean atmosphere is supplied to the mask in sufficient quantity to provide the wearer with an adequate respiratory supply. There is little or no resistance to inhalation, and the surplus air usually provides a cooling and refreshing effect. The only disadvantage is the necessity of trailing a hose that is connected to the air source. The hose has a tendency to become entangled; on the other hand, lifelines and signal cords have the same inclination.

Emergency egress units with small portable compressed air bottles that provide sufficient time to escape from a contaminated atmosphere are generally provided.

Filter Masks

There are very few, if any, fire departments or rescue organizations still equipped with filter-type respirators because they have two serious deficiencies that have caused many fatalities: these masks do not supply oxygen and they are not capable of offering complete protection in high concentrations of carbon monoxide and other toxic gases and vapors.

Filter masks were advocated and used by emergency personnel for many years. Under the right conditions, these masks offer adequate protection to the wearer, but are severely restricted in their versatility. A large number of fire fighters and other emergency personnel who depended on these units for respiratory protection have been killed when the atmosphere did not contain an adequate amount of oxygen, there was an excessive concentration of carbon monoxide, or there were other toxic gases and vapors present for which the breathing apparatus was not designed. Authorities have now specified that only self-contained breathing apparatus be used for protection from the dangers of toxic gases or oxygen-deficient atmospheres encountered by the fire and rescue services.

However, these units are still popular in industry, and it is possible to encounter them in a factory, ice plant, or warehouse location. Canisters are still being supplied to offer protection against one certain type of respiratory hazard, such as acid gas, organic vapor, ammonia, or carbon monoxide. It is strongly recommended that fire fighting and rescue personnel confine their use of breathing apparatus to the various self-contained types.

RESCUE FROM HAZARDOUS MATERIALS

QUESTIONS FOR REVIEW

1. Which one of the following has had the greatest effect on the number of fire fighters killed by burns and structural collapse each year?
   a. New construction materials.
   b. Self-contained breathing apparatus.
   c. Lack of sufficient fire fighting personnel.
   d. Hiring of women into the fire service.

2. Modern respiratory protection equipment is least valuable for allowing rescue personnel to:
   a. Better search for victims in burning buildings.
   b. More effectively attack the blaze.
   c. Allow more efficient interior ventilation tactics.
   d. Permit wearers to exceed the time limitations of the equipment.

3. When rescue personnel are preparing to enter a burning structure, which one of the following is the least dependable for establishing whether a respirable atmosphere exists within the building?
   a. The five human senses, especially smell.
   b. The presence of thick black smoke.
   c. The absence of bright roaring flames.
   d. The absence of heat.

4. Normal atmospheric air contains which one of the following gases in the greatest volume?
   a. Oxygen.
   b. Carbon dioxide.
   c. Nitrogen.
   d. Ozone.

5. The least practicable method of extending the duration of self-contained breathing apparatus is by:
   a. Reducing physical activity.
   b. Adopting a calm emotional state.
   c. Not connecting tubes until ready to enter toxic atmosphere.
   d. Taking extremely shallow breaths.

CHAPTER 20

Managing Hazardous Gases, Liquids, and Chemicals

During their everyday duties, emergency service personnel are exposed to a wide variety of toxic, flammable, and explosive gases, vapors, liquids, and solids. Very often they are required to enter structures and other areas where they are exposed to the extremes of heat and frigid temperatures, to an atmosphere that is deficient in oxygen, to lethal gases created by a fire, and to the solid irritating tarry particles that are a component of smoke. The physical exertion, excitement, and anxiety involved in many emergency incidents add to the health problems by increasing the breathing and heart rates. Therefore, among the occupational hazards of being a fire fighter, police officer, ambulance attendant, emergency medical technician, lifeguard, or in any other of the many public service occupations are the severe demands placed on the respiratory, cardiovascular, and nervous systems. All personnel must be well trained in the hazards they may encounter so that duties may be performed with maximum safety under even the most adverse conditions (Figures 20-1 and 20-2).

HAZARDOUS MATERIALS

One of the great concerns of emergency service personnel is the hazardous nature of materials they may be called upon to deal with. Many people believe that they should be concerned with these dangers only when responding to an incident in a chemical factory or warehouse. They are under the impression that occupancy names or signs will adequately warn them of the hazards. These beliefs are no longer valid because we live in a chemical environment. A large variety of common and exotic chemicals are being encountered more frequently, in larger quantities, and under quite unexpected circumstances. Chemicals that a few years ago were laboratory curiosities and were handled with the most elaborate of safety precautions are now common items of industry and commerce. They are being

RESCUE FROM HAZARDOUS MATERIALS

Fig. 20-1. This warning sign did not prevent a disastrous tanker explosion that killed many persons and demolished valuable property. The 810-foot-long tank ship had just completed unloading crude oil and was preparing to sail from the harbor when heat from an unknown source ignited the large volume of flammable vapors remaining in the empty tanks. Some authorities believe faulty venting of the flammable vapors from the cargo tanks to be the primary cause of the blast.

Fig. 20-2. A tanker loaded with a blend of butane and other highly flammable liquids exploded and continued to burn for several days. Several persons were killed instantly and the wharves and warehouses on both sides of this harbor channel were engulfed by the flames and destroyed. The wharf from which this photograph was taken was not totally destroyed because a large fireboat blasted a path through the floating concentration of flames with large master streams of water. This enabled the crew to make a massive attack on the blaze and stop its destructive advance. (Courtesy of the Los Angeles City Fire Department.)

transported over the highways, railroads, and through the air in increasingly large quantities. They may be encountered anywhere in the country; no place is so remote that it can be considered to be perfectly safe and unconcerned. Many materials whose toxicity and flammability are extraordinarily high are found in such diverse occupancies as agricultural supply stores, farms, hardware stores, and medical research centers.

Although hazardous liquids and gases are the most common, consideration must also be devoted to flammable solids when thinking of dangers; many of these resemble flammable liquids in the caution that should be observed. Others, such as organic peroxides, are potential explosives. Still others, particularly the finely divided metal powders, resemble grain or plastic dusts in their behavior in a fire situation; they burn explosively.

While the flammable characteristics of materials are usually given the paramount consideration when judging the hazardous properties of chemicals, an equally important factor is the toxic nature of the material or of its decomposition products. Even a very small fire or an unexpected combination of chemicals can be disastrous if a cloud of toxic gas or vapor is generated.

Placards and Labels

The Department of Transportation has adopted a uniform marking of hazardous materials in transit to better convey important information about dangerous chemicals and other commodities. The placards and labels can be very useful in identifying the material and its principal hazards. However, an absence of any warning signs should not be taken as an assurance that the substance is safe. Nicely labeled and uncontaminated, fresh in their containers, most chemicals are quite innocuous. When the label is burned or washed off after exposure to unknown conditions of heat and moisture, mixed on the floor with unfamiliar and strange substances, or floating in a pool of water, these same chemicals can be a tremendous hazard. Chemicals that are harmless by themselves can be turned into a deadly combination when mixed together. Unless the commodity can be positively identified as having no explosive, flammable, combustible, corrosive, poisonous, or irritating property, it should be treated with caution.

Even with better controls and legislation on hazardous material cargoes, incidents still occur and when they do, emergency operations must be prepared to minimize the danger. Fire, police, and medical services personnel should know how to properly handle a variety of potentially hazardous situations. Each incident is different, but if trained personnel can recognize the hazard and react quickly to it, many injuries and fatalities can be eliminated.

Trucks that carry potentially dangerous cargo can be identified by placards placed on the front, rear, and both sides of the vehicle. These placards are required by federal law on any vehicle carrying 1,000 pounds (450 kilograms) of any potentially dangerous material or any amount of extremely dangerous material. The eleven placard categories include: Explosive A (maximum explosive hazard), Explosive B (flammable hazard), Flammable, Flammable Gas, Compressed Gas, Corrosives, Poisons, Oxidizers, Radioactive, Cargo Fire—Avoid Water, and Dangerous.

The following definitions of terms utilized on storage and shipping placards and labels will aid in determining the individual dangers of the commodity:

Class A Explosive; a substance or material which detonates to create a maximum hazard.

Class B Explosive; in general, functions by rapid combustion.

Class C Explosive; includes certain types of manufactured articles containing Class A or Class B Explosives, or both, as components but in restricted quantities. Creates a minimum hazard.

Combustible Liquid; any liquid with a flash point from 100°F to 200°F.

Corrosive Material; any liquid or solid that causes destruction of human skin tissue or a liquid that has a severe corrosion rate on steel.

Etiologic Agent; a viable microorganism, or its toxin, which causes or may cause human disease.

Explosive; any chemical compound, mixture, or device whose purpose is to substantially and instantaneously release gas and heat.

Flammable Gas; any compressed gas meeting the lower flammability limits as specified.

Flammable Liquid; any liquid with a flash point less than 100°F.

Flammable Solid; any solid material, other than an explosive, which is liable to cause fires through friction, absorption of moisture, spontaneous chemical changes, retained heat from manufacturing, or which can be ignited readily and when ignited burns so vigorously and persistently as to create a serious hazard.

Hazardous Material; a substance or material which has been determined to be capable of posing an unreasonable risk to health, safety, and property.

Irritating Material; a liquid or solid substance which upon contact with fire or when exposed to air gives off dangerous or intensely irritating fumes.

Nonflammable Gas; any compressed gas other than a flammable compressed gas.

Oxidizer; a substance such as chlorate, permanganate, inorganic peroxide, nitro carbo nitrate, or a nitrate that yields oxygen readily to stimulate the combustion of organic matter.

Poison A; extremely dangerous poisons of such a nature that a very small amount of the gas, or vapor of the liquid, mixed with air is dangerous to life.

Poison B; less dangerous poisons which are known to be so toxic to man as to afford a hazard to health.

Pyroforic Liquid; any liquid that ignites spontaneously in dry or moist air at or below 130°F (54°C).

Radioactive Material; any material, or combination of materials, that spontaneously emits ionizing radiation.

Chem-Cards

Chem-Cards (Figure 20-3) are an excellent way to determine the hazards of a chemical and the best methods of abating the dangers. These Chem-Cards are prepared by the Manufacturing Chemists' Association and many drivers transporting hazardous materials carry these informational cards in the cabs of their vehicles. Each Chem-Card provides a brief description of a specific hazardous material, a statement concerning the hazardous nature of the chemical, and general guidelines for coping with an emergency situation in case of a fire, leak, spill, or exposure to humans.

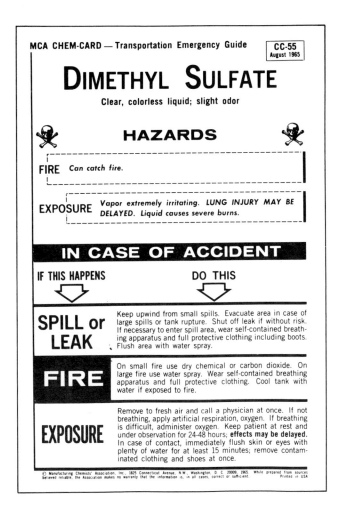

Fig. 20-3. Facsimile of a typical Chem-Card, which briefly describes, states the hazardous nature of, and offers general guidelines for coping with an emergency incident involving a hazardous material. (Courtesy of the Manufacturing Chemists' Association.)

Emergency Action Guide

The U.S. Department of Transportation has published a booklet entitled *Hazardous Materials—Emergency Action Guide* to help emergency service personnel during the first 30 minutes of an incident in which a spill or fire involves the most commonly encountered volatile, toxic, gaseous, and/or flammable materials that are shipped. General and specific safety procedures that should be followed are provided. Example pages from that book are shown in Figures 20-4 and 20-5.

Each left-hand page identifies a specific hazardous material, outlines its potential hazards, and provides immediate action information for fires, spills, and emergency medical care of the injured. The page also lists certain functions and services for those with appropriate resources and equipment.

Each right-hand page specifies recommended evacuation areas and distances for protecting the public from dangerous concentrations of toxic vapors and explosions. Where applicable, necessary water pollution controls are provided. This page also lists procedures to follow when assistance is needed or when appropriate resources and equipment are not available.

RESCUE FROM HAZARDOUS MATERIALS

Chlorine
(Nonflammable Gas, Poisonous)

Potential Hazards

Fire: —Cannot catch fire.
—May ignite combustibles.

Explosion: —Container may explode due to heat of fire.

Health: —Contact may cause burns to skin or eyes.
—*Vapors may be fatal if inhaled.*
—Runoff may pollute water supply.

Immediate Action

—Get helper and notify local authorities.
—If possible, wear self-contained breathing apparatus and full protective clothing.
—Keep upwind and estimate *Immediate Danger Area.*
—Evacuate according to *Evacuation Table.*

Immediate Follow-up Action

Fire: —Move containers from fire area if without risk.
—Cool containers with water from *maximum distance* until well after fire is out.
—Do not get water inside containers.
—Do not use water on leaking container.
—Stay away from ends of tanks.

Spill or Leak: —Do not touch spilled liquid.
—Stop leak if without risk.
—Use water spray to reduce vapors.
—Isolate area until gas has dispersed.
—Do not get water inside containers.

First Aid: —Remove victim to fresh air. Call for emergency medical care. *Effects of contact or inhalation may be delayed.*
—If victim is not breathing, give artificial respiration. If breathing is difficult, give oxygen.
—If victim contacted material, immediately flush skin or eyes with running water *for at least 15 minutes.*
—Remove contaminated clothes.
—Keep victim warm and quiet.

For Assistance Call Chemtrec toll free (800) 424-9300
In the District of Columbia, the Virgin Islands, Guam, Samoa, Puerto Rico and Alaska, call (202) 483-7616.

Additional Follow-up Action

—For more detailed assistance in controlling the hazard, call Chemtrec (Chemical Transportation Emergency Center) toll free (800) 424-9300. You will be asked for the following information:
- Your location and phone number.
- Location of the accident.
- Name of product and shipper, if known.
- The color and number on any labels on the carrier or cargo.
- Weather conditions.
- Type of environment (populated, rural, business, etc.)
- Availability of water supply.

—Adjust evacuation area according to wind changes and observed effect on population.

Water Pollution Control

—Prevent runoff from fire control or dilution water from entering streams or drinking water supply. Dike for later disposal. Notify Coast Guard or Environmental Protection Agency of the situation through Chemtrec or your local authorities.

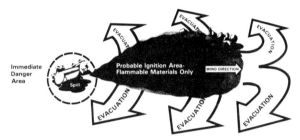

Evacuation Table — Based on Prevailing Wind of 6-12 mph.

Approximate Size of Spill	Distance to Evacuate From Immediate Danger Area	For Maximum Safety, Downwind Evacuation Area Should Be
200 square feet	20 yards (24 paces)	528 feet long, 528 feet wide
400 square feet	30 yards (36 paces)	528 feet long, 528 feet wide
600 square feet	40 yards (48 paces)	1,056 feet long, 528 feet wide
800 square feet	45 yards (54 paces)	1,056 feet long, 528 feet wide

In the event of an explosion, the minimum safe distance from flying fragments is 2,000 feet in all directions.

Fig. 20-4. Emergency Action Guide for controlling an incident involving chlorine gas. Information contained on the two pages describes the potential hazards, actions to be taken, and the recommended method of evacuating residents from the affected area. (Courtesy of CHEMTREC.)

This publication was compiled from sources representing the best opinions on the subject at the time of its preparation. It is recommended that a copy of this booklet be obtained from the National Highway Safety Administration, U.S. Department of Transportation, and kept in the vehicles of all emergency service personnel, such as fire, police, medical, and maintenance, who may encounter emergency incidents involving hazardous materials.

Procedures at Transportation Incidents

It is impossible to provide a concrete sequence of actions that emergency personnel should follow upon the arrival at the scene of a hazardous material transportation incident, but the following procedures will assist in controlling the situation:

1. Approach all incidents that may involve hazardous materials from upwind and upslope, if possible. Warn other responding units of the hazard and recommend upwind routes of approach.
2. Park vehicles at least 1,500 feet (450 meters) upwind of the scene.

Managing Hazardous Gases, Liquids, and Chemicals

Hydrocarbon Fuels
(Flammable Liquid)

Potential Hazards
Fire: — Highly flammable.
— Flammable vapors may spread from spill.

Explosion: — Container may explode due to heat of fire.
— Runoff may create fire or explosion hazard in sewer system.

Health: — Vapors indoors may cause dizziness or suffocation.

Immediate Action
— Get helper and notify local authorities.
— If possible, wear full protective clothing.
— Eliminate all open flames. No smoking. No flares.
— Keep upwind. Isolate hazard area.
— Evacuate by *at least* 2,000 feet.

Immediate Follow-up Action
Fire: — **Small Fire:** Dry chemical or CO_2.
— **Large Fire:** Water spray or fog.
— Move containers from fire area if without risk.
— Cool containers with water from *maximum distance* until well after fire is out.
— Stay away from ends of tanks.

Spill or Leak: — Stop leak if without risk.
— Use water spray to reduce vapors.
— **Large Spills:** Dike for later disposal.
— **Small Spills:** Take up with sand, earth or other noncombustible, absorbent material.

First Aid: — Remove victim to fresh air.
— Use standard first aid procedures.

For Assistance Call Chemtrec toll free (800) 424-9300
In the District of Columbia, the Virgin Islands, Guam, Samoa, Puerto Rico and Alaska, call (202) 483-7616.

Additional Follow-up Action
— For more detailed assistance in controlling the hazard, call Chemtrec (Chemical Transportation Emergency Center) toll free (800) 424-9300. You will be asked for the following information:
- Your location and phone number.
- Location of the accident.
- Name of product and shipper, if known.
- The color and number on any labels on the carrier or cargo.
- Weather conditions.
- Type of environment (populated, rural, business, etc.)
- Availability of water supply.

Water Pollution Control
— Prevent runoff from fire control or dilution water from entering streams or drinking water supply. Dike for later disposal. Notify Coast Guard or Environmental Protection Agency of the situation through Chemtrec or your local authorities.

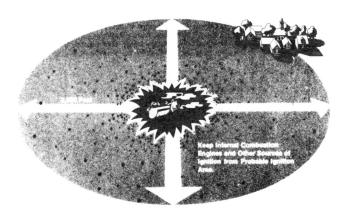

In Case of Explosion, the Minimum Safe Distance from Flying Fragments is 2,000 Feet In All Directions.

Fig. 20-5. Emergency Action Guide for controlling an incident involving a flammable liquid. Information on the two pages describes the potential hazards, actions to be taken, and the minimum safe distance in case of an explosion. (Courtesy of CHEMTREC.)

3. Identify the type of cargo involved. Determine from the plant supervisor, truck driver, or other knowledgeable person what the material is and how it reacts. Check for labels and placards.
4. Wear full protective clothing, including self-contained breathing apparatus.
5. Evacuate the unnecessary people from the vicinity if there is a toxic or explosive hazard.
6. Protect the structural exposures and confine the fire if the material or vehicle is burning.
7. Remove the victims upwind to obtain fresh air immediately. Treat the victims if an accurate diagnosis can be obtained.

If an accurate diagnosis is not available, personnel can obtain necessary information on disposal of the material and specific medical treatment by contacting CHEMTREC, The Chemical Transportation Emergency Center in Washington, D.C. Their toll free number, 800-424-9300, is available for emergency calls 24 hours a day.

CHEMTREC maintains detailed files on chemical products, including trade names and personnel contacts at manufacturers and

RESCUE FROM HAZARDOUS MATERIALS

shippers. The CHEMTREC representative on duty can readily find the necessary data to provide immediate information on the nature of the product and the appropriate way to handle it. Before calling CHEMTREC, personnel should gather the following information: the name of the shipping company, the location of the incident, the products being shipped, the destination of the vehicle, and the nature of the incident.

GASES

Gases are classified according to their physical and chemical properties and their effect on human functioning. When considering the relative hazards of gases, parts per million of the gas in the normal atmosphere is usually given. To give some idea of what a light concentration of gas this amounts to, one part per million (ppm) is roughly equivalent to 1 ounce (28 grams) of vermouth dissolved in 8,000 gallons (30,400 liters) of gin.

Irritant Gases

Gases that injure the air passages or lungs, inflaming the surfaces of the respiratory tract, are called irritant gases. They are divided into three groups.
1. Gases highly soluble in water that affect the upper respiratory tract. These include ammonia and hydrogen chloride.
2. Gases that affect the middle and upper respiratory tract. These are more serious as their action is deeper and harder to treat. Examples are chlorine and sulfur dioxide.
3. Gases that primarily affect the lungs, causing them to fill with fluid (edema). Phosgene and nitrous fumes are examples.

Asphyxiants

Asphyxiants can generally be divided into two classes:
1. The simple asphyxiants such as carbon dioxide, nitrogen, helium, hydrogen, and methane. These are nontoxic and act mechanically by excluding oxygen.
2. Chemical asphyxiants. These are divided according to their site of action:
    a. Asphyxiants that combine with the hemoglobin in the blood and prevent it from carrying oxygen from the lungs to the body tissues. Carbon monoxide is an example.
    b. Asphyxiants that act on the body tissues and prevent the utilization of the oxygen brought to them by the blood. The cyanide derivatives act in this manner.

Anesthetic Gases and Vapors

Anesthetic gases and vapors, which produce a loss of sensation and are generally heavier than air, are classified as follows:
1. Petroleum vapors such as gasoline, kerosene, butane, propane, etc., usually do not cause serious aftereffects in concentrations too low to cause unconsciousness. Others in this class are acetylene, ethyl ketone, and acetone.
2. Carbon tetrachloride, trichlorethylene, toluene, methyl alcohol, and carbon disulfide injure the liver, kidneys, brain, and other internal organs after long exposure.

HEAT

The physiological effects of exposure to heated air are many. Dehydration, heat exhaustion, blockage of the respiratory tract, and burns are the immediate results. Body heat is regulated by a heat-controlling center in the brain that operates by stimulating perspiration through the sweat glands, the evaporation cooling the body. Rescue personnel should not enter atmosphere exceeding 120 to 130°F (49 to 54°C) without special protective clothing and masks; even trained fire fighters would not be able to inhale more than one or two breaths of moisture saturated air at these temperatures without serious consequences. The cooling effect of skin moisture evaporation may counteract the skin effect of heat in dry air.

Skin surface temperatures may increase in heated air until blisters or actual charring occur. Studies have shown that second degree burns and unbearable pain may occur in 20 seconds at a temperature of 131°F (54°C) and in one second at 158°F (70°C). However, before the human skin can absorb sufficient heat to raise the skin surface temperature, other heat-absorbing properties of the body will react. These include the heat absorbed in the evaporation of perspiration and the heat convected away from the skin by the blood circulation (Figure 20-6).

Because of the low specific heat properties of heated air, respiratory injuries usually do not extend very far down the victim's airway; therefore, tissue damage is restricted to the upper passages. The one exception to this may be steam burns because steam has a greater ability to carry heat. Fire fighters can be seriously injured if they inhale the steam that is created when they apply a stream of water into a superheated structural atmosphere.

Fig. 20-6. Shield is used to protect fire fighters from the devastating heat of a blazing oil tank. (Courtesy of Los Angeles City Fire Department.)

SMOKE

Many fire fighters and other emergency personnel have the mistaken belief that if they can breathe and the air is not uncomfortable, the atmosphere must be tenable. This is not necessarily true as some of the most lethal by-products of combustion are not discernible by odor, taste, or color. Emergency service personnel should adopt the policy of using self-contained breathing apparatus at every structure fire and whenever there is a gas or chemical leak or spill where there is the slightest chance that toxic gases, vapors, or dusts could be encountered. Continued emphasis on the use of breathing apparatus must be exercised, because more fire fighters are placed out of action by smoke inhalation than by any other single injury (Figure 20-7).

Composition of Fire Gases

The composition of the fire gases will vary, depending on the materials burning. Since the variety of materials that can be involved in a fire is infinite, there is no limit to the combination of gases that can kill. But what complicates things even more is that the combustion products from a given fuel will vary, depending on the level of oxygen and the temperatures involved.

There is also a synergistic effect. Two gases in combination can be more deadly than either singly. Put several gases together and a more serious problem is compounded. Three components of combustion gases are universally present and represent the majority of the dangers: (1) carbon monoxide which dulls the senses and kills, (2) heat, and (3) suspended particulate matter that reduces vision, irritates the eyes, damages respiratory passages, and causes panic.

In practically all fire disasters where there is a large loss of life, the primary cause of death is the inhalation of heated, toxic, and oxygen deficient atmospheres. The composition of gases produced by combustion depends on the:

1. Chemical composition of the burning material.

Fig. 20-7. Fire fighters open roof with axes to ventilate the heat and smoke from a structure fire. (Courtesy of the New York City Fire Department.)

2. Rate of heating.
3. Oxygen content of the atmosphere at or near the burning surface.
4. Temperature of the evolved gases.

Many toxic gases are released by the combustion of common substances. Carbon monoxide (CO) and carbon dioxide (CO_2) are formed whenever any substance containing carbon molecules is burned. The following irritating and/or toxic gases are also released by the materials listed:

Wood, cotton, and paper: formaldehyde, formic acid, carbolic acid, methyl alcohol, acetic acid, methane, and acetaldehyde.

Petroleum products: acrolein, in addition to those generated by the burning of wood.

Wool and silk: sulfur dioxide, hydrogen sulfide, hydrogen cyanide, and ammonia.

Plastics: hydrogen chloride, aldehydes, ammonia, cyanide, nitrogen oxides, phosgene, and chlorine.

Rubber: sulfur dioxide.

Knowledge of the products of combustion from common materials under different burning conditions will help in understanding the effect of smoke on health. In all fires, smoke is produced with varying amounts of dusts, fibers, fumes, mists, vapors, and gases. Smoke is made up of the volatized products of combustion. It is the small particles of carbon and the tarry specks that make the gases visible. The fire gases rendered visible by these particles are what is generally defined as smoke; in some instances, condensed steam contributes to the visibility of fire gases.

Toxic Residue After Fire Is Out

The overhaul and cleanup stage of fire fighting tends to be the most hazardous because the crew often believes that the dangers are over when the flames are extinguished. They remove their breathing apparatus to be more comfortable; they are unaware that heated and wet materials continue to generate toxic gases and vapors long after the blaze has been extinguished. A wise officer will allow the crew an opportunity to withdraw from the building into the cool clean air after the fire has been knocked down. This will serve two purposes—the crew can rest and the structure will have a chance to become ventilated.

SMOKE INHALATION

Smoke inhalation is actually three different problems; hypoxia, systemic poisons, and local irritants that attack the bronchopulmonary tree. While any one of these three could be disabling or fatal, the combined effects of two or more is synergistic; that is, the combined effects of two or more of the inhalants is more toxic than the total effect would be if each were inhaled separately.

Hypoxia

Hypoxia is any condition in which there is an inadequate supply of oxygen for the body tissues. It occurs in fires when the slow burning of materials reduces the oxygen content of a closed room to dangerously low levels. Open flame usually vanishes as the oxygen concentration drops, but smoldering with the release of heat and smoke may continue.

RESCUE FROM HAZARDOUS MATERIALS

When the oxygen content of the atmosphere falls from its usual level of about 21 to 15 percent, muscular skill is diminished; when it drops to between 14 and 10 percent, a person may still be conscious but has faulty judgment. This is not obvious to the individual, who becomes fatigued very quickly. A level between 10 and 6 percent will cause collapse, but the person can be revived by fresh air and oxygen. Increased oxygen demands during periods of exertion may result in oxygen deficiency symptoms at much higher percentages.

Death from hypoxia occurs because of a lack of oxygen to vital organs, and the subsequent breakdown in cellular metabolic functions. Even if hypoxia is reversed early, it may result in some permanent disability; this depends on the organs involved.

Hypoxia is a common type of injury among persons who have been overcome while asleep and when fire fighters without breathing apparatus linger in a fire area after the fire has been knocked down. As a rule, emergency personnel rarely are subjected to hypoxia injury because of the almost universal use of self-contained breathing apparatus. Air usually enters the structure with the fire fighters. Effective fire departments simultaneously launch ventilation activities with the fire attack, and well-trained fire fighters will leave the fire area after the blaze has been knocked down to allow the room time to air out and cool down.

Pulmonary Irritants

The most common irritants found in fire gases include: acetic acid, acrolein, acetaldehyde, formic acid, formaldehyde, ammonia, furfural, tar, and sulfur dioxide. They arise in various combinations in fires involving wood, paper, coal, petroleum products, textiles, animal fats, and oils, wool, silk, rubber, photographic film, celluloid; plastics also produce a great amount of smoke and toxic gases.

Irritant gases and vapors generally give sufficient warning of their presence. Under some conditions, irritant materials may condense on small smoke particles and be inhaled deep into the lungs. These gases irritate the mucous membranes of the eyes and respiratory tract, producing pain and discomfort. They are, however, of secondary importance in causing quick deaths. Usually the earliest symptoms of pulmonary irritants are a burning sensation beneath the breastbone and a cough without phlegm. Often there is a metallic taste in the mouth. The burning can cause serious concern and may mimic a heart attack, blood clot in the lung, or rupture of the aorta. Usually after the acute symptoms, there may be a silent period with no discomfort, followed by shortness of breath, cough, chest discomfort and painful breathing, asthmatic wheezing, reduced breath sounds, rapid breathing, rapid heartbeat, and increased bluishness (cyanosis) around the mouth and mucous membranes.

Pulmonary injury to the body may not be immediately apparent; damage to the trachea, bronchus, and lung may not be evident until about three hours after exposure. Some common side effects of the pulmonary irritants are bronchitis from chemical burns, acute emphysema from the destruction of lung tissues, areas where hemorrhage occurs in the lungs, and infiltration of inflammatory cells into the alveolar spaces of the lungs. Secondary infections such as bronchopneumonia sometimes follows. If the exposure is excessive with severe damage to the lungs, permanent loss of pulmonary function may result. As healing occurs, a proliferation of scar tissue within the lungs may lead to permanent disability.

Managing Hazardous Gases, Liquids, and Chemicals

As an example of the hazardous properties of fire gases, one of the most common irritants produced by combustion is sulfur dioxide. Not all of the pulmonary irritants have identical effects at the same concentrations as sulfur dioxide, but in general, similar symptoms will appear.

Sulfur dioxide will usually cause nasal irritation at concentrations of 15 parts per million (ppm). Above 28 ppm, bronchial constriction and increasing resistance to breathing begins to appear. Once the concentration is greater than 40 ppm, the victim suffers constriction of the bronchial and pulmonary veins. By the time it has reached 400 ppm, fluid may accumulate in the lungs, causing pulmonary congestion. Internal bleeding in the lungs may occur at levels between 600 and 1,000 parts per million (ppm).

Systemic Poisons

There are basically three common poisonous gases that disable and kill smoke inhalation victims: carbon monoxide, hydrogen sulfide, and hydrogen cyanide. By far the most significant fire fighting problem is carbon monoxide, because more fire deaths occur from carbon monoxide than from any other single poison. Carbon monoxide is generated in every fire, but the less complete the combustion, the greater the production of carbon monoxide. As a general rule, the darker the smoke, the higher the carbon monoxide level, because black smoke is high in particulate carbon as a result of incomplete combustion.

Carbon Monoxide

Carbon monoxide is the most important of all the gases encountered in fire fighting and ranks first among the causes of fire deaths. It is colorless and tasteless; the gas has a slight garliclike odor that is hardly noticeable. The density is 0.967, as compared with air, and is, therefore, slightly lighter than air; when it is heated, it is considerably lighter than air and will rise rapidly. Unlike most common poisonous gases, carbon monoxide can almost never be detected by odor, and never by taste or irritation of the respiratory tract.

This gas is formed by the incomplete combustion of materials containing carbon. The amount of carbon monoxide formed depends on several variables:
1. The amount and type of materials involved in the combustion.
2. The amount of oxygen that is available.
3. The temperature of the fire.

Complete combustion of carbonaceous materials gives off carbon dioxide gas (CO_2). As the oxygen content of the atmosphere decreases there is less oxygen available for the oxidation (burning) of the material, thus producing a larger amount of carbon monoxide (CO). The explosive range of the gas is from 12.5 to 74 percent, which makes it extremely dangerous.

Hemoglobin has an affinity for carbon monoxide 250 to 300 times as great as its affinity for oxygen. Thus carbon monoxide causes asphyxiation by combining with the hemoglobin of the blood and preventing it from transporting its oxygen from the lungs to the tissues. Relatively small amounts of carbon monoxide, even when present in an atmosphere that is adequate in oxygen, may produce effects which are dangerous to life in a short time. An undetectable concentration of carbon monoxide in the atmosphere could be rapidly fatal. A 50 percent carbon monoxide level in the blood can occur within 30 to 90 seconds after exposure to a 5 percent carbon monoxide atmosphere. Two or three good breaths of this concentration may kill a victim outright. These concentrations are easily reached in average residential fires. A one percent room air concentration of carbon monoxide will cause a 50 percent blood level in 2 1/2 to 7 minutes.

Symptoms. The symptoms of carbon monoxide poisoning make their appearance when the blood is about 25 percent saturated, unconsciousness at about 40 to 50 percent, and death when the saturation reaches 65 percent. A concentration of 0.64 percent of carbon monoxide in the atmosphere will give a blood saturation of 80 percent in one or two minutes; this causes a headache, dizziness, weakness, unconsciousness, and death. A person who is exposed to 1.28 percent concentration of carbon monoxide in air will become unconscious after two or three breaths and probably will die in one to three minutes.

The victim may resemble a case of acute alcoholism in which judgment is lost. A cherry-red discoloration of the skin *may* be noted, though this is far more often described than actually observed. Many variables, such as physical condition, amount of exertion, and presence of other toxic gases, affect the amount of carbon monoxide that can be tolerated without causing death.

Treatment. Ventilation of the lungs with 100 percent oxygen will permit the blood gradually to overcome the carboxyhemoglobin state (the product of the chemical union of hemoglobin and carbon monoxide) and to return to a normal oxyhemoglobin (oxygen and hemoglobin) state.

Some medical authorities advocate the use of an oxygen-carbon dioxide mixture (carbogen) with a resuscitator or inhalator. The purpose of this is to stimulate deeper spontaneous ventilation by elevating above normal the carbon dioxide pressure in the blood. An increase in the depth of ventilation will help to reoxygenate the blood more rapidly. However, ventilation can be easily assisted or controlled with 100% oxygen, without added carbon dioxide.

The patient should be kept warm and treated for shock. Prolonged exposure to carbon monoxide may be followed by pneumonia.

Hydrogen Sulfide

Hydrogen sulfide is a significant contributor to death under certain circumstances. It is a flammable, highly toxic gas that is four times more toxic than carbon monoxide. It has a characteristic rotten egg odor and is slightly heavier than air. Its explosive range is 4.3 to 45 percent.

At 0.2 percent concentration of the gas, the sense of smell is paralyzed and the characteristic odor is not distinguishable. In large concentrations it will paralyze the respiratory center of the central nervous system, resulting in suffocation. Since concentrations of 0.1 percent in air are presumed to be rapidly fatal, it is vitally important to take protective action at the first indication of the odor of hydrogen sulfide. Frequently, persons on the outside of a building may recognize the hazard before those working inside, because of the lower concentration of gas outside.

Hydrogen sulfide is found around oil refineries, dye works, tanneries, chemical laboratories, and sewage disposal plants; it may be produced by the combustion of hair, hides, wool, silk, or rubber.

Symptoms. Hydrogen sulfide acts both as an asphyxiant and as an irritant. Low concentrations are indicated by headache, weariness, and dizziness. Higher concentrations produce pains in the chest, depression, delirium, convulsions, coma, and death. Respiration often fails before the heart action is stopped.

Treatment. Treat for shock and administer oxygen.

Hydrogen Cyanide

Hydrogen cyanide is the gas that fumes from hydrocyanic acid and is extremely poisonous. It is commonly found in the by-products of combustion. There is a synergistic effect when hydrogen cyanide is mixed with carbon monoxide as the hydrogen cyanide increases a person's respiratory rate and causes more carbon monoxide to be inhaled.

It is formed by the action of water and acid on cyanides, and may be the product of incomplete combustion of certain nitrogen containing materials such as wool, silk, and some plastics. Hydrogen cyanide is a vermin fumigant and therefore presents a serious life hazard where buildings are being fumigated. Exposure to 0.01 percent even for a short time is said to be very dangerous. The characteristic odor is similar to bitter almonds; it sometimes warns of the presence of hydrogen cyanide, but this characteristic is not reliable since it may be masked by other odors or the sense of smell may be quickly impaired. Being lighter than air, it will rapidly diffuse upwards and is very soluble in water.

The action of this gas is very rapid; it is the lethal gas used in death chambers for capital punishment. The gas may be absorbed rapidly through unbroken skin, so no type of breathing equipment offers full protection. Hydrogen cyanide acts on the body tissues, preventing the utilization of the oxygen brought to them by the blood hemoglobin. The result is a rapid paralysis of the respiratory center.

Symptoms. Small amounts may cause dizziness, slow and shallow breathing, palpitation of the heart, nausea, vomiting, constriction of the throat, foaming of the mouth, convulsions, and unconsciousness. Sublethal symptoms are increased heart rate to over 100 beats per minute, gasping respirations, and muscle spasms. The skin may be rosy red in color as in carbon monoxide poisoning, and may produce an itching rash over the entire body.

Treatment. Treat for shock. Remove to a hospital immediately and give oxygen en route. If possible, notify the hospital so that the proper antidote can be ready and waiting.

Carbon Dioxide

Carbon dioxide is produced by the complete combustion of carbonaceous materials. It is odorless and colorless, nonflammable, and nontoxic. It is one and one-half times as heavy as air, so it may be encountered in low places such as basements, wells, and mines.

The principal danger of carbon dioxide is that it acts as an anesthetic in high concentrations. Concentrations of 2 to 5 percent stimulate the breathing, making it uncomfortable and laborious. A 10 percent concentration, even with sufficient oxygen present, may result in unconsciousness and death.

Symptoms. Symptoms may include headache, dizziness, shortness of breath, weakness, and loss of consciousness. Cyanosis will be evidenced by a blue or dusky gray coloring, particularly noticeable in the lips, cheeks, and ears.

RESCUE FROM HAZARDOUS MATERIALS

Treatment. Ventilation of the lungs will rapidly remove the carbon dioxide from the blood. This should be done under controlled conditions permitting a reduction in the carbon dioxide pressure in steadily increasing steps, since rapid, unmonitored, uncontrolled ventilation can result in serious derangements of the acid-base balance of the cerebrospinal fluid.

Acrolein

Acrolein (acrylic aldehyde) is an irritating and toxic gas produced during combustion of petroleum products, fats, oils, and many other common materials. Concentrations in the range of 0.0153 to 0.024 percent are presumed to be dangerous to life during an exposure period of 30 to 60 minutes.

Nitrous Fumes

Nitrous gas fumes were responsible for a large number of deaths in the Cleveland Clinic Disaster in 1929 when fire consumed a large amount of cellulose nitrate X-ray film and the toxic smoke and vapors spread throughout the building. Even though nonhazardous cellulose acetate safety film is now used, there is still a large amount of old cellulose nitrate film stored in film libraries and medical record sections.

The color of the fumes may vary from colorless to deep red or reddish-brown, depending on the mixture of the various oxides of nitrogen. The fumes are heavier than air and are extremely poisonous. The fumes are given off when nitric acid is exposed to air and when materials such as lacquers, pyroxylin articles, explosives, rayon, and other nitrogen compounds are consumed by fire. Nitrogen dioxide is one of the most dangerous gases; it is widely used in many industrial processes.

Symptoms. Symptoms may not appear until several hours after exposure, since it is possible to inhale a fatal amount without a great deal of discomfort. Headache, dizziness, nausea, and vomiting may occur. Death is usually caused by pulmonary edema (accumulation of fluid in the lungs).

Treatment. The patient must be kept quiet and should be treated for shock. Persons exposed to nitrous fumes should be kept under close medical supervision and watched for latent effects for at least 24 hours, even though they show no symptoms and feel well. Inhalation of 100 percent oxygen is recommended.

COMMONLY ENCOUNTERED HAZARDS

Gases from Burning Plastics

When burned, some plastics emit some particularly deadly gases and vapors. A few years ago fire fighters were called to extinguish a small fire that involved some plastic articles; no one bothered to wear breathing apparatus for such a minor blaze. Immediately after the fire, one of the crew who was the most involved with the fire fighting became ill and suffered a tightness in the chest, a burning sensation that seemed to close off the throat, a headache in the front of the head, shortness of breath, dizziness, and nausea. At work the next morning, the fire fighter had a seizure resembling epilepsy, but regained consciousness after the seizure. An hour later the victim became unconscious, was taken to the hospital, and died.

Others of the crew experienced some of the same symptoms, but they attributed them to the normal breathing of irritating smoke and hard work. No one considered the symptoms serious enough to warrant seeking medical attention.

The autopsy of the victim who died revealed a severe accumulation of fluid in the lungs (pulmonary edema) and hemorrhage due to inflammation of the lungs (chemical pneumonitis) caused by exposure to chemical smoke and fire. There was an incidental finding of hardening of the coronary arteries supplying the heart.

Polyvinyl chloride (PVC) plastic is commonly used as insulation for electrical wiring and many other industrial uses. When heated, PVC produces hydrogen chloride and several other hazardous chemicals. When inhaled as a gas or combined with water vapor, hydrogen chloride acts as an irritant to the lining of the eyes and the breathing surfaces of the body. A small concentration causes local irritation to the throat after a short exposure; more severe exposure results in pulmonary edema, as well as in spasms of the voice box. High concentrations are dangerous even for a very brief exposure.

When gases such as hydrogen chloride enter the lungs, they react chemically with water to produce strong acids or alkalies. The violent inflammatory reaction which results can damage the lungs. The typical effect of hydrogen chloride on the breathing tube (trachea) is destruction of the superficial lining. Electrocardiograms performed after exposure are apt to reveal premature heartbeats or other heart irregularities.

Petroleum Vapors

The flammable hazards of petroleum products is well known, but other dangers are largely ignored. Petroleum vapors have an anesthetic effect on the human body that is similar to that caused by chloroform, ether, and alcohol. Concentrations far below the explosive range may rapidly be lethal if inhaled. These vapors are very dangerous and the line between unconsciousness and death is very narrow.

Symptoms. Symptoms may include burning of the eyes, headache, dizziness, restlessness, and intoxication with confusion, disorders of speech, hearing, and sight resulting. Exposure to petroleum vapors sometimes terminates in convulsions, coma, or death.

Treatment. Treat for shock and administer oxygen.

Carbon Tetrachloride

Carbon tetrachloride is a clear, colorless liquid with a peculiar odor. The vapor density, as compared with air, is 5.32; this makes it considerably heavier than air. It is nonflammable and has been widely used as a fire extinguishing agent, although its toxicity has made it illegal for this use in many states. It is also used as a cleaning agent and as a fumigant.

In the presence of excess moisture or high temperatures, such as occur in open flames, it decomposes into hydrochloric acid, phosgene gas, and other toxic gases. Carbon tetrachloride is potentially toxic by inhalation of its fumes, by contact with the skin or mucous membranes, or by ingestion. Poisoning may result from a single, brief exposure to a high concentration of vapor or from a prolonged, excessive, or repeated exposure to a low concentration. The vapors are toxic in concentrations of 50 parts per million (ppm) for continued exposure. No one should enter a confined area where carbon tetrachloride has been used on a fire until it has been thoroughly ventilated.

Symptoms. Many cases of carbon tetrachloride poisoning are undiagnosed, the condition usually being treated as nephritis, heart disease, or hepatitis. It acts first as an anesthetic, affecting the central nervous system and producing mental confusion, depression, headache, dizziness, fatigue, nausea, and vomiting. The secondary action affects the liver, kidneys, heart, skin, lungs, and nervous system. Pulmonary edema may occur.

Treatment. Treat for shock and administer oxygen. Do not allow patients to exert themselves.

Methyl Bromide

Methyl bromide is a chlorinated hydrocarbon, as is carbon tetrachloride; it is used as a fumigant and a fire extinguishing agent. Its extreme toxicity limits its use in fire extinguishers to fixed systems, such as the engine nacelles of aircraft, where there is no life hazard. It is odorless except in high concentrations. When used as a fumigant, small amounts of a warning gas (chloropicrin) are often added. Dangerous amounts may be absorbed through the skin, even though breathing apparatus is utilized.

Symptoms. Generally the symptoms are similar to those evidenced in carbon tetrachloride poisoning, though there is often a delayed reaction after exposure.

Treatment. Treatment is the same as for carbon tetrachloride poisoning. Persons who have been exposed, but show no symptoms, should be carefully watched for latent effects for at least 24 hours.

Cyanides

Cyanides are found in electroplating plants, heat treating areas, fumigation, and other industrial applications. The cyanides are odorless when dry; when acting upon by acids or moisture, some give off hydrogen cyanide gas. The cyanides are extremely poisonous and should not be handled with the bare hands or taken internally.

Symptoms and treatment are the same as for hydrogen cyanide.

Selenium

Selenium is a nonmetallic element related to sulfur and it resembles sulfur in many of its properties. It is steel gray in color. In fires involving considerable amounts of selenium, the formation of selenium dioxide is recognized by the reddish-brown or rusty-red smoke produced. The smoke is extremely pungent and intolerable, producing eye and nasal irritation. It has a definite garliclike odor resembling carbon disulfide. Selenium compounds should not be handled because toxic amounts may be transferred to the mouth or absorbed through a skin abrasion or injury.

Selenium is used in various photoelectric devices, paints and pigments, radio and TV rectifiers, and other industrial applications. Selenium rectifiers are used extensively for the conversion of alternating current to direct current.

Symptoms and treatment are similar to those for sulfur dioxide poisoning.

Ammonia

Ammonia is a colorless, transparent gas with a penetrating, pungent odor. It is extremely irritating to the eyes, nose, throat, lungs, and moist parts of the body; for this reason persons normally will not remain in an atmosphere containing ammonia long enough to suffer serious consequences. Exposure to 0.5 to 1 percent ammonia in air for one-half hour is sufficient to cause death or serious injury.

The explosive range of ammonia is between 16 and 25 percent by volume in air, with an ignition temperature of 1,204°F (651°C). It is 0.598 times as heavy as air. Ammonia is widely used as a refrigerant and is also formed when matter containing nitrogen burns. It is very soluble in water, which will absorb about 900 times its own volume of vapor at ordinary temperatures. This makes it easy to disperse ammonia with a fog water spray.

Symptoms. Irritation of the eyes, respiratory tract, and moist parts of the body under the arms, between legs, and back of neck are typical symptoms, accompanied by convulsive coughing and difficult breathing caused by a spasm of the throat.

Treatment. Give oxygen. Wash the affected parts, including the eyes, with cool running water. Treat for shock.

Sulfur Dioxide

Sulfur dioxide is commonly used as a refrigerant in small refrigeration units. It is colorless, suffocating, and extremely irritating. It is inconceivable that anyone would remain voluntarily in a dangerous concentration of the gas. Sulfur dioxide is a little more than twice as heavy as air and is formed by burning sulfur; it is also used as a bleaching agent and as a fumigant.

Symptoms and treatment are the same as for ammonia.

Chlorine

Chlorine is classified by the Department of Transportation as a nonflammable compressed gas; in containers it has both a liquid and a gaseous phase. It is used for treating sewage, bleaching, and for water purification. It is commonly encountered in industrial plants. It was one of the World War I antipersonnel gases.

Chlorine is one of the chemical elements. Neither the gas nor the liquid is explosive or flammable; both react chemically with many substances. Chlorine is only slightly soluble in water. The gas has a characteristically pungent, irritating odor and is greenish-yellow in color. It is about two and one-half times as heavy as air and is very irritating to the lungs.

As the gas is heavier than air, when it escapes it will seek the lowest level in the building or area in which the leak occurs. Although dry chlorine does not react with (corrode) many metals, it is very reactive (strongly corrosive) when moisture is present.

As soon as there is any indication of the presence of chlorine in the air, immediate steps should be taken to remedy the condition. Chlorine leaks always get worse if they are not promptly corrected. If the leak is extensive, all persons in the path of the fumes must be warned to leave the area. Keep upwind of the leak and above it.

Never use water on a chlorine leak. Chlorine is only slightly soluble in water, but if water is applied to a leaking container, moisture will increase the corrosive action. This will enlarge the hole and escalate the release of the chemical.

Symptoms. Chlorine gas will irritate the mucous membranes, the respiratory system, and the skin. Large amounts cause irritation of eyes, coughing, and labored breathing. Exposure to high concentrations result in retching and vomiting, followed by difficult breathing. Death can result from suffocation.

Treatment. Treatment is symptomatic; there is no known antidote for chlorine gas inhalation. Administer oxygen and wash the affected parts of the body. If even minute quantities of liquid chlorine enter the eyes, or if the eyes have been exposed to strong concentrations of the gas, they should be flushed immediately with copious quantities of running water.

Chlorine Bleach

Unfortunately, few people realize that mixing commonly encountered substances may release irritating or toxic gases and fumes. Many housewives have been killed or made severely ill while using an ordinary toilet bowl cleaner. Not satisfied with the way it was removing stains, they would add some household bleach or other substance and stir it with a brush.

When the widely used household chlorine bleach, which is a sodium hypochlorite solution, is combined with an acid or acid producing substance, such as a toilet bowl cleaner or vinegar, there is a sudden release of chlorine gas. Likewise, when a chlorine bleach is mixed with ammonia, lye, or other alkaline substance, the action will liberate a highly irritating gas.

People should not make the mistake of thinking that because certain household products are good and useful, the combination of two or more of them will do a better job than one alone.

Calcium Hypochlorite

While only moderately toxic to health, calcium hypochlorite is a very strong irritant. Its primary hazard is chemical burns to the eyes, skin, and mucous membranes. Upon decomposition, fumes of chlorine are emitted which are capable of causing irritations to the respiratory tract. This chemical is not combustible and is stable under ordinary conditions. It is, however, a very reactive and powerful oxidizer, particularly when moistened. Danger is present when a small quantity of water is poured onto the chemical, and most particularly if the chemical is premixed with any organic substances. Fires in swimming pool supply stores have caused many casualties among fire fighters when this material became moistened.

Calcium hypochlorite is employed for the clarification of water and controlling the growth of organisms such as algae. It is found in swimming pool supply houses in large quantities.

PESTICIDES

All pesticide chemicals should be handled with care. Not all pesticides and insecticides are poisonous; many are not required to be labeled with the word *Poison* and to display the skull and crossbones symbol. Extreme care is required when handling poison pesticides because of the potentially hazardous nature of these products. While some pesticides may not be as dangerous as others, some products can cause serious illness or death if allowed to remain on the skin for a short period of time or inhaled into the lungs.

A small amount of some poison pesticides taken into the body can cause serious illness or death if harmful amounts are absorbed through the skin, inhaled while breathing, or swallowed while eating or drinking.

Pesticides are applied to crops by aircraft, spray rigs, and small hand sprayers. The concentrates are usually found in chemical warehouses and agricultural insecticidal spray companies throughout the world. In agricultural areas they may be encountered while being applied or while being stored in bulk quantity in the field.

Emergency personnel can receive lethal pesticide poisoning by coming into contact with the chemical itself, by being exposed to the gases and vapors released by a fire, or by wading in contaminated water runoff while combating a fire.

The widespread use of organophosphates in agriculture is now on the increase; this is a group of very toxic insecticides that are closely related chemically to the wartime nerve gases. Various chemicals, such as parathion, malathion, chlorthion, diazinon, and disulfoton belong to this group.

Vaporization greatly increases the potential danger of these chemicals because of more rapid access to the nervous system following either inhalation or absorption through the skin and mucous membranes. As is typical of organophosphates, these chemicals exert their lethal effects by inhibiting a vital enzyme, cholinesterase, found in nerve tissue. While this enzyme is tied up by the pesticide, much of the nervous system can no longer function properly and poisoning thus occurs.

These insecticides are only slightly irritating and give little immediate warning. They impair the nervous system of the body and poisoning may result from one heavy exposure or from frequent small exposures because the effects are cumulative. Contact with the bare skin must be avoided. Self-contained breathing apparatus should be worn, together with full protective clothing, whenever any of the organophosphates are encountered at a fire or other emergency incident. High or sustained heat, 200 to 240°F (93 to 116°C), can break down the product to form a highly toxic gas.

Symptoms of Pesticide Poisoning

Pesticide poisoning symptoms will vary a great deal; they resemble those of heat prostration, smoke inhalation, and a variety of other ills. Generally speaking, signs or symptoms of any unusual nature and discomforts of any kind by those handling the products or combating a fire may indicate possible poisoning.

Symptoms of poisoning include headache, dizziness, nausea, diarrhea, increased salivation, flow of tears, and sweating. The organophosphates cause the pupils of the eyes to become smaller (constricted) and vision is impaired. The muscles of the eyelids and tongue may twitch, followed by muscle spasms over the entire body. Unusual amounts of mucus may result in difficult breathing. Coma is noted in cases of severe poisoning, and death is primarily due to respiratory failure and cardiac arrest.

Treatment of Pesticide Poisoning

The product labels of all poisonous substances contain first aid instructions that clearly outline the procedure to be followed in case of contact or exposure. The statements on these labels are specific for the products on which they are used; *read the label.*

In organophosphate poisoning, treatment must be started immediately; obtaining medical help is imperative. Treatment must be rapid, and is presently accomplished with large doses of atropine, sometimes in conjunction with another drug known as pralidoxime. If poisoning is not too severe, prompt treatment will usually assure

RESCUE FROM HAZARDOUS MATERIALS

recovery, but acute intoxication with pesticides of this type can also be fatal. Other treatment includes:

1. Eyes: Contact with the eyes may cause temporary blindness. Flush with water immediately and continue for at least 15 minutes.
2. Skin: Immediately remove contaminated clothing and scrub skin with soap and water for 15 minutes.
3. Internal: If swallowed, make patients vomit by giving an emetic such as Syrup of Ipecac.
4. Inhalation: Remove victim from the contaminated atmosphere and keep the patient under continuous observation. Obtain medical aid immediately. Be prepared to administer artificial respiration.

Fires in Pesticide Storage

Fires in warehouses and other storage areas containing poison pesticides can be very dangerous because they add the possibility of poisoning to the usual fire fighting hazards. If the fire fighting procedure is not handled properly, the water or chemicals used to combat the blaze could very easily spread contamination over a wide area. Sometimes, even the heat of the fire or air currents created by the fire vaporize certain insecticides and cause poison particles to become airborne. Consequently, fire fighters and other emergency personnel engaged in the fire fighting operation must avoid breathing fumes and smoke resulting from the fire in a poison pesticide area and must prevent skin contact with water or debris in or from these areas. During a pesticide fire, action must be taken to protect fire fighters, safeguard the public, fight the fire, and neutralize toxic waste.

When fire fighters and other emergency personnel fight a chemical fire, the emergency can be compounded by inappropriate action. It is usually better to allow pesticides already involved in a fire to burn if extinguishment will not save the building. Heat detoxifies the chemicals. They will probably not be usable anyway, and the fire reduces the amount of contaminated waste to be disposed of later. In general, large amounts of water should not be used on a pesticide fire. It merely spreads contamination. Fire fighters should concentrate their efforts on protecting the population and surrounding area.

General guidelines to follow when combating a fire that involves agricultural and garden chemicals are given in Table 20-1.

Whenever possible, avoid working in areas downwind of the fire. If there is a residential section immediately downwind of the fire, evacuate the residents until the fire is brought under control.

Avoid heavy fire streams, if possible, since the force of the stream of water spreads contamination and causes dusts to become airborne. Airborne dusts may present an explosion danger as well as a toxic hazard.

Containers may rupture violently if they become overheated; therefore, keep a safe distance from a fire in which these are present. If drums are not leaking, they may be cooled with a water spray to prevent overheating. If possible, dike the runoff water from a fire to prevent it from entering sewers or streams.

Cleanup

Response to any incident where a pesticide or insecticide may have been encountered requires thorough cleanup afterwards. Clothing worn at the time of the incident should be regarded as contaminated and not be worn again until it has been laundered. If clothing has been badly contaminated, it should be destroyed by burial or burning. Personnel should clean themselves by thoroughly washing with soap and water as soon as possible.

Contaminated tools and vehicles should be washed with a solution of one quart of sodium hypochlorite and one cup of detergent in 2 gallons (7.6 liters) of water, or a 5 percent sodium carbonate (soda ash) solution plus detergent, or a 5 percent trisodium phosphate solution plus detergent. Everything must be scrubbed thoroughly and rinsed with clean water.

Table 20-1.

FIRE FIGHTING TACTICS	
FOR FIRE DEPARTMENTS	
Contact facility operator	Determine type, quantity and hazards of products. Determine if fire should be fought after weighing fire fighting & postfire hazards vs. possible salvage.
Notify physician to stand by	Physicians may obtain poison control information by contacting Chevron Chemical Co. Many other manufacturers will provide similar information.
Contact Chevron Chemical Company	Maintain liaison for specialized information, particularly during a large fire. Many other manufacturers will provide similar information.
Evacuate downwind & isolate area	Patrol area to keep out spectators.
Wear personal protective equipment	Wear rubber or neoprene gloves, boots, turn-outs & hat. If contact cannot be avoided (such as entering an unventilated building for rescue) also wear self-contained breathing apparatus (Air Paks).
Attack fire from upwind & from a safe distance	Bottles, drums, metal & aerosol cans are not vented and may explode.
Contain fire and protect surroundings	Prevent spread of fire by cooling nearby containers to prevent rupture (move vehicles & rail cars if possible). Burning chemicals cannot be salvaged.
Use as little water as possible & contain run-off	Contaminated run-off can be the most serious problem. Water spreads contamination over a wide area. Construct dikes to prevent flow to lakes, streams, sewers, etc. Cooling effect of water retards high-temperature decomposition of the chemicals to less toxic compounds.
Use water fog spray not straight stream	Fog spray is more effective for control. Avoid breaking bottles and bags: adds fuel and contamination. Straight streams spread fire and contamination.
Poisoning—Avoid product, smoke, mist and run-off	In case of contact or suspected poisoning, leave site immediately. Any feeling of discomfort or illness may be a symptom of poisoning. Symptoms may be delayed up to 12 hours. Chemicals may poison by ingestion, absorption through unbroken skin, or inhalation. Wash face and hands before eating, smoking or using toilet. Do not put fingers to mouth or rub eyes.

Source: Chevron Chemical Company

FUMIGATIONS

A rescue call to a building being fumigated should be carefully evaluated, taking into consideration the length of time the victim has been overcome and the toxicity of the gas. If the fumigant is very toxic, or if the victim has been exposed very long, death has undoubtedly already occurred. It would be foolhardy to jeopardize rescuers' lives to retrieve a corpse.

The first step would be to call the fumigation operators for assistance and advice. Where there is no possibility that the victim is still alive, personnel wearing self-contained masks should open the doors and windows from the outside and thoroughly ventilate before entering.

RIOT CONTROL AGENTS

Vomiting, tear, and psycho gases produce unpleasant symptoms, usually only for a short time, and do not cause death when properly used. They are employed to control riots, to force people out of buildings and caves, and for personal protection. They are often used for breathing apparatus and gas mask drills.

Vomiting Gases

Inhaling vomiting gases makes a person sick. The general symptoms are: a sense of fullness in the nose, a severe headache, intense burning in the throat, and tightness and pain in the chest. These are followed by uncontrollable coughing, violent sneezing, nausea, and finally, by vomiting.

The symptoms may be delayed several minutes. Persons inhaling vomiting gas before getting a protective mask on may still get sick later on. It would be natural, then, to think that the mask is leaking and to take it off. If this were done, however, the victim would be exposed to more gas and it might be disastrous. The protective mask should be worn in spite of coughing, sneezing, salivation, or nausea. The mask may be lifted from the face briefly, if necessary, to permit vomiting or to drain saliva from the facemask. Clear the mask each time it is adjusted to the face before resuming breathing. Carrying on duties as vigorously as possible will lessen and shorten the symptoms.

Tear Gases

Tear gases are not very toxic; this makes them useful in civil disorders to disperse crowds, to squelch prison riots, to drive a barricaded person with a gun out of a building. The vapor of tear gases produces a sharp, irritating pain in the eyes, resulting in an abundant flow of tears. There is usually no permanent damage to the eyes and the effects wear off quickly. For a short time the victims may not be able to see. A mask, put on before any tear gases get into the eyes, will give complete protection. Some of the new tear gases also cause a runny nose, severe chest pains, and, sometimes, vomiting. These effects also wear off quickly.

For protection, put on a mask and clear it. Force the eyes to stay open so tears form; when the vision clears, carry out assigned duties. When it is safe to remove the mask, blot away the tears but do not rub the eyes. If a liquid or solid agent has entered the eyes, which is a rare occurrence, force the eyes open and flush them with clean water.

Psycho Gases

Psycho gases produce mental symptoms and may also result in physical symptoms such as a staggering gait, dizziness, and blurred vision. Some of the gases cause fainting spells and some will cause a severe muscular weakness. The mental symptoms often resemble alcoholic drunkenness; persons act silly, giggle, or become angry and belligerent. Psycho agents sometimes cause hallucinations. These

Managing Hazardous Gases, Liquids, and Chemicals

gases may become widely used in civil disorders; they do not kill, but they will render a person ineffective.

By the time a victim of psycho gas poisoning realizes that something is wrong, there may be too much mental confusion to do anything about it. If many persons are affected, it may be necessary to confine them temporarily under guard to prevent accidents. While the administration of oxygen is generally helpful in instances of gas poisoning, little specific information is available concerning treatment for persons who have been exposed to gases designed to disrupt the mental processes.

QUESTIONS FOR REVIEW

1. Placards and labels to convey important warnings and other information have been adopted by the:
    a. Department of Commerce.
    b. Department of Transportation.
    c. Department of Labor.
    d. Department of Health, Welfare, and Education.

2. When practicable, victims who are being removed from a toxic cloud of gas or vapor should be evacuated:
    a. Upwind.
    b. Downwind.
    c. Upslope.
    d. Downslope.

3. Unless absolutely necessary, rescue personnel should not enter an environment where the temperature exceeds:
    a. 120°F (49°C).
    b. 110°F (43°C).
    c. 100°F (38°C).
    d. 90°F (32°C).

4. Which one of the following gases is the primary cause of fatalities at structure fires?
    a. Hydrogen cyanide.
    b. Hydrogen sulfide.
    c. Carbon dioxide.
    d. Carbon monoxide.

5. If a farm worker has been poisoned by being sprayed on the body with an organophosphate pesticide, which of the following will likely offer the most relief from the symptoms?
    a. A rapid and hot shower.
    b. Inhalation of 100 percent oxygen.
    c. Treatment for shock and administration of aspirin to relieve headache.
    d. Injection of a large dose of atropine.

CHAPTER 21

Handling Nuclear Radiation Incidents

Since World War II, the applications of radioactive materials for peaceful as well as national defense purposes have increased tremendously. With few exceptions, these applications in many different fields are a direct benefit to human society. The attainment of these increased beneficial effects, however, creates additional requirements for the shipment of radioactive substances to more places, over more lanes of transportation, and through more shipping terminals. Despite strict regulations designed to ensure the safe handling and shipping of such materials, accidents have occurred, and the prospects are for greater numbers in the future. There is a distinct possibility that emergency service organizations may be called to the scene of a radioactive materials incident at any time.

Many experts may be necessary to deal with radiation accidents. The initial emergency work of fire fighters, police officers, rescue squads, ambulance attendants, and other emergency service personnel at the accident site may be supplemented later by physicians and nurses when the exposed or injured victims are received at a medical facility. The vast majority of accidents involving radioactive materials occurs in facilities that use these materials daily. There will almost always be professional advice available. Request guidance and follow it.

This chapter is concerned with the peaceful radiation hazards. Although this brief treatment of nuclear radiation cannot take the place of a complete training course, it is designed to provide fundamental and reliable information that any emergency service personnel can use for guidance in taking immediate action in case of an accident. The number of users of radioactive materials has increased in recent years. Indications are that present expansion will continue so that radioactive materials will be even more prevalent in the future. In order to deal with these problems intelligently, a firm understanding of nuclear radiation and radiation protection is necessary.

RADIATION

We have always been exposed to some degree of radiation through cosmic rays and naturally occurring radioactive materials. With the development of atomic energy for peaceful uses, many people now come into contact with radioactive materials through their work. Radiation is a form of energy transmission. Anything that blocks any radiation or stops the waves of radiation absorbs the energy from the wave. The absorption of energy can cause much damage to living tissue. There are many forms of radiation; all can be harmful.

Radiation affects the body in different ways. Because radiation is a form of energy, it is capable of doing work. Formerly it was thought that very low doses of radiation would not hurt the body, but now it is known that ionizing radiation may be harmful even in small doses. Basically, these rays of energy are absorbed by cells and cause changes within the cells. Because life is dependent on billions of individual cells contained in the body, the destruction of a large number of these cells results in sickness and sometimes in death. This is called *radiation sickness.*

The degree of radiation sickness depends on the type of radiation and the part of the body exposed to it. Large doses of radiation to an arm may cause the loss of the arm but have only a limited effect on the rest of the body. Most important is the quantity of radiation received by the whole body or by the important organs of the body. A victim may receive two or more types of radiation exposure simultaneously.

External Radiation

A victim who has received external radiation exposure to all or a part of the body is no more a hazard to rescuers or the environment than is a person who has received radiation therapy or a diagnostic X-ray. A harmful dose from radioactive material, a reactor, an accelerator, or other nuclear facility may cause body damage, but the patient is not radioactive and does not present a hazard to attendant personnel. Other than treating this victim as necessary for wounds and other injuries, there is nothing further to be done except transporting the patient to a hospital where specialized treatment may begin.

External Contamination

External contamination of body surfaces and/or clothing by liquids or by solid particles may represent some degree of hazard to personnel handling the victim. Special precautions must be taken in this case to prevent the transfer of contamination from the skin or clothing while transporting the patient to a place for proper decontamination and treatment for radiation exposure. All personnel and equipment that come in contact with this type of patient must be considered contaminated until a radiation survey shows otherwise. Materials and crew used in handling such a victim should be segregated until a radiation survey is made by competent personnel.

External Contamination with Wounds

When external contamination is complicated by a wound, care must be taken not to cross-contaminate surrounding surfaces from the wound and vice versa. The main consideration here is to treat the patient for personal injuries as appropriate, but avoid spreading the contaminating radioactive material by handling the victim as little as possible. Cover the stretcher, including the pillow, with a blanket to contain the contamination. Qualified emergency medical care personnel can clean the wound and surrounding surfaces of the skin and seal off the cleaned area. When crushed, dirty tissue is involved, early wet debridement following wound irrigation may be indicated to remove radioactive material contamination. Further debridement and more definitive therapy should await sophisticated measurement and consultant guidance by a physician.

Wounds should be treated in a manner that will avoid further contamination from other parts of the body. Compresses should be fixed in place with elastic bandages, not adhesive tape.

Internal Contamination

An individual who has received internal contamination by inhalation or ingestion is no hazard to rescuers, other persons, or the environment. Following the cleansing of any contaminated substance deposited on the skin from airborne material, the patient is similar to a chemical poisoning victim. Body wastes should be collected and saved so that estimates of the amount of radioactive material present in the victim's body can be made to assist in determining the appropriate therapy.

Radiation Survey

The only method of detecting gamma radiation hazards is by using a Geiger counter, ionization chamber, or other instrument designed to detect radiation signals. A Geiger counter measures the rate of radiation, which is expressed as roentgens per hour. Should rescue personnel be called upon to provide emergency medical care for a patient who is exposed to radiation, they should proceed with a knowledge of the dangers they will face. Information gained from witnesses or experts in the area and the use of a survey instrument are helpful in determining what the risk will be during the procedure.

Whole body exposure up to 100 roentgens (100 r) in 1 hour, which would be the same as 50 r per hour for 2 hours or 200 r per hour for 30 minutes, has little or no effect. If a reading of 500 r per hour were on the meter, a rescuer could go into the area for 12 minutes and receive exposure equivalent only to 100 r per hour. If the risk is small, or if it must be taken to carry out a rescue operation, the operation should proceed, keeping in mind the three most important factors that determine how much radiation a person would receive:

1. Length of time exposed.
2. Distance from the source of radiation.
3. Amount of shielding from the radiation source.

A rescuer taking a calculated risk in entering an area of radioactivity should remain as briefly as possible and should not enter again without clearance from an appropriate authority.

Emergency service personnel should possess and carry beta-gamma radiation-monitoring survey meters, or know where and from whom they can be readily obtained. This information must be kept up to date. With these survey meters, a well-trained operator can determine whether:

1. There is or is not beta or gamma radiation.
2. The radiation is low-level; the readings do not exceed the maximum on a low-range Geiger counter-type radiation survey meter.

Handling Nuclear Radiation Incidents

3. The radiation is high-level; it can be measured only on a high-range meter.
4. The radioactive material is or is not scattered around the area and has or has not contaminated shoes, clothing, or uncovered areas of the skin.

Beta-gamma radiation survey meters will not register alpha or neutron radiation. Therefore, if there is any evidence that the radioactive material is an alpha or neutron emitter, as on a package label or shipping document, precautions for radioactive contamination should be taken pending the arrival of an expert radiological assistance team.

SEARCH AND RESCUE

Whenever there is an indication that rescuers might be subjected to radioactive exposure, some thought must be devoted to surveying the area to determine the hazard. Personnel would never be sent into a lethal concentration of toxic gases without respiratory protection; likewise, rescuers at nuclear incidents should first determine what the potential dangers are before they risk injury or death. Whenever there is any hint that radioactive materials or any form of radioactivity might be present, full protective clothing and self-contained breathing apparatus should be worn (Figure 21-1).

Radioactive particles may be carried by dust or smoke particles, and every precaution must be taken to avoid the problem of cross-contamination. The best protection against radiation contamination from dust is adequate protective clothing and self-contained breathing apparatus. Several layers of clothing, including hat and gloves, will protect a person from radioactive dust.

Rescue Action

The primary objective of a rescue squad should be to protect the victims from additional injuries, including exposure to radioactive materials. After this has been accomplished, attention can be devoted to emergency medical care, preventing the spread of radioactivity, and decontamination.

The speed and priority of rescue depend largely on the dangers to which the victims are exposed. If the victims are found in a highly contaminated area, they should be moved to a less exposed section immediately. Get in and get out should be the rule. Rescuers should remove the patient and themselves from the hazardous area as soon as possible, even if it means violating some of the other rules of emergency medical care. Radiation is as great a danger to the rescuer and the patient as a burning building, poisonous gas, or an explosive. If there is reason to suspect that the rescuer and the victim are carrying radioactive material in their clothing or shoes, they should stop at the edge of the exposed area and remove as much contaminated clothing as possible. The rescuers should wash both themselves and the patient if they are badly contaminated. By carrying out this simple procedure, they will greatly reduce the hazards to themselves and fellow crew members. The length of time of radiation exposure will be cut down and, with a little care, contamination will not be carried elsewhere. Water used to wash with becomes contaminated itself and should be kept in a covered container for proper disposal.

If a resuscitator or other breathing aid is to be used, internal contamination may be minimized by wiping the victim's face, but

RESCUE FROM HAZARDOUS MATERIALS

Fig. 21-1. When combatting a fire or offering aid at an accident, remain alert to the possibility that radioactive materials could be present. Whenever there is the slightest indication that any form of radioactivity might be involved, stay upwind and wear full protective clothing and self-contained breathing apparatus.
(Courtesy of the Los Angeles City Fire Department. Photo by Rock.)

being careful to stroke away from the nose, mouth, and eyes.

Notify the hospital by radio that a radiation patient is being brought in. The danger in transporting a victim of radiation is only in the amount of radioactive materials on the body or clothing to which others might be exposed.

Decontamination

Decontamination requires removing the clothing; careful peeling or cutting of the clothing from a victim's body should remove most of the contamination. Clothing and other belongings of the patient should be placed in plastic bags, sealed, and properly identified. A shower will remove external contamination, with particular attention devoted to the hair and parts of the body that may rub together. A tub bath is not as good as a shower because it does not remove the particles as well.

All ambulance equipment, linens, cots, and fittings may be contaminated. Ask the radiological survey team to check all equipment before it is used again. Washing will take care of this, but it must be carried out under competent supervision. Remember that the ambulance itself should also be thoroughly washed inside and out to remove any radioactive dust before it is used again.

Handling Nuclear Radiation Incidents

TRANSPORTATION INCIDENTS

One of the areas of real and justified concern to emergency personnel in the development of peaceful uses of atomic energy is the possibility of fires or accidents during the transportation of radioactive materials. These problems cannot possibly be anticipated except in a most general way. Because accidents occasionally happen in circumstances where professional guidance may not be readily available, personnel should know how to evaluate the dangers involved in a fire fighting or rescue operation.

Authorities recommend that a radioactive survey should be made whenever there is a serious fire in a transportation facility such as a truck terminal, an express depot, a freight warehouse, a transit shed on a pier, an airplane, the hold of a ship, or other place in which radioactive material in transit may be located and the nature of the contents involved is not known. High- and low-range radiation detection instruments should be used to determine the dangers before any close-up operations, such as overhauling, are commenced (Figure 21-2).

Fig. 21-2. Authorities recommend that whenever there is a serious fire in a transportation facility which may house radioactive material in transit, a radioactive survey should be made before any close-up operations are started. (Courtesy of the Los Angeles City Fire Department.)

Radioactive Material Shipments

At the present time in the private segment of the nuclear industry in the United States, the majority of the current shipments of radioactive materials consist of small or intermediate quantities of substances in relatively diminutive packages. Most of these parcels contain radioisotopes intended for medical diagnostic or therapeutic applications by thousands of doctors and hospitals throughout the country. Many such materials are quite often of very short half-life; therefore, they must be supplied by the producer to the user via the most rapid available means of transportation. It follows, therefore, quite logically that the majority of these parcels will be shipped by air freight or air express.

All radioactive substances are kept in labeled and shielded containers when not in use. Radioactive material shipped in interstate commerce is dispatched under regulations of the Interstate Commerce Commission (ICC). These shipments are marked with a bright purple propeller on a yellow background. Certain types of radiation materials must be accompanied by trained personnel at all times during transit.

Radiation Hazards

Probably the only real hazard of injury from radiation in a transportation accident would occur if a relatively large, powerful, radiation source, such as that producing gamma radiation, became unshielded and persons unaware of the danger remained in the radiation field. The most likely way this could happen would be in an intense fire, where a container might rupture and the resulting smoke and vapor would constitute a sort of miniature fallout.

At an accident, so long as the container remains intact, there is no hazard, since the maximum radiation level at the surface is kept to an inconsequential amount. If the fire should be extremely severe, and sufficient heat is generated to melt the lead container, the shielding might be displaced. Molten lead is as good a shield for radiation as solid lead, but the melting of the lead may damage the container, or allow the radioactive material to relocate in the container so that it is no longer in the center of the lead shielding. If such a container is exposed to fire, it is important to keep it cooled down; this will minimize damage to the container (Figure 21-3).

If a vehicle transporting radioactive materials is on fire at the time of arrival of fire fighting units, little or nothing would be accomplished even if the fire could be extinguished instantaneously. Normal procedures provide for approaching such a fire with caution. Chemicals that react violently with one another, or with water, may be present. All such fires demand caution, and close-in fire fighting operations should not be undertaken unless there is some definite benefit to be gained. In the absence of radiation-detection equipment, the only evidence of a cargo of radioactive materials might be the possible discovery of quantities of melted lead. While it is quite possible that the lead in the shipment may be simply an item being transported, the possible presence of a gamma emitter should be suspected, and a withdrawal to a safe distance made if radiation-detection instruments are not available for an immediate evaluation of the situation.

Fire fighting techniques on any truck fire, the contents of which are unknown, should provide for the shielding of personnel, either by the emergency vehicles, or the use of a ditch or other natural protection. This shielding is needed more for protection from a possible explosion than from radiation. A solid stream of water should be used from the maximum range that will allow the water to fall down over the fire. If there is no violent reaction between the water and the contents, fire extinguishing efforts can be continued, but no overhauling should be undertaken close to the scene until more information is available. If a violent reaction results from the use of water, as might happen if sodium were in a shipment, use of water should be discontinued and the fire allowed to burn out, unless the vehicle directly exposes other property. In such an event, it is best to attempt to tow the burning vehicle to a safer location.

Sequence of Emergency Procedures

If confronted with an accident in which radioactive material is present, remain calm and do not panic. The mere fact that one or more containers at an accident scene are labeled radioactive does not mean that they have suddenly become dangerous. However, if any of them are broken open or appear to be leaking, seek expert advice before attempting to handle them. Once again it must be emphasized that there is no substitute for good judgment and common sense.

It might be difficult to decide what to do first if a person was injured or trapped in the accident and the first rescuer was alone and first on the scene. The decision to act must then be made with the knowledge that help will be needed as soon as possible and the sooner authorities are notified, the quicker help will arrive. Nevertheless, the saving of human life and care of the injured is the first consideration.

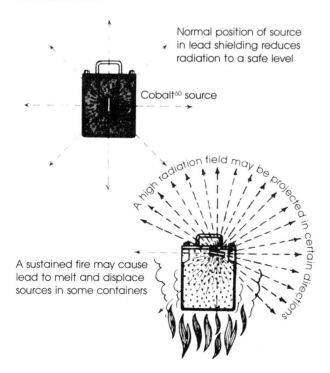

Fig. 21-3. A sustained fire may melt the lead shielding of a container, displacing the radiation source.

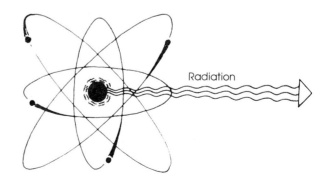

Fig. 21-4. Radiation originates from the nucleus of an atom.

RESCUE FROM HAZARDOUS MATERIALS

Following are the proper procedures in a radioactive materials incident:

Keep away from the wreckage, material, container, or other material at the scene, except to rescue people, when there is an indication that nuclear radiation materials might be present.

Report the accident as quickly as possible. If radiological material is suspected, notify the nearest military installation or office of the U.S. Energy Research & Development Administration.

Warn sightseers and bystanders to stay well away, 500 yards (450 meters) or more if possible.

Stay out of smoke, mist, dust, or other visible substances which are airborne. It is best to stay upwind and on high ground to be safest.

People who have been in the area affected by the accident may have become contaminated and should be held in some nearby place for examination before they are released to go their way. If they will not stay, their names, addresses, and telephone numbers should be obtained.

Do not permit people to handle debris or take souvenirs from the accident scene.

Turn over control at the accident scene only to properly identified authorities.

ATOMIC WEAPON HAZARDS, PRECAUTIONS, AND PROCEDURES

Atomic weapons may be transported by aircraft, truck, train, and naval vessel. Stringent safety measures have been incorporated in the design of all atomic weapons, as well as the handling and transportation procedures, to preclude a nuclear explosion in the event of any accident, even of a military aircraft in operational exercises. For all atomic weapons in combat aircraft, a specific sequence of positive action is required to ready them for a nuclear detonation.

Possible Hazards

Even though nuclear weapons are so designed as to prevent a nuclear yield in the event of accidental detonation, there is still the hazard commensurate with conventional weapons and materials. The two components of a nuclear weapon that constitute the most probable hazard in the case of an accident are the high explosives and the plutonium. Other components may produce hazards, but they are of such nature that precautions taken against explosives and plutonium are more than sufficient for their control. It should be kept in mind that accidents in which nuclear weapons or components are present will usually take in other materials in more widespread use, such as gasoline or other volatile and explosive fuels. If fire occurs, acrid, suffocating, and toxic fumes and smoke will probably be generated by the combustion of the surrounding materials. In that event, normal procedures and precautions applicable to this type of fire should be taken.

High Explosives. Most atomic weapons will contain conventional high explosives in varying amounts up to many hundreds of pounds. These high explosives comprise the major hazard associated with accidents involving atomic weapons. Due to the probable presence of high explosives in any atomic weapons shipment, accidents or fires relating to such shipments must be treated as accidents or fires involving conventional high explosives. The following is a summary of knowledge concerning high explosives and their danger and should be applied to atomic weapons where appropriate.

In any accident involving a high explosive there is some possibility of a detonation occurring. The detonation may range from a very small one to one of considerable magnitude, or it may be a series of small explosions. The breakup of the weapon due to impact or a small explosion will probably result in the local scattering of small pieces of high explosives. It is unwise for anyone other than trained demolition personnel to attempt clearing an area of broken high explosives.

If a nuclear weapon is enveloped in the flame of a gasoline fire, the high explosives are likely to ignite, burn, and in most cases, detonate. These detonations may range from one large explosion to several small ones. It is extremely difficult to extinguish large quantities of burning high explosives. Whenever burning high explosives are confined, as in an intact weapon, detonation may occur at any time. When high explosives burn, torching (jets of white flame coming out of the weapon) may be observed, but torching is not always evident before detonation. High explosives may melt at comparatively low temperatures, flow out of the weapon, and resolidify. In this state, they are extremely sensitive to shock. If unconfined, high explosives may burn, producing toxic gases and leaving a poisonous residue. Ignition or detonation of the high explosives in a nuclear weapon involved in a fire can be prevented if the temperature of the explosives is kept below 300°F (149°C).

While it is not feasible to predict the exact effect of an accident involving high explosives, the possibility of the accidental nuclear explosion of an atomic weapon is so remote as to be negligible.

Plutonium may be dispersed as small particles if detonation of the high explosives occurs, or as fumes if a fire occurs.

Body Effects of Plutonium. When small particles of plutonium are suspended in the air, it is possible to inhale them and thus deposit plutonium in the lungs. They may also be swallowed, but in such cases only a small percentage is retained by the body, since the plutonium is in a highly insoluble form. Cuts in the skin provide a third source of entry through which plutonium may enter the blood stream. Plutonium is not a hazard if it remains outside the body, because it is an alpha emitter. The alpha particles have a very short range and lack the ability to penetrate the skin. This characteristic of plutonium radiation makes its behavior markedly different from that of fallout from an atomic explosion in that it does not emit the more penetrating beta and gamma radiation. Conventional survey meters are of little use in detecting alpha radiation, and only special teams trained for handling nuclear accidents are capable of evaluating the radiological situation at the scene of an accident.

Field experiments indicate that the principal potential source of intake of plutonium into the body is by inhalation during the passage of the cloud resulting from the detonation of the explosive, or from a fire. Whereas it is always desirable to reduce to a minimum the intake of plutonium, one may enter or remain in a highly contaminated open area for short periods of time (up to several hours) after passage of the cloud.

Emergency Procedures

Immediate assistance to personnel must be rendered where possible. However, except for the saving of lives, personnel should keep

RESCUE FROM HAZARDOUS MATERIALS

away from the accident. There is always the danger of detonation of conventional high explosives. The nearest military installation or Atomic Energy Commission office must be notified of the accident. The area is to be cleared of all nonessential personnel, to a distance of at least 2,000 feet (600 meters) or more.

If there is a fire and *if the weapon is not burning or engulfed in flames,* rescuers may attempt to extinguish the fire in the normal manner (from the upwind side only). They must keep the weapon cool, but be prepared to stop the water if it accelerates the burning. The high explosive will not explode if the temperature is kept below 300°F (149°C). Foam must be used with caution on a bomb as it may act to trap heat inside; foam is an insulator and may retain sufficient heat to detonate the weapon.

If the weapon is engulfed in flames or if the high explosive is burning (torching), the area should be cleared to at least 1,500 feet (450 meters) of all personnel; do not attempt to fight the fires. Smoke should be avoided and the downwind area cleared. But if dense smoke must be encountered for long periods of time, dust-filtering masks, goggles, or breathing apparatus should be used. These are not needed for short stays in the smoke and nonavailability should never hold up rescue efforts. Personnel who have entered smoke in the crash area should report to the special team for monitoring, and, if necessary, decontamination after the initial action is over.

After the burning has subsided, and if the AEC team has not yet arrived, the following actions should be taken: do not attempt to clean up the scene of the accident; refrain from touching, removing, or examining any items in the vicinity of the explosion; do not pocket or retain as souvenirs any object found in the accident area; do not permit reentry into the scene of the accident by anyone.

NUCLEAR WEAPONS EXPLOSIONS

The atomic (or nuclear) bomb is similar to the more conventional (or high explosive) type of bomb insofar as its destructive action is due mainly to blast or shock. However, apart from the fact that nuclear bombs can be thousands of times more powerful than the largest TNT bombs, there are other more basic differences. First, a fairly large proportion of the energy in a nuclear explosion is emitted in the form of light and heat, generally referred to as *thermal radiation.* This is capable of causing skin burns and of starting fires at considerable distances. Second, the explosion is accompanied by highly penetrating and harmful, but invisible, rays called the *initial nuclear radiation.* Finally, the substances remaining after a nuclear explosion are radioactive, emitting similar radiation over an extended period of time. This is known as the *residual nuclear radiation,* or *residual radioactivity.*

Most of the material damage caused by an air burst nuclear bomb is due, directly or indirectly, to the shock (or blast) wave which accompanies the explosion. The combination of high peak overpressure and long duration of the compression phase of the blast wave results in mass distortion of buildings, similar to that caused by earthquakes. An ordinary explosion will usually damage only part of a large structure, but the nuclear blast can surround and destroy whole buildings.

Thermal Radiation

One third of the total energy of a nuclear explosion is emitted in the form of thermal (heat) radiation. Because of the enormous amount of energy liberated in an atomic bomb, very high temperatures are attained. These may be of the order of several millions of degrees, compared with a few thousand degrees in the case of a TNT explosion. Although blast is responsible for most of the destruction caused by a nuclear air burst, thermal radiation will contribute to the overall damage by igniting combustible materials. These fires may spread rapidly among the debris produced by the blast. In addition, thermal radiation is capable of causing skin burns on exposed persons at distances so far from the explosion that the effects of the blast and of the initial nuclear radiation are not significant.

Thermal radiation can cause burn injuries either directly by absorption of the heat by the skin, or indirectly, as a result of fires started by the radiation. The direct burns are often called *flash burns*, since they are caused by the flash of thermal radiation from the ball of fire. The indirect (or secondary) burns are referred to as *flame burns* and are identical with skin burns caused by any large fire, no matter what its origin. Burns, irrespective of their cause, are generally classified according to their severity and depth.

Initial Nuclear Radiation

One of the unique features of a nuclear explosion is the fact that it is accompanied by the emission of nuclear radiation. These radiations are quite different from thermal radiation and consist of gamma rays, neutrons, beta particles, and a small proportion of alpha particles.

The range of alpha and beta particles is comparatively short and they cannot reach the ground from an air burst. The initial nuclear radiation may be regarded as consisting only of gamma rays and neutrons produced during a period of one minute after the nuclear explosion. Both of these nuclear radiations, although different in character, can penetrate considerable distances through the air and through most solid materials. Although the energy of the initial gamma rays and neutrons is only about 3 percent of the total explosion energy, compared with some 33 percent thermal radiation, the nuclear radiations can cause a large proportion of the bomb casualties.

Residual Nuclear Radiation and Fallout

Residual nuclear radiation is defined as that emitted after one minute from the instant of a nuclear explosion. This radiation arises mainly from the bomb residues (the fission products) and, to a lesser extent, from the uranium and plutonium that have escaped fission. About 1.75 ounces (49 grams) of fission products are formed for each kiloton of bomb explosion; this is comparable to the radioactivity of a hundred thousand tons of radium. One kiloton of nuclear bomb energy is equivalent to 1,000 tons (900 metric tons) of TNT explosion. These fission products will contaminate a large area in a fallout contour pattern downwind that may cause casualties for hundreds of miles (kilometers) from the explosion.

ATOMIC STRUCTURE

It is useful for rescue personnel to have an understanding of basic forms of matter and certain terminology relating to the structure of matter. The basic knowledge of atomic structure is essential in that it provides a foundation upon which a person may build a clearer concept of the "whys" of radiation and its associated hazards.

All materials are made up of atoms, the structure of which is similar to the structure of our solar system. The solar system is composed of a sun, or central body, around which the planets revolve. In an atom there is a central body, or nucleus, around which revolve a number of smaller bodies called electrons. Here the similarity ends because the solar system occupies a vast amount of space whereas that occupied by an atom is infinitely small.

Components of the Atom

The atom consists of a heavy, dense core, called a nucleus, which contains practically all of its weight. This nucleus is surrounded by electrons that whirl around the nucleus at tremendous speeds. Electrons are very small and have practically no weight (Figures 21-4 and 21-5).

The nucleus consists of little individual balls of matter which are about the same size in all atoms. Some of these little balls of matter have a positive electrical charge, and are called protons. Others, called neutrons, have no electrical charge and are said to be electrically neutral. Both a proton and a neutron are much larger and heavier than an electron.

For each positively charged proton in the nucleus, there is a negatively charged electron in orbit around the nucleus. The number of protons in the nucleus, therefore, determines the number of electrons in the orbit. The number of protons also determines the nature of the element. An element is formed when large numbers of identical atoms are bound together.

Elements

Elements are found in nature with the number of protons ranging all the way from 1 proton (hydrogen) to 92 protons (uranium). Each time the number of protons is changed an entirely different element is formed (Figure 21-6). The periodic chart of the elements places 92 natural elements in numerical succession and in periods or groups.

Synthetic Elements

In recent years, nuclear scientists have created additional elements; as far as is known they are not found in nature. Their atomic numbers range from 93 to 104. They are known as the trans-uranic elements of which element Number 94, plutonium, is an important member. All are radioactive.

Molecules

Atoms group together in certain numbers and combinations to form molecules. Many molecules consisting of identical atoms make up an element. Molecules composed of combinations of different atoms form compounds. Many of these compounds are familiar in everyday life such as water (H_2O), sugar ($C_{12}H_{22}O_{11}$), or salt (NaCl).

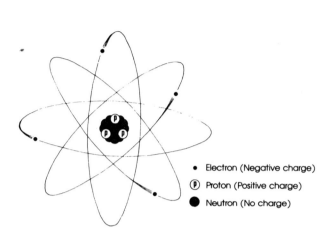

Fig. 21-5. The elementary particles that make up an atom.

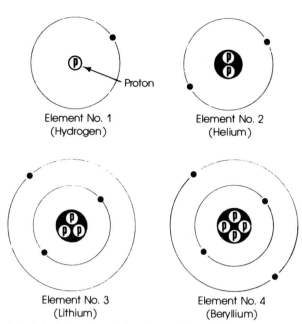

Fig. 21-6. A comparison of the atoms of the first four elements.

Isotopes

The number of neutrons in the nucleus may range from zero to almost 150. In certain elements it is found that different atoms of the same element have the same number of protons but vary in the number of neutrons. The organic chemist will treat these variants as the same element because the organic chemist works with orbital electrons, and since all these atoms have the same number of protons, they will have the same number of electrons in orbit. Since they have a different number of neutrons in the nucleus, however, the various atoms of the same element will not all weigh the same (Figure 21-7). To the nuclear physicist and physical chemist, they are different substances of the same element that vary in their atomic weight. These are called isotopes. Many of the isotopes are unstable, and, therefore, radioactive. Some of the isotopes are stable and not radioactive.

RADIATION

The nuclei of radioactive atoms contain excess energy and therefore are said to be in an excited state. They rid themselves of this excitation by emitting the excess energy in the form of radiation. All atomic radiation falls into two general categories: that which consists of rays and that which consists of subatomic bits of matter (particles).

All radioactive materials emit ionizing radiation. The term *ionizing* covers those forms of radiation that directly or indirectly cause rearrangement of orbital electrons in the atoms of a substance through which the radiation passes. In living tissue this can cause biological damage.

There are a number of very common and familiar types of

RESCUE FROM HAZARDOUS MATERIALS

radiation, such as radio and radar waves, heat, and light, which are non-ionizing in character. They do not pack enough energy to affect orbital electrons. Damage resulting from some of these forms of radiation is generally confined to the outer layers of the body. Sunburn and thermal burns are two common examples. Fortunately their effects are usually quickly apparent and the exposed individual can take protective action before serious damage can result. However, this is not the case with the ionizing types of radiation because the electron rearrangement within the body cannot be felt by the exposed individual, and the resulting biological damage does not become apparent until it may be too late. There are three common types of ionizing radiation to which an emergency worker may be exposed during an accident or other emergency (Figure 21-8).

Gamma Radiation. Gamma radiation consists of extremely short electromagnetic waves of pure energy having no mass or weight. These waves are similar to X-rays. Gamma radiation travels at the speed of light and has a relatively long range and great penetrating ability.

Beta Radiation. Beta radiation consists of elementary particles carrying negative electrical charges. Beta particles are ejected at high velocities but have shorter range (usually several feet in air), and have less penetrating ability than gamma radiation. Beta radiation is often referred to as corpuscular radiation.

Alpha Radiation. Alpha radiation consists of relatively heavy atomic particles each containing two protons and two neutrons bound together and identical with the nucleus of a helium atom. Alpha particles have extremely short range (several inches in air) and very little penetrating ability. Alpha radiation is also referred to as corpuscular radiation.

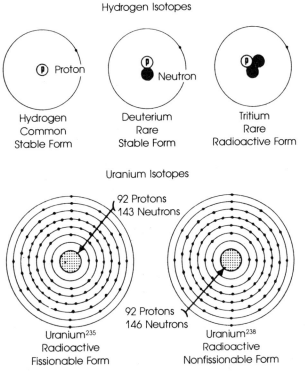

Fig. 21-7. Isotopes of the elements hydrogen and uranium.

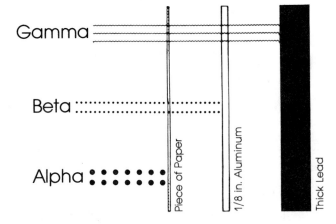

Fig. 21-8. The three common types of radiation.

There are several other types of radiation that are not as common as those just discussed. Protective measures taken for gamma, beta, and alpha radiation will usually suffice for the less common types of radiation.

Radioactive Decay

Excess energy ejected in the form of radiation is called radioactive decay or disintegration. Some radioisotopes decay directly into a stable state in one step. Others decay through a series of steps or chains, forming different radioactive elements called daughter products before finally reaching a stable state.

Gamma radiation originates when the discharge of one of the particles from a nucleus does not take sufficient energy along with it to leave the nucleus in a contented state. If the particle leaving the nucleus does not take with it all of the energy that the atom would tend to lose in this particular disintegration, it throws off some of the energy in the form of gamma radiation. Therefore, in addition to beta or alpha radiation, gamma radiation can be given off by many radioactive materials.

Radioactive Half-Life

It is impossible to look at any one atom and predict exactly when the disintegration process will take place; but for a large number of atoms, it can be statistically determined with accuracy that half of the atoms will disintegrate in a given period of time. This is the so-called *half-life* of a radioactive material (Figure 21-9) and may range from fractions of a second up to millions of years. The following radioactive elements demonstrate this wide range of half-lives: argon 41, 109 minutes; iodine 131, 8 days; radium 226, 1620 years; plutonium 239, 24,000 years.

A useful rule of thumb is that the passage of seven half-lives will bring a radiation level down to a little below 1 percent of its original level, and in ten half-lives the level will be down to less than 0.1 percent. The shorter the half-life, the more highly radioactive the material will be.

Isotopes with very short half-lives are not very useful because they decay so rapidly that they cannot be put to effective use. Isotopes with half-lives measured in hours or days can be put to practical use, but this requires close scheduling of the use of the isotopes with their manufacture. It is for this reason that many short half-life radioisotopes used for medical purposes are shipped by air express.

THE FISSION PROCESS

The terms *radioactive* and *fissionable* are not equivalent. Fissionable defines those atoms that can be split apart by nuclear bombardment, thus releasing large amounts of energy in the process. Just because a material is radioactive, it is not necessarily fissionable, and most of the radioisotopes handled in commerce and industry are not fissionable. The material must be capable of keeping up and multiplying the chain reaction process so that a large number of fissions can be made to take place in a relatively short time.

Certain elements have the characteristics of being able to emit bits of matter (neutrons) which, when they strike other atoms of the

RESCUE FROM HAZARDOUS MATERIALS

same matter, cause those atoms to break open and emit more bits of matter (neutrons) that continue into the chain reaction process. When these atoms are split open, a certain amount of energy, which has been holding the atom together, is released. This energy can be harnessed. This is fission energy, and should not be confused with radiation energy (Figure 21-10).

Only two materials that are readily available have this property. One is uranium 235, which is an isotope of uranium comprising about 1/140th of the uranium found in nature, the balance being mostly uranium 238, which is not fissionable. The other fissionable material is plutonium. Plutonium is not a natural element, but a synthetic one created from nonfissionable uranium 238 in a nuclear reaction brought about in an atomic reactor.

The nucleus of each atom is held together by a force called binding energy. Some of this energy is released when a nucleus is split in two by the fission process. This is due to the fact that it takes less binding energy to hold together the two fragments that resulted from the splitting of the nucleus than it took to hold the original nucleus before fission (Figure 21-11).

PROBLEMS OF RADIATION

Radiation materials emit energy that has the power to damage living tissue if received in sufficient quantity. Within certain limits, however, damage can be repaired by the body so that there is no apparent effect.

The human body has varying tolerances to all kinds of damage to its mechanism. This, in general, comes about through a complicated biological process known as mitosis, that is, cell division in which worn out or damaged cells are replaced. From the prenatal state to maturity, this cell division process is rapid but continues at a slower pace after a person has reached maturity.

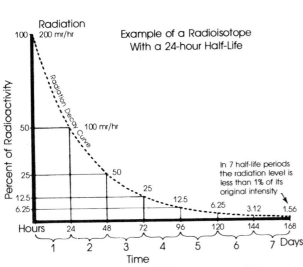

Fig. 21-9. Radioactive half-life.

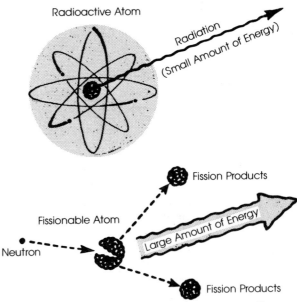

Fig. 21-10. The difference between a radioactive atom and a fissionable atom.

Radiation is like any other form of damage to the body, whether it be from excessive heat, poison, electrical shock, disease, or wounds. There are various injury limits that the body can stand, depending upon the type, severity, and duration of the damage. If the injury is sudden and massive the body's repair mechanism is overwhelmed and death results quickly. There is also a wide middle ground in which continued damage of a lesser nature, when prolonged over a period of time, can gradually overcome the body's ability to repair itself and will cause the individual to sicken and eventually succumb.

Then there is that area in which damage to the body is small enough to cause no demonstrable effects over an extended period of time because of the body's ability to continuously overcome and repair the small amount of damage being produced.

The effects of excessive radiation exposure on the body are manifested in several ways:

Radiation Sickness. Radiation sickness is produced by a massive overdose of penetrating external gamma radiation, which causes nausea, vomiting, diarrhea, malaise, hemorrhage, and a lowering of the body's resistance against disease and infection, and, if serious enough, death.

Radiation Injury. Radiation injury consists of localized injurious effects, generally from overdoses of the less penetrating beta radiation and most often to the hands because contact is usually with the hands. This can cause injuries such as burns, loss of hair, and skin lesions. Genetic damage is also a form of radiation injury, usually of permanent nature.

Radioactive Poisoning. Radioactive poisoning is illness resulting when dangerous amounts of certain types of radioactive materials enter the body; it may cause such diseases as anemia and cancer. The alpha radiation emitters are the most dangerous in this respect.

EXTERNAL AND INTERNAL RADIATION

Rescue personnel can get into trouble with radiation by two entirely different means: one, by radiation originating from a radioactive source located outside the body from which the radiation comes at the body like a continuous shower of tiny, invisible bullets; the other, by exposure of internal body organs to radioactive material which has been taken into the body and which may have collected in these body organs. It should be obvious that precautions against one type of hazard will not be particularly helpful in protecting against the other type of hazard, and that the radiation problem is made up of two separate problems.

This is indeed the case. As a matter of fact, certain radioactive materials are no hazard at all outside the body. However, if these same materials got inside the body in sufficient quantities, they could produce a serious case of radioactive poisoning.

Therefore, it is important for rescue personnel to understand that the radiation problem is not one problem, but two problems: that of external radiation exposure, and that of internal radioactive poisoning. Of course, it is possible in a given situation that both external and internal hazards are present.

There are two types of external radiation hazards: long range, highly penetrating external radiation called gamma; and short range, less penetrating radiation called beta.

RESCUE FROM HAZARDOUS MATERIALS

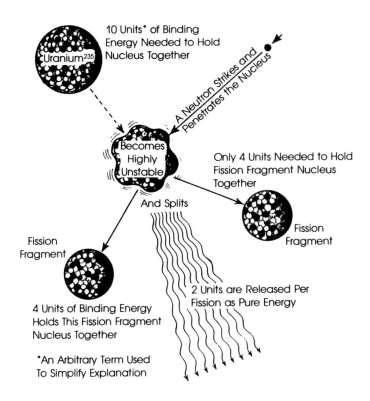

Fig. 21-11. The release of binding energy from a fissionable atom.

The long range, highly penetrating external radiation is similar to X-rays and consists of very short waves of pure energy having no mass or weight. These rays originate from certain radioactive materials which are usually located outside the body, and so the rays come at the body like a continuous shower of tiny, invisible bullets (Figure 21-12). In order to visualize this, it should be conceived as coming one ray at a time. Each ray can be considered a bundle of energy. It may penetrate the body to some depth before it does damage and the energy of the ray is spent.

Some rays pass entirely through the body without hitting anything; these do no harm. Very few of the rays have enough energy to penetrate the nucleus of the atom and cause any change; they do, however, hit the electrons that are spinning around the nucleus and the energy of the ray is spent. This results in a transfer of energy from the rays to those electrons that are ejected from the atoms. This process is called *ionization*.

Atoms in their normal state are electrically neutral; that is, an atom contains the same number of electrically negative electrons in orbit as it contains electrically positive protons in its nucleus. When radiation causes displacement and rearrangement of these electrons, positive and negative ions are created (Figure 21-13). There are several ways in which ionization can take place, but the end result is that ion pairs are formed, which, if produced in sufficient quantities, can cause complex changes in the body chemistry that result in varying degrees of sickness or death, depending on the amount of ionization present. When an electron is knocked off an atom, the cell of which the atom is a part is damaged. The body's

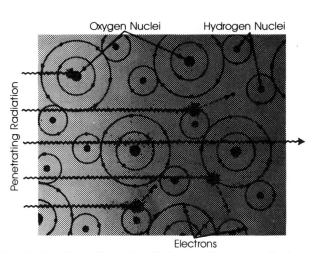

Fig. 21-12. The effect of radiation on the atoms that compose the body.

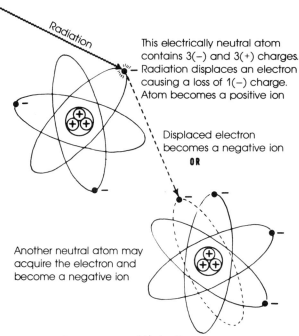

Fig. 21-13. The process of ionization.

repair mechanism swings into action to repair the damaged cell. Exposure to excessive radiation can cause sufficient damage to bring about death.

The Roentgen

It is possible to have radiation exposure so small that it has no apparent effect, and, on the other hand, sufficient radiation exposure to cause death. It is therefore necessary that some unit of measurement be used with which to measure radiation exposure. This will allow a means of determining what the acceptable limits are and of being sure that those limits are not exceeded.

The unit of measurement for penetrating external radiation exposure is the roentgen (abbreviated r), named in honor of the man who discovered the X-ray in 1895. It is an arbitrary unit of measurement. Its exact definition is of significance only to experts; the primary importance for the lay person lies in how many roentgens will injure living cells. It must be remembered that the roentgen is used only to measure gamma radiation and X-rays passing through air. As a roentgen is a rather large amount of radiation, a subunit measurement of the roentgen is used for more convenience when dealing with small amounts of radiation. This is the milliroentgen (abbreviated mr), or one-thousandth of a roentgen.

Since the roentgen measures gamma or X-ray radiation in air only, it cannot be used to define the biological effects in an exposed individual. The term used for this purpose is REM, which means Roentgen Equivalent Man. One roentgen of penetrating external radiation produces one rem of body damage.

Single Exposures

It is impossible to say how many roentgens it will take to kill any specific individual because each person varies in resistance to any attack upon the body, whether it be by radiation, electricity, poison, injury, or disease. It is certain, however, that no person can survive exposure of 1,000 roentgens of total body radiation delivered in a short space of time (Table 21-1).

Table 21-1 EFFECTS RESULTING FROM SINGLE WHOLE-BODY EXPOSURE TO EXTERNAL RADIATION.

SINGLE DOSE OVER WHOLE BODY		EFFECT
In Roentgens	In Milliroentgens	
Less than 25	Less than 25,000	Clinically not detectable
25 - 100	25,000 - 100,000	Blood changes but no illness expected
100 - 300	100,000 - 300,000	Slight to severe illness
300 - 500	300,000 - 500,000	Illness and possible death
500 - 1000	500,000 - 1,000,000	Survival possible
Over 1000	Over, 1,000,000	Survival improbable

The effect of 1,000 roentgens of radiation delivered to the whole body is by no means the same thing as 1,000 roentgens delivered to a small portion of the body any more than a third-degree burn of the palm of the hand is the same thing as a third-degree burn of a large area of the body.

Similarly, the time element is an important part of the definition. A short space of time is defined as 24 hours, or less. The ability of the body to withstand any damage is, of course, increased if the same amount of injury given to the body is spread out over a longer period of time. The dose it takes to kill one specific individual is not good measure of the fatal dose to others because of differences in each individual's body metabolism.

Nuclear Radiation Injury

After the discovery of X-rays and radium toward the end of the nineteenth century, serious and sometimes fatal exposure to radiation was sustained before the dangers were realized. The harmful effects of radiation appear to be due to the ionization (and excitation) produced in the living tissue cells, thus damaging or destroying some of the constituents essential to their normal functioning.

Exposure to radiation makes striking and characteristic changes in the body: the total number of white cells increases sharply the first two days, and then decreases below normal levels. This increases susceptibility to infection. The amount of platelets (a blood clotting constituent) decreases and the possibility of hemorrhage is greater. The red blood cells decrease, resulting in anemia. The possibility of leukemia, a disease associated with an overproduction of white blood cells, is increased. The healing of all wounds is retarded.

Radiation Tolerance Levels

In emergency operations where appreciable amounts of radiation are present, one should not hesitate to accept a total-body exposure of twenty-five roentgens in a single day. If it becomes necessary to make decisions concerning repeated exposures, the following statement may be used as a "rule of thumb" guide: exposure of 25r per day at weekly or longer intervals for a total of eight exposures (200 r) may be experienced without serious loss of efficiency due either to immediate illness or significant general deterioration in health and ability. Serious illness from repeated exposures over a period of weeks is improbable if before each re-exposure, the degree of radiation damage already produced and that to be expected are evaluated by medical personnel. Training activities should involve no more than 0.3 r per week. The cardinal principle must be: *avoid all unnecessary exposure.*

Protection from External Radiation

The means of protecting personnel from external radiation exposure are a combination of three things: time, distance, and shielding. Protection is provided by controlling the length of time of exposure, controlling the distance between the worker and the source of the radiation, and placing a shielding material between the worker and the source of the radiation. It is not generally feasible to use only one factor of protection. The factor of time is always involved; that is, time is usually used in combination with distance, shielding, or both. In order to understand each of these factors of protection, it is helpful to discuss them separately.

Time. The effect of time on radiation exposure is easy to understand. If a person is in an area where the radiation level from penetrating external radiation is 100 milliroentgens per hour, then in one hour exposure would be 100 mr, and in two hours, 200 mr, and so on.

Time is used as a safety factor by keeping the time of exposure down to the absolute minimum. For instance, if work must be done in a high radiation area, the work to be done should be carefully preplanned outside the hazard area so that the minimum time is used within the radiation area to accomplish the work. The basic principle is to limit the exposure of personnel to radiation in all cases to the minimum time necessary to accomplish the task, and to use the minimum number of personnel so that the accumulated units of exposure will be the absolute minimum.

Distance. The effect of distance on radiation exposure is quite startling. The effect is measured by the inverse square law. That is to say, the intensity of radiation lessens by the square of the distance from the source (Figure 21-14).

For example, standing 1 foot (30 centimeters) from a source of radiation giving off 1,000 roentgens per hour of penetrating external radiation, a person would be exposed to 1,000 r per hour, but would receive only 250 r per hour at two feet, which would double the distance. The effect on the radiation level is to reduce it to one-half squared, or one-fourth. If the distance is tripled, the level is reduced to one-third squared, or one-ninth; this would be 111 r per hour.

Shielding. In order to stop a high proportion of the rays before they reach a person, a material which has many electrons in its makeup must be placed between the person and the source of the radiation. The more electrons in the atoms of the material, the more radiation will be stopped.

Figure 21-15 shows lead and water in a rough comparison of their atomic makeup. Notice that lead has more electrons in the orbits of each atom than does water. Therefore lead, with 82 electrons per atom, makes a better shield than water.

Figure 21-16 shows the relative efficiency of various shielding materials. Lead, iron, concrete, and water are efficient in about the proportions shown in the illustration in stopping the same amount of radiation. Lead is quite compact and is suitable where space

RESCUE FROM HAZARDOUS MATERIALS

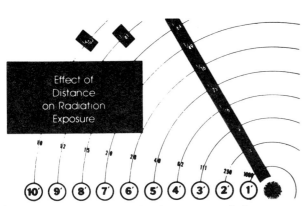

Fig. 21-14. The effect of distance on radiation exposure.

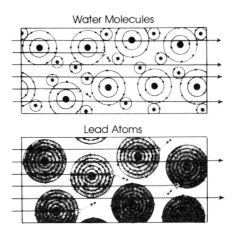

Fig. 21-15. Comparison of the atomic structures of water and lead.

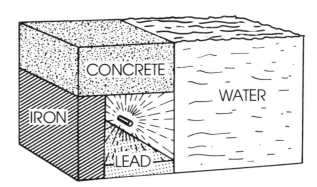

Fig. 21-16. The relative efficiency of various materials for shielding against gamma radiation.

requirements are a factor. It must be realized that no material can be an absolute barrier to penetrating external radiation. Regardless of the thickness of the material, some radiation can get through the material without hitting any electrons and, therefore, without being absorbed.

Where there is an obviously shielded area, no attempt should be made to disrupt the shielding or to get behind it to perform fire fighting or a rescue until adequate advice can be obtained.

It is well known that lead melts at very low temperatures. When the lead is molten it will have the same shielding properties as before; however, the radioactive source in the container's cavity could conceivably float to the highest point in the container and the shielding will be greatly reduced (see Figure 21-3).

Any lead shielding containers that have been exposed to fire, therefore, should be approached with caution. Lead is relatively low in value and costly to transport. Therefore, any melted lead found at a fire in a transportation situation, such as an airplane, truck, railway express car, truck terminal, or freight depot, should be viewed with suspicion. It is possible that the lead was actually being used as shielding for a radiation source, and that the source may be exposed, causing high radiation.

Protection against Beta Radiation

Thus far, the discussion has been concerned primarily with long range, penetrating external gamma radiation. There is another less hazardous type called beta radiation. This radiation is less penetrating, has a shorter range, and represents an external hazard to personnel only if they come into close contact with the radioactive materials by handling them or by not promptly washing off any material that may have gotten on their bodies.

Radiation Detection and Monitoring

Grave damage may result to emergency crews unless the extent and the intensity of radioactive contamination can be determined. It is possible to accumulate harmful doses in a short time, without any visible effect, and with possibly irreparable damage.

Ionizing radiation emitted from radioactive materials cannot be detected by any of the human senses. A person receiving a large overexposure of external radiation would not be aware of it until sometime later when radiation sickness developed. Therefore radiation must be detected by mechanical, chemical, or electrical means. Recent research, however, indicates that a person exposed to a massive overdose of external radiation, in the order of 5,000 r or more, would feel the effects almost immediately.

The radiation received by a person exposed to a radioactive locality is described in two ways. The first, called a *total dose*, describes the total cumulative quantity of radiation (measured in roentgens) received over the period of exposure. The second, called the *dose rate* (measured in roentgens per hour) is the amount received by a person at a given instant.

Because ordinary methods will not detect radiation, it becomes necessary to use special detection devices, such as geiger counters, ion chambers, film badges, pocket dosage meters, and other detection instruments.

Internal Radiation and Protection

The internal radiation exposure problem, in contrast to that of external radiation, is much more complicated and involves many factors. Monitoring alpha radiation is more difficult than monitoring other types of radiation. Certain instruments are available that can evaluate alpha contamination only on surface areas. This is of little benefit to rescue personnel as it does not tell them how much is in the air. Other more specialized types of monitoring equipment require time and special techniques usually not available at the scene of an emergency.

The principal hazard associated with internal radiation is due to the nature of alpha particles. Alpha radiation cannot penetrate beyond the outer layer of skin; therefore, it presents no radiation hazard when kept outside of the body. The greatest internal radiation hazard to rescue personnel is that of alpha-emitting radioactive material getting into the body. There are four possible ways for radioactive materials to get into the body: by breathing, by swallowing, through breaks in the skin, and by absorption through the skin. The most likely way for radioactive materials to enter the body of personnel is by breathing and swallowing, and to a lesser degree, through wounds.

RESCUE FROM HAZARDOUS MATERIALS

The emergency services are accustomed to dealing with toxic concentrations so deadly that almost immediate death will result if personnel are exposed to them for even short periods of time. Such gases as carbon monoxide, oxides of nitrogen, and others that are often encountered are almost instantly recognized as being fatal when present in high concentrations. This, however, is not the nature of the hazards of radioactive materials. Rather, the hazard is the subtlety of radioactive materials which, over a period of time, may prove detrimental to the individual because of steady damage caused by the radiation emitted. These materials cannot be sensed and may not, necessarily, be accompanied by smoke or irritating fumes which by their very nature are self-protecting in that they tend to drive the rescuers out of the hazardous area before they can receive a dangerous amount of toxic material.

The fundamental rule is to require that all personnel entering a radiation area during a fire or other emergency wear the self-contained type of mask equipment until it is determined that no airborne hazard exists. Smoking and eating are usually forbidden in areas where radioactive materials are handled or may be present, because of the danger of transferring radioactive contamination from the hands to the cigarette or food, and then to the mouth and into the body. The introduction of radioactives through wounds is avoided by standard safety techniques to prevent injury. The hazard of absorption through the skin, where it exists, is handled by the provision of suitable airtight protective clothing and gloves.

PROTECTIVE AND DECONTAMINATION PROCEDURES

In order to minimize injuries to personnel from residual radioactivity, it is essential that members of rescue squads understand the principles of protection and decontamination.

Protection

The following practices should be followed when working near radioactive material:

1. A gamma survey meter should be carried by one member of each working party to measure the intensity of radiation to which personnel are exposed.
2. Respiratory protection will be required for all personnel entering a contaminated area. The degree of protection should be governed by the type of contamination and the method of decontamination employed. Rescue personnel entering a dangerous area should wear an approved, self-contained breathing apparatus; if these are not available, the minimum of an all-service mask containing a filter approved for toxic smokes should be used.
3. When the usual rubber boots, coats, helmets, and gloves are used in a contaminated area, they should be washed down with a clean stream of water and monitored. Exposure can be reduced somewhat by frequent washing down of equipment and apparel; care must be exercised, however, not to drive radioactive material into the mouth, nose, ears, and eyes. Canvas clothing, when not of the discardable type, should be

laundered in special equipment. Every precaution against the spread of contamination should be taken during a change of clothes.

4. Any rescuer who receives a cut or abrasion, or who inadvertently handles radioactive contamination with bare hands, should report the incident, have the exposure evaluated, and receive the necessary medical aid.

5. No smoking, eating, or drinking should be permitted in radiologically "hot areas" until they are declared safe. These precautions should be followed until adequate air and surface contamination surveys show the hazard to be negligible.

6. All personnel engaged in operations within the contaminated area should take a thorough shower bath, washing several times with soap and water, as soon as facilities become available. Violent scrubbing with a brush should be avoided to prevent the possibility of breaking the skin and thus contaminating the blood stream. Particular attention should be given to hair washing, skin folds and creases, and cleaning under the fingernails. Soap and water should be kept clear of the eyes and other body openings.

Decontamination

Decontamination may be defined as the removal of radioactive contamination from a person or object followed by the safe disposal of radioactive waste. It must be constantly borne in mind that radioactive material cannot be destroyed, but can be removed and will decay of its own accord. The easiest and sometimes the only way of preventing radiation casualties is to avoid the contamination. If this is not possible, decontamination may have to be carried out in the following manner:

1. Wet contamination on smooth surfaces—flush with water and detergents followed by scrubbing and flushing. Steam under pressure is very helpful. Treatment is more difficult if the contamination has been allowed to dry.

2. Contamination in the form of dust—use vacuum cleaning. If further treatment in necessary, brush and vacuum again.

3. Contamination of greasy surfaces—remove greasy material with dry-cleaning solvents. If necessary, follow by scouring with water, soaps, and detergents.

4. Deeply absorbed contamination—removal of the surface is necessary; at the same time, further penetration of the contaminant should be prevented. Remove the top few inches (centimeters) of earth, use caustic paint removers, sand or grind the surface, or use special compounds.

5. Firmly held contamination—cover with a suitable thickness of sealing material, or dispose of contaminated objects or clothing by deep burial in the earth or at sea in weighted, sealed containers.

   Decontamination may have to be carried out in several stages, and a combination of methods may be necessary.

QUESTIONS FOR REVIEW

1. Which one of the following statements best describes the hazards of nuclear radiation?

RESCUE FROM HAZARDOUS MATERIALS

    a. Radiation dangers are most serious in research laboratories.
    b. Radiation hazards may be encountered in any medical laboratory.
    c. Trucks, aircraft, and other transportation under the control of the Department of Transportation are prohibited from carrying hazardous materials that may emit radiation.
    d. Everyone has been exposed to some degree of radiation throughout her/his life.

2. Nuclear radiation damages the human body to the greatest degree by affecting the:
    a. Cells.
    b. Muscle tissue.
    c. Brain.
    d. Blood.

3. A person may be most seriously affected if radioactive materials are deposited in/on:
    a. Bare skin.
    b. Food or drink.
    c. Lungs.
    d. Stomach.

4. Rescue personnel respond to a truck accident and discover that a truck that is plainly identified as transporting radioactive materials is burning. The best method of safely combating this hazardous situation is to:
    a. Launch a massive fire fighting attack with spray streams.
    b. Protect exposed structures, withdraw personnel, and let it burn.
    c. Use a tow truck to remove the vehicle to a safer location and then extinguish the blaze.
    d. Use common fire fighting procedures as no extraordinary hazards are created.

5. A military aircraft that has been identified by the crew as carrying atomic weapons is burning with great ferocity. The hazards of combating this blaze are:
    a. Possibly catastrophic, as the weapon may fission at any moment.
    b. Serious, as the weapon will fission if the temperature exceeds 300°F (149°C).
    c. Moderate, as there is only a 50-50 chance that the weapon will fission.
    d. Serious, as the weapon will not fission but conventional explosives may detonate.

CHAPTER 22

Handling Bomb Threats and Explosives

Bomb threats and the use of explosives are becoming more prevalent throughout the entire country; no section can be considered safe from this type of calamity. Most bombs appear to be harassment devices. They are used to threaten, to invoke fear, or for monetary gain. Threats of bombing and actual explosions are sometimes employed as much for their psychological, terroristic effect as for any physical damage. Most bombers in this country do not appear to desire to kill anyone, though there are often fatalities.

Explosions and the threat of bombing have created a need for practical knowledge to cope with the increasingly violent activities of people who represent segments of unrest in our society. Repeated criminal acts in which the use of or threat to use explosives against educational institutions, industry, law enforcement, and the general public place a most urgent responsibility on all emergency service organizations. To ensure a safe place to live and work, everyone concerned with these problems should be prepared. Forethought on the many possible dilemmas is necessary. With organized standard operating procedures, most bomb threat incidents will be resolved with minimal personal injury and property damage (Figure 22-1).

What should be done if a bomb threat is received? Should it be ignored or would it be best to always evacuate the building? If the callers realize that they can disrupt the organization every time they make a telephoned bomb threat, an epidemic of calls may occur; if the personnel are not evacuated and an explosion does occur, there could be many persons killed or maimed. An official could be damned for evacuating the structure every time a threat is made and damned for ignoring the warning when an explosion occurs. A strategy should be formulated and a standard operating procedure adopted so all concerned officials and agencies will handle the incident in the safest and most cooperative manner.

Handling Bomb Threats and Explosives

Fig. 22-1. Explosive reactions are not always predictable. Two university students were using a case of dynamite as a target for their indoor rifle range, believing that blasting caps were necessary for detonation. They were both badly injured.

PURPOSE OF BOMB THREATS

The only two reasonable explanations for a telephone call reporting that a bomb is to go off in a particular installation are:
1. The caller has definite knowledge or believes that an explosive or incendiary device has been or will be placed and wants to minimize personal injury or property damage. The caller may be the person who placed the device or someone else who has become aware of such information.
2. The caller wants to create an atmosphere of anxiety and panic that will, in turn, possibly result in a disruption of the normal activities at the institution where the device is purportedly located.

Creating Panic

When a bomb threat call has been received, there will be a reaction to it. If the call is directed to an installation where a vacuum of leadership exists or where there has been no organized advance planning to handle such threats, the call will result in panic.

Panic is one of the most contagious of all human emotions. Panic is defined as a "sudden, excessive, unreasoning, infectious terror." Panic is caused by fear—fear of the known or the unknown. Panic can also be defined in the context of a bomb threat call as the ultimate achievement of the caller.

Once a state of panic has been reached, the potential for personal injury and property damage is dramatically increased. Emergency and essential facilities can be shut down or abandoned and the community denied their use at a critical time.

Abandonment

Another possible motive is achieving abandonment of the premises where the bomb has been planted. Leaving facilities unattended can lead to destruction of the facility and the surrounding area. Large chemical manufacturing plants, power generators, steam boilers, and other such facilities require the attention of operating personnel.

DEVELOPING A PLAN FOR HANDLING BOMB THREATS

The importance of being prepared with a plan to handle bomb threats cannot be overemphasized. Here is where rescue and emergency service organizations can do the community a great service. Community leaders, business people, merchants, and civil organizations should be made aware of the growing possibility of bombing being used to terrorize or to achieve certain ends.

Effects of not being prepared or not having an organized plan to handle bomb threat calls can result in a lack of confidence in the leadership of a company or organization. This will be reflected in lower productivity, apprehensive personnel, and a reluctance to continue employment at a location that is being subjected to bomb threat calls.

Following are some guidelines for the proper procedures in telephoned and written threats. These procedures should be made a part of educational efforts by rescue groups, publicized in the media, and presented at civic and business club meetings to make the community conscious of preparedness.

Telephoned Threats

When an anonymous caller reports that a bomb has been placed on the premises, the procedures outlined here should be followed. In all cases, the report should be treated as genuine until it has been investigated and the area searched.

1. The person receiving the call should attempt to keep the caller on the line as long as possible. Ask the caller to repeat the message. Record every word spoken by the person. If so equipped, the switchboard operator should be instructed to turn on a tape recorder to make a transcription of the conversation. This record would be invaluable later to apprehend and convict the caller.
2. If the caller does not indicate the location of the bomb or the time of possible detonation, ask for this information. Try to learn when the bomb is set to explode, where it is, what kind

it is, what it looks like, who placed the bomb and why, and what the explosive is. Ask about the motives behind the bomb threat.
3. The caller should be informed that the building is occupied and the detonation of a bomb could result in death or serious injury to many innocent people.
4. Pay particular attention to peculiar background noises such as motors running, airplanes, traffic, background music, and any other noise that may give a clue as to the location of the caller.
5. Listen closely to the voice characteristics: male or female; young, middle age, or old; accents; speech impediments; tone; and loudness. Note the voice quality; was the person calm or excited? Did the caller appear to be familiar with the location involved? Was the person's voice familiar?
6. This information should be reported immediately to the police; fire department; FBI; and the Department of the Treasury, Bureau of Alcohol, Tobacco, and Firearms.

If more than one or two threats are made in a month, consideration should be given to requesting the telephone company to monitor the telephone calls for a period of time. Because of the number of demands for this service, tracing devices can be left at a given location for only a short time.

Written Threats

When written threats of bombing are received, the person receiving the message should be instructed to save all materials, including any envelope or container. Once the message is recognized as a bomb threat, further unnecessary handling should be avoided. Every possible effort must be made to retain evidence such as fingerprints, handwriting or typewriting, paper, and postal marks, which are essential to tracing the threat and identifying the writer.

While written messages are usually associated with generalized threats and extortion attempts, a written warning of a specific device may occasionally be received. It should never be ignored. With the growing use of voice print identification techniques to identify and convict telephone callers, there may well be an increase in the use of written warnings and calls to third parties.

Authorities should be notified immediately and given all the particulars available.

EVACUATION OF THE PREMISES

Fire and police departments should always recommend evacuation in response to bomb threats; however, authority for evacuating a building is usually the responsibility of the officials in charge of the structure. Fire and police vehicles responding to a bomb threat should cease using their red lights and sirens before they near the area. Their arrival should be as inconspicuous as possible.

The decision to evacuate or not to evacuate is one of the most serious decisions to be made by the management. To help the administrator decide, consideration should be made as to how the caller made statements and answered questions. This will help determine whether to take the threat seriously, evacuate the premises, and initiate a search.

RESCUE FROM HAZARDOUS MATERIALS

The three elements common to all bomb emergencies are disorganization, confusion, and fear. For these reasons, the bomb should be contained or removed with the least amount of disruption and commotion.

Management may pronounce a carte blanche policy in the event of a bomb threat; evacuation will be effected immediately. This decision circumvents the calculated risk and gives prime consideration for the safety of personnel in the building. This can result in production down-time, and can be costly, if the threat is a hoax. The alternative is for management to make the decision on the spot at the time of the threat. There is no magic formula that can produce the proper decision.

In the past, the vast majority of bomb threats turned out to be hoaxes. However, today more of the threats are materializing. Thus, management's first consideration must be for the safety of people. It is practically impossible to determine immediately whether a bomb threat is real.

TERRORIST ACTIVITIES

Investigations have revealed that the targets for terrorist bombings are not selected at random. The modus operandi for selecting the target(s) and planting the explosives appears to follow this pattern. The target is selected because of political or personal gain to the terrorist. It is then kept under surveillance to determine the entrances and exits most used, and when. This is done to determine the hours when very few people are in the building. The idea is that the intent is not to injure or kill people, but to destroy the structure. Reconnaissance of the building is made to locate an area where a bomb can be concealed, do the most damage, and where the bomber is least likely to be observed.

A test, or dry run, of the plan is often made. After the dry run and at a predetermined time, the building is infiltrated by the bomber(s) to deliver the explosives or incendiary device. The device may be fully or partially preset prior to planting. If it is fully set and charged, it is a simple matter for one or two of the group to plant the device in a preselected concealed area. This can be accomplished in a minimum of time. If the device is not fully set and charged, one member may act as a lookout while others arm and place the device. Most devices used for the destruction of property are usually of the time-delay type. These contrivances can be set for detonation to allow sufficient time for the bomber(s) to be a considerable distance away before the bomb threat call is made and the device detonated.

BOMB SEARCH

Regardless of whether the building is evacuated, a search of the premises should be made. Persons working in the building should help search for the bomb. They are familiar with the area and can more easily spot something out of place. To be proficient at checking a building, search personnel must be thoroughly familiar with all hallways, restrooms, false ceiling areas, and every location where an explosive or incendiary device may be concealed. If the emergency service personnel are to scrutinize the premises, the contents and the floor plan will be strange to them unless they have previously reconnoitered the building. Thus, it is extremely important that maintenance personnel or other employees who are thoroughly familiar with the premises be assigned to aid them. When a room or particular facility has been searched, it should be marked or the room sealed with a piece of tape and reported to the group supervisor.

Precautions in Search

The evacuation or search unit should be trained only in evacuation and search techniques and not in the methods of neutralizing, removing, or otherwise having contact with the device. If a bomb or other suspicious object is located, it should not be disturbed but a string or paper tape may be run from the device location to a safe distance and used later as a guide to the device.

Do not touch a strange or suspicious object. Its location and description should be reported to the person in charge. The removal and disarming of a bomb or suspicious object must be left to the professionals in explosive ordnance disposal. Prior to any bomb threat, every official who could become confronted with this type of problem should determine what local agencies maintain bomb squads. There is no uniformity; some cities delegate this responsibility to the police department, others depend on the fire department, and areas close to military installations receive this service from that organization. Every bomb threat standard operating procedure should include a list of emergency telephone numbers for summoning assistance.

If the danger zone is located, the area should be blocked off or barricaded with a clear zone of at least 300 feet (90 meters) until the object has been removed or disarmed.

During the search of the building, a rapid two-way communication system is of utmost importance. Such a system can be readily established through existing telephones. *Caution*—the use of radios during the search can be dangerous. The radio transmission energy can cause premature detonation of an electric initiator (blasting cap).

During the search, emergency medical personnel should be on the scene and standing by in case of an accident caused by an explosion of the device. Fire fighting units should be alerted and prepared to combat any blaze which may result.

Room Search

The following technique is based on the use of a two-person searching team. There are many minor variations possible in searching a room. Only the basic methods are discussed in the following pages.

Listening. When the two-person search team enters the room to be searched, they should first move to various parts of the room and stand quietly with their eyes shut and listen for a clockwork device. Frequently, a clockwork mechanism can be quickly detected without the use of special equipment. Even if no clockwork mechanism is detected, the team is now aware of the background noise level within the room itself.

Background noise or transferred sound is always disturbing during a building search. When checking a structure, if a ticking sound is heard but cannot be located, one might become unnerved. The ticking sound may come from an unbalanced air conditioner fan several floors away or from a dripping sink down the hall. Sound will transfer through air conditioning ducts, along water pipes, and

RESCUE FROM HAZARDOUS MATERIALS

through walls. One of the worst types of buildings to work in is one that has steam or water heat. This type of structure will constantly thump, crack, chatter, and tick due to the movement of the steam or hot water through the pipes and the expansion and contraction of the pipes. Background noise may also be outside traffic sounds, rain, wind, etc.

Detailed Examination. The person in charge of the team should determine how the room is to be divided for searching and to what height the first sweep should extend. The initial check should cover all items resting on the floor up to the selected height.

The room should be divided into two equal parts. The equal division should be based on the number and type of objects in the room, not the size. An imaginary line is then drawn between two objects to indicate the zones.

To select the initial height of the first search, look at the furniture or objects in the room and determine the average height of the majority of items resting on the floor. In an average room this height usually includes table or desk tops, chair backs, and similar objects. The first searching height usually covers the items in the room up to hip height.

After the room has been divided and a searching height has been selected, both searchers go to one end of the room division line and start from a back-to-back position. This is the starting point, and the same point will be used on each successive searching sweep. Each searcher now starts searching around the room, working toward the other member, checking all items resting on the floor around the wall area of the room. When the two searchers meet, they will have completed a wall sweep and should then work together and check all items in the middle of the room up to the selected hip height. It is important to thoroughly check the floor under carpets and rugs. The first searching sweep should also include those items which may be mounted on or in the walls, such as air conditioning ducts, baseboard heaters, built-in wall cupboards, etc., if these fixtures are below hip height. The first searching sweep usually consumes the most time and effort. During all searching sweeps, use an electronic or medical stethoscope on walls, furniture, floors, and other items which could conceal an explosive device.

The second room searching sweep usually covers the height from the hip to the chin or top of the head. The two searchers should return to the starting point and repeat the searching techniques at the second selected searching height. This sweep usually covers pictures hanging on the walls, built-in bookcases, tall table lamps, etc.

The third searching sweep usually covers the area from the chin or top of the head up to the ceiling. This sweep includes high mounted air conditioning ducts, hanging light fixtures, etc.

If the room has a false or suspended ceiling, the fourth sweep involves investigating this area. Check flush or ceiling mounted light fixtures, air conditioning or ventilation ducts, sound or speaker systems, electrical wiring, structural frame members, etc.

Post a conspicuous sign or marker indicating that the search has been completed. If the use of signs is not practical, use a piece of colored tape across the door and doorjamb approximately 2 feet (60 centimeters) above floor level. The method of indicating that an area has been checked should be uniform among all team members, and understood by all. It would not be wise to adopt a method that could allow a saboteur an opportunity to keep everyone out of the room in which the bomb is planted by hanging a sign on the doorknob.

The room searching technique can be expanded. The same basic methods can be used to check a convention hall or airport terminal. Encourage the use of common sense or logic. If a guest speaker at a convention has been threatened, common sense would indicate searching the speaker's platform and microphones first, but always return to the searching technique. Do not rely on random or spot checking of only logical target areas. The bomber may not be a logical person.

Outside Areas. When outside areas are checked, pay particular attention to street drainage systems, manholes in the street, and in the sidewalk. Thoroughly check trash receptacles, garbage cans, dumpsters, incinerators, parked cars and trucks, mail boxes, and other places large enough where a device can be concealed.

Schools. School bombings are usually directed against nonstudent areas. Find out which teachers or staff members are unpopular and where they work. The problem areas in schools are student lockers and the chemistry laboratory.

Student lockers are locked; no accurate record of the combinations are available. Every other locker seems to tick or click. School authorities should be requested to open the lockers so they can be checked.

Chemistry labs should be treated with caution. Each year some student tries to make an explosive mixture or rocket fuel in the classroom, gets scared, and phones in a bomb call. The best procedure is to ask the chemistry teacher to inspect the classroom, laboratory, and chemical storage area. The teacher will know most of the items and where they should be.

If repeated bomb threats are received at schools, it is likely that a student wants to disrupt the classes and force an early school dismissal. It will tend to reduce the number of this type of bomb scare if the school authorities will conduct makeup classes on Saturday or lengthen the school day to regain the lost class time.

Office Buildings. The biggest problem in office buildings is many locked desks; a repair of desk locks is an expensive item. There are many other areas to inspect, such as filing cabinets, storage closets, wall lockers, etc. Many companies have a security system to protect fashions and industrial secrets; these are usually guarded with alarm systems that ring large bells when a pressure mat, electric eye, microswitch, or other device is disturbed.

Auditoriums, Amphitheaters, and Convention Halls. Here, thousands of seats must be checked on hands and knees. Look for cut or unfastened seats with a bomb inserted into the cushion or back. Check out the stage area, which has tons of equipment on it; also the speaker's platform and the microphones. The area under the stage generally has crawlways, tunnels, trapdoors, dressing rooms, and storage areas. The sound system is extensive and the air conditioning system is unbelievable. The entire roof area in a theater frequently has one huge storage room and maintenance area above it. Check all hanging decorations and lighting fixtures.

Airport Terminals. These structures combine all the problems covered under schools, office buildings, and auditoriums. In addition, there are outside areas and aircraft.

Aircraft. The complexities of aircraft design make it unlikely that

RESCUE FROM HAZARDOUS MATERIALS

even a trained searcher will locate any but the most obvious explosive or incendiary device. Therefore, detailed searches of large aircraft must be conducted by maintenance and crew personnel who are entirely familiar with the construction and equipment of the plane. In emergency situations where searches must be conducted by public safety personnel without the aid of aircraft specialists, the following general procedures should be used:

1. Evacuate the area and remove all personal property.
2. Check the area around the craft for bombs, wires, or evidence of tampering.
3. Tow the aircraft to a remote area.
4. Starting on the outside, work toward the plane's interior.
5. Begin searching at the lowest level and work up.
6. Remove freight and baggage, then search cargo areas.
7. Check out rest rooms and lounges.
8. Be alert for small charges placed to rupture the pressure hull or cut control cables. The control cables usually run underneath the center aisle.
9. With special attention to refuse disposal containers, check food preparation and service areas.
10. Search large cabin areas in two sweeps.
11. Check the flight deck.
12. Simultaneously, search the baggage and freight in a safe place under the supervision of airline personnel. If passengers are asked to come forward to identify and open their baggage for inspection, it may be possible to quickly focus in upon unclaimed luggage.

Elevator Wells and Shafts. Elevator wells are usually 1 to 3 feet (30 to 90 centimeters) deep with grease, dirt, and trash that must be probed by hand. To check elevator shafts, get on top of the car with flashlights and inspect the shaft while slowly moving the car upward. Remember that the counterweights are descending as the car is rising; check the counterweights thoroughly. Often there are strong winds in the shaft; do not stand too close to the edge of the car.

SUSPICIOUS OBJECT LOCATED

Types of bombs used in terrorist activities include explosive, chemical, and fire. Most bombs encountered in other than a war situation will be crude and, fortunately, usually limited in size and force. Any bomb, no matter how insignificant looking, has the power to kill or maim instantly anyone who is foolish enough to handle it without adequate training.

It is imperative that personnel taking part in the search be instructed that their mission is only to search for and report suspicious objects, not to move, jar, or touch the article or anything attached thereto. The removal or disarming of a bomb must be left to the professionals in explosive ordnance disposal. Remember that bombs and explosives are made to explode, and there are no absolutely safe methods of handling them. Following are the proper procedures:

1. Report the location and an accurate description of the object to the official in charge of the incident.
2. Evacuate the building and keep all unnecessary people at least 90 meters (300 feet) away from the structure.

3. Place sandbags or mattresses, not metal shield plates, around the object. Do not attempt to completely cover the device.
4. Identify the danger area, and block it off with a clear zone of at least 300 feet (90 meters); include the area below and above the object.
5. Check to see that all doors and windows are open to minimize the primary damage from the blast and reduce the secondary damage from fragmentation.
6. Do not permit reentry into the structure until the device has been removed or disarmed, and the building declared safe for reentry.

LETTER AND PACKAGE BOMBS

Letter and package bombs are not new. While the latest incidents have involved political terrorism, such bombs are made for a wide variety of motives. The particular form of these bombs varies in size, shape, and components. They may have electric, nonelectric, or other sophisticated firing systems.

Management personnel in all areas of business and industrial organizations should be educated to make sure that personnel at all levels be prepared for the possibility of letter and package bombs.

Mail handlers should be instructed to be alert to recognize suspicious looking items. Although there is no approved, standard detection method, the following precautions are suggested:
1. Look at the sender's address; is it a familiar one?
2. Is correspondence from the sender expected? Do the characteristics of the envelope or package resemble the expected contents?
3. If the item is from another country, is it expected? Are there any friends or relatives traveling in that area? Was some item purchased from a business associate, charitable or religious organization, or other group?

If a suspicious-looking letter or package is received, do not attempt to open it. Isolate it and evacuate everyone in the vicinity to a safe distance. Notify the police and await their arrival.

BOMB EXPLOSION

In the event of an explosion, personnel must be alert for possible additional bombs or devices. Terrorist groups have initially exploded a small charge of explosives and later, after the emergency personnel have arrived, exploded a larger bomb that killed and injured police and fire fighters. If an explosion occurs, the scene should be secured immediately.

It is proper to emphasize that the extrication of the injured may be very dangerous for those trying to help. Protection for emergency care personnel is of primary importance in explosion emergencies. Falling debris and fire hazards exist for a long time after such an event. Rescue must be conducted with these hazards in mind.

BOMB DISPOSAL

Explosive ordnance disposal units composed of personnel who are specially trained in handling explosives will help curb the fear and

RESCUE FROM HAZARDOUS MATERIALS

confusion that bomb threats instill by preventing the destruction that the explosives create.

The personnel in these units should not take any unnecessary chances. Each incident demands extreme caution to protect the lives of those involved. Standard operating procedures, top efficiency, and sophisticated equipment are the prime ingredients necessary to disarm, dismantle, or destroy a bomb package.

Explosive ordnance disposal units should have on their vehicle the equipment that has the capability of identifying the contents of a bomb package. These devices would include an X-ray unit, listening devices with contact microphones that can be immersed in water, and a closed-circuit television. The primary intent is to allow the personnel an opportunity to conduct an inspection while spending a minimum amount of time near the bomb. The X-ray machine puts an image of the contents of the bomb package on a fluoroscope screen; the television camera projects the image from the camera to the closed circuit set in the squad truck for easy and safe viewing.

Standard body armor should be worn any time the crew is in close proximity to any explosive device.

Once it has been determined what type of explosive device was planted—clock- and radio-controlled explosives and Molotov cocktails are the most common—the team must then decide on the best way to destroy or remove it.

Before attempting to disarm a bomb, sandbags or other protective barriers should be placed around the device. Then, if an explosion should occur before the disarming process is completed, the sandbags will channel the explosion in a safe direction.

The primary method of rendering a bomb safe at the scene is by disarming it. The two most common disarming devices are the water cannon and a shotgun. The water cannon destroys the contents of a package by shooting a charge of water into it. Powder inside the cannon drives the water into the package at great force and high velocity. The shotgun disarmer consists of shooting either a buckshot, chilled lead shot, or a rifle deerslug from a 12-gauge shotgun. The bullets provide an opening in the package so that the contents can be inspected or they may destroy the fusing system to prevent the bomb from detonating.

Dismantling a bomb package may be one of the methods considered, but this is more hazardous and is utilized only when necessary.

EXPLOSIVES

An explosion is the rapid release of energy. The amount of material that releases this energy, the type of material, and, to some extent, the size of the space and degree of confinement of an explosion will contribute to its overall magnitude. Most people hear only the sound of an explosion. Very few see the fiery flash of the energy release. A person who is too close to see this flash is usually burned.

At the same time as the flash occurs, a shock wave starts outward in all directions from the point of the explosion. Two types of damaging pressure are exerted by this wave, and they occur almost simultaneously. One of these is known as overpressure. It is the squeeze effect on objects within its range. The pressure is measured in pounds per square inch (psi) over atmospheric pressure. It has nothing to do with motion.

At the same time as the shock wave, a second type of pressure, called dynamic pressure, is exerted. This is motion and is the equivalent of a very high wind. It pushes, tumbles, or tears apart objects in its path. It is difficult to determine the existence of these two pressures separately. Assume that they occur together.

Shock waves exert the same general effect on all objects in their path, whether the object is a building or the human body. First, the pressure around the object rises very sharply (overpressure) as the wave hits it, bends around it, and tends to crush it inward. Simultaneously, a rush of wind (dynamic pressure) will strike the object and tend to push it over, tear it apart, and hurl its debris through the air. Next, there is a slight decrease in pressure to below normal (suction), as the air turns and rushes back in a reverse direction. Less, but still considerable, damage is done during the suction phase. At this point, fires caused by the explosion that were not extinguished by the blast will break out.

Injuries due to explosions are caused by the blast itself (overpressure and dynamic pressure), by flying objects that penetrate the body, by heat and fire from the blast, and by other related causes such as inhalation of smoke. Any injuries suffered in an explosion can be complicated by the amount of time that may pass before rescue can be effected; during this time lapse, uncontrolled hemorrhage, shock, and the effects of inhaling smoke and fumes can be compounded.

Blast Force

Within the blast area, direct injuries resulting from the squeeze and tearing of the blast forces are predominant. These include rupture of the eardrums, rupture of internal hollow or solid organs, internal hemorrhage, and, particularly, contusion of the lungs. Lung contusions, analogous to skin bruises, are accompanied by edema and hemorrhage into the lungs, which may occur to such an extent that those portions of the lung no longer oxygenate the passing blood. A patient may experience severe hypoxia (insufficient oxygen) from this injury. All such trauma as those mentioned should be expected from explosions.

Heat

Within the blast area and close to the point of explosion, much heat is liberated, which is rapidly dissipated. Victims close to this point may be burned. The extent of injury will depend on the amount of skin exposed and the nearness of the person to the explosion. Burning agents may be hot liquids or vapors, but the usual cause of the injuries is a dry hot gas from the explosion itself.

Flying Objects

Persons near the blast area, but not trapped, may suffer indirect injuries caused by flying debris, collapse of buildings, or by being blown about by the wind of the blast. These injuries will include abrasions, lacerations, contusions, concussions, and fractures. There may be penetrating wounds of the body and internal organs. The speed of flying objects, their size and shape, and what they consist of determine the degree of injury they inflict.

Falling Objects

Collapse of buildings or other structures may add injuries caused by falling objects to the foregoing injuries.

Being Blown About

If a person is blown about or caused to fall any distance, he can expect injuries directly related to this violence. This trauma is usually lacerations, contusions, fractures, or any of the internal complications of blunt injury.

Defining Terms

The term *explosion* is an effect, rather than a cause, which generally results from a vigorous reaction with a sudden release of a large amount of energy due to rapid production of gases and the liberation of heat. An explosion exerts enormous pressures at a rate sufficient to have destructive effects. Such action is accompanied by a loud report with objects flying off in different directions.

A *detonation* is regarded as an extremely rapid and violent explosion, with a practically instantaneous release of chemical energy. This phenomenon has specific characteristics in that it is associated with a shock wave which is traveling faster than the speed of sound. It is characterized by a most powerful and sudden disruptive effect, generally evidenced by high fragmentation or great shattering (brisance). This is also referred to as a high order explosion. Explosives that produce this effect are dynamite, TNT, sensitized ammonium nitrate, and similar bursting compounds.

A *deflagration* is a type of explosion that ordinarily occurs in a relatively slow, pushing manner. These are referred to as low-order explosives and are commonly utilized as propellants. In this group are black powder, smokeless powder, flammable gases and vapors, and dusts.

Causes of Explosions

Explosions are caused by many substances other than the chemical compounds that are labeled as explosives. Following are the most common causes of explosions:

1. Gases: natural gas, sewer gas, and other flammable gases.
2. Vapor from flammable liquids: gasoline, solvents, cleaning fluids, and other low flash point flammable liquids.
3. Dusts: Some of the most destructive explosions have resulted from the ignition of a cloud of flammable dust. Among these are combustible metals, grain, starch, plastic, and carbonaceous dusts.
4. Explosives and blasting agents of a commercial variety. Bombs and incendiary devices of a military nature are also deadly.
5. Unstable or explosive commercial chemicals.
6. Steam, air, mechanical, or electrical explosions.

The rates of transformation of explosives have been found to vary greatly. One group that includes smokeless powder and black powder undergoes autocombustion at rates that vary from a few inches (centimeters) per minute to approximately 1,320 feet (400 meters) per second. These are known as *low explosives*. Being categorized as a low explosive does not mean low in hazard, for explosives so classified are among the most dangerous; they are commonly utilized as propellants. Black powder has been responsible for some of the most devastating explosions in history.

A second group, which includes TNT and nitroglycerine, has been found to undergo detonation at rates from 3,300 to 28,050 feet

(1,000 to 8,500 meters) per second; such materials are known as *high explosives*. This group is used for their brisance, or shattering effects.

From the standpoint of emergency personnel, any explosive is a potential danger when involved in a fire, accident, or other incident. The principal advantage in attempting to classify explosives is that the handling personnel can segregate initiating explosives from the detonating materials, as well as highly flammable substances from explosives that are detonated by heat.

HOW EXPLOSIVES REACT TO HEAT OR FIRE

Although many tests have been conducted by civil and military authorities, and much has been learned of value, it should be remembered that the action of explosives in a fire or violent accident is not always predictable. The first job of any emergency service personnel is to see that conditions are maintained in any area where explosives are handled so as to prevent any fire from starting. But if a blaze should erupt, a knowledge of the possible reactions of various explosives, so far as this information is known, should prove helpful.

Small Arms Ammunition

Small arms ammunition, which includes cartridges with bullets less than 1 inch (0.4 centimeter) in diameter, is not considered an explosive hazard. Its principal danger in connection with storage is its involvement in fire from an outside source.

When involved in fire, the principal danger is from flying bullets and cases, which, being light and not having much velocity, seldom go over 200 yards (180 meters). Application of hose streams from behind whatever protection may be afforded is good practice in order to avoid flesh wounds from possible missiles. According to military authority, such a fire may be controlled and extinguished by flooding or spraying with large amounts of water.

Fixed or Semifixed Ammunition

This type of ammunition resembles the ordinary bullet in appearance, having the propelling charge case attached to the projectile but capable of being removed to vary the amount of the charge. This type also includes the large calibers.

The danger when involved in fire is similar to that of small arms ammunition; a fire may be fought in a magazine or cargo containing boxes of fixed ammunition until it appears to be getting out of control. There would appear to be little likelihood of a whole pile of such ammunition detonating at once if the boxes in which it is stored catch fire.

Black Powder

Black powder, or gunpowder, is a mixture of potassium nitrate, charcoal, and sulfur. It is considered to be one of the most dangerous explosive hazards as it is ignited easily by heat, friction, or sparks. It is used in propellant charges, igniters, primers, and as a bursting charge for hand grenades. It is commonly found in sporting good stores, gun shops, and in the homes and garages of people who reload their own ammunition.

When ignited, black powder burns so fast and with such

intense heat that the effect is an instantaneous explosion. All that can be done is to attempt to keep the fire in other materials from reaching the containers of black powder. If it appears that the efforts will fail, no attempt should be made to combat such a fire; fire fighters should either get away and seek cover or, if there is insufficient time, fall flat on the ground to avoid injury.

Blasting Caps, Fuses, Detonators

The two primary hazards in these devices are shock and the possibility of fire. Electric blasting caps may be ignited by radar, radio transmission, and certain other stray electrical impulses; they are not likely to detonate en masse from any of these dangers, while other types might. The total quantity of explosive materials in these items is usually limited, so structural damage will remain light. Small missiles having slight range would be formed.

Blasting Gelatin

Blasting gelatin is nitroglycerine stiffened by having 7 percent collodion cotton dissolved in it. It is among the most powerful of all the explosives in common nonmilitary use; it may detonate when near or in a fire. No attempt should be made to combat a blaze involving blasting gelatin, but efforts should be directed toward covering exposures.

Bombs

Military bombs are of several types. The practice, or dummy, bomb offers no hazard as it is stored empty. The boxes containing other types of bombs are marked to indicate whether they contain chemical agents, high explosives, etc.

Fires occurring in fragmentation bombs, because they only have a small amount of TNT or amatol, are apt to cause intermittent detonations similar to the reaction described for fixed ammunition.

Demolition bombs, since they have a thin case and a very high percentage of explosive content, tend to detonate en masse if a fire occurs in a magazine in which they are stored. This could also happen in a fire or other accident involving trucks or railroad cars during transportation. Fire fighters and rescue personnel summoned to such a situation should keep at least 1,000 feet (300 meters) away, using available cover, and stand by to extinguish fires which may start as a result of the explosion.

Chemical bombs include incendiaries, smoke producers, toxic gas releasers, etc. If these are present in or close to a fire, their small explosive charges are likely to burst them and release the toxic gases. Emergency personnel should wear self-contained breathing apparatus and work from the windward side when combating fires involving such materials. All occupants of buildings to the leeward should be notified of escaping gases.

Dynamite

Dynamite, which consists of various proportions of nitroglycerine mixed with wood meal and oxidizing agents such as sodium and ammonium nitrates, usually comes packed in wooden or fiberboard boxes of 20 to 50 pounds (9 to 22.5 kilograms) capacity.

If unconfined, it may burn without an explosion, or it may explode with a high-order detonation. Therefore, it is not recommended to combat a fire in which dynamite has been a factor.

Pyrotechnics and Fireworks

The principal hazard of pyrotechnics and fireworks is the potential of fire. Some pyrotechnics may ignite spontaneously if exposed to moisture or high temperature; under similar conditions, most fireworks are difficult to ignite. Most types burn with intense heat and without serious explosions.

Authorities recommend that water be immediately applied in large quantities if these materials are present in a fire. Steam or fog is not considered to be as effective as water in deluge amounts. Fire fighters should work from behind barriers and not expose themselves unnecessarily.

Smokeless Powder

When unconfined, smokeless powders will burn fiercely with little smoke or ash and without explosion. At fires in which large quantities of this powder are a factor, even a deluge of water applied from close range would not extinguish the flames until all of the explosive had been consumed. However, hose lines are beneficial to extinguish fires in other materials and prevent the blaze from involving the explosives.

Many small arms powders with a cellulose nitrate base are nearly as sensitive to friction as black powder and should be treated with the same care. Smokeless powder dust is especially hazardous; loose powder may be ignited by sparks, friction, or intense heat; it burns rapidly with excessive heat.

Burning powder in a ship's hold or other confined space may be quite serious; it may explode and produce considerable structural damage and debris acting as missiles. Any smokeless powder may become unsafe to handle after exposure to high temperatures because the material starts to decompose.

If water can be applied immediately and in large quantity, the smokeless powder of the large grain size may be controlled in a manner comparable to cellulose nitrate. However, if it gets beyond the incipient stage and a large quantity is threatened, fire fighters should take cover until after the big flare occurs; they should then take action to extinguish any secondary fires which have been started.

TNT and Other High Explosives

High explosives in bulk form and demolition blocks have relatively similar hazardous characteristics. Although stable in storage, they can be ignited by spark and friction and be detonated by shock. When ignited, they will burn vigorously. In large quantities, such as a bulk shipment likely to be found on board vessels, trucks, and railroad cars, they would be very likely to detonate.

TNT, as stored in bulk, is in the form of finely divided white or yellow crystals that can be ignited by friction or spark; it also burns like tar. Detonation should always be expected.

Amatol is TNT mixed with ammonium nitrate; it has characteristics and hazards similar to TNT. It is not recommended to combat a fire involving amatol. Precautions should be taken to protect against injury from an explosion.

SIZE-UP OF INCIDENTS RELATING TO EXPLOSIVES

When sizing up a hazardous situation, the officer in charge should not forget the duty of protecting personnel and equipment. To rush in foolishly and attempt to combat a fire involving a large quantity of highly flammable, toxic, or explosive materials would be a breach of duty. The officer would not be considering either the public or emergency personnel because after the whole crew was wiped out, there could be a long delay in getting other companies to the scene to fight the secondary fires or control the other dangers at the scene. Factors to be considered are:

1. Will lives be lost unless the fire is extinguished? If property loss is the only consideration, then extreme caution in exposing personnel is warranted.
2. What are the dangers of the hazardous materials involved? Is there a likelihood of an en masse detonation? If there is a blast, would missiles and heavy repeated explosions result? What are the possibilities of intense heat and flash flames exposing or igniting adjacent properties?
3. What are the possibilities of extinguishment before highly dangerous explosives become involved? If the possibilities look good, either because the explosives can be removed or the fire is so far away from the hazardous materials that hose streams can quickly control the situation, then fighting the blaze is indicated. If the chances are slim of controlling the fire, then efforts should be directed toward evacuating the area, placing the emergency personnel behind protective cover, and standing by at a safe distance to be in readiness to extinguish fires caused by subsequent explosions.

FIRE FIGHTING RULES

1. Attempt to determine the type of explosive or other hazardous material, and the possible dangers.
2. Evacuate residents and bystanders from the surrounding area. If toxic gases or vapors are being emitted, start on the downwind side first.
3. Protect exposures to prevent spread of the fire.
4. Keep fire fighting apparatus at a safe distance.
5. Do not attempt to combat a fire involving hazardous materials at close range unless certain that an explosion or release of toxic gas or vapor will not result from exposure to intense heat.
6. When approaching the fire, utilize all available cover for protection, such as ditches, dikes, structures, etc.
7. If it appears that an explosion is imminent, all personnel should fall flat on the ground rather than attempting to run; this will reduce the dangers of being struck by fragments or a blast of heat and gases.
8. Self-contained breathing apparatus should be worn at any fires or other incidents involving chemical munitions or toxic materials.

QUESTIONS FOR REVIEW

1. The principal purpose of most bombing incidents in the United States is considered to be the intent to:
   a. Create fatalities.
   b. Promote a war.
   c. Injure and discredit police organizations.
   d. Harass.

2. Bomb squads usually recommend that when a threat that explosives have been planted in a building has been received, the threat should be treated as:
   a. A hoax.
   b. A hoax, unless some proof is offered.
   c. Authentic.
   d. Authentic, if some proof is offered.

3. The most contagious of all human emotions is generally regarded to be:
   a. Fear.
   b. Love.
   c. Panic.
   d. Hate.

4. Fire and police departments _____ evacuation in response to bomb threats. The missing term is:
   a. Always demand.
   b. Normally require.
   c. Discreetly suggest.
   d. Always recommend.

5. The initial search of a room to find an explosive device will normally cover:
   a. All items resting on the floor up to a selected height.
   b. Closets and under beds.
   c. Behind large articles of furniture.
   d. Vases, boxes, and other closed containers.

GLOSSARY OF RESCUE EQUIPMENT AND TECHNIQUES

Air Lift Bags Laminated fabric bags that may be inflated to raise vehicles and other heavy objects.

Axe, Flat-Headed A long-handled axe that can be used for both cutting and pounding.

Axe, Pick-Headed A long-handled axe with a cutting blade on one side and a pointed pick on the other. This is probably the most versatile tool used by fire fighters and rescuers. It can be used for cutting, prying, and forcing.

Block and Tackle Device used to move or lift heavy objects. Blocks that consist of frames and pulleys allow the lifting power to be multiplied in ratio to the number of sheaves used.

Breathing Apparatus Self-contained respiratory protection which will allow rescuers to enter and survive in atmospheres that have excessive amounts of smoke and toxic gases, or which have a low concentration of oxygen.

Chain Hoist A mechanism that is used to increase the mechanical advantage of a person so heavy weights may be raised.

Chisel, Pneumatic A chisel powered by compressed air that is extensively used for extricating trapped victims from crashed vehicles.

Claw Tool A lever with a curved claw on one end that will enable workers to multiply their human forces. Extensively used in forcible entry.

Come-Along A ratchet device which enables a chain or cable to create a considerable force to raise or move objects.

Cribbing Blocks of wood used to support or stabilize heavy objects.

Cutting Torch An oxygen-acetylene torch that may be used to cut through metals. Must be used with caution around crashed vehicles to prevent igniting spilled fuel.

Hook, Bail Also called a **Hay Hook.** A long curved hook with a hard wood "T" handle. Used to move baled hay, cargo, and is useful to pull out the glass in vehicle windows.

Hurst Power Tool A hydraulic spreading and pulling tool that is commonly called the *Jaws of Life*. This device is widely used for extricating trapped victims from crashed vehicles and collapsed structures.

Hydraulic Rescue Equipment A remotely controlled hydraulic jack with a wide range of rams and attachments that will allow pulling, pushing, and spreading. This tool is widely used for extricating trapped victims from crashed vehicles and collapsed structures.

Jack A type of machine for lifting heavy loads. There are three types: ratchet, screw, and hydraulic.

Knots A way of fastening a rope to itself or to another rope or object.

Lever A lever consists of a rigid bar that is capable of rotating about some point of support known as the fulcrum or the axis. Levers permit multiplying force so heavy objects may be raised or moved.

Glossary of Rescue Equipment and Techniques

Manual Lifting Force that a human being can generate by using the leg and back muscles. The average person should not attempt to lift more than 80 percent of his own weight.

Mini Quic Bar® A small forcible entry tool that is useful for prying and forcing. It is mainly utilized for forcible entry.

Pry Bar A type of metal lever used to force locked doors and windows or to move heavy objects.

Quic Bar® A multipurpose rescue tool that consists of a lever that can pry or force objects. Mainly used for forcible entry and to extricate victims from wrecked vehicles.

Quic K-Bar-T® Rescue Kit A rescue tool that consists of a slide hammer that drives either a chisel or a cutting blade. Mainly used to extricate victims from crashed vehicles.

Rope A fibrous line that is normally made of manila or synthetic fibers.

Saw, Chain A saw powered by a gasoline engine that is extremely useful for cutting lumber and other building materials.

Saw, Power-Driven Commonly called a rescue saw and is driven by a gasoline engine. A variety of circular blades will allow rapid cutting of metals, construction materials, bricks, masonry, and a wide range of other substances.

Snatch Block A block that is used to change the hauling direction of a rope or cable.

Wedges Tapered blocks of wood used to fill thin spaces or steady an object that is being stabilized.

Winch A drum that is driven by either an electric motor or a vehicle engine. Cable wound on the drum will exert considerable pulling force.

Wrecking Bar A type of long metal lever which enables rescuers to multiply their human forces so heavy objects may be raised or moved.

Appendix

Answers to the Questions for Review

Chapter 1	Chapter 7	Chapter 13	Chapter 19
1.-**b**	1.-**c**	1.-**b**	1.-**b**
2.-**a**	2.-**a**	2.-**a**	2.-**d**
3.-**d**	3.-**d**	3.-**d**	3.-**a**
4.-**d**	4.-**b**	4.-**a**	4.-**c**
5.-**c**	5.-**d**	5.-**d**	5.-**d**

Chapter 2	Chapter 8	Chapter 14	Chapter 20
1.-**a**	1.-**a**	1.-**d**	1.-**b**
2.-**d**	2.-**d**	2.-**a**	2.-**a**
3.-**c**	3.-**a**	3.-**b**	3.-**a**
4.-**b**	4.-**c**	4.-**c**	4.-**d**
	5.-**b**	5.-**a**	5.-**d**

Chapter 3	Chapter 9	Chapter 15	Chapter 21
1.-**c**	1.-**a**	1.-**a**	1.-**d**
2.-**b**	2.-**b**	2.-**c**	2.-**a**
3.-**d**	3.-**c**	3.-**d**	3.-**c**
4.-**b**	4.-**d**	4.-**b**	4.-**b**
5.-**d**	5.-**c**	5.-**c**	5.-**d**

Chapter 4	Chapter 10	Chapter 16	Chapter 22
1.-**c**	1.-**a**	1.-**a**	1.-**d**
2.-**d**	2.-**c**	2.-**d**	2.-**c**
3.-**a**	3.-**b**	3.-**b**	3.-**c**
4.-**d**	4.-**b**	4.-**c**	4.-**d**
5.-**b**	5.-**d**	5.-**c**	5.-**a**

Chapter 5	Chapter 11	Chapter 17	
1.-**d**	1.-**d**	1.-**a**	
2.-**b**	2.-**a**	2.-**c**	
3.-**d**	3.-**b**	3.-**b**	
4.-**c**	4.-**c**	4.-**a**	
5.-**a**	5.-**c**	5.-**d**	

Chapter 6	Chapter 12	Chapter 18	
1.-**b**	1.-**c**	1.-**d**	
2.-**a**	2.-**b**	2.-**d**	
3.-**a**	3.-**a**	3.-**a**	
4.-**d**	4.-**c**	4.-**c**	
5.-**c**	5.-**d**	5.-**a**	

Index

A

Abandonment, 12-13
ABCs (of prehospital emergency care), 4, 68, 71
Access: to aircraft wreckage, 113-114; through crashed auto's roof, 94-96; to highway accident victims, 87-100
Accident death rate, 2
Acetylene, 28-29
Acrolein, 398
Aerial ladders, 221-225
Aerodontalgia, 283
Aerotitus, media, 283
Air: composition of, 357; conservation of, 358-360; laws of, 278-280
Air ambulance operations (helicopter), 345-347
Aircraft: bomb threats to, 443-444; fixed-wing, 341-342; location of overdue, lost, or downed, 312; use in searches of, 313-314. See also Helicopter rescue
Aircraft abandonment, cold water survival after, 304-306
Aircraft accidents, 106-118; approaching, 107; crash hazards of, 110-111; extricating occupants of, 111-112; fire fighting at, 107-116, 350; forcible entry at, 113-114; governmental agencies and, 112-113; high- and low-impact crashes, 110, 111; military, 114-117
Air cylinders, recharging, 375-376
Air embolism, 283-284
Air lift bags, 26-27

Airline masks, 381
Airport terminals, bomb threats to, 443
Air stations, respiratory, 376-378
Alarm bell (on breathing apparatus), 367-368, 375
Alpha radiation, 422, 425, 431-432
Ambulance services, 3
Ammonia, 400-401
Ammunition, 449
Amperes, 165, 170-172
Amphitheaters, bomb threats to, 443
Anchors, 297-298
Anesthetic gases and vapors, 390
Anoxia, 284
Apartment houses, searching for victims in, 215
Area search procedures, 325, 327, 331-332
Armament, and military aircraft accidents, 114, 117
Army litter/stretcher, 61
Arson, 124-125
Asphyxiants, 390
Atomic structure, 420-421, 426-427, 430
Atomic weapons: emergency procedures for, 417-418; explosions of, 418-419; hazards of, 416-417
Attitude, professional, 8, 10
Auditoriums, bomb threats to, 443
Automobiles. See Motor vehicles
Automotive accidents. See Highway accidents
Avalanches, 328-332
Axes, 27

Index

B

Backboards, 56-58, 73, 74, 101. See also Basket stretcher
Backdraft explosions, 200, 201-203; preventing, 203-205
Backpack (for breathing apparatus), 375
Baskets: helicopter rescue, 348, 349; mesh screened (for water rescue), 307-308
Basket stretcher, 84, 221, 223, 262
Batteries (aircraft), disconnecting, 112
Batteries (vehicle): disconnecting, 103; fires in, 80-81
Beckets, 35, 36
Being blown about, by explosions, 448
Bends: in diving, 284-286; of rope, 34-36
Beta radiation, 422, 425, 431
Bight, 34, 35, 36
Black powder (gunpowder), 448, 449-450
Blanket drag, 55, 56
Blanket stretchers, 61, 62
Blast force, 447
Blasting materials, 450
Bleeding control, 71, 72
Blizzards, 233
Block and tackle, 22-23, 24
Boards (for water/ice rescue), 290. See also Backboards
Boats: grappling from, 296-297; for ice/water rescue, 290, 291, 295
Bombs, 117, 450; disposal of, 445-446
Bomb threats, 436-446
Boom hazards, from electrical contact, 177
Borrowed servant concept, 13
Bottom guidelines, 297-298
Bowline, 35, 36
Boyle's Law, 279
Bracing. See Shoring techniques
Breaking methods, 91, 93-94
Breathing: buddy, 360; control of, 359-360
Breathing apparatus, 197, 201; cleaning, 373-374; closed-circuit, 361; compressed air, 360-361; emergency operation of, 372; filter, 361; inspecting before each use of, 372-373; for mines, tunnels, pits, holes, 264, 265, 267-268; principles of operation, 371-372; for respiratory protection, 354-381; routine inspection of, 374-375; self-contained underwater (SCUBA), 278-283; service life of, 358-359; testing and approval of, 368; types and use of, 368-381
Buddy breathing, 360
Buddy lines, 298
Buddy system, uses of, 298-299, 365
Buildings: construction types of, 127-128; damage effects on, 128-129; evacuation of, 351; high-rise, 218-219; landslides and, 259; removing victims from, 222-225; rescue from burning, 196-219; rescue from collapsed, 120-144; searching for bomb threat in, 440-444; signs of collapse of, 123-126; stages of rescue from, 126-127; void formation in, 127, 128, 129-130
Buoyancy, human body's, 293
Buoys, in water rescue, 295, 297
Burning structures, rescue from, 196-219
Burns, electrical, 170, 172-173
Bus accidents, 102-104

C

Caisson disease, 284-286
Calcium hypochlorite, 402
Canopy jettison areas, 114, 115-116
Canyons, rescue from, 186-188, 225. See also High places
Carbon dioxide, 397-398
Carbon monoxide, 395-396
Carbon tetrachloride, 399-400
Cardiac evaluation, 72
Cardiopulmonary resuscitation (CPR): to electrical shock victims, 172, 184-185; during transportation, 46

Index

Carry techniques: chair, 53; fire fighters, 47-49; pack-strap, 51; and personnel safety, 41-42; seat, 50; three-person lift and, 52-53
Casualties. See Victims
Cave-ins, 258-268
Cave rescue, 332-335
Chain hoists, 23
Chain saws, 31
Chair carry, 53
Chem-Cards, 386-387
Chemical bombs, 450
Chemicals, hazardous, 383-390
Chemox oxygen breathing apparatus, 380
CHEMTREC, 389-390
Children, trapped, 181-182
Chill index, wind, 233-234
Chlorine, 401-402
Chlorine bleach, 402
Circular sweep pattern, 299, 301
Civil liability, 11-17
Claw tools, 20
Cleanliness, personnel's, 7
Cleanup, after pesticide encounter, 404-405
Cliffs. See Canyons; High places
Closed-circuit breathing apparatus, 378-380
Clothing: drag technique using, 55, 56; safety, 6, 121; uniform, 6-7
Clove-hitch, 34-35
Cold water survival, 304-307
Come-alongs, 24, 25, 99
Communications, 313
Compressed air cylinders, 281
Compression, 266-267; or respiratory air, 376-377
Conduct: ethical/moral, 6; professional, 7-11
Confidentiality, 15
Confined areas, victim transportation in, 47
Consent, from victim/patient, 14-15
Construction accidents, 4
Contour search procedure, 325, 326
Convention halls, bomb threats to, 443
Conversation, rescue personnel's, 9-10
Cooperation, 5, 11
Cribbing, 22
Crowbars, 37

Cutting tools, 27-31
Cutting torch, 28-29
Cyanides, 397, 400
Cyclones, 245-249
Cylinder valve, on breathing apparatus, 375

D

Dead shore, 140-142
Debris, 127, 143; tunneling through, 132-137
Decay, radioactive, 423
Decompression, 267, 268
Decompression sickness, 284-286
Decontamination, of radioactive materials, 412
Deep water rescue, 307-308
Deflagration, 448
Demand-type breathing apparatus, 369-378
Demolition, 124
Demolition bombs, 450
Desert survival and rescue, 336-339
Detonation, 448
Diesel engines, stopping, 103
Diesel fuel, flammability of, 79
Digging operations, 261
Discipline, of rescue personnel, 5-6
Disentanglement: from crashed motor vehicles, 87, 88-100; from machinery, 180-181; procedures of, 90-91; techniques, tools, and equipment for, 89, 90, 91-100
Displacement method, 91
Distortion method, 91
Distribution circuits, 167
Dividing. See Severing methods
Diving: accidents, 274-276; diseases, 283-288; preparations for rescue, 298-299; SCUBA, 277-288; tenders, 298
Dogs, in avalanche rescue, 331
Doors, electrical devices protecting, 168-169
Doors, opening: aircraft, 113-114; automobile, 91-93; building, 37-39, 168-169, 181-182; bus, 102-103; elevator, 153-155
Dosage, nuclear radiation, 431

Dragging techniques, water rescue, 295-297
Drag hooks, 295
Drag techniques, 53-56
Drill sessions, for breathing apparatus use, 361-362, 363
Driving liability, rescue personnel's, 17
Drowning, 271-273; in cold water, 273; under ice, 291-293
Drownproofing, 303, 306
Duty, rescue personnel's, 16
Dynamic pressure, 447
Dynamite, 450

E

Ears: anatomy of, 281-282; accidents/diseases of, 283
Earthquakes, 252-256; elevator damage from, 161-162; safety procedures for, 255-256; tsunami waves and, 249-250, 256
Education (of personnel). See Training (of public), 3, 270, 312-313
Egress systems, military aircraft, 115-116
Electric shock, medical effects of, 170-173, 174
Electrical burns, 170, 172-173
Electrical contact, rescue from, 174-175
Electrical generating stations/substations, 169
Electrical power, elevator rescue and, 150-151, 152, 160
Electrical wires, hazards of, 131-132, 165, 167-170
Electricity: currents of, 164-165, 170-172, 323-324; helicopter's static charge of, 349; from power companies, 166-167; shutting off, 169-170, 180; terms of, 165-166
Electrified booby traps, 168-169
Elements, natural and synthetic, 420, 421
Elevating platforms, 222, 225-226
Elevators, 145-163; bomb threats to walls/shafts of, 444; cables for, 151-152; call buttons on, 149; communication with passengers in, 148, 152-153; construction of, 147-150; doors/locks on, 153-155; earthquake damage and, 161-162; electrical power and, 148, 150-151, 152, 160; emergency exits on, 157-162; emergency unlocking devices, 149-150; fire in/around, 159-161, 219; forcible entry into, 155-157; freight, 153, 158-159; hydraulic, 148, 152; lighting in, 148-149; locating stalled car, 151; mechanics for, 147; passenger cooperation in, 152-153; psychological care of passengers in, 145-147, 153; removing occupants of, 155; victim pinned between hoistway and, 162
Emergency Action Guide, for hazardous materials, 387-388
Emergency medical care: at aircraft accidents, 112; at diving accidents, 274-276; at drownings, 272-273; of electrical burns, 173; at highway accidents, 70-74, 82, 88-89; liability for, 11-12; during pole top rescue, 185-186; priorities/sequence, 71, 83
Emergency Medical Technicians (EMTs), 3-5
Emotions, coping with, 8, 9-10, 362
Emphysema, 286
Engine compartment, vehicle fires in, 79-80
Engines, aircraft, 114-115
Environmental factors, 85-86
Equipment, 19-40; for extrication from crashed motor vehicle, 89, 90, 91-100; fire apparatus, 221-226; for high places rescue, 186-190, 191; for ice (water) rescue, 289-291; life-support, 46-47; for mountain rescue, 327-328; transportation by helicopter of, 351; for water rescue, 294-295, 297-298. See also specific kinds, e.g., Breathing apparatus; Rope; etc.
Ethical standards, 6
Eustachian tube, 281-282

Index

Evacuation, in response to bomb threats, 439-440, 444. *See also* Removal of victim
Evaluation, 316-320. *See also* Size-up
Exhaustion factor, 321, 338
Exploration, of collapsed buildings, 126
Explosions: causes of, 448-449; defined, 448; in mines, 263
Explosives, 436, 437, 445, 446-452; in nuclear weapons, 416-417; reaction to heat or fire of, 449-451. *See also* Bombs
Exposure factor, 321
Exposure protection, 209
External contamination (by radiation), 409-410
External radiation exposure, 409, 425-431
Extrication: from aircraft, 111-117; from collapsed buildings, 121; from crashed vehicles, 82-105. *See also* Removal of victims

F

Facepiece, of breathing apparatus, 363-365, 374, 381
Falling objects, from explosions, 447
Fallout, 419
Fatalities, aircraft accident, 112
Federal Aviation Administration (FAA), 112
Figure-of-eight knot, 34
Filling station, for respiratory air, 378
Filter masks, 381
Fire apparatus, 221-226
Fire fighters: carry technique of, 47-49; drag technique of, 53-54; respiratory protection for, 354-381
Fire fighting hazards, of electrical contact, 175-177
Fire fighting strategy, 198-205
Fire gases, composition of, 392-393
Fire rescue: helicopter use in, 349-351; by police officers, 206-207
Fires: aircraft accident, 107-113, 114, 117; automotive, 77-81; behavior of, 198-205; in buildings, 124-125, 196-219; in buses, 103; elevator, 159-161; explosive-related, 452; during floods, 239; in mines, 263; in pesticide storage, 404, 405; in pressurized tunnels, 268; radioactive materials in, 414-415; signal (in wilderness rescue), 315; tsunamis and oil, 250
Fire stream techniques, 109-110, 176, 199-200
Fireworks, 451
First aid care, liability for, 12
Fission process, 423-424, 426
Fission products, of nuclear explosions, 419
Flammable liquids, at highway accidents, 69, 70, 79-80
Flares, at highway accident site, 75-76
Flash floods, 238, 239, 240-241
Floating the victim (of diving accidents), 275-276
Floats, 295, 297
Floods: desert, 336; hurricane, 243; river, 238-241, 277. *See also* Water, rescue from
Floor (building), collapse of, 124, 125, 129-130, 136-137
Floor (motor vehicle), access through, 96, 97
Flying objects, from explosions, 447
Flying shore, 140, 141
Forcible entry, 37-40
Forecasting. *See* Prediction
Forepole method, 134, 135
Fractures, at highway accidents, 73, 74, 100
Frames, for tunnels, 134-136
Freezing rain, 232
Freight elevators, 153, 158-159
Fuel: aircraft, 108; diesel, 79. *See also* Gases and vapors
Fulcrum. *See* Levers
Fumigations, 405-406
Funnels (tornado), 246-247

G

Gamma radiation, 422, 423, 425-430

Garbage disposals, 182
Gases, laws of, 278-280
Gases and vapors: at aircraft accidents, 113-114; around burning structures, 197, 201, 202-203; around collapsed buildings, 131, 132, 137, 138-139; combustible, 357; explosions caused by, 448-449; at highway accidents, 69, 70, 79-80, 87; managing hazardous, 383-407; in mines, 263; oxygen-acetylene cutting torch and, 28-29; respiratory protection against, 354-381; riot control by, 406-407; skin penetration by, 365; in tanks, pits, holes, 264-266; toxic, 357; in tunnels, 137; types of, 390
Gasoline washdowns, 69, 70
Genetic damage, from radiation, 425
Gibbs (on elevators), 156
Glass. *See* Breaking methods; Windows; Windshield removal
Good Samaritan laws, 13-14
Government agencies: aircraft crashes and, 112-113; in wilderness search and rescue operations, 311-312, 315-316
Grab-and-pull technique, 101
Grappling techniques, 296-297
Ground currents, from lightning, 323-324
Gunpowder, 448, 449-450
Guns. *See* Armament

H

Hail, 238
Half-hitch, 34
Half-life, radioactive, 423, 424
Hand tools, 260
Hasps, forcing, 40
Hazard control, 86-87
Hazardous areas, on buses, 103-104
Hazardous gases. *See* Gases and vapors
Hazardous materials, 383-390; *Emergency Action Guide* for, 387-388; labeling/defining, 385-386, 387; procedures at transportation incidents, 388-390. *See also* Radioactive materials
Hazards: aircraft crash, 110-111; armament, 117
Heart: arrhythmias, and drowning, 272; electrical shock damage to, 171, 172, 184-185
Heart-lung resuscitator, 46-47
Heat: from explosions, 447; physiological effects of, 391. *See also* Fires
Helicopter rescue, 341-351; from deep water, 307-308; medical problems of, 346-347; from wilderness areas, 320
Hemorrhage control, priority of, 71, 72
High places, rescue from, 184-195, 315. *See also* Buildings; Mountain rescue
High-pressure atmosphere, respiratory protection in, 360-361
High-pressure bottles, recharging, 376
High-rise structures, fire in, 218-219
High-voltage wires, downed, 87
Highway accidents, 66-81; extrication from vehicles at, 82-105
Highway conditions, winter, 232
Highway Safety Act (1966), 3
Hinge pins, door, 38
Hitches, 34-35, 36
Hoisting/hoists, 186-187, 225, 266, 349. *See also* Carry techniques; Lifting Holders, of breathing apparatus, 387
Holes, rescue from, 264-266
Horse collar sling, 348-349, 350
Hose lines: breathing apparatus, 374, 375; fire fighting, 199-200, 251
Hose masks, 381
Hotels, searching for victims in, 215
Hot sticks, 175
Human chain, for water/ice rescue, 291
Hurricanes, 241-245
Hurst power tool, 24-25; uses of, 93, 94, 96, 98
Hydraulic components, vehicle fires in, 80

Index

Hydraulic elevators, 148, 152
Hydraulic jack, 21
Hydraulic rescue equipment, 25-26; uses of, 93, 94, 100
Hydrogen cyanide, 397
Hydrogen sulfide, 396-397
Hypothermia, 304
Hypoxia, 284, 393-394, 447

I

Ice rescue, 288-293
Ice storms, 232
Illness, discussing, 9, 10
Immediate action, in search and rescue operations, 314-316
Immobilization, victim, 57, 58, 73, 74, 100-102
Impaled victims, 181
Impedance, 165
Improvisation, stretcher, 61, 62
In loco parentis, 11
Incline drag, 55
Initial nuclear radiation, 418, 419
Injuries: discussing, 9, 10; nuclear radiation, 425, 428
Insecticides, 402-405
Institutionalized liability, 13
Intensity, earthquake, 253
Internal contamination (by radiation), 410, 425-429, 431-432
Intravenous (IV) equipment, 47
Ionization, 426, 427
Ionizing, 421-422
Irritant gases, 390
Isotopes, 421, 422, 423

J

Jacking, buses, 103-104
Jacks, 20-21; hydraulic, 25-26; rules for use of, 21-22
Jammed doors (vehicle), opening, 91-93
Jumping (from high places), 189-190

K

Kinetic energy chart, 189
Knots, 34-36

L

Labels, for hazardous materials, 385-386, 387
Labyrinth travel, 334
Ladder leg-lock, 192
Ladders, 191-194; as bridging between buildings, 194, 195; carrying, 191; climbing, 191-192; hazards of, 177; helping people down, 192-194; as sheerleg, 262; in water/ice rescue, 289, 290
Laminated glass, 93-94
Landing sites, helicopter, 342-345
Landslides, 258-268
Laser radiation, from aircraft, 115
Law enforcement personnel, 7, 12, 206-207
Legal terms, 13
Letter bombs, 445
Levers, 19, 20
Liability: civil, 11-17; driving, 17; institutionalized, 13; releases to avoid, 15
Life jackets/preservers, 289, 295
Lifelines, 264, 366-367
Lifesaving care. *See* Emergency medical care
Lifesaving nets, 188-190
Life-support equipment, 46-47
Lifting: devices for, 20-27; manual, 27. *See also* Carry techniques; Hoisting/hoists
Lightning, 173, 236-238, 323-324
Lights, underwater, 298
Limbs, injured, 44
Liquid petroleum gas (LPG), flammability of, 79
Liquids, hazardous, 383-390
Liquefied oxygen, (LOX), 108
Listening, in bomb threat search, 441-442
Litters. *See* Stretchers
Live wires. *See* Electrical wires
Locating victims. *See* Searching for victims
Locked doors. *See* Doors
Log rolling, 57-58
Looting, 125
Lowering the victim, 186
Lungs: diving accidents and, 283-284; heart- resuscitator, 46-47

M

Machinery accidents, 179-181
Machines, simple rescue, 19
Magnitude, earthquake, 253-254
Manual rescue tools, 20, 27, 37
Maps, wilderness travel, 313
Markers, water rescue, 295
Marking system, for searched areas, 213, 214, 216-217, 442-443
Marsh rescue, 339
Masks, breathing apparatus, 363-364, 374, 381
Mechanical aptitude, 5, 19
Medical opinions, of rescue personnel, 9, 10
Medical personnel, cooperation with, 11
Metal detector, in avalanche rescue, 331
Methyl bromide, 400
Military aircraft, 114-117
Mine accidents/rescue, 262-264, 335-336
Missing persons, search for, 324-327
Molecules, 420
Moral standards, 6
Motor vehicles: electrical contact with, 174; extrication from crashed, 82-105; submerged (in water), 273-274. See also Highway accidents
Mountain rescue, 327-332, 344
Muscles, for lifting and moving, 41-42

N

NASA breathing apparatus, 370-371
National Transportation Safety Board (NTSB), 112
Natural disasters, 230-256
Negligence, 15-17. See also Liability
Nets: lifesaving, 188-190; water rescue, 308
Neurological status, of auto accident victims, 74
Nitrogen absorption (bends), 284-286
Nitrogen narcosis (rapture of the depths), 286-287
Nitrous fumes, 298
Nuclear radiation. See Radiation
Nuclear weapons. See Atomic weapons

O

Office buildings, bomb threats to, 443
Ohm, 166
Ohm's law, 166
Open-circuit (demand-type) breathing apparatus, 369-378
Openings, See Doors; Windows
Overboard, person, 307
Overexertion, of SCUBA divers, 277
Overhand knot, 34
Overpressure, 446
Oxygen, administering, 46
Oxygen-acetylene cutting torch, 28-29
Oxygenation. See Respiration control
Oxygen deficiency: atmospheric, 356-357; in SCUBA diving, 284
Oxygen poisoning, 287-288
Oxygen systems, aircraft, 108, 116-117

P

Package bombs, 445
Pack-strap carry, 51
Pancake collapse, 129
Panic: bomb threats creating, 437-438; in burning structures, 197; in elevator rescue, 146; by SCUBA divers, 277, 281
Paramedics, 4, 5
Patient roll, 57-58
Patients. See Victims
Pedals, motor vehicle, extrication from, 100
Penetrator hoist, 349
Personnel, 2-18; civil liability of, 11-17; conduct of, 7-11; muscles of, 41-42; qualifications of, 5-7; untrained, 85
Pesticides, 402-405
Petroleum vapors, 399

Index

Physiological aspects: of diving, 281-283; of heat, 391
Pike pole hook, 297
Pilot rescue, 116-117
Pits, 264-266
Placards, for hazardous materials, 385-386, 387
Planks, for water/ice rescue, 290
Plaster, falling, 126
Plastics, gases from burning, 398-399
Plutonium, body effects of, 417
Pneumatic chisel, 31
Pneumatic components, vehicle fires in, 80
Pneumatic rescue equipment, 30, 31; uses of, 94-95, 98, 100
Pneumothorax, 287
Poisons/poisoning: oxygen, 287-288; pesticide, 403-404; radioactive, 425; systemic, 395-397
Pole and blanket stretchers, 61
Poles, for water/ice rescue, 291
Pole top rescue, 184-185
Police. See Law enforcement personnel
Porto-Power units, 25
Precautions: avalanche rescue, 329-330; ice rescue, 288-289
Prediction: of floods, 240-241; of hurricanes, 243-244; of tornadoes, 247-248; of tsunamis, 251
Prehospital emergency medical care, 3, 4, 68, 71
Pressure: changes of, with water depth, 280-281; from explosions, 446-447
Pressure gauges, on breathing apparatus, 375
Pressurized tunnels, 266-268
Priorities, rescue: from burning structures, 208-209; at highway accident scene, 67
Pry bars, 20, 21
Psycho gases, 406-407
Psychological aspects of diving, 281
Psychological care, of elevator passengers, 145-147, 153
Pulley systems, 24
Pulmonary irritants, in fire gases, 394-395
Purification, of respiratory air, 377
Pyrotechnics, 451

Q

Quagmires, 339-340
QUIC BARs, 37, 38
Quicksand, 339-340

R

Radiation, 421-423; aircraft radar 115; decontamination procedures, 433; detection and monitoring of, 431; external, 409, 425-431; incidents, 408-433; injury, 425, 428; internal, 410, 425-429, 431-432; laser, 115; problems of, 424-432; protection from, 429-431, 432-433; sickness, 409, 425; survey, 410-411; tolerance levels, 428-429
Radioactive decay, 423
Radioactive half-life, 423, 424
Radioactive materials, 413-416
Radioactive poisoning, 425
Rain: freezing, 232; thunderstorms, 235-238
Raking shore, 140
Rapid search procedures, 325-326, 331
Rappeling, 187-188
Ratchet jack, 20-21
Reasonable care, 12
Recovery of victims: of cave-ins, 261-262; of drowning, 297-299; of under-ice drowning, 291-293
Regulator, breathing apparatus, 374
REM, 427
Removal of victims: from aircraft, 111-117; from burning structures, 216; by helicopter, 348-349, 351; from high places, 191; from highway accidents, 100-101, 104; from water, 276-277. See also Extrication
Rending. See Severing methods
Reports, of highway accidents, 70
Rescue basket, 348, 349
Rescue coil, 36-37
Rescue saws, 29-30
Residual nuclear radiation/ radioactivity, 418, 419
Resistance (electrical), 166

Respiration, 356-358; control of, 71-72
Respiratory air stations, 376-378
Respiratory hazards, 355-357
Respiratory protection, 354-381
Respondet Superior, 13
Ring buoy, 290
Riot control gases, 406-407
River floods, 238-241
Rockets, military, 117
Rock slides, 332
Roentgen, 427-429
Roof (building), helicopter landing on, 350, 351
Roof (motor vehicle), access through, 94-96, 104
Rope, 31-37; care of, 33-34; comparison charts, 32; electrical contact rescue using, 174-175, 176-177; inspection of, 32-33; knots, hitches, bends, 34-35; lifeline, 264, 366-367; in rescue coil, 36-37; for water/ice rescue, 290-291; wire, 37
Route finding, in cave rescue, 333-334
Routes, to accidents, 66
Rubble removal, 127
Rural helispots, 345

S

Saddle-back carry, 50
Safe load, of block and tackle, 23
Safe places, inside collapsed buildings, 127, 128, 136, 137
Safety education, for wilderness areas use, 312-313
Safety lines, 298
Safety precautions: for breathing apparatus use, 365-368; around collapsed buildings, 122-123; for flood areas, 241; for hurricane areas, 244-245; in thunderstorm, 237-238
Safety procedures, earthquake, 255-256
Safety rules: around helicopters, 342, 343; in snowstorm, 234; tornado, 249
Safety switches, on elevators, 150-151
Saws, 29-31

Scanning procedures, from helicopter, 347-348
School bombings, 443
School buses, accidents of, 104
Scoop stretcher, 60
Screw jack, 21
SCUBA diver rescue, 277-288
Search and rescue operations (SAR), 310-340; helicopter use in, 347-349; in radiation incidents, 411-412
Searching techniques, in bomb threats, 440-444
Searching for victims: in burning structures, 196, 199, 208-217; at highway accidents, 67-68, 87; procedures, in wilderness areas, 325-327; suspension or termination of, 320-321; under ice, 291-293; under water, 293-303; wilderness areas, 310-340
Seat carry, 50
Seat catapult (ejection systems), 116
Seats, motor vehicle, removing or forcing, 97-98
Selenium, 400
Self-generating oxygen breathing apparatus, 380-381
Self-rescue, from icy water, 289
Semicircular sweep pattern, 299-300, 301
Service circuits, 167
Service lines (gas, water, electric), tunneling, and avoiding, 133
Severing methods, 91
Sewers, broken, 132
Shafts. *See* Tunneling techniques
Sheet bends, 35, 36
Shielding, from nuclear radiation, 429-430
Ship abandonment, cold water survival after, 304-305
Shock: electric, 170-173, 174; at highway accident sites, 73-74
Shock waves: of explosions, 446-447, 448; of nuclear bombs, 418
Shoring operations, 261
Shoring techniques, 139-141
Sinusitis, 288
Size-up: of aircraft accidents, 107-108, 111; of collapsed building, 120, 121; of explosive-related incidents, 451-452; of fireground, 207-209; 350-351; of highway

Size-up (continued):
accidents, 67, 86-87; of landslides/cave-ins, 260
Skin resistance, 171
Sled stretchers, 63
Slings, rescue, 307-308, 348-349, 350
Slough rescue, 339
Small arms ammunition, 449
Smoke, 392-393. *See also* Backdraft explosions; Fires; Gases and vapors
Smoke inhalation, 392, 393-398
Smokeless powder, 451
Snatch block, 23
Snow, 232-233
Spinal injuries, 44; at highway accidents, 73, 74, 101
Square knot, 35, 36
Stabilization: of fractures, 73, 100; of vehicles or cargo, 86-87; of victims, 43-44, 45
Stairwells, of high-rise structures, 218-219
Standards: of care, 16; ethical/moral, 6
Steel structures, fire's effect on, 125
Steering wheel, motor vehicle, 98-99
Sternal compression. *See* Cardiopulmonary resuscitation
Stokes stretcher, 61, 348, 349
Storage, of respiratory air, 377-378
Storms, winter, 231-234
Straight sweep search patterns, 300-302
Stretchers, 58-63; for helicopter rescue, 348, 349; for water/ice rescue, 290
Structural elements, raising/supporting, 139-143. *See also* Floors; Walls
Strutting techniques, 141-142
Sulfur dioxide, 401
Summoning aid, to wilderness areas, 315
Sumps, 335
Survival: of avalanches, 330; in cold water, 304-307; in deserts, 336-338; in mine accident, 263-264; in structure fire, 205-206
Suspension, of search and rescue operations, 320-321

Suspicious object located, procedures when, 444-445
Swamp rescue, 339
Sweep patterns, in water rescue search, 299-303
Swinging doors, 38-39
Systemic poisons, 395-397

T

TNT, 451
Tackle, block and, 22-23, 24
Tanks, 264-266
Teamwork, 5, 122
Tear gases, 406
Techniques: of breaching walls, 138; carry, 47-53; drag, 43, 53-56; elevator rescue, 151-157; for moving victims, 42-46; for personnel safety, 41-42; of raising/supporting floors and walls, 139-143; trenching, 137-138; tunneling, 132-137; of ventilation, 138-139
Tempered glass, 94; doors of, 39
Termination, of search and rescue operations, 320-321
Terminology: legal, 13; for rescue personnel, 5
Terrain, rough, transportation in, 62-63
Terrorist activities, 440
Thermal radiation, 418, 419
Three-person lift and carry, 52-53
Thumb knot, 34
Thunder, 238
Thunderstorms, 235-238
Tides, hurricanes and high, 243
Timber hitch, 35, 36
Timbering tunnels, 133, 134-137
Time factor, in search and rescue operations, 318-319, 324
Tire fires: aircraft, 108, 115; automobile, 81
Torching, 417
Tornadoes, 245-249
Tort law, 16
Tow trucks, training personnel of, 84
Toxic residue, after fire is out, 393
Traffic control: bystanders and, 75, 77; with flares, 75-76;

using hand signals, 76; at highway accidents, 68-69, 71, 74-77
Training: for breathing apparatus use, 361-365; for law enforcement personnel, 7; of rescue personnel, 3-5, 6; in vehicle rescue operations, 83-85
Transmission circuits, 166
Transportation, 41-63, 73, 74, 83, 84; from cave-ins, 268; from crashed motor vehicles, 101-102; of hazardous materials, 388-390; by helicopter, 351; methods of, 56-63; of radioactive materials, 413-416; techniques of, 47-56; from wilderness areas, 316, 317. See also Removal of victims
Trapped victims, 179-182; in collapsed structures, 121-122, 126; in motor vehicles, 78
Travel vibration, of helicopters, 346
Travois (stretcher), 63
Treatment: rapidity of, 43; right to refuse, 15
Trenching, 137-138
Truck accidents, 69
Trunks, automobile: access through, 96-97; fires in, 80
Tsunamis, 249-251, 256
Tunneling techniques: debris, 132-134; timbering and lining, 133, 134-137
Tunnels: collapse of, 262-264; pressurized, 266-268
Twisters, 245-249

U

Unconsciousness, victim, 44-45
Underwater search and recovery, 291-303
Uniforms, 6-7
Utilities, damaged, hazards from, 130-132

V

V-type collapse, 129, 130
Van accidents, 69

Vapors. See Gases and vapors
Vehicles (motor). See Motor vehicles
Vehicles (rescue), positioning of, 68-69
Ventilation: in fire and rescue service, 200-201, 202, 351, 356; helicopter's role in, 351; of tanks, pits, holes, 264-265; during transportation, 46; of tunnels, under debris, 138-139
Vertical shore, 140-142
Victims: accessibility of, 184; care of, 42-43; complications in handling, 43-47; consent of, 14-15; hard-to-control, 45-46; right to refuse treatment of, 15; stabilization of, 43-44, 45; transporting, 41-63. See also Recovery of victims; Removal of victims; and type of accident, e.g., Highway; Mine; etc.
Visibility, traveling with poor, 205
Voice, tone of, 10
Voids: as safe places, 127, 128, 136, 137; tunneling to connect, 132
Volcanoes, 251-252
Voltage, 166, 170-172
Volume, changes of, with water depth, 280-281
Vomiting, victim's, 45
Vomiting gases, 406

W

Walking, techniques of desert, 338
Walls (building): breaching, 138; collapse of, 124, 129-130; removing, 142-143; shoring, 139-141; strutting, 141-142
Walls (elevator), breaching, 156-157
Warning devices, of breathing apparatus, 367-368, 375
Warnings, weather, 232
Warning signs, about hazardous materials, 385-386, 387
Watch, weather, 232
Water, rescue from, 270-308, 350

Index

Water hazards: in caves, 334-335, 336; of collapsed buildings, 131; electricity and, 166
Waterspouts (tornado), 245
Waves, shock, 418, 446-447, 448
Waves, tsunami, 249-251
Weather, 230-231
Weather conditions, and wilderness travel, 313, 322-324
Wedges, 22
Weighted line, search with, 302, 303
Wilderness search and rescue, 310-340
Winches, 24, 25
Wind chill factor, 233-234; in cold water survival, 306

Windows, breaking/opening: aircraft, 113, 114; automobile, 93-94; building, 39-40, 168-169; bus, 103, 104
Windows, electrical devices on, 168-169
Windshield removal (crashed vehicles), 93, 99, 104
Windstorms, 247. *See also* Tornadoes
Winter conditions, in wilderness areas, 317, 319
Winter storms, 231-234
Wire rope/cable, 37
Wires, *See* Electrical wires
Wrecker trucks, training personnel of, 84
Wrecking bars, 20
Written bomb threats, 439